TEUBNER-TEXTE zur Mathematik · Band 131

Symposium "Analysis on Manifolds with Singularities", Breitenbrunn 1990

Edited by Bert-Wolfgang Schulze and Hans Triebel

Springer Fachmedien Wiesbaden GmbH 1992

ISBN 978-3-663-11578-6 ISBN 978-3-663-11577-9 (eBook)
DOI 10.1007/978-3-663-11577-9

Ursprünglich erschienen bei B.G. Teubner Verlagsgesellschaft Stuttgart • Leipzig 1992

The present Teubner-Text continues a series of publications on analysis of the Karl-Weierstrass-Institute for Mathematics, Berlin (cf. the volumes **96** (1985/86), **106** (1986/87), **112** (1988), **117** (1990)). The volume contains contributions from the International Workshop "Analysis in Domains and on Manifolds with Singularities" in Breitenbrunn, Germany, April 30 - May 5, 1990.

Der vorliegende Teubner-Text setzt eine Serie von Publikationen zur Analysis des Karl-Weierstraß-Instituts für Mathematik, Berlin, fort (vgl. die Bände **96** (1985/86), **106** (1986/87), **112** (1988), **117** (1990)). Der Band enthält Beiträge der Internationalen Konferenz "Analysis in Domains and on Manifolds with Singularities" in Breitenbrunn, Deutschland, 30. April - 5. Mai 1990.

Le Teubner-Text présent est le prolongement d'une serie des publications sur analyse mathématique par l'Institut de Mathématiques Karl-Weierstrass, Berlin (cp. les tomes **96** (1985/86), **106** (1986/87), **112** (1988), **117** (1990)). Le volume contient des articles du Congrès International "Analysis in Domains and on Manifolds with Singularities" à Breitenbrunn, Allemagne, 30. avril - 5. mai 1990.

Данный текст продолжает серию публикаций института математики им. Карла Вейерштрасса г. Берлина в области анализиса (ср. т. **96** (1985/87гг.), **106** (1986/87гг.), **112** (1988г.), **117** (1990г.)). Он содержит доклады международной научной конференции "Analysis in Domains and on Manifolds with Singularities" в г. Брейтенбрунн, Германия, 30 апреля по 5 мая 1990г.

PREFACE

The present Teubner-Text contains the contributions from the International Workshop "Analysis in Domains and on Manifolds with Singularities", Breitenbrunn, Germany, 30. April - 5. May 1990.

In recent years the analysis on manifolds with singularities became more and more interesting, not only because of the progress in solving corresponding singular problems in partial differential equations but also of the new relations to other parts of mathematics such as geometry, topology and mathematical physics. Other motivations come from concrete models in engineering and applied sciences which lead to partial differential equations in domains with a piece-wise smooth geometry (conical points, edges, corners, ..., higher singularities), piece-wise smooth data or boundary and transmission conditions, degenerate coefficients, and so on. There are natural relations to the numerical analysis where also the asymptotics of solutions close to the singularities play a role.

As for the smooth cases it is necessary to develop structure principles and unified theories that cover as much as possible the huge variety of concrete situations, often being treated by individual papers under very specific assumptions.

Such a framework in the elliptic (and the parabolic) theory is the calculus of pseudo-differential operators, with the parametrix constructions on symbolic level and the asymptotics of solutions as an element of the regularity. The pseudo-differential calculus is at the same time link to corresponding extensions of the index theory. Moreover, the micro-local analysis in general is responsible for the non-elliptic singular problems, similarly to the classical theory of Fourier integral operators in the case of hyperbolic (or more general classes of) equations.

Most of the basic things is yet to be done. For instance, a general index theory on piece-wise smooth or non-compact spaces is far from being achieved in a satisfying form. But it is already obvious by the present knowledge that the structures will be much more complex than in the classical case.

Also the solvability of elliptic boundary value problems in domains with higher corners, the functional analysis of spaces with asymptotics in those cases, the problem of branching of exponents in the asymptotics along edges with complicated model cones, the nature of singular terms for corners, and many other topics from that area seem to be on a beginning, in spite of the stream of papers and so much inside and experience in particular cases. In other words the prospects in the analysis on manifolds with singularities are actually exciting.

The conference in Breitenbrunn was a meeting of outstanding specialists in this field. So this collection of contributions corresponds to the general intention of this series of publications.

The editors are indebted to the colleagues from the Freie Universität Berlin, Dr. B. Schomburg, Ms A. Link and Mr R. Regitz for technical assistance in preparing this volume.

Berlin, Jena October 1991

B.-W. Schulze, H. Triebel

CONTENTS

REFLECTION AND REFRACTION OF SINGULARITIES FOR WAVE EQUATIONS WITH INTERFACE CONDITIONS GIVEN BY FOURIER INTEGRAL OPERATORS[1]

Felix ALI MEHMETI
Mainz, Fed. Rep. of Germany

0 Introduction

Cauchy problems for hyperbolic operators often have the property, that the singularities of the initial data propagate along the bicharacteristic strips of the operator (cf. e.g. [13]). We consider, in the linear case, the situation where the bicharacteristics hit transversally a spacelike interface, which is 'active' in the sense that the interface condition is given via certain Fourier integral operators. Taking the identity, we obtain classical transmission conditions. A suitable functional analytic setting is furnished by the interaction concept [3], [6], [7], which covers very general mutual influences of evolution phenomena on different domains.

We state the laws of reflection and refraction of singularities at the interface in terms of the canonical relations of the given Fourier integral operators (most proofs are only sketched). These results are obtained using a modified parametrix construction of J. Cl. Nosmas [28], the calculus of Fourier integral operators [13], [14], [21], arguments involving the Boutet de Monvel algebra [17], [30], [33] and a nonrecursive formula for the compatibility conditions for abstract inhomogeneous wave equations [5]. The latter is based on the theory of T. Kato on time dependent evolution equations [22] and generalizes a regularity result of Lions-Magenes [24]. The structural similarities on the functional analytic level to interaction in quantum field theory are underlined by the observation that in both cases new singularities (particles) are created in the interaction zone.

Our situation is a model for active interfaces; it arises naturally as result of coordinate transformations of transmission problems; it gives structural insights of the following kind: classical transmission or boundary operators are all defined via the trace operator, whose canonical relation governs the laws of reflection and refraction of singularities. The latter can only be changed by introducing further Fourier integral operators with nontrivial canonical relations. Especially it cannot be changed using standard pseudodifferential operators.

Interaction problems are models for evolutionary or stationary phenomena on several domains with various kinds of mutual influence (cf. [3], [6], [7]). This concept is

[1]This research was supported by the Deutsche Forschungsgemeinschaft under grant Al 289/1-1

inspired by methods of B. Gramsch in [15], [16], where he derives a rough asymptotic estimate for eigenvalues of elliptic transmission or boundary value problems. Important special cases are transmission problems on ramified spaces (i.e. families of domains, where certain parts of the boundaries are identified, cf. G. Lumer [26], S. Nicaise [27], J.v. Below [9], and many other contributions of these authors, who considered especially the parabolic case). I have treated mainly *hyperbolic* transmission and interaction problems (cf. [1], ..., [5]) from the viewpoint of evolution equations, obtaining results on local and global existence and regularity. Here arises also naturally the question of propagation of singularities. This complex has been investigated for interface problems for wave equations by S. Hansen [18], J.Cl. Nosmas [28], V. Petkov [29], M. Taylor [34].

Consider two half spaces $X_1 := \{x \in I\!R^n : x_n > 0\}$, $X_2 := \{x \in I\!R^n : x_n < 0\}$ and the interface $\Sigma := \{x \in I\!R^n : x_n = 0\}$. Let $u_i : [0,\infty) \times X_i \to I\!R$ be a distributional solution of the following Cauchy problem for wave equations with Fourier integral interface conditions:

$$\begin{array}{lll} (W) & \frac{\partial^2}{\partial t^2} u_i(t,x) \; - \; a_i \Delta_x u_i(t,x) \; + \; c_i u_i(t,x) \; = \; 0 & \\ & \forall t \in [0,\infty], x \in X_i, \; i = 1,2 & \\ (Init) & u_i(0,x) \; = \; x_{0,i}, \; \frac{\partial}{\partial t} u_i(0,x) \; = \; y_{0,i} \;\; \forall x \in X_i, \; i = 1,2 & (1) \\ (I_0) & F_1^{\Sigma}\big(u_1(t,\cdot)\big|_{\Sigma}\big) \; = \; F_2^{\Sigma}\big(u_2(t,\cdot)\big|_{\Sigma}\big) \;\; \forall t \in [0,\infty) & \\ (I_1) & F_1^{\Sigma}\big(a_1 \frac{\partial}{\partial x_n} u_1(t,\cdot)\big|_{\Sigma}\big) \; = \; F_2^{\Sigma}\big(a_2 \frac{\partial}{\partial x_n} u_2(t,\cdot)\big|_{\Sigma}\big) \;\; \forall t \in [0,\infty) \, , & \end{array}$$

where, for simplicity, $a_i > 0$, $c_i > 0$ are constants, $x_{0,i} \equiv 0, i = 1,2$, $y_{0,1} \in H^1(X_1)$ has compact support, $y_{0,2} \equiv 0$ and F_i^{Σ} are certain Fourier integral operators associated with local canonical graphs (cf. assumption 1.2). The (I_i) are called interaction conditions.

What will happen to a wave hitting the interface Σ? Our main results (theorems 5.4, 5.7, 5.9) show that the canonical relation of the Fourier integral operator describing the refraction from X_1 into X_2 is a composition of canonical relations:

$$\mathcal{R} := Bichar(P_2)\big|_{I\!R \times X_2} \circ {}^{\dagger}B \circ {}^{\dagger}C_2^{e,\Sigma} \circ C_1^{e,\Sigma} \circ B \circ (\mathcal{E}_1\big|_{X_1}) \qquad (2)$$

where $\mathcal{E}_i$ corresponds to the pure Cauchy problem for the wave equation with subscript i, B corresponds to the trace operator on $I\!R \circ \Sigma$, $C_i^{e,\Sigma}$ can be defined in terms of the canonical relation of F_i^{Σ} and $Bichar(P_i)$ is the bicharacteristic relation of P_i. We have

$$WF(u_2) \subseteq \mathcal{R} \circ WF(y_{0,1}) \, . \qquad (3)$$

A similar inclusion holds for $WF(u_1)$ describing the reflection. (2) and (3) mean, that a singularity in the initial conditions in X_1 is propagated along a bicharacteristic strip in the light cone in $I\!R \times X_1$, hits then Σ, undergoes a modification of its projection to $T^*\Sigma$ according to the canonical relations of the F_i^{Σ} and is finally sent in a bicharacteristic strip in the light cone of $I\!R \times X_2$ (further, it is reflected back in $I\!R \times X_1$).

In this construction we considered the nondiffractive case, i.e. the case that the bicharacteristics hit Σ transversally. S. Hansen proved propagation of singularities in the diffractive case (where a parametrix construction is no longer possible) for F_i^{Σ} = identity. M. Lorenz [25] investigated the reflection of jumps at boundaries.

Propagation of singularities for nonlinear hyperbolic problems without interfaces has been treated by M. Beals [8], J.M. Bony [10], M. Sablé-Tougeron [31] (nonlinear boundary conditions) and others. A future development might be the extension to interaction problems.

I am grateful to Professor B. Gramsch for valuable support. Further I thank Professor S. Hansen and Dr. E. Schrohe for many discussions.

1 Notations and Assumptions

NOTATION 1.1

(i) $X_1 := \{x \in I\!R^n : x_n > 0\}$, $X_2 := \{x \in I\!R^n : x_n < 0\}$, $\Sigma := \{x \in I\!R^n : x_n = 0\}$.

(ii) Elements of $I\!R \times X_{1,2}$ are denoted by

$$\bar{x} := (x_0, x', x_n),$$

where $x_0 \in I\!R$ is the time variable and $x_n \in I\!R$ is the space variable measuring the distance of (x', x_n) to Σ. $x' = (x_1, \ldots, x_{n-1})$ are the other space variables.

(iii) Write $D_i := \frac{1}{\sqrt{-1}} \frac{\partial}{\partial x_i}$, $i = 0, \ldots, n$; $D := (D_1, \ldots, D_n)$.

(iv) Concerning Fourier integral operators, canonical relations and graphs we use the notations of [35], vol. II, chapter VIII, for wave front sets cf. [35], vol. I, chapter VI. For pseudo-differential operators we use also the notations of [35], vol. I.

Further we consider the following operators and forms:

ASSUMPTION 1.2

(i) Consider real numbers $d_0, d_1 > 0$ and realvalued functions $a_i^{lj} \in C^\infty(I\!R^n)$, $l, j = 1, \ldots, n$, $i = 1, 2$, such that all derivatives are bounded, with the following properties: strong ellipticity

$$\sum_{l,j=1}^{n} a_i^{lj}(x)\xi_l\xi_j \geq d_0 \sum_{j=1}^{n} | \xi_j |^2 \quad \forall(\xi_1, \ldots, \xi_n) \in I\!R^n, x \in X_i, i = 1, \ldots, n;$$

and symmetry

$$a_i^{lj}(x) = a_i^{jl}(x) \ \ \forall\, l, j = 1, \ldots, n; \ x \in X_i; \ i = 1, 2.$$

Assume $a_i^{jn} \equiv 0, j = 1, \ldots, n-1; i = 1, 2$ (this is a technical assumption used only in section 3 which might be released using better variational inequalities). Consider further

$$c_i \in C^\infty(\overline{I\!R^n}) \text{ with } c_i \geq d_1 \quad \text{for } i = 1, 2\ .$$

All derivatives of c_i are assumed to be bounded.

Assume that all these functions are constant outside a compact subset of $I\!R^n$.

(ii) Consider the differential operator of second order

$$Q_i(x, D) := \sum_{l,j=1}^{n} D_j a_i^{lj}(x) D_l + c_i(x)$$

$$= -\sum_{l,j=1}^{n} \frac{\partial}{\partial x_j} a_i^{lj}(x) \frac{\partial}{\partial x_l} + c_i(x), \quad \forall x \in I\!R^n, \quad i = 1,2 ,$$

with its principal symbol

$$Q_{i,2}(x,\xi) = \sum_{l,j=1}^{n} a_i^{lj}(x)\xi_l\xi_j \ , x \in I\!R^n, i = 1,2.$$

(iii) Consider the differential operator of first order

$$\partial_i(x,D) \ := \ \sum_{l,j=1}^{n} a_i^{ln}(x)\sqrt{-1}\ D_l$$

$$= \ a_i^{nn}(x)\sqrt{-1}D_n \ = \ a_i^{nn}(x)\frac{\partial}{\partial x_n} \ , \quad x \in I\!R^n, \quad i = 1,2 ,$$

with its symbol

$$\partial_i(x,\xi) := \sqrt{-1}\sum_{l=1}^{n} a_i^{ln}(x)\xi_l \ = \ a_i^{nn}(x)\sqrt{-1}\xi_n \ , \quad x \in I\!R^n, i = 1,2.$$

(iv) Define $P_i(x, D_0, D) \ := \ -D_0^2 \ + \ Q_i(x,D) \ = \ \frac{\partial^2}{\partial x_0^2} \ + \ Q_i(x,D) \quad , \ x \in I\!R^n, i = 1,2.$

(v) Let $a_i : H^1(X_i) \times H^1(X_i) \to I\!R$ be the bilinear form associated with Q_i, i.e.

$$a_i(u,v) := \int_{X_i} \Big[\sum_{l,j=1}^{n} a_i^{lj}(x)\Big(\frac{\partial}{\partial x_l}u(x)\Big)\Big(\frac{\partial}{\partial x_j}\overline{v(x)}\Big)\Big] d^n x \ + \ \int_{X_i} c_i(x)u(x)\overline{v(x)} d^n x \ ,$$

(vi) Define

$$\tilde{a} : \prod_{i=1}^{2} H^1(X_i) \times \prod_{i=1}^{2} H^1(X_i) \to \mathbb{C}$$

by

$$\tilde{a}((u_i),(v_i)) \ := \ \sum_{i=1}^{2} a_i(u_i,v_i)$$

LEMMA 1.3

(i) $Q_{i,2}(x,\xi) \ \geq \ d_0 > 0 \ \ \forall \ |\xi| = 1, \ x \in I\!R^n.$

(ii) $\tilde{a}$ *is coercive (i.e.* $\exists \alpha > 0 : \ \tilde{a}(u,u) \ \geq \ \alpha \parallel u_i \parallel^2_{\prod_{i=1}^2 L^2(X_i)} \ \forall u \in \prod_{i=1}^2 H^1(X_i)$*).*

(iii) $\tilde{a}((u_i),(v_i)) \ - \ ((Q_i(x,D)u_i),(v_i))_{\prod_{i=1}^2 L^2(X_i)}$

$= \sum_{i=1}^2 (-1)^{i-1} \int_\Sigma (b_0\partial_i(x,D))(b_0 v_i)(x) dx_1 \ldots dx_{n-1} \ \forall (u_i) \in \prod_{i=1}^2 H^2(X_i), \ (v_i) \in \prod_{i=1}^2 H^1(X_i),$ *where* b_0 *is the restriction operator to* Σ *(the factor* $(-1)^{i-1}$ *is due to the negative* x_n- *coordinates in* X_2*).*

Proof:

(i) Strong ellipticity of Q_i.

(ii) Gårding's inequality.

(iii) Green's formula on halfspaces.

□

ASSUMPTION 1.4

(i) Nondiffraction condition:

Every bicharacteristic strip of $Q_{i,2}(x,\xi)$ starting from $T_0^*X_1$ or $T_0^*X_2$, hitting the hyperplane Σ, intersects it transversally and in a unique point (note that the Hamiltonian flow of Q_i is globally defined, due to 1.2 (ii)).

(ii) Changement of the index:

$\sum_{l,j=1}^{n-1} a_1^{lj}(x)\xi_l\xi_j/a_1^{nn}(x) \; - \; \sum_{l,j=1}^{n-1} a_2^{lj}(x)\xi_l\xi_j/a_2^{nn}(x) \; > \; 0 \; \forall x \in \Sigma$ and $\| \xi' \| = 1$.

(Thus X_1 is the medium with the smaller 'index'. It will carry the support of the initial conditions of our problem in section 4).

Further we consider certain Fourier integral operators:

ASSUMPTION 1.5

(i) Let $F_i^\Sigma \in I^0(\Sigma, \Sigma, C_i^\Sigma), i = 1, 2$, be Fourier integral operators associated with the local canonical graphs $C_i^\Sigma \subseteq T^*\Sigma \times T^*\Sigma$.

(ii) Assume further for $i = 1, 2$

$$F_i^\Sigma \in \mathcal{L}(H^s(\Sigma)) \;\; \forall s \in I\!R,$$

and that F_i^Σ is unitary in $\mathcal{L}(L^2(\Sigma))$.

REALIZATIONS 1.6 Let us give a realization for the Fourier integral operator F_j^Σ following [35], vol. II, section VIII, 8. Let A_j^Σ denote a classical pseudo-differential operator of order one on Σ with the principal symbol a_j^Σ, which is realvalued. Let $\{\Phi_j^\Sigma(t), t \leq T\}$ be the Hamilton flow of a_j^Σ. Assume that A_j^Σ is symmetric and call $\tilde{A}_j^\Sigma$ its Friedrichs extension, which is selfadjoint. Then $\{U_j(t) | t \in I\!R\}$ with $U_j(t) \; := \; \exp it\tilde{A}_j^\Sigma$ is a group of unitary operators on $L^2(\Sigma)$. Further, by [35], section VIII, 8, $U_j(t)$ is a unitary Fourier integral operator in $\mathcal{L}(L^2(\Sigma))$ for all $t \in I\!R$.

If $|t|$ is sufficiently small, the graph of the map

$$(x, \xi) \mapsto \Phi_j^\Sigma(x, \xi)$$

defines a local canonical graph $C_{t,j}$ which is the canonical relation of $U_j(t)$. The amplitude of $U_j(t)$ can be constructed by transport equations (cf. [35], equations (8.7), ..., (8.13)). Choose a convenient $t^* \in I\!R$ and define $F_j^\Sigma := U_j(t^*)$. Simple examples are coordinate rotations and translations.

NOTATION 1.7

(i) Consider the trace operators

$b_0 : C^\infty(\overline{X_i}) \to C^\infty(\Sigma)$ defined by $b_0 u \; := \; u\big|_\Sigma$,

$\beta_0 : C^\infty(\overline{I\!R \times X_i}) \to C^\infty(I\!R \times \Sigma)$ defined by $\beta_0 u \; := \; u\big|_{I\!R\times\Sigma}$,

$\gamma_{0,t} : C^\infty(\overline{I\!R \times X_i}) \to C^\infty(\overline{X_i})$ defined by $\gamma_{0,t} u \; := \; u\big|_{\{x_0=t\}} \; = \; u(t, \cdot)$,

$\gamma_{0,t}^\Sigma : C^\infty(\overline{I\!R \times \Sigma}) \to C^\infty(\overline{\Sigma})$ defined by $\gamma_{0,t}^\Sigma u \; := \; u\big|_{\{x_0=t, x_n=0\}}$.

These operators are Fourier integral operators

$$b_0 \;\in\; I^{1/4}(\Sigma, X_i, B_0),$$
$$\beta_0 \;\in\; I^{1/4}(I\!R \times \Sigma, I\!R \times X_i, B),$$
$$\gamma_{0,t} \;\in\; I^{1/4}(\{t\} \times X_i, I\!R \times X_i, G_{0,t}),$$
$$\gamma_{0,t}^\Sigma \;\in\; I^{1/4}(\{t\} \times \Sigma, I\!R \times \Sigma, G_{0,t}^\Sigma),$$

with the corresponding canonical relations

$$\begin{aligned} B_0 &= \Big\{\big((x',0,\xi',0),(x',0,\xi',\xi_n)\big) : (x',\xi') \in T^*I\!R^{n-1}, \xi_n \in I\!R\Big\} \\ B &= \Big\{\big((x_0,x',0,\xi_0,\xi',0),(x_0,x',0,\xi_0,\xi',\xi_n)\big) : \\ &\quad (x_0,x',\xi_0,\xi') \in T^*I\!R^n, \xi_n \in I\!R\Big\} \\ G_{0,t} &= \Big\{\big((t,x,0,\xi),(t,x,\xi_0,\xi)\big) : (x,\xi) \in T^*I\!R^n, (t,\xi_0) \in T^*I\!R\Big\} \\ G^{\Sigma}_{0,t} &= \Big\{\big((t,x',0,\xi'),(t,x',\xi_0,\xi')\big) : (x',\xi') \in T^*I\!R^{n-1}, (t,\xi_0) \in T^*I\!R\Big\} . \end{aligned}$$

Write shortly $\gamma_0 := \gamma_{0,0}$ and $\gamma_1 := \frac{\partial}{\partial x_0} \circ \gamma_{0,0}$.

We shall use the special imbeddings in these descriptions for compositions of canonical relations.

(ii) For an open subset $Y \subseteq I\!R^p$ with $C^\infty-$boundary we define the admissible distributions:

$$\begin{aligned} \mathcal{D}'_a(Y) &:= \{u \in \mathcal{D}'(Y) : u \text{ has an extension } \tilde{u} \in \mathcal{D}'(I\!R^n) \\ &\qquad \text{such that } \mathrm{WF}(\tilde{u}) \cap N^*_{\partial Y} = \emptyset\} \end{aligned}$$

($N^*_{\partial Y}$ means the conormal bundle of ∂Y). Note, that for $u \in \mathcal{D}'_a(I\!R \times X_i)$, $\beta_0 u \in \mathcal{D}'(I\!R \times \Sigma)$ is defined (cf. [13], section 5).

2 A Cauchy Problem with Fourier Integral Interaction Conditions

The main objective of this paper is the study of the singularities of the solution of the following problem:

For given $f \in \mathcal{E}'(X_1)$ find $u = (u_1, u_2) \in \prod_{i=1}^2 \mathcal{D}'_a(I\!R \times X)$ with

$$\begin{aligned} &(W) \quad P_i(\cdot, D_0, D)u_i = 0,\ i = 1,2 \\ &(Init) \quad \gamma_0 u_i = 0,\ i = 1,2 \\ &\qquad \gamma_1 u_1 := \tfrac{\partial}{\partial x_0} \circ \gamma_0 u_1 = f;\ \gamma_1 u_2 = 0 \qquad\qquad (4) \\ &(I_0) \quad F^{\Sigma}_1 \circ \gamma^{\Sigma}_{0,t} \circ \beta_0\, u_1 = F^{\Sigma}_2 \circ \gamma^{\Sigma}_{0,t} \circ \beta_0\, u_2 \quad \forall t \in I\!R \\ &(I_1) \quad F^{\Sigma}_1 \circ \gamma^{\Sigma}_{0,t} \circ \beta_0 \circ \partial_1(\cdot, D)\, u_1 = F^{\Sigma}_1 \circ \gamma^{\Sigma}_{0,t} \circ \beta_0 \circ \partial_2(\cdot, D)\, u_1 \quad \forall t \in I\!R . \end{aligned}$$

Conditions $(I_i), i = 1,2$ are called interaction conditions. For $F^{\Sigma}_i =$ identity, $i = 1,2$, problem (4) is a model for propagation of waves in the half spaces X_i and their reflection and refraction along the interface Σ (a transmission problem). The F^{Σ}_i allow the description of somehow active interfaces with a modified law of refraction. Note that we consider only problems with the same Fourier integral operators in (I_0) and (I_1). The γ_i and $\gamma^{\Sigma}_{0,t}$ will always be applicable due to the hyperbolicity of P_i.

Our aim is to find a pair of admissible distributions solving (4) modulo $C^\infty-$functions. Such an approximate solution can be obtained by a parametrix:

DEFINITION 2.1 Let $K \subseteq X_1$ be a compact subset. A parametrix of the problem (4) is a mapping

$$\mathcal{E}'_K \to \prod_{i=1}^{2} \mathcal{D}'_{F_i}(I\!R \times X_i)$$

$$f \mapsto (u_1, u_2)$$

with

$$\begin{aligned}
h_i &:= P_i(\cdot, D_0, D)u_i \in C^\infty(\overline{I\!R \times X_i}),\ i = 1,2 \\
g_{0,1} &:= \gamma_0 u_1 \in C^\infty(\overline{X_1}) \\
g_{1,1} &:= \gamma_1 u_1 - f \in C^\infty(\overline{X_1}) \\
g_{0,2} &:= \gamma_0 u_2 \in C^\infty(\overline{X_2}) \\
g_{1,2} &:= \gamma_0 u_2 \in C^\infty(\overline{X_2}) \\
g_0 &: t \to F_1^\Sigma \circ \gamma_{0,t}^\Sigma \circ \beta_0\, u_1 - F_2^\Sigma \circ \gamma_{0,t}^\Sigma \circ \beta_0\, u_2 \text{ in } C^\infty(\overline{I\!R \times \Sigma}) \\
g_1 &: t \to F_1^\Sigma \circ \gamma_{0,t}^\Sigma \circ \beta_0\, \partial_1(\cdot, D)u_1 - F_2^\Sigma \circ \gamma_{0,t}^\Sigma \circ \beta_0\, \partial_2(\cdot, D)u_2 \text{ in } C^\infty(\overline{I\!R \times \Sigma})
\end{aligned}$$

Further the functions h_i, g_{ij}, g_j, $i = 1,2$, $j = 0,1$, are assumed to satisfy the compatibility conditions of infinite order with respect to the interaction conditions (I_i). This notion is defined for our situation in [5].

3 The Interaction Operator with Fourier Integral Interaction Conditions

In this section we transform problem (4) of section 2 in a functional analytic formulation using the interaction concept described in [3]:

DEFINITION 3.1 Define

$$V := \{(u_1, u_2) \in \prod_{i=1}^{2} H^1(X_i) : F_1^\Sigma \circ b_0 u_1 = F_2^\Sigma \circ b_0 u_2\}$$

REMARK 3.2 By assumption 1.5, V is a closed subspace of $\prod_{i=1}^2 H^1(X_i)$.

DEFINITION 3.3

(i) Define $a : V \times V \to I\!R$ by

$$a((u_i),(v_i)) := \tilde{a}((u_i),(v_i)) = \sum_{i=1}^{2} a_i(u_i, v_i)\ .$$

(ii) Define $H := \prod_{i=1}^2 L^2(X_i)$. Clearly $V \hookrightarrow H$. Endow V with the structure induced by $\tilde{V} := \prod_{i=1}^2 H^1(X_i)$ and identify H with H'. Thus

$$V \hookrightarrow H \cong H' \hookrightarrow V'\ .$$

(iii) Let $A : D(A) \to H$ be the selfadjoint positive operator constructed from a and V by Friedrichs extension, i.e.

$$\begin{aligned}
&\tilde{A} : V \to V',\ \tilde{A}(v) := a(u,v)\ \ \forall\, u, v \in V \\
&D(A) := \{u \in V : \tilde{A}u \in H' = H\} \\
&A := \tilde{A}|_{D(A)}\ .
\end{aligned}$$

The assumptions are satisfied due to Gårdings inequality. It follows $D(A^{1/2}) = V$.

(iv) Call V interaction space and A interaction operator.

As we shall see, the domain of the interaction operator A is characterized by the interaction conditions (I_i), $i = 0, 1$:

THEOREM 3.4

$$\begin{aligned} D(A) &= \{(u_1, u_2) \in \prod_{i=1}^{2} H^1(X_i) : Q_i(\cdot, D)u_i \in L^2(X_i), \\ F_1^\Sigma \circ b_0 u_1 &= F_2^\Sigma \circ b_0 u_2 \quad in\ H^{1/2}(\Sigma), \\ F_1^\Sigma \circ b_0 \partial_1(\cdot, D)u_1 &= F_2^\Sigma \circ b_0 \partial_2(\cdot, D)u_2 \quad in\ H^{-1/2}(\Sigma)\} \end{aligned}$$

REMARK 3.5 The second equation in 3.4 is a priori an equation between distributions which are not necessarily in $L^2(X_i)$. Theorem 3.6 shows, that they are in $L^2(X_i)$.

Proof of theorem 3.4:

Let $V_0 := \prod_{i=1}^{2} H_0^1(X_i)$, $\varrho : V' \to V_0'$ be the restriction to V_0 and $\tilde{A} : V \to V'$ be defined by $\tilde{A}u(v) := a(u, v)$. Define $A_f \in \mathcal{L}(V, V_0')$ by $A_f := \varrho \tilde{A}$ (the 'formal' operator, cf. [32], [3]). Then holds for $u = (u_1, u_2) \in D(A_f) := \{u \in V : A_f u \in H\}$:

$$u \in D(A) \Leftrightarrow a(u, v) - (A_{fu}, v)_H = 0 \quad \forall\, v \in V .$$

By Green's formula for halfspaces follows:

$$\begin{aligned} a(u, v) - (A_f u, v)_H &= \tilde{a}((u_i), (v_i)) - ((Q_i u_i), (v_i))_{\prod_{i=1}^{2} L^2(X_i)} \\ &= \langle b_0 \partial_1 u_1, b_0 v_1 \rangle - \langle b_0 \partial_2 u_2, b_0 v_2 \rangle , \end{aligned}$$

where $\langle \cdot, \cdot \rangle$ is the duality bracket in $(H^{-1/2}(\Sigma), H^{1/2}(\Sigma))$. Therefore: for $u \in D(A_f) \cap \prod_{i=1}^{2} \mathcal{S}(X_i)$ holds by [32]:

$$\begin{aligned} u \in D(A) &\Leftrightarrow a(u, v) - (A_f u, v)_H = 0 \Leftrightarrow \\ 0 &= \langle b_0 \partial_1 u_1, b_0 v_1 \rangle - \langle b_0 \partial_2 u_2, b_0 v_2 \rangle \\ &= \langle b_0 \partial_1 u_1, b_0 v_1 \rangle - \langle b_0 \partial_2 u_2, (F_2^\Sigma)^* F_1^\Sigma b_0 v_1 \rangle \\ &= \langle b_0 \partial_1 u_1, b_0 v_1 \rangle - \langle ((F_2^\Sigma)^* F_1^\Sigma)^* b_0 \partial_2 u_2, b_0 v_1 \rangle \quad \forall\, v_1 \in H^1(X_i) \end{aligned}$$

(note that $F_1^\Sigma \circ b_0 v_1 = F_2^\Sigma \circ b_0 v_2$ and the F_i^Σ are unitary). v_1 can be chosen, such that $v_1 \neq 0$ only arbitratily close to some $x \in \Sigma$. Therefore

$$u \in D(A) \Leftrightarrow b_0 \partial_1 u_1 = ((F_2^\Sigma)^* F_1^\Sigma)^* b_0 \partial_2 u_2 \text{ on } \Sigma,$$

i.e.

$$F_1^\Sigma b_0 \partial_1 u_1 = F_2^\Sigma b_0 \partial_2 u_2 \text{ on } \Sigma.$$

By continuity, this equation holds in $H^{-1/2}(\Sigma)$ if we consider an arbitrary $u \in D(A_f)$. □

THEOREM 3.6 $\qquad D(A) \subseteq \prod_{i=1}^{2} H^2(X_i)$

Idea of the proof:

We construct a parametrix of the stationary version of the interaction problem (4) in the Boutet de Monvel algebra. This is carried out transforming the problem into a transmission problem via similarity and proving the existence of a parametrix in the simplified situation (using methods of [17], [30], [33]). This proof has been suggested to me by E. Schrohe. □

4 Exact Solution in the Hilbert Space Setting

Consider the interaction problem (4) in section 2. If the initial condition f is in $L^2(X_1)$, (4) can be reformulated putting $x(t) := (u_i(t,\cdot))$: Find $x(\cdot) \in C^2([0,\infty), H)$ with

$$\begin{cases} \ddot{x}(t) + Ax(t) = 0 & \forall\, t \in \mathbb{R}, \\ x(0) = (0,0) =: x_0,\ \dot{x}(0) = (f,0) =: y_0, & \\ x(t) \in D(A) & \forall t \in \mathbb{R}. \end{cases} \tag{5}$$

A solution $x(\cdot)$ of (5) represents an *exact* solution of the interaction problem (4). It is indicated in the subsequent lemma. A parametrix of (4) defines only a solution modulo C^∞, but it gives geometrical information on the singularities which cannot be obtained in the Hilbert space setting.

LEMMA 4.1 *For every $k \in \mathbb{N}, x_0 \in D(A^{(k+1)/2}), y_0 \in D(A^{k/2})$ there is one and only one solution $x(\cdot)$ of (5). It satisfies*

$$x(\cdot) \in C^{\,k+1-j}(\mathbb{R}, D(A^{j/2})),\ j = 0,\dots,k+1$$

and

$$x(t) = \left((\sqrt{A})^{-1}\sin(\sqrt{A}t)\right)x_0 + \left(\cos(\sqrt{A}t)\right)y_0 \quad \forall\, t \in \mathbb{R}\,.$$

Proof: Cf. [3] □

5 Construction of the Parametrix

We use the parametrix for the Cauchy problem without interface given e.g. in [13]: let E_i be the exact solution operator for the Cauchy problem for $P_i(\cdot, D_0, D)$ in $\mathbb{R}^{n+1}$, i.e.

$$E_i : f \mapsto u_i(\cdot,\cdot),$$

where u_i is the solution of

$$\begin{aligned} P_i(\cdot, D_0, D)u_i &= 0 \quad \text{on} \quad \mathbb{R}^{n+1} \\ \gamma_0 u_i &= 0 \quad \text{on} \quad \{x_0 = 0\} \\ \gamma_1 u_i &= f \quad \text{on} \quad \{x_0 = 0\}\,. \end{aligned}$$

Writing $E_i(t) : f \mapsto u_i(t,\cdot)$ for $t \in \mathbb{R}$, $\{E_i(t) : t \in \mathbb{R}\}$ is a strongly continuous group in $\mathcal{L}(L^2(\mathbb{R}^n))$, whose existence follows from section 4. By [13], E_i it is a Fourier integral operator:

$$E_i \in I^{-1-1/4}(\mathbb{R}^{n+1}, \mathbb{R}^n, \mathcal{E}_i)\,,$$

where $\mathcal{E}_i = \Big\{\big((y,\eta),(x_0,\xi_0)\big) : (y,\eta)$ is on the bicharacteristic strip through (x_0,ξ) with $\xi\big|_{T^*\{x_0=0\}} = \xi_0$ and $P_{i,2}(x_0,\xi) = 0\Big\}$. $\mathcal{E}_i$ describes the 'light cones' starting from the points of $\{x_0 = 0\}$.

DEFINITION 5.1

(i) $\mathcal{G}_j^i := B \circ \mathcal{E}_i\big|_{X_j} \subseteq T^*\{x_n = 0\} \times T^*\{x_0 = 0\}$, $i, j = 1, 2$.

(ii)

$$C_i^{e,\Sigma} := \Big\{\big((x_0, x', \xi_0, \xi'), (\tilde{x}_0, \tilde{x}', \tilde{\xi}_0, \tilde{\xi}'),\big) : x_0 = \tilde{x}_0, \xi_0 = \tilde{\xi}_0,$$
$$\big((x', \xi'), (\tilde{x}', \tilde{\xi}'),\big) \in C_i^{\Sigma}\Big\} , i = 1, 2.$$

LEMMA 5.2

(i) Under condition 1.4 (i), $\mathcal{G}_j^i, i = 1, 2$ is a local canonical graph.

(ii) $C_i^{e,\Sigma}, j, i = 1, 2$ is a local canonical graph.

Proof:

(i) Lemma 2.4 in [28].

(ii) Let C_i^{Σ} be locally given by the graph of the symplectomorphism

$$f_i^U : U \subseteq T^*\Sigma \to T^*\Sigma$$

(where Σ is identified canonically with $\mathbb{R}^n$ by $\Sigma \ni (x', 0) \leftrightarrow x' \in \mathbb{R}^{n-1}$). Then $C_i^{e,\Sigma}$ is described locally by the graph of

$$f_i^{e,U}(x_0, x', \xi_0, \xi') := (x_0, \xi_0, f_i^U(x', \xi')).$$

Clearly, $f_i^{e,U}$ is also a symplectomorphism for $i = 1, 2$.

□

DEFINITION 5.3 $\mathcal{R} := \mathcal{E}_2 \circ {}^t(C_2^{e,\Sigma} \circ \mathcal{G}_1^2) \circ (C_1^{e,\Sigma} \circ \mathcal{G}_1^1)$, $\mathcal{S} := \mathcal{E}_1 \circ {}^t\mathcal{G}_2^1 \circ \mathcal{G}_1^1$

These formulae can easily be restated as indicated in (2) in the introduction.

THEOREM 5.4 *$\mathcal{R}$ and $\mathcal{S}$ are conic, lagrangian submanifolds of $T_0^*\mathbb{R}^{n+1} \times \Omega_1$ with respect to $\sigma_{\mathbb{R}^{n+1}} - \sigma_{X_1}$.*

(Here $\Omega_i := \{(x, \xi) \in T_0^* X_{1,2}$: the bicharacteristic strip of Q_i issued from (x, ξ) hits $\Sigma\}$ which is an open subset of $T_0^*\mathbb{R}^n$ by [28] lemma 2.1.1.)

Idea of the proof: The $\mathcal{G}_j^i, i, j = 1, 2$ and $C_i^{e,\Sigma}, i = 1, 2$ are local canonical graphs by 5.2. The $\mathcal{E}_i, i = 1, 2$ are imbedded, closed, conic, lagrangian submanifolds of $(T^*\mathbb{R}^{n+1} \times T^*\mathbb{R}^n) \setminus 0$. So $\mathcal{R}, \mathcal{S}$ are conic, lagrangian subsets, because these properties are conserved under composition. They are also submanifolds, because the composition of a submanifold of a product and a local graph is again a submanifold. □

Another expression for $\mathcal{R}$ and its interpretation can be found in the introduction of this paper.

DEFINITION 5.5 $\mathcal{R}_\varepsilon := \Big\{\big((y, \eta), (x.\xi)\big) \in \mathcal{R} : y_0 \geq |t(x, \xi)| - \varepsilon$ or $y_0 \leq -|t(x, \xi)| + \varepsilon\Big\}$ where $t(x, \xi)$ is the time which the bicharacteristic of P_1 starting in $(x, \xi) \in \Omega_1$ needs to reach Σ (uniquely determined by assumption 1.4 (i)). Define $\mathcal{S}_\varepsilon$ analogously.

LEMMA 5.6 *For $\varepsilon > 0$, $\mathcal{R}_\varepsilon$ and $\mathcal{S}_\varepsilon$ are closed conic lagrangian submanifolds of $T_0^*I\!R^{n+1} \times \Omega_1$ with respect to $\sigma_{\mathbb{R}^{n+1}} - \sigma_{X_1}$.*

THEOREM 5.7 *For every compact subset $K \subseteq X_1$ there exists $\varepsilon = \varepsilon(K)$ and Fourier integral operators*

$$S = S_K \in I^{-1-1/4}(I\!R^{n+1}, X_1, \mathcal{S}_\varepsilon)$$

$$R = R_K \in I^{-1-1/4}(I\!R^{n+1}, X_1, \mathcal{R}_\varepsilon)$$

such that their distribution kernels K_S and K_R satisfy

$$WF(K_S) \subseteq \check{\mathcal{S}}_\varepsilon\,, \quad WF(K_R) \subseteq \check{\mathcal{R}}_\varepsilon$$

and such that the mapping

$$\mathcal{E}'_K \to \prod_{i=1}^{2} \mathcal{D}'_a(I\!R \times X_i) \tag{6}$$

$$f \mapsto \left((E_1 f - Sf)\big|_{\mathbb{R}\times X_1}, Rf\big|_{\mathbb{R}\times X_2}\right) \tag{7}$$

is a parametrix of the problem (4).

Idea of the proof: First we translate the interaction problem (4) into equations involving the Fourier integral operators R, S, E_i, F_i^Σ and various trace operators. Then we prove that the choice of the canonical relations $\mathcal{R}_\varepsilon$ and $\mathcal{S}_\varepsilon$ and is compatible with these equations. The amplitudes of R and S are constructed with an iteration scheme. Here we use the calculus of Fourier integral operators, especially the theorems on products, products with vanishing principal symbol, and on asymptotic sums (cf. [13], [14], [21], [35]). □

COROLLARY 5.8 *Take the situation of 5.7 with $F_i^\Sigma \in \Psi^0(\Sigma)$ satisfying 1.5 (ii). Then the conclusion of 5.7 holds, taking the closed homogeneous canonical relations $\mathcal{R}_\varepsilon$ and $\mathcal{S}_\varepsilon$ as given in [28], section 2 (corresponding to $\mathcal{R}_\varepsilon$ and $\mathcal{S}_\varepsilon$ in our notation, if we take $C_i^\Sigma :=$ diag $T^*(\Sigma) := \{((x',\xi'),(x',\xi')) \,:\, (x',\xi') \in T^*(\Sigma)\}$.)*

THEOREM 5.9 *Let $K \subseteq X_1$ be a compact subset. Consider $f \in H^1(X_1)$ with ess sup $f \subseteq K$. Let (u_1, u_2) be the exact solution of (4) with the initial condition f as indicated in section 2. Then:*

$$WF(u_1) \subseteq \mathcal{E}_1 \circ WF(f) \cup \mathcal{S} \circ WF(f)|_{\mathbb{R}\times X_1}$$

$$WF(u_2) \subseteq \mathcal{R} \circ WF(f)|_{\mathbb{R}\times X_2}$$

Idea of the proof: A big difference to the treatment of transmission conditions in [28] is, that the proof of this propagation result on the basis of the parametrix is more complicated. It uses the H^2–regularity stated in 3.6 and can be reduced to a regularity theorem for inhomogeneous interaction problems with inhomogeneous interaction conditions. Here the theory of T. Kato [22] and a trick of Lions-Magenes [24] are involved. The regularity theorem is worked out in [5].

One finds, that the difference of the exact solution and the approximate solution defined by the parametrix is C^∞. The assertion is then a consequence of the fact, that a Fourier integral operator transports the wave front set of a distribution via its canonical relation. □

EXAMPLE 5.10 For $i = 1, 2$ take any $f_i \in C^\infty(\Sigma)$ such that $f_i(\xi')^{-1} = \overline{f_i(\xi')}$, then Op f_i is unitary. Assume further $f_i(\xi') = 1$ outside a compact subset $M_i \subseteq \Sigma$ and $|f_i(\xi')| = 1$ in M_i. Then $f_i \in S^0_{1,0}(\Sigma)$ and thus Op $f_i \in \Psi^0(\Sigma)$. Clearly, Op f_i respects all Sobolev spaces. For $F_i^\Sigma :=$ Op f_i, the conclusion of 5.7 is thus valid with $C_i^{e,\Sigma} =$ diag $T^*(I\!R \times \Sigma)$. In general, the laws of reflection and refraction of singularities are not changed if we replace the transmission conditions in the classical transmission problem (T) in [28] by the interaction conditions (I_i), if the F_i^Σ are pseudo-differential operators.

Further admissible choices for F_i^Σ are translations and rotations and other Fourier integral operators with nontrivial canonical relations which can defined as in 1.6.

REFERENCES

[1] ALI MEHMETI, F.
Lokale und globale Lösungen linearer und nichtlinearer hyperbolischer Evolutionsgleichungen mit Transmission, Dissertation, Joh. Gutenberg-Universität, Mainz (1987).

[2] ALI MEHMETI, F.
Global Existence of Solutions of Semilinear Evolution Equations with Interaction, *Symp. 'Part. Diff. Equ.', Holzhau 1988*, Teubner-Texte zur Mathematik 112, Leipzig 1989.

[3] ALI MEHMETI, F.
Regular Solutions of Transmission and Interaction Problems for Wave Equations, *Math. Meth. Appl. Sci.*, **11**, 665-685 (1989).

[4] ALI MEHMETI, F.
Global Existence for Semilinear Evolution Equations with Applications to Interaction Problems, Preprint, 1989.

[5] ALI MEHMETI, F.
Existence and Regularity of Solutions of Cauchy Problems for Inhomogeneous Wave Equations with Interaction, to appear in the Proceedings "Workshop on Matrix and Operator Theory", Rotterdam, June 26-29, 1989.

[6] ALI MEHMETI, F. , NICAISE, S.
Compact Imbeddings and Interaction Problems, to appear in the Proceedings of the workshop on "Operator Semigroups and Evolution Equations" in Blaubeuren, 1989.

[7] ALI MEHMETI, F., NICAISE, S.:
Some Realizations of Interaction Problems, to appear in the Proceedings "2nd International Conference on Trends in Semigroup Theory and Evolution Equations", Delft, september 25-29, 1989.

[8] BEALS, M.
Propagation and Interaction of Singularities In Nonlinear Hyperbolic Problems (Birkhäuser, Boston, Basel, Berlin, 1989).

[9] BELOW, J.V.
Classical Solvability of Linear Parabolic Equations on Networks, *J. Diff. Eq.* **72**, 316-337 (1988).

[10] BONY, J.M.
Calcul symbolique et propagation des singularités pour les edp nonlinéaires, *Ann. Sc. Ec. Norm. Sup., Série 4*,**14**,(1981) 209-246.

[11] BONY, J.M.
Microlocal Analysis in Nonlinear Partial Differential Equations, lectures at the second C.I.M.E. session, Montecatini Terme, Italy (1989).

[12] BREZIS, H.
Monotonicity Methods in Hilbert Spaces and Some Applications to Nonlinear Partial Differential Equations, Contributions to Nonlinear Functional Analysis, E. Zarantonello (Ed.), 1971.

[13] DUISTERMAAT, J.J.
Fourier Integral Operators, Courant Inst. of Math. Sci., New York University, 1973.

[14] DUISTERMAAT, J.J., HÖRMANDER, L.
Fourier Integral Operators II, *Acta Math.***128**,(1972) 183-269.

[15] GRAMSCH, B.
Zum Einbettungssatz von Rellich bei Sobolevräumen, *Math. Zeitschr. 106* (1968) 81-87.

[16] GRAMSCH, B.
Asymptotik der Eigenwerte stark elliptischer Operatoren auf kompakten verzweigten Räumen; unpublished (1979).

[17] GRUBB, G.
Functional Calculus of Pseudo-Differential Boundary Problems, Birkhäuser, Boston, Basel, Stuttgart, 1986.

[18] HANSEN, S.
Singularities of Transmission Problems, *Math Ann.*, **268**, 233-253 (1984).

[19] HANSEN, S.
Fortpflanzung von Singularitäten bei Transmissionsproblemen, 2 lectures held in Mainz in august 1989.

[20] HANSEN, S.
Solution of a Hyperbolic Inverse Problem by Linearization, preprint Paderborn (1990).

[21] HÖRMANDER, L.
Fourier Integral Operators I, *Acta Math.***127**,(1971) 79-183.

[22] KATO, T.
Linear and Quasilinear Equations of Evolution of Hyperbolic Type, *C.I.M.E., II ciclo, Cortona,* 125-191 (1976).

[23] LIONS, J.L.
Optimal Control of Systems Governed by Partial Differential Equations, Springer Grundlehren Bd. 170, Berlin, Heidelberg, New York (1971).

[24] LIONS, J.L., MAGENES, E.
problèmes aux limites nonhomogenes et applications, vol. 2, travaux et recherches math. 18, Dunod, Paris, 1968.

[25] LORENZ, M.
Reflexion von diskreter Asymptotik für streng hyperbolische Randwertaufgaben, preprint 1989.

[26] LUMER, G.
Espaces ramifiés et diffusions sur les réseaux topologiques, *C.R. Acad. Sc. Paris,* t. 291, Série A (1980) 627-630.

[27] NICAISE, S.
Problèmes de Cauchy posés en norme uniforme sur les espaces ramifiés élémentaires, *C.R. Acad. Sc. Paris, t. 303,* Série I, no. 10 (1986) 443-446.

[28] NOSMAS, J.CL.:
Paramétrix du problème de transmission pour l'equation des ondes; *J. Math. Pures et Appl.* **56**, 423-435, 1977.

[29] PETKOV, V.
Inverse scattering problem for transparent obstacles; Math. Proc. Camb. Phil. Soc. 92, 361-367 ,1982.

[30] REMPEL, S., SCHULZE, B.-W.:
Index Theory of Elliptic Boundary Problems, Berlin, Akademie-Verlag, 1982.

[31] SABLÉ-TOUGERON, M.
Régularité microlocale pour des problèmes aux limites nonlinéaires, *Journées 'Equations aux Dérivées Partielles', St. Jean-De-Monts 1984,* Conf. No. 1.

[32] SHOWALTER, R. E.
Hilbert Space Methods for Partial Differential Equations (Pitman, London, San Franzisco, Melbourne, 1977).

[33] SCHROHE, E.
Randwertprobleme auf nichtkompakten Mannigfaltigkeiten, unpublished manuscript, 1990.

[34] TAYLOR, M.
Diffraction Effects in the Scattering of Waves, in *Singularities in Boundary Value Problems,* Nato Advanced Study Institute Series (H.G. Garnir, Ed.) D. Reidel Publishing Company, Dordrecht, Boston, London, 1981.

[35] TRÈVES, F.
Introduction to Pseudodifferential and Fourier Integral Operators, Plenum Press, New York and London, 1980.

F. Ali Mehmeti
FB Mathematik
Johannes-Gutenberg-Universität
Saarstraße 21
W-6500 Mainz, Deutschland

On the Solutions for a Class of Elliptic Equations Degenerate at the Boundary

L.A. Bagirov
Moscow Civil Engin. Institute
USSR Moscow 113114 Jaroslawskoe sh. 26.

1 Degeneration at all points of the boundary.

Let $G \subset \mathbb{R}^n$ be a domain with the boundary Γ which is an $(n-1)$-dimensional C^∞-manifold. Let us consider an equation

$$A(x,D)u(x) = \sum_{|\alpha|\le m, |\beta|\le m} D^\alpha (a_{\alpha\beta}(x)\rho^{\tau-2m+|\alpha|+|\beta|}(x) D^\beta u(x)) = f(x) \tag{1}$$

where

$$\alpha = (\alpha_1, \alpha_2, \ldots, \alpha_n), \ \beta = (\beta_1, \beta_2, \ldots \beta_n),$$

$$|\alpha| = \alpha_1 + \alpha_2 + \ldots + \alpha_n, \ D^\alpha = D_1^{\alpha_1} \ldots D_n^{\alpha_n}, \ D_k^{\alpha_k} = (-i)^{\alpha_k} \partial^{\alpha_k} / \partial x_k^{\alpha_k};$$

$\rho(x)$- is the regularized distance from the point $x \in G$ to the boundary Γ; $\tau \in \mathbb{R}$.

We will consider the eqaution (1) under the following two conditions.

Condition 1 *For* $\forall x \in \overline{G}$, $\forall \xi \in \mathbb{R}^n$

$$|\sum_{|\alpha|=|\beta|=m} a_{\alpha\beta}(x)\xi^{\alpha+\beta}| \ge \delta|\xi|^{2m}, \ \delta > 0, \ \delta = \text{const}.$$

Condition 2 *For* $\forall x \in \overline{G}$ *and for all numbers* $\{\xi_\alpha\}$, $|\alpha| \le m$, $\xi_\alpha \in \mathbb{C}$

$$Re \sum_{|\alpha|\le m, |\beta|\le m} a_{\alpha\beta}(x)\xi_\alpha\overline{\xi}_\beta \ge \delta_0 \sum_{|\alpha|=m} |\xi_\alpha|^2, \ \delta_0 = \text{const} > 0.$$

We shall consider solutions of the equations (1) from the class of functional spaces $V_\sigma^s(G)$, which have for integer $s \ge 0$, $\sigma \in \mathbb{R}$ the following norm

$$||f : V_\sigma^s(G)||^2 = \sum_{|\alpha|\le s} \int_G \rho(x)^{2[\sigma - s + |\alpha|]} |D^\alpha f|^2 dx.$$

We will also need the space $V_\sigma^s(G)$ for integers $s < 0$ namely $V_\sigma^s(G) = (V_{-\sigma}^{-s}(G))'$.

The space $V_\sigma^s(G)$ for integers $s < 0$ may be effectively described (see [1] Theorem 1.2.7).

If $f \in V_\sigma^s(G)$, $s < 0$, $\sigma \in \mathbb{R}$, then

$$f = \sum_{\alpha \le m} D^\alpha f_\alpha(x), \tag{2}$$

where

$$f_\alpha(x)\rho(x)^{-s+\sigma-|\alpha|} \in L^2(G),\ |\alpha| \le m,\ ||f : V^s_\sigma(G)|| = \inf \sum_{|\alpha|\le m} ||f_\alpha(x)\rho(x)^{-s+\sigma-|\alpha|} : L^2(G)||,$$

the infinum is carried out for all possible representations of the form (2).

If $a_{\alpha\beta}(x) \in C^\infty(\overline{G})$, then the operator

$$A(x,D) : V^{s+m}_{\sigma+\tau}(G) \to V^s_\sigma(G) \tag{3}$$

is a bounded operator for all integers s and for all σ.

If $s = m$ then the operator (3) is bounded for every $\sigma \in \mathbb{R}$ and all measurable bounded coefficients $a_{\alpha\beta}(x)$.

Theorem 1 *Let the condition 2 be satisfied, $a_{\alpha\beta}(x)$ be bounded measurable functions on G; $\tau \le 2m$, $\tau \ne 1,3,\dots,2m-1$. The there exists σ_0 such that for $|\sigma+\tau/2| < \sigma_0$ and for every $f \in V^{-m}_\sigma(G)$ there exists a unique solution of the equation (1) from the space $V^m_{\sigma+\tau}(G)$ and the following estimate holds*

$$||u : V^m_{\sigma+\tau}(G)|| \le c||Au : V^{-m}_\sigma(G)||. \tag{4}$$

This theorem is proved in [1] under some additional conditions on σ_0, in [1] a review of papers devoted to the equation (1) is also to be found.

We will sketch here the proof. It differs from the one given in [1] and is more natural.

Proof of Theorem 1

Introduce a bilinear form

$$a(u,v) = \sum_{|\alpha|\le m, |\beta|\le m} \int_G \rho(x)^{\tau-2m+|\alpha|+\beta|} a_{\alpha\beta}(x) D^\beta u(x) \overline{D^\alpha v(x)} dx.$$

As usual, we will call the function $u \in V^m_{\sigma+\tau}(G)$ the solution of the equation (1) with $f \in V^{-m}_\sigma(G)$, if

$$a(u,v) = \sum_{|\alpha\le m} \int_G f_\alpha(x) \overline{D^\alpha v(x)} dx = <f,v> \tag{5}$$

for $\forall v \in C_0^\infty(G)$.

From the Condition 2 it is easy to get the inequality

$$Re\ a(u,u) \ge \delta_0 \sum_{|\alpha|=m} \int_G \rho(x)^\tau |D^\alpha u(x)|^2 dx \tag{6}$$

for $\forall u \in C_0^\infty(G)$.

Using the inequality proved in the monograph [2] (see p. 319) from (6) we get for $\tau \ne 1,3,\dots 2m-1$

$$Re\ a(u,u) \ge \delta_1 ||u(x) : V^m_{\tau/2}(G)||^2,\ \delta_1 > 0. \tag{7}$$

Inequality (7) is valid for $\forall u \in V^m_{\tau/2}(G)$ since in this space the set $C_0^\infty(G)$ is dense.

Consider the auxiliary bilinear form

$$b_\gamma(u,v) = a(u\rho^{-\gamma}, v\rho^\gamma),\ \gamma - \tau/2 = \sigma.$$

It is easy to see, that

$$Re\ b_\gamma(u,u) \ge \delta_3 ||u : V^m_{\tau/2}(G)||^2 - C|\gamma|^2 ||u\rho^{\tau/2-m} : L^2(G)||^2.$$

Then if $|\gamma| < (\delta_2/c)^{1/2} = \sigma_0$ for $\forall u \in V^m_{\tau/2}(G)$

$$Re\, b(u,u) \geq \delta_3 ||u : V^m_{\tau/2}(G)||^2,\ \delta_3 > 0. \tag{8}$$

Since the form $b_\gamma(u,v)$ is coercive for $|\gamma| < \sigma_0$ there exists $w \in V^m_{\tau/2}(G)$ being the unique solution of the integral identity

$$b_\gamma(w,\Phi) = < f_1, \Phi > \quad \forall \Phi \in C_0^\infty(G), \tag{9}$$

where f_1 is determined by the inequality $< f_1, \Phi > = < f, \Phi\rho^\gamma >$, $f_1 \in V^{-m}_{\tau/2}(G)$.

Denote $u = w\rho^{-\gamma}$, $\Psi = \Phi\rho^\gamma$. From (9) it follows that

$$a(u,\Psi) = a(w\rho^{-\gamma}, \Phi\rho^\gamma) = b_\gamma(w,\Phi) = < f_1, \Phi > = < f, \Psi > .$$

It means that u is the unique solution of the equation (1) from the space $V^m_{\gamma+\tau/2}(G)$ an the estimate (4) holds.

Consider now the equation (1) with the the condition 1 and assume that $a_{\alpha\beta}(x) \in C^\infty(\overline{G})$. Then the equation (1) can be rewritten in the form

$$A(x,D)u(x) = \sum_{|\alpha| \leq 2m} a(x)\rho(x)^{\tau-2m+|\alpha|} D^\alpha u(x) = f(x), \tag{10}$$

where $a_\alpha(x) \in C^\infty(\overline{G})$.

We will investigate the equation (10) in the spaces $V^s_\sigma(G)$ with $s \geq 0$. Related equations were studied in [3]-[5].

Let $z \in \Gamma$. Let us make a change of variables $x \to y = (y', y_n)$ so that y_n axis point along the inner normal to Γ at the point z and the coordinates y' belong to the tangent plane to Γ at z.

Rewrite the operator

$$A(z,\rho,D_x) = \sum_{|\alpha| \leq 2m} \rho^{-2m+\tau+|\alpha|} a_\alpha(z) D^\alpha_x$$

in new variables and denote the obtained operator by $\hat{A}(z, y_n, D_{y'}, D_{y_n})$, Let us introduce an operator

$$\hat{A}_0(z, y_n, D_{y_n}) = y^{2m-\tau} \hat{A}(z, y_n, 0, D_{y_n}).$$

Condition 3 *For $\forall z \in \Gamma$ the polynomial $\hat{A}_0(z,\mu)$ has no roots on the line $Im\ \mu = \tau + \sigma - (s + 2m) + 1/2$.*

Let us introduce a space $E^s_\sigma(\mathbb{R}_+)$, of functions $v(y_n)$, $y_n \in \mathbb{R}_+ = \{y_n : y_n > 0\}$ such that

$$||v : E^s_\sigma(\mathbb{R}_+)||^2 = \sum_{k=0}^{s} y_n^{2\sigma}(1 + y_n^{2[k-s]})|D^k_{y_n} v(y_n)|^2 dy.$$

It is easy to prove that the operator

$$\hat{A}(z, y_n, \Theta, D_{y_n}) : E^{s+2m}_{\sigma+\tau}(\mathbb{R}_+) \to E^s_\sigma(\mathbb{R}_+) \tag{11}$$

for $\forall z \in \Gamma$, $\forall \Theta \in \mathbb{R}^{n-1}$, $|\Theta| = 1$ is continuous.

Theorem 2 *Under the condition 1 the operator (11) has finite-dimensional kernel and cokernel if and only if the condition 3 is satisfied.*

Condition 4 *$\exists \sigma_-, \sigma_+ \in \mathbb{R}$ such that $\forall z \in \Gamma$, $\forall \Theta \in \mathbb{R}^{n-1}$, $|\Theta| = 1$ and $\forall \sigma, \sigma_- < \sigma < \sigma_+ dim\ Ker\hat{A} = dim\ Coker\hat{A} = 0$.*

Theorem 3 *Let the conditions 1, 3, 4 be satisfied. Then for $\forall \sigma, \sigma_- < \sigma < \sigma_+$ the operator has finite-dimensional kernel and cokernel.*

The scheme of the proof of this theorem is given in [1].

The condition 4 seems rather hard to check. But in some cases it is possible to make it more explicit, i.e. to calculate σ_-, σ_+.

Example 1 *Let $r = (x^2 + y^2)^{1/2}$, $\Phi = \arctan y/x$. Consider in a circle $K = \{(x, y) : r < 1\}$ of the radius 1 the equation*

$$D_x((1-r)^\tau D_x u) + D_y((1-r)^\tau D_y u) = f(x, y). \tag{12}$$

Introducing polar coordinates $x = r\cos\Phi, y = r\sin\Phi$ we get

$$D_r[(1-r)^\tau D_r(v)] - ir^{-1}(1-r)^\tau D_r v + r^{-2}(1-r)^\tau D_\Phi^2 v = F(r, \Phi), \tag{13}$$

where $v(r, \Phi) = u(r\cos\Phi, r\cos\Phi)$, $F(r, \Phi) = f((r\cos\Phi, r\cos\Phi)$.

If $z \in \Gamma = \{(x, y) : r = 1\}$, then the operator $\hat{A}(z, y_n, \Theta, D_{y_n})$ has the form

$$\hat{A}(D_t(t^\tau D_t v) + t^\tau v \tag{14}$$

where $t = 1 - r$, and $\hat{A}_0(z, tD_t) = (tD_t)^2 - i(\tau - 1)tD_t$.

Condition 3 means that $\sigma \neq s + 0.5$, $\sigma \neq s - \tau + 0.5$. The operator (14) is the Bessel operator and elementary analysis gives $\sigma_+ = s + 1 - 0.5(\tau + |\tau - 1|)$, $\sigma_- = s + 1 + 0.5(|\tau - 1| - \tau)$.

Theorem 1 states that there exists such $\sigma_0 > 0$ that for $|\sigma + \tau/2| < \sigma_0$ the exists a unique solution of the equation (1) from the space $V^m_{\sigma+\tau}(G)$. The number σ_0 is not given explicitely by this theorem. But for some equations its value can be obtained.

Example 2 *Consider the equation (12) in the circle $K = \{(x, y) : r < 1\}$. The funtionals $a(u, v)$ and $b_\gamma(u, v)$ have the form*

$$\begin{aligned} a(u, v) &= \int_K (1-r)^\tau [D_x u \overline{D_x v} + D_y u \overline{D_y v}] dx dy; \\ b_\gamma(u, v) &= \int_K (1-r)^\tau \left[D_x((1-r)^{-\gamma} u)\overline{D_x((1-r)^\gamma v)} + D_y((1-r)^{-\gamma} u)\overline{D_y((1-r)^\gamma v)}\right] dx dy. \end{aligned}$$

It is easy to see that

$$b_\gamma(u, v) = a(u, v) - \gamma^2 \int_K (1-r)^{\tau-2} |u|^2 dx dy. \tag{15}$$

For all $u \in C_0^\infty(K)$

$$a(u, u) \geq \mu_0 \int_K (1-r)^{\tau-2} |u|^2 dx dy, \tag{16}$$

where the constant $\mu_0(\tau) > 0$ is the minimal eigenvalue of the problem:

$$D_x((1-r)^\tau D_x u) + D_y((1-r)^\tau D_y u) = \lambda(1-r)^{\tau-2} u, \ u|_{r=1} = 0.$$

This problem is reduced to the solution of the hypergeometric equation, which we do not give here for the sake of brevity. From (15) and (16) we get for this example $\sigma_0 = (\mu_0)^{-1/2}$.

2 Degeneration on a part of the boundary ($n = 2$)

Let G be a domain of the plane, whose boundary consists of the interval $\Delta = (a, b)$ of the OX_1 axis and a smooth curve joining the points $(a, 0)$ and $(b, 0)$ and lying in the half plane $x_2 > 0$.

Let us consider in this domain an equation

$$A(x, D)u(x) = \sum_{|\alpha|\leq 2m} a_\alpha(x) x_2^{\tau-2m+|\alpha|} D^\alpha u(x) = f(x). \tag{17}$$

Here $\tau \in \mathbb{R}$ is the degeneration index, $\alpha = (\alpha_1, \alpha_2)$, $|\alpha| = \alpha_1 + \alpha_2$, $a_\alpha(x) \in C^\infty(\overline{G})$.

Condition 5 *For* $\forall x \in \overline{G}$, $\forall \xi \in \mathbb{R}^2$

$$|\sum_{|\alpha|=2m} a_\alpha(x)\xi^\alpha| \geq \delta|\xi|^{2m}, \quad \delta > 0.$$

Let us pose the boundary condition on Γ_0:

$$B_j(x', D_x)u(x)|_{\Gamma_0} = 0 \quad (j = 1, 2, \ldots, m), \quad x' \in \Gamma_0. \tag{18}$$

Here B_j are differential operators of order $m_j \leq 2m - 1$ with smooth coefficients.

Condition 6 *For every point* $x' \in \overline{\Gamma}_0$ *the Shapiro-Lopatinsky conditions are satisfied.*

We shall seek for the solution of the problem (17), (18) in the space $V^s_{\gamma_1,\gamma_2,\sigma}(G)$ which has the following norm

$$\|u : V^s_{\gamma_1,\gamma_2,\sigma}(G)\|^2 = \sum_{|\alpha|\leq s} \int_G |D_x^\alpha u|^2 x_2^{2[\sigma-s+|\alpha|]} r_a^{2\gamma_1} r_b^{2\gamma_2} dx,$$

where $s \geq 0$ is an integer, $\gamma_1, \gamma_2 \in \mathbb{R}$, $\sigma \in \mathbb{R}$, $r_a = ((x_1 - a)^2 + x_2^2)^{1/2}$, $r_b = ((x_1 - b)^2 + x_2^2)^{1/2}$.

We do not pose any condition on Δ, because the set of functions from C^∞ with $u|_\Delta = 0$ is dense in $V^s_{\gamma_1,\gamma_2,\sigma}(G)$.

Let us denote

$$\mathcal{V}^s_{\gamma_1,\gamma_2,\sigma+\tau}(G) = \{u : u \in V^s_{\gamma_1,\gamma_2,\sigma}(G), B_j u|_{\Gamma_0} = 0, \ j = 1, 2, \ldots, m\}$$

It is easy to see that the operator

$$A(x, D) : \mathcal{V}^{s+2m}_{\gamma_1,\gamma_2,\sigma+\tau}(G) \longrightarrow V^s_{\gamma_1,\gamma_2,\sigma}(G) \tag{19}$$

is continuous.

Let us evaluate the conditions under which the problem (17), (18) is solvable and let us investigate the behaviour of solution in the neighbourhood of the points $(a, 0), (b, 0)$.

If $m = 2$, then the sharp edge plate deformation problem is reduced to the equation of the form (17) (see [6]). When $m = 1$, $\tau = 1$, then the Euler-Poisson-Darboux equation $y\Delta u + au x + bu y = 0$, which is often used in applications, also belongs to the class in consideration. The investigation of boundary problems for the equations (17) with $m = 1$ was begun by Keldysh and continued by many mathematicians. The main results are given in the monographs [7]-[9] and the review [10], where a full bibliography may be found.

Condition 7 *In a small neighbourhood of the points* $(a,0)$ *and* $(b,0)$ *the curve* Γ_0 *is given by the equations*

$$x_2 = k_a(x_1 - a) \quad \textit{and} \quad x_2 = k_b(x_1 - b),$$

respectively

It is obvious that when a generalized solution is constructed, no requirements on the character of the intersection of Γ_0 and Δ have to be imposed, but when a strong solution is constructed or the smoothness of the generalized solution is investigated, it is not so.

At all points $z = (x_2, 0)$ it is necessary to impose the conditions analogous to the Conditions 3 and 4. We will not go into details here.

Let us formulate additional conditions on the operator A at the points $(a,0), (b,0)$ It is sufficient to do it for the point $(a,0)$. For the point $(b,0)$ they are quite analogous. Let $K_a = \{x : 0 < \arctan(x_2/(x_1-a)) < \Phi_a = \arctan k_a\}$, $r = r_a = ((x_1-a)^2 + x_2^2)^{1/2}$, $\Phi = \arctan(x_2/(x_1-a))$.

Consider in K_a the problem

$$\begin{aligned} \sum_{|\alpha|\le 2m} a_\alpha(a,0)x_2^{\tau-2m+|\alpha|}D^\alpha u &= f, \quad x \in K_a \\ B_j^0((a,0),D)u|_{\Phi=\Phi_a} &= 0, \quad j = 1,\ldots,m, \end{aligned}$$

where B_j^0 is the leading part of the operator B_j.

The change of variables $x - a = r\cos\Phi$, $y = r\sin\Phi$ transforms it into

$$\begin{aligned} P(\Phi, rD_r, D_\Phi)v(r,\Phi) &= r^{2m-\tau}f(r,\Phi), \\ Q_j(\Phi, rD_r, D_\Phi)v(r,\Phi)|_{\Phi=\Phi_a} &= 0, \quad j = 1,\ldots,m. \end{aligned} \tag{20}$$

With the problem (20) it is natural to connect a polynomial pencil

$$P(\Phi,\lambda,D_\Phi)w(\Phi) = g(\Phi), \quad Q_j(\Phi,\lambda,D_\Phi)w(\Phi)|_{\Phi=\Phi_a} = 0, \quad j = 1,\ldots,m. \tag{21}$$

Let $\omega_a = (0,\Phi_a)$ be an interval of $\mathbb{R}$. Introduce the space $E_\sigma^s(\omega_\alpha)$ with the norm

$$||w : E_\sigma^s(\omega_a)||^2 = \sum_{|\alpha|\le s} \int_{\omega_a} (\sin\Phi)^{2(\sigma-s+|\alpha|)}|\lambda|^{2\alpha_1}|D_\Phi^{\alpha_1}w(\Phi)|^2 d\Phi;$$

the subspace $\mathcal{E}_\sigma^s(\omega_a) = \{w : w \in E_\sigma^s(\omega_a), Q_j w|_{\Phi=\Phi_a} = 0,\ j = 1,\ldots,m\}$; and the set $\mathcal{K}_{N,\Psi} = \{\lambda : \lambda \in \mathbb{C}, |\arg\lambda - \kappa\pi| < \Psi < \pi/2,\ \kappa = 0,1, |\lambda| > N\}$.

Theorem 4 *If the Conditions 17, 18 and the Conditions 3, 4 are satisfied at the point* $(a,0)$, *then for* $\forall\sigma, \sigma_- < \sigma < \sigma_+$, $\exists N > 0$ *and* $\varphi < \pi/2$ *such that the operator* $P(\Phi,\lambda,D_\Phi) : \mathcal{E}_{\sigma+\tau}^{s+2m}(\omega_a) \to E_\sigma^s(\omega_a)$ *is an isomorphism for* $\forall\lambda \in \mathcal{K}_{N,\Psi}$.

Corollary 1 *If the conditions of the Theorem 4 are satisfied, then the operator* $P^{-1}(\Phi,\lambda,D_\Phi) = P^{-1}(\lambda)$ *is a meromorphic function of* λ.

Denote the inverse operator of the problem (21) by $P_a^{-1}(\lambda)$, and the inverse operator of the analogous problem for the point $x = (b,0)$ by $P_b^{-1}(\lambda)$.

Theorem 5 *Let the Conditions 17, 18 and the Conditions 3, 4 be satisfied at all points of the interval* $\overline{\Delta}$. *Let the operator* $P_a^{-1}(\lambda)$ *has no poles on the line* $Im\ \lambda = \gamma_1 + \sigma + \tau - s - 2m + 1$, *and the operator* $P_a^{-1}(\lambda)$ *have no poles on the line* $Im\ \lambda = \gamma_2 + \sigma + \tau - s - 2m + 1$.

Then for $\forall\sigma, \sigma_- < \sigma < \sigma_+$ *the operator given by the equality (19) is a Noether operator.*

Theorems 4, 5 enable us to obtain the asymptotics of the solution of the problem (17), (18) when $x \to (a,0)$ or $x \to (b,0)$ analogously to elliptic problems in the domains with corner points (see [11]). The form of these asymptotics depends on the poles of the operators $P_a^{-1}(\lambda)$, $P_b^{-1}(\lambda)$ in some strip and on their eigenfunctions and associated eigenfunctions.

Example 3 *Consider the problem*

$$D_{x_1}(x_2^\tau D_{x_1}u) + D_{x_2}(x_2^\tau D_{x_2}u) = f(x),\ x \in G,\ u|_{\Gamma_0},\ u \in V^{s+2}_{\gamma_1,\gamma_2,\sigma+\tau}(G). \tag{22}$$

As we have seen in Example 1 the numbers σ_- and σ_+ for this problem can be evaluated explicitely and the Conditions 3 and 4 are satisfied if $\sigma \neq s+0.5$, $\sigma \neq s-\tau+0.5$,

$$\sigma_- = s+1+0.5(|\tau-1|-\tau) < \sigma < \sigma_+ = s+1-0.5(\tau+|\tau-1|).$$

Let $a = 0$. Calculate the poles. To do it we have to find such λ that the problem (22), in which the domain G is replaced by the angle $K = \{x : 0 < \Phi < \Phi_0 \leq \pi\}$, has the solution of the form $r^{i\lambda}w(\Phi)$.

For the definition of λ and $w(\Phi)$ we get the problem

$$(\sin\Phi)^\tau(D_\Phi^2 + \lambda^2)w(\Phi) - i\tau(\sin\Phi)^{\tau-1}(\lambda\sin\Phi + \cos\Phi D_\Phi)w(\Phi) = 0,$$
$$w(\Phi_0) = 0,\quad w(\phi) \in \mathcal{E}^{s+2}_{\sigma+\tau}(\omega_0),\quad \omega_0 =]0,\Phi_0[.$$

In the case when $\Phi_0 = \pi$ the condition $w(\Phi_0 = 0)$ is canceled.

This problem is reduced to the evaluation of some solutions of the hypergeometric equation. We will restrict ourselves to the case $\tau \neq 2n-1$. Simple calculations give us transcendental equations for λ:

$$\begin{aligned} F(\tau+i\lambda; -i\lambda; 0.5(\tau+1); (\sin(\Phi_0/2))^2) &= 0, \quad \tau > 1; \\ F(0.5(\tau+1)+i\lambda; 0.5(1-\tau)-i\lambda; 0.5(3-\tau); (\sin(\Phi_0/2))^2) &= 0, \quad \tau \leq 1. \end{aligned}$$

Here F is the hypergeometric Gauss function.

If $\Phi_0 = \pi/2$ or $\Phi_0 = \pi$, the equations obtained can be solved explictely. Their roots are

$$\begin{array}{lll} \Phi_0 = \pi,\ \tau \geq 1 & \lambda_k' = -i(k+0.5(1-\tau)), & \lambda_k'' = i(k+0.5(1+\tau)); \\ \Phi_0 = \pi,\ \tau < 1 & \lambda_k' = -i(k+1-\tau), & \lambda_k'' = i(k+1); \\ \Phi_0 = \pi/2,\ \tau \geq 1 & \lambda_k' = i(2k+1+\tau), & \lambda_k'' = -i(2k+1); \\ \Phi_0 = \pi/2,\ \tau < 1 & \lambda_k' = i(2k+2), & \lambda_k'' = -i(2k+2-\tau);\ k = 0,1,2,\ldots \end{array}$$

A part of the results given in this report can be found in the paper [1].

References

[1] S.M. Nicol'skij, P.I. Lizorkin, N.V. Miroshin. Weighted functional spaces and their applications to boundary value problems for degenerate elliptic equations, lzv. VUZ'ov Matematika, (1988), N8,pp. 4-30.

[2] H. Triebel, Theory Function Spaces Differential Operators, VEB Deutscher Verlag der Wissenschaften, Berlin, 1978.

[3] M.I. Vishik, V.V. Grushin. Boundary-value problems for elliptic equations degenerate at the boundary, Mat. Sb., (1969), Vol. 80, N4, pp. 455-491.

[4] P.Bolley, J.Camus. Reqularité pour une classe de problèmes aux limites elliptiques dégénérés variationnels, Boll. Unione mat Ital., (1977), v.14, N1, pp.77-100.

[5] P.Bolley, J.Camus. Sur une classe d'opérateurs elliptiques et dégénérés à plusieurs variables, Boll.Soc. Math.France, Suppl.Mem., (1973), 34, p. 55-140.

[6] S.G. Mikhlin. Variational Methods in Mathematical Physics, Nauka, Moscow, 1966.

[7] M.M. Smirnov. Degenerate Elliptic and Hyperbolic Equations, Nauka, Moscow, 1966.

[8] A.V. Bitsadze. Some Classes of Partial Differential Equations, Nauka, Moscow, 1973.

[9] O.A. Oleinik, E.V. Radkevich. Second-order Equations with Non-negative Characteristics Form, VINITI, Moscow, 1971.

[10] V.P. Glushko, Y.B. Savchenko. Degenerate elliptic equations of high order: spaces, operators, boundary value problems. Itogi Nauki i Tech., Math. Anal., (1985),v.23, pp.125-218.

[11] V.A. Kondrat'ev. Boundary value problems for elliptic equations in domains with conical or corner points. Trudy Mosk.Mat. Obshch., (1967), v.16, p.209-292.

[12] L.A. Bagirov. On the solutions of boundary value problems for elliptic equation degenerate at the part of the boundary. mat.Zametki, (1990), v.47, N1, pp.29-38.

EDGE ASYMPTOTICS ON A SKEW CYLINDER

Martin Costabel and Monique Dauge

Heidelberg and Nantes

1. POSITION OF THE PROBLEM

1.a Origin. This problem was posed as a question to the first author by I. Babuška (Maryland).

It is well known that singularities of the domain give rise to a loss of regularity for the solutions of any elliptic boundary value problem. The situation is rather well understood when the singularities are isolated points of the boundary and are of conical type (see [4], [8]).

When a conical singularity is tensorized with an affine space, one gets an edge. Regularity results are rather complete in that case [6], [11]. If the operator is translation invariant along the edge, the asymptotics can be derived in a direct way from the asymptotics on the corresponding conical domain [1].

But for physical examples in the ordinary three-dimensional space, when a bounded domain has edges and no corners, then the edges are necessarily curved. The simplest example is a cylinder with circular basis, cut orthogonally to its generating lines. But this example is very particular: The opening of the edge is everywhere $\pi/2$ and the curvature of the edge is constant. If one cuts the cylinder by a plane which is skew with respect to the generating lines, then the edge is elliptic and the opening angle is varying. This gives rise to difficulties for the precise analysis of the structure of the solution, due to the fact that the asymptotics in the corresponding two-dimensional domains depend in a discontinuous way on the opening parameter. In particular, the coefficients of the singular functions along the edge (stress intensity factors etc.) will blow up at certain points. Such a behavior causes difficulties also for numerical approximations.

1.b Geometry of the domain. Let B be an analytic bounded domain in $\mathbb{R}^2$. Let Ψ be an affine function $(x_2, x_3) \mapsto x_1 = \Psi(x_2, x_3)$. We assume that

$$\text{for all } (x_2, x_3) \in \overline{B}, \quad \Psi(x_2, x_3) > 0.$$

We introduce

$$\Omega = \left\{(x_1, x_2, x_3) \in \mathbb{R}^3 \mid (x_2, x_3) \in B,\ 0 < x_1 < \Psi(x_2, x_3)\right\}.$$

This is our skew cylinder. We denote by M the top edge and by $\partial_1\Omega$ the side of the cylinder:

$$\partial_1\Omega = \left\{(x_1, x_2, x_3) \in \mathbb{R}^3 \mid (x_2, x_3) \in \partial B,\ 0 < x_1 < \Psi(x_2, x_3)\right\}.$$

The union of the top and the bottom of the cylinder is denoted by $\partial_2\Omega$.

For what we are going to do, a more general class of domains is admissible: The piecewise analytic domains with edges with the property that in each point of an edge, the domain is locally analytically diffeomorphic to a straight wedge.

For instance, if Ψ_1 and Ψ_2 are analytic on $\mathbb{R}^2$ and such that

$$\forall (x_2, x_3) \in \overline{B}, \quad \Psi_1(x_2, x_3) < \Psi_2(x_2, x_3),$$

we get a cylinder with curved ends:

$$\Omega = \left\{(x_1, x_2, x_3) \in \mathbb{R}^3 \mid (x_2, x_3) \in B,\ \Psi_1(x_2, x_3) < x_1 < \Psi_2(x_2, x_3)\right\}.$$

Such a situation is useful in reaction-diffusion, for instance.

Interesting extensions could be:

1. Include the situation where an isolated point on the edge has an opening angle equal to π. Such a situation appears for instance when two isometric skew cylinders are glued together along their skew tops.
2. Include piecewise C^∞ manifolds with edges.
3. Include higher order singularities (corners of any dimension).

1.c Boundary value problems. We choose two simple examples: a mixed problem and the Neumann problem for the Laplace operator. Let us denote by $(\mathcal{P}_I)$, resp. $(\mathcal{P}_{II})$ these two problems. We write

$$(\mathcal{P}_I) \quad \begin{cases} \Delta u = f & \text{on } \Omega \\ u = 0 \quad \text{on } \partial_1\Omega, & \frac{\partial u}{\partial n} = h_2 \quad \text{on } \partial_2\Omega. \end{cases}$$

$$(\mathcal{P}_{II}) \quad \begin{cases} \Delta u = f & \text{on } \Omega \\ \frac{\partial u}{\partial n} = h_1 \quad \text{on } \partial_1\Omega, & \frac{\partial u}{\partial n} = h_2 \quad \text{on } \partial_2\Omega. \end{cases}$$

We take the Laplace operator because it is the simplest possible example of an elliptic operator. Moreover, it is physically relevant. We choose the mixed Dirichlet-Neumann problem because the singularities appear at a low regularity level (in a generic way, $u \notin H^2(\Omega)$). We also consider the Neumann problem, because it does not satisfy the isomorphism condition in weighted Sobolev spaces which is used in [6], [9]. Lastly, the behavior of such problems in a two-dimensional angle is well known [2].

Other boundary value problems behave in a similar way as $(\mathcal{P}_I)$ and $(\mathcal{P}_{II})$. As interior operator, we can take any elliptic second order operator with real analytic coefficients. With such operators, we would in a natural way consider any Dirichlet-Neumann boundary problems (including pure Dirichlet and pure Neumann problems), the Neumann condition being defined using the conormal derivative associated with the interior operator. In this way we obtain, with our above class of piecewise analytic domains, a class of boundary value problems which is invariant with respect to analytic diffeomorphisms.

It is also possible for us to consider mixed oblique derivative problems (semi-variational problems) in a framework related to that in [7].

1.d Exponents of the singularities.

Let us first recall the result in a two-dimensional angle G with opening ω. We use polar coordinates (r, θ) so that G corresponds to $r > 0$, $0 < \theta < \omega$. For any integer $j \geq 1$, we set

$$\begin{aligned}
\nu_j &= \begin{cases} (2j-1)\dfrac{\pi}{2\omega} & \text{for the mixed problem} \\[2ex] \dfrac{j\pi}{\omega} & \text{for the Neumann problem} \end{cases} \\
\sigma_j(r,\theta) &= r^{\nu_j} \cos \nu_j \theta \\
S_j(r,\theta) &= r^{\nu_j} (\log r \cos \nu_j \theta - \theta \sin \nu_j \theta) \quad \text{if } \nu_j \in \mathbb{N}.
\end{aligned}$$

Proposition 1.1 *Let $s \in \mathbb{R}$, $s > 1/2$. Let us assume that the interior datum in G has H^{s-1} regularity and that the Neumann data have $H^{s-\frac{1}{2}}$ regularity. Then the variational solution w of the mixed Neumann-Dirichlet problem, resp. Neumann problem on G, admits the following decomposition,*

$$w = w_{\text{reg}} + w_{\text{sing}}$$

where

$$w_{\text{reg}} \in H^{s+1-\varepsilon}(G) \quad \forall \varepsilon > 0$$

$$w_{\text{sing}} = \sum_{1 \leq \nu_j < s,\, \nu_j \notin \mathbb{N}} c_j \sigma_j + \sum_{1 \leq \nu_j < s,\, \nu_j \in \mathbb{N}} c_j S_j.$$

Remark 1.2 If $\forall j \in \mathbb{N}$ there holds $\nu_j \neq s$, then $w_{\text{reg}} \in H^{s+1}(G)$. On the other hand, if $\exists j \in \mathbb{N}$ such that $\nu_j = s$, then w_{reg} does not belong to $H^{s+1}(G)$, in general. If we remain in the scale of ordinary Sobolev spaces, it is impossible to remove this ε. If one characterizes the preimage, one finds a space in which H^{s+1} is not closed. Conversely, the image of H^{s+1} is not closed in $H^{s-1}(G) \times H^{s-\frac{1}{2}}(\partial G)$.

Let us return now to our skew cylinder. We assume that for a fixed real $s > 1/2$

$$\begin{aligned} f &\in H^{s-1}(\Omega) \\ h_j &\in H^{s-\frac{1}{2}}(\partial_j\Omega), \quad j = 1, 2 \end{aligned}$$

with the usual compatibility condition for the Neumann problem. Then there exists a variational solution $u \in H^1(\Omega)$.

For any $x \in M$, let $\omega(x)$ be the opening angle of Ω in x. We set

$$\nu(x) = \begin{cases} \dfrac{\pi}{2\omega(x)} & \text{for} \quad (\mathcal{P}_I) \\ \dfrac{\pi}{\omega(x)} & \text{for} \quad (\mathcal{P}_{II}) \end{cases}$$

If $s < \min_{x \in M} \nu(x)$, then $u \in H^{s+1}(\Omega)$. Since, anyway, $\min_{x \in M} \nu(x) > 1/2$, we have $u \in H^{\frac{3}{2}+\varepsilon}(\Omega)$ and the Neumann conditions make sense. From now on we assume that

$$\min_{x \in M} \nu(x) < s.$$

Then u can be split into two parts for any $\varepsilon > 0$,

$$u = u_{\text{reg}} + u_{\text{sing}}$$

where $u_{\text{reg}} \in H^{s+1-\varepsilon}(\Omega)$ and u_{sing} is an asymptotics. Such a decomposition generally depends on $\varepsilon > 0$. Our aim is to describe the structure of such splittings. In particular, we want to separate as much as possible the roles of the different variables, the abscissa y along the edge, the distance r from the edge and the angular variable θ. More precisely, we intend to separate what comes from the geometrical framework (domain and boundary value problem) and what comes from the data (f, h_j). The part that comes from the geometrical framework will be described as an analytic bundle. The part that comes from the data will be described as $H^{s-\varepsilon}$ sections of this bundle. Only the second notion depends on ε. By contrast, the bundle is an object which is associated to the pair (domain, boundary value problem) in a canonical way.

2. SIMPLE ASYMPTOTICS

2.a A simple result. The exponents of the singularities which appear in the asymptotics along the edge depend on the edge parameter y. They are the same as in a two-dimensional problem for the Laplace operator with lower order terms and the opening $\omega(y)$, which is the opening of Ω at the point y. We can enumerate them using a double index $k = (k_1, k_2)$, where $k_1, k_2 \in \mathbb{N}$ and

$$\nu_{(0,k_2)}(y) = k_2$$

and for $k_1 \geq 1$,

$$\nu_k(y) = \begin{cases} (2k_1 - 1)\dfrac{\pi}{2\omega(y)} + k_2 & \text{for } (\mathcal{P}_I) \\ \dfrac{k_1 \pi}{\omega(y)} + k_2 & \text{for } (\mathcal{P}_{II}). \end{cases}$$

We have to include all positive integers in those exponents due to the possible interaction between polynomials and singularities.

What can be expected as asymptotics along the edge is

$$\sum_{k,q,\beta} c_{k,q,\beta}(y)\, r^{\nu_k(y)} \log^q r\, \varphi_{k,q,\beta}(y,\theta) \tag{2.1}$$

where only the $c_{k,q,\beta}$ depend on the data (f, h_j). Actually, such an asymptotics in tensor product form is not convenient in general, since the $c_{k,q,\beta}$ are not regular enough: The best which can be expected is $H^{s-\nu_k(y)}$. Such an asymptotics could be valid only if $s = \infty$, or if the 'regular part' has only half the regularity of the data. Therefore we introduce a function $\Phi(y, r)$ such that its partial Fourier transform satisfies

$$\mathcal{F}_{y\to\xi}\Phi(\xi, r) = \phi(r|\xi|)$$

where ϕ is a rapidly decreasing function, has a Fourier transform with compact support, and satisfies for a sufficiently large N

$$\phi(0) = 1, \quad \frac{d^l}{ds^l}\phi(0) = 0 \quad (l = 1, \ldots, N).$$

We define the convolution with respect to y,

$$(\Phi * c)(y, r) := \int \Phi(y - y', r)\, c(y')\, dy'.$$

The following theorem describes our result on the simple edge asymptotics.

Theorem 2.1 *Let J be an open segment in M and $\tilde{J}$ any open set in M such that $\overline{J} \subset \tilde{J}$ holds and let $\mathcal{U}_J$ be a sufficiently small neighborhood of J in $\overline{\Omega}$. Let χ be an analytic diffeomorphism which defines on $\mathcal{U}_J$ local cylindrical coordinates (y, r, θ) with $0 < \theta < \omega(y)$. In local coordinates, we write I and $\tilde{I}$ for J and $\tilde{J}$. We assume that for some $\varepsilon_0 \geq 0$ there holds*

$$\text{for all } k \text{ we have} \quad \forall y \in \tilde{I},\ \nu_k(y) < s \quad \text{or} \quad \forall y \in \tilde{I},\ \nu_k(y) \geq s - \varepsilon_0 \tag{2.2}$$

$$\text{for all } y \in \tilde{I}, \quad \text{if } \nu_k(y) = \nu_{k'}(y) < s \quad \text{then } k = k'. \tag{2.3}$$

To each k there exists a finite set of indices (q, β) and analytic functions $\varphi_{k,q,\beta}(y, \theta)$ such that any solution u of problem $(\mathcal{P}_I)$ or $(\mathcal{P}_{II})$ can be decomposed into

$$u = u_{\text{reg}} + u_{\text{sing}}.$$

Here $u_{\text{reg}} \in H^{s+1-\varepsilon}(\mathcal{U}_J)$ $\forall \varepsilon > \varepsilon_0$ and

$$u_{\text{sing}} = \chi^{-1}\left(\sum_{k,q,\beta} (\Phi * c_{k,q,\beta})(y,r)\, r^{\nu_k(y)} \log^q r\, \varphi_{k,q,\beta}(y,\theta)\right). \tag{2.4}$$

The coefficients $c_{k,q,\beta}(y)$ are defined on $\tilde{I}$ and satisfy $c_{k,q,\beta} \in H^{s-\nu_k(y)-\varepsilon}(I)$ for all $\varepsilon > 0$. The sum extends over those k for which $\nu_k < s$ holds on $\tilde{I}$.

Remark 2.2 If $k_1 = 0$ then the logarithmic terms are absent and for any fixed $y \in M$, the function $r^{\nu_k(y)} \varphi_{k,0,\beta}(y,\theta)$ is a polynomial in cartesian variables. The corresponding term in the asymptotics can be put into the regular part. We include such terms in our asymptotics as a preparation for the study of crossing points.

Remark 2.3 Assume that $k_2 = 0$ and $\nu_k \notin \mathbb{N}$. Then $q = 0$ and β has only one value. We have

$$\varphi_{k,0,1}(y,\theta) = \cos \nu_k(y)\theta.$$

Remark 2.4 Only the first terms (with $k_2 = 0$) arise directly from the conormal principal symbol of the operator. The other terms come from other parts of the operator and there exists no simple relation between the symbol of the operator and the singularities. For instance, one obtains a logarithmic term by differentiating $r^{\nu_k(y)}$ with respect to y.

2.b The question of optimality. Results related to Theorem 2.1 have been proven for the Dirichlet problem by Kondratev [5] concerning the first singularity, Nikishkin [10] for all singularities, Maz'ya and Roßmann [9] for more general operators. We have to mention some differences between our result and those of the above authors: The first one (but not essential) is our use of spaces with non-integer exponents. The second one is that our results do not need the isomorphism assumption in weighted Sobolev spaces; such an assumption excludes the Neumann problem, for instance.

The third difference lies in our localization. All the above authors make the following assumption

$$\forall k,\ \forall y \in M, \quad \nu_k(y) \neq s. \tag{2.5}$$

In such a case ε_0 may be taken as 0 in our theorem. It is interesting to realize that such an assumption means a severe restriction on the possible range for the couple (α, s) where α describes the slope of the top of the skew cylinder: $0 < \alpha < 1$ and $\beta = \alpha\pi/2$ is the angle between the top of the cylinder and the horizontal (x_2, x_3)-plane.

Let us illustrate this restriction. The maximal opening is $(1+\alpha)\pi/2$ and the minimal one is $(1-\alpha)\pi/2$. It is easy to check that if

$$\nu_{(1,0)}\left((1-\alpha)\frac{\pi}{2}\right) \geq \nu_{(1,1)}\left((1+\alpha)\frac{\pi}{2}\right) \tag{2.6}$$

then the union of the numerical ranges of all the ν_k's on M is a full half axis. The condition (2.6) is fulfilled in the example of the Neumann problem $(\mathcal{P}_{II})$ if

$$\frac{2}{1-\alpha} \geq \frac{2}{1+\alpha} + 1,$$

i. e., if $\alpha \geq \sqrt{5} - 2 \simeq 0.236$.

As we already explained for two-dimensional problems, if the condition (2.5) does not hold, there is no optimal regularity for the regular part (see Remark 1.2). Even if condition (2.5) holds, it is still difficult to obtain the optimality. Only [9] claim the optimality when there is no logarithmic term is the asymptotics and moreover $s - \nu_{(1,0)} < 1$.

Concerning the regularity of the coefficients, even in the straight edge case with a translation invariant operator, when logarithmic terms appear, there also appear coefficients with less regularity than $H^{s-\nu_k}$. Such coefficients have the property that for a certain number $Q > 0$

$$\mathcal{F}^{-1}_{\xi\to y}\hat{c}(\xi)\,\log^{-Q}(|\xi|+2) \in H^{s-\nu_k}.$$

We think that this problem of optimality has a similar level of complexity as the considerations about the localization of coefficients in §16 of [1].

2.c The crossing of exponents. In Theorem 2.1 we made the assumption (see (2.3)) that there is no crossing of exponents, i.e., there are no points y such that for some k, k' with $k \neq k'$ there holds $\nu_k(y) = \nu_{k'}(y)$. All the authors quoted above also require this condition. We will see that such a crossing of exponents in general induces the blowing up of coefficients in the expansion (2.4).

For our problem of the skew cylinder, it is impossible to avoid such crossings. For y_0 such that $\omega(y_0) = \pi/2$ (there always exist two such points), we have $\nu_k(y_0) = \nu_{k'}(y_0)$ for

$$\begin{array}{llll} k = (1,0) & \text{and} \quad k' = (0,1) & \text{for} & (\mathcal{P}_I) \\ k = (1,0) & \text{and} \quad k' = (0,2) & \text{for} & (\mathcal{P}_{II}). \end{array}$$

The points where crossing of exponents will eventually appear (for large s) are dense in M, so this phenomenon occurs in a generic way.

Our situation has one special feature, however: It is possible to choose the exponents as analytic functions; here this choice $\nu_k(y)$ is obvious, as it is also obvious for the oblique derivative problems. There are other cases where such an analytic choice is not obvious but still possible. Here is an example.

Let us consider a domain $\Gamma \subset \mathbb{R} \times \mathbb{R}^n$ with an edge and such that

$$\Gamma = \{(y,z) \in \mathbb{R} \times \mathbb{R}^n \mid z \in K(y)\}$$

where for each y, $K(y)$ is a cone. Let $G(y)$ denote $K(y) \cap S^{n-1}$. We suppose that the family $(G(y))_y$ is analytic. And we consider, for instance, the Neumann problem for Δ on Γ. Then the exponents will be

$$\nu_{(k_1,k_2)}(y) = 1 - \frac{n}{2} + \sqrt{(1-\frac{n}{2})^2 + \lambda_{k_1}(y)} + k_2$$

where $k_1, k_2 \in \mathbb{N}$ and $(\lambda_{k_1}(y))_{k_1 \in \mathbb{N}}$ is the increasing sequence of eigenvalues of the Neumann problem for the (positive) Laplace-Beltrami operator on $G(y)$. As a consequence of a result of [3], there exists an analytic choice for the eigenvalues , i. e., analytic functions $\tilde{\lambda}_{k_1}$ such that for any y, the sequence $(\tilde{\lambda}_{k_1}(y))_{k_1 \in \mathbb{N}}$ is an enumeration of the eigenvalues with repetition according to the multiplicities. In general, this does not coincide with the enumeration in increasing order. Let us quote as an example the case when the $G(y)$ are spherical caps with opening $\alpha(y)$ where $0 < \alpha(y) < \pi$ (see §18 of [1]).

Our results can be applied to such a geometrical situation, with the corresponding appropriate choice of the exponents.

2.d Motivations. In Section 3, we will present the main results of this paper. Our motivations for their presentation are the following.

1. To give an asymptotics in the neighborhood of crossing points which is as explicit and as simple as possible
2. To eliminate as many technical hypotheses as possible.

To achieve these aims, we have chosen to treat in a first stage a class of problems which is restricted by the following two requirements:

1. Analyticity for the coefficients and the faces of the domain
2. No bifurcation points (see below).

This class of problems is sufficiently large to contain the examples described above, in particular the skew cylinder problems.

As already said, for second order operators with real coefficients, there exist analytic choices for the exponents. Indeed, this is also true for general elliptic second order operators with complex coefficients, because the poles of the resolvent of the associated operator pencil are always simple (see [1], §14 for instance). But such an analytic choice is generically impossible for fourth order operators such as the bilaplacian. The basic problem is the expression of the roots of a polynomial whose coefficients depend analytically on a parameter. The roots are algebraic but, in general, non-analytic functions of the parameter.

Such situations of bifurcations are studied in [12]. It would be interesting to give the actual structure of asymptotics for general elliptic boundary value problems. In the general case there appear combinations of both crossings and bifurcations. We think that even then it will be possible to reach the aims we described at the end of the first section, i. e., to separate all that can be separated.

3. ASYMPTOTICS AT CROSSING POINTS

3.a Ordering the exponents. Let y_0 be a crossing point, i. e., a point where there exist distinct k and k' such that

$$\nu_k(y_0) = \nu_{k'}(y_0) < s. \tag{3.1}$$

Since we assume that our cylinder is actually skew, crossing points are isolated, so there exist open intervals I and $\tilde{I}$ with $y_0 \in I$, $\overline{I} \subset \tilde{I}$, and there is no other crossing point in $\overline{\tilde{I}}$.

If the opening angle along the edge is constant (as it happens for the base of our cylinder or as it would be in the case of a plane circular crack), then if condition (3.1) is satisfied, it holds along the entire edge. In such a case we have a superposition and not a crossing, and the simple asymptotics of Theorem 2.1 is valid.

Let $\mathcal{K}_{y_0}$ be the set of indices,

$$\mathcal{K}_{y_0} := \{k = (k_1, k_2) \mid \nu_k(y_0) < s\}.$$

We denote by $\mu_1, \ldots, \mu_{j_0}$ the distinct elements of the set

$$\{\nu_k(y_0) \mid k \in \mathcal{K}_{y_0}\}.$$

Since y_0 is a crossing point, the cardinality of $\mathcal{K}_{y_0}$ is strictly larger than j_0. For each j, let $\mathcal{K}_{y_0,j}$ be the subset of $\mathcal{K}_{y_0}$,

$$\mathcal{K}_{y_0,j} := \{k \in \mathcal{K}_{y_0} \mid \nu_k(y_0) = \mu_j\}.$$

The μ_j are either crossing exponents (if $\#\mathcal{K}_{y_0,j} > 1$) or simple exponents (if $\#\mathcal{K}_{y_0,j} = 1$).

For each k, we call multiplicity of ν_k the maximal power of $\log r$ which appears in the asymptotics (2.4) along with the term $r^{\nu_k(y)}$ for $y \in I \setminus \{y_0\}$. Then we denote by $(k_j^q)_{1 \le q \le q_j}$ an enumeration of $\mathcal{K}_{y_0,j}$, repeating each term according to its multiplicity.

Finally, we set for $y \in \tilde{I}$:

$$\mu_j(y) := \max_{k \in \mathcal{K}_{y_0,j}} \nu_k(y). \tag{3.2}$$

3.b Direct formulation of asymptotics. What essentially changes from the simple asymptotics (2.4) is the behavior of the functions of r. Instead of having separately the terms $r^{\nu_l(y)} \log^p r$, we have now special combinations of these terms which cannot be separated. Let us introduce these combinations.

Definition 3.1 Let $q \geq 1$ an integer and $\nu_1, \ldots, \nu_q$ be complex numbers, not necessarily distinct. Let γ be any simple curve surrounding $\nu_1, \ldots, \nu_q$ in the complex plane. Then we define

$$S[\nu_1, \ldots, \nu_q; r] = \frac{1}{2\pi i} \int_\gamma \frac{r^\lambda}{(\lambda - \nu_1) \cdots (\lambda - \nu_q)} \, d\lambda.$$

Here are some examples. We assume that ν_1 is different from ν_2.

$$S[\nu_1; r] = r^{\nu_1} \tag{3.3}$$

$$S[\nu_1, \nu_1; r] = r^{\nu_1} \log r \tag{3.4}$$

$$S[\nu_1, \nu_2; r] = \frac{r^{\nu_1} - r^{\nu_2}}{\nu_1 - \nu_2} \tag{3.5}$$

$$S[\nu_1, \nu_1, \nu_2; r] = \frac{r^{\nu_1} \log r}{\nu_1 - \nu_2} - \frac{r^{\nu_1} - r^{\nu_2}}{(\nu_1 - \nu_2)^2} \tag{3.6}$$

When all the ν_l are distinct, we obtain

$$S[\nu_1, \ldots, \nu_q; r] = \sum_{l=1}^{q} \frac{r^{\nu_l}}{\prod\limits_{\substack{k=1\\k\neq l}}^{q} (\nu_l - \nu_k)}. \tag{3.7}$$

Remark 3.2 Example (3.5) gives (3.4) as a limit case for $\nu_2 \to \nu_1$. More generally, the function S is analytic in all its arguments on $\mathbb{C}^q \times (0, \infty)$. On the other hand, example (3.7) shows that the coefficients of the powers r^{ν_l} blow up near the points where two ν_l's coincide.

Theorem 3.3 *Let J, $\tilde{J}$, I and $\tilde{I}$ be defined as in Section 3.a and in Theorem 2.1. Let also $\mathcal{U}_J$ and χ be as in Theorem 2.1. We still assume that for some $\varepsilon_0 \geq 0$ condition (2.2) holds. To each $j = 1, \ldots, j_0$ and to each $q = 1, \ldots, q_j$, there exists a finite set of indices γ and analytic functions $\psi_{j,q,\gamma}(y, \theta)$ such that any solution u of problem $(\mathcal{P}_I)$ or $(\mathcal{P}_{II})$ can be decomposed into*

$$u = u_{\text{reg}} + u_{\text{sing}}.$$

Here $u_{\text{reg}} \in H^{s+1-\varepsilon}(\mathcal{U}_J)$ $\forall \varepsilon > \varepsilon_0$ and

$$u_{\text{sing}} = \sum_{j=1}^{j_0} v_j$$

with

$$v_j = \chi^{-1} \left(\sum_{q,\gamma} (\Phi * d_{j,q,\gamma})(y, r)\, S[\nu_{k_j^1}(y), \ldots, \nu_{k_j^q}(y); r]\, \psi_{j,q,\gamma}(y, \theta) \right). \tag{3.8}$$

The coefficients $d_{j,q,\gamma}(y)$ are defined on $\tilde{I}$ and satisfy $d_{j,q,\gamma} \in H^{s-\mu_j(y)-\varepsilon}(I)$ for all $\varepsilon > 0$.

Remark 3.4 If there is no crossing in $\tilde{I}$, then this statement yields the same result as Theorem 2.1. Indeed, the sets $\mathcal{K}_{y_0,j}$ are all reduced to one element and the functions in r are all of the form $S[\nu, \ldots, \nu; r]$, i. e. $r^\nu \log^q r$.

Remark 3.5 In a generic way, the coefficients $d_{j,k,\gamma}$ do not vanish at y_0 (see the example in Subsection 3.d below). As a consequence, if $\mathcal{K}_{y_0,j}$ has more than one element, the coefficients $c_{k,q,\beta}$ for $k \in \mathcal{K}_{y_0,j}$ tend to infinity at y_0, in general.

3.c Bundle formulation of asymptotics. As in Subsection 3.a, we denote by I a neighborhood of the crossing point y_0. For any $y \in I \setminus \{y_0\}$ and $k \in \mathcal{K}_{y_0}$, let $B_k(y)$ be the vector space spanned by the functions of (r, θ) which occur in the simple asymptotics (2.4) in the terms corresponding to the exponents $\nu_k(y)$:

$$B_k(y) := \operatorname{span}\left\{r^{\nu_k(y)} \log^q r \, \varphi_{k,q,\beta}(y, \theta)\right\}.$$

The $y \mapsto B_k(y)$ define analytic bundles over $I \setminus \{y_0\}$. The following theorems give another description of what happens at the crossing points.

Theorem 3.6 *The sum*

$$y \mapsto \bigoplus_{k \in \mathcal{K}_{y_0}} B_k(y)$$

which is defined on $I \setminus \{y_0\}$, extends as an analytic bundle on I.

It is possible to consider smaller sums, each of which corresponds to a single crossing exponent μ_j: Define C_j by

$$C_j(y) := \bigoplus_{k \in \mathcal{K}_{y_0, j}} B_k(y).$$

Theorem 3.7 *For any $j = 1, \ldots, j_0$, the bundle $y \mapsto C_j(y)$ extends as an analytic bundle on I.*

Remark 3.8 It is an open problem whether the bundles B_k themselves extend to analytic bundles on I. We have solved this problem in one special case: Let B_{k_1} and B_{k_2} be one-dimensional bundles over $I \setminus \{y_0\}$, given by

$$B_{k_l}(y) = \operatorname{span}\left\{r^{\nu_{k_l}(y)} \varphi_{k_l}(y, \theta)\right\} \quad (l = 1, 2).$$

If their sum extends to an analytic bundle on I, then both B_{k_1} and B_{k_2} are analytic on I.

Even if we knew that the bundles B_k extend to analytic bundles on I, the statements of the Theorems 3.6 and 3.7 would not be trivial: The sum of two analytic bundles is not always an analytic bundle. There may occur a collapse in dimension. For instance, if $B_1(y)$ is generated by $r^{\nu_1(y)}\varphi(\theta)$ and $B_2(y)$ by $r^{\nu_2(y)}\varphi(\theta)$, and if $\nu_1(y) = \nu_2(y)$ only in $y = y_0$, then $B_1 + B_2$ collapses in y_0. But a sum of analytic bundles can always be extended as an analytic bundle. In our example, $B_1(y_0) + B_2(y_0)$ has to be replaced by $\operatorname{span}\left\{r^{\nu_1(y)}\varphi(\theta),\ r^{\nu_1(y)} \log r \, \varphi(\theta)\right\}$. A trivialization of this extension is given by

$$\left\{r^{\nu_1(y)}\varphi(\theta),\ \frac{r^{\nu_1(y)} - r^{\nu_2(y)}}{\nu_1(y) - \nu_2(y)} \varphi(\theta)\right\}.$$

This extension property does not hold, in general, for C^∞-bundles. Let us give a simple example of bundles with values in $\mathbb{R}^3$:

$$\begin{aligned} B(y) &= \operatorname{span}\{(1,0,0)\} \\ B'(y) &= \operatorname{span}\left\{(1, e^{-1/y^2} \cos \frac{1}{y}, e^{-1/y^2} \sin \frac{1}{y})\right\}. \end{aligned}$$

The sum $B(y) + B'(y)$ cannot be extended to a C^∞-bundle in $y = 0$.

We start now from the analytic extension $\tilde{C}_j(y)$ of C_j. Let $y \mapsto X_{j,\alpha}(y,\cdot,\cdot)$ for $\alpha = 1,\dots,A_j$ define a trivialization of $\tilde{C}_j$.

As a consequence of the form of the B_k's, we get the following lemma.

Lemma 3.9 *There exist analytic functions $\psi^{j,\alpha}_{q,\gamma}(y,\theta)$ such that*

$$X_{j,\alpha}(y,r,\theta) = \sum_{q,\gamma} S[\nu_{k^1_j}(y),\dots,\nu_{k^q_j}(y);r]\,\psi^{j,\alpha}_{q,\gamma}(y,\theta).$$

All the objects discussed in this subsection up to now are only linked with the geometrical framework (domain and boundary value problem). The singular parts v_j in Theorem 3.3 can now be written as a kind of regularized $H^{s-\varepsilon}$-sections of the bundles $\tilde{C}_j$:

$$v_j = \chi^{-1}\left(\sum_{\alpha=1}^{A_j}(\Phi * b_{j,\alpha})(y,r)\,X_{j,\alpha}(y,r,\theta)\right)$$

with $b_{j,\alpha}(y) \in H^{s-\mu_j(y)-\varepsilon}(I)$.

3.d An example. Let us illustrate our statements by a simple example. We consider the mixed problem $(\mathcal{P}_I)$. We take $s \in (1, 2/(1+\alpha))$, where α is the "obliquity" of the skew cylinder (see Section 2.b). Since there is one zero Dirichlet condition, $\nu_{(0,0)}$ does not appear. With that choice of s, only $\nu_{(1,0)}$ and $\nu_{(0,1)}$ are relevant. For simplicity, let us denote

$$\begin{aligned}\nu_1(y) &:= \nu_{(1,0)}(y) = \frac{\pi}{2\omega(y)}\\ \nu_2(y) &:= \nu_{(0,1)}(y) = 1.\end{aligned}$$

There are exactly two points $y \in M$ where $\nu_1(y) = \nu_2(y)$ holds. These are the two points y_0, y_0' where $\omega(y) = \pi/2$. On $M \setminus \{y_0, y_0'\}$, the simple asymptotics (2.4) holds. Here $q = 0$ and only one value of β is required. We write l instead of $(l, 0, 1)$. Then we can choose

$$\begin{aligned}\varphi_1(y,\theta) &= \cos\nu_1(y)\theta\\ \varphi_2(y,\theta) &= \sin(\omega(y)-\theta).\end{aligned}$$

Here it is possible to compute $c_2(y)$ since it depends only on the pointwise value of the boundary datum h_2 on the edge:

$$c_2(y) = \frac{h_2(y,0)}{\cos\omega(y)}.$$

We have the precise regularity result $c_2 \in H^{s-1}_{\text{loc}}(M \setminus \{y_0, y_0'\})$.

Concernig the bundle representation of the asymptotics, we observe the following facts.

$B_1(y)$ is generated by $r^{\nu_1(y)} \cos \nu_1(y)\theta$.
$B_2(y)$ is generated by $r \sin(\omega(y) - \theta)$.
When $y = y_0$ or y_0', then $B_1(y) = B_2(y)$.

A basis of the analytic extension of $B_1 + B_2$ is given by

$$\begin{aligned}
X_1(y, r, \theta) &= r^{\nu_1(y)} \cos \nu_1(y)\theta. \\
X_2(y, r, \theta) &= \frac{r \sin(\omega(y) - \theta) - r^{\nu_1(y)} \cos \nu_1(y)\theta}{1 - \nu_1(y)} \\
&= \frac{r - r^{\nu_1(y)}}{1 - \nu_1(y)} \sin(\omega(y) - \theta) \\
&\quad + r^{\nu_1(y)} \frac{\sin(\omega(y) - \theta) - \cos \nu_1(y)\theta}{1 - \nu_1(y)} \\
&= \frac{r - r^{\nu_1(y)}}{1 - \nu_1(y)} \cos \nu_1(y)\theta \\
&\quad + r \frac{\sin(\omega(y) - \theta) - \cos \nu_1(y)\theta}{1 - \nu_1(y)}.
\end{aligned}$$

The different forms of X_2 correspond to different possible orders of enumeration in the representation of Lemma 3.9.

If we want to get the direct representation of Theorem 3.3, we need 3 basis functions, for instance:

$$\begin{aligned}
S[\nu_1(y); r]\, \psi_{1,1}(y, \theta) &= X_1(y, r, \theta) \\
S[\nu_1(y); r]\, \psi_{1,2}(y, \theta) &= r^{\nu_1(y)} \frac{\sin(\omega(y) - \theta) - \cos \nu_1(y)\theta}{1 - \nu_1(y)} \\
S[\nu_1(y), \nu_2(y); r]\, \psi_{2,1}(y, \theta) &= \frac{r - r^{\nu_1(y)}}{1 - \nu_1(y)} \sin(\omega(y) - \theta).
\end{aligned}$$

Now we can compare the three representations of a singular part, namely the "simple asymptotics" of Theorem 2.1, the "direct representation" of Theorem 3.3, and the "bundle representation" with the basis X_1, X_2. Assume that we have

$$\begin{aligned}
c_1\, r^{\nu_1} \varphi_1 + c_2\, r^{\nu_2} \varphi_2 &= d_{1,1}\, S[\nu_1; r]\, \psi_{1,1} + d_{1,2}\, S[\nu_1; r]\, \psi_{1,2} + d_{2,1}\, S[\nu_1, \nu_2; r]\, \psi_{2,1} \\
&= b_1\, X_1 + b_2\, X_2.
\end{aligned}$$

Then there hold the following relations between the coefficients.

$$\begin{aligned}
b_1 &= c_1 + c_2 \\
b_2 &= c_2(1 - \nu_1)
\end{aligned}$$

$$
\begin{aligned}
c_1 &= b_1 - b_2/(1-\nu_1) \\
c_2 &= b_2/(1-\nu_1) \\
d_{1,1} &= b_1 \\
d_{1,2} &= d_{2,1} = b_2.
\end{aligned}
$$

These relations clearly display the blow-up of the coefficients in the simple asymptotics at the points y_0, y_0' and also the necessity for the introduction of the exponents $\mu_j(y)$ in (3.2).

Acknowledgement. This work was done while both authors were supported by the DFG Forschergruppe KO 634/32-1 at the TH Darmstadt and the Unité de Recherche Associée 758 of the CNRS at the University of Nantes.

REFERENCES

[1] M. DAUGE. *Elliptic Boundary Value Problems in Corner Domains — Smoothness and Asymptotics of Solutions.* Lecture Notes in Mathematics, Vol. 1341. Springer-Verlag, Berlin 1988.

[2] P. GRISVARD. *Boundary Value Problems in Non-Smooth Domains.* Pitman, London 1985.

[3] T. KATO. *Perturbation Theory for Linear Operators.* Springer-Verlag, Berlin - Heidelberg - New York 1976.

[4] V. A. KONDRAT'EV. Boundary-value problems for elliptic equations in domains with conical or angular points. *Trans. Moscow Math. Soc.* **16** (1967) 227–313.

[5] V. A. KONDRAT'EV. Singularities of a solution of Dirichlet's problem for a second order elliptic equation in a neighborhood of an edge. *Differential Equations* **13** (1970) 1411–1415.

[6] V. G. MAZ'YA, B. A. PLAMENEVSKII. L^p estimates of solutions of elliptic boundary value problems in a domain with edges. *Trans. Moscow Math. Soc.* **1** (1980) 49–97.

[7] V. G. MAZ'YA, B. A. PLAMENEVSKII. On boundary value problems for a second order elliptic equation in a domain with edges. *Vestnik Leningrad Univ. Mathematics* **8** (1980) 99–106.

[8] V. G. MAZ'YA, B. A. PLAMENEVSKII. Estimates in L^p and in Hölder classes and the Miranda-Agmon maximum principle for solutions of elliptic boundary value problems in domains with singular points on the boundary. *Amer. Math. Soc. Transl. (2)* **123** (1984) 1–56.

[9] V. G. MAZ'YA, J. ROSSMANN. Über die Asymptotik der Lösungen elliptischer Randwertaufgaben in der Umgebung von Kanten. *Math. Nachr.* **138** (1988) 27–53.

[10] A. NIKISHKIN. Singularities of the solution of the Dirichlet problem for a second order equation in a neighborhood of an edge. *Moscow Univ. Math. Bull.* **34**(2) (1979) 53–64.

[11] S. REMPEL, B. W. SCHULZE. ***Asymptotics for Elliptic Mixed Boundary Problems.*** Akademie-Verlag, Berlin 1989.

[12] R. SCHMUTZLER. About the structure of branching asymptotics for elliptic boundary value problems in domains with edges. Lecture given at the International Workshop "Analysis in Domains and on Manifolds with Singularities", Breitenbrunn, GDR. 1990.

Martin Costabel
Institut für Angewandte Mathematik
Universität Heidelberg
Im Neuenheimer Feld 293
D-6900 Heidelberg

Monique Dauge
Département de Mathématiques
Université de Nantes
2 rue de la Houssinière
F-44072 Nantes Cedex 03

A COLLOCATION METHOD FOR A SCREEN PROBLEM IN $\mathbb{R}^3$

Martin Costabel, Frank Penzel and Reinhold Schneider

Heidelberg and Darmstadt

1. INTRODUCTION

Let $\Omega := (-1,1) \times (-1,1) \subset \mathbb{R}^2$. The first kind integral equation on Ω,

$$Vu(x) := \frac{1}{4\pi} \int_\Omega u(y) \frac{dy}{|x-y|} = f(x) \quad (x \in \Omega) \tag{1.1}$$

gives the solution of the Dirichlet problem for the Laplace equation in $\mathbb{R}^3 \backslash \overline{\Omega}$ with Dirichlet data f given on Ω ("screen problem") (see [6], [16]). For the numerical approximation of the solution u of (1.1), we introduce a grid $\square^N$ on $\mathbb{R}^2$,

$$\square^N := \{x_k^N := \frac{k}{N} = (\frac{k_1}{N}, \frac{k_2}{N}) \mid k \in \mathbb{Z}^2\}. \tag{1.2}$$

As trial space we use the space $S^d(\square^N)$ of tensor product splines of degree d on $\square^N$ (see (2.4)). Then our collocation scheme is the following: Find a function $u^N \in S^d(\square^N)$ whose nodal values vanish outside $\overline{\Omega}$ such that

$$Vu^N(x_k^N) = f(x_k^N) \quad \text{for } x_k^N \in \square^N \cap \overline{\Omega}. \tag{1.3}$$

We show the stability and convergence of this method in the Sobolev space $\tilde{H}^{-1/2}$, see Theorem 3.1. Deatailed proofs for the case when the degree d of the splines is odd, in particular the practically important case $d = 1$, can be found in [4]. A stability proof for the even degree case has recently been found (see [3]). Thus we have convergence also for the simplest case $d = 0$, i. e., midpoint collocation with piecewise constant trial functions. In [4] we also discuss the relation of our work with other recent papers on spline collocation methods. See also [9] for a review.

Our basic observation is that the collocation matrices are related to finite sections of a fixed infinite Toeplitz matrix (This was observed in the one-dimensional case in [10]). The symbolic calculus for Toeplitz matrices, based on Fourier series, then shows that this infinite matrix is positive definite and symmetric. These properties are inherited by the collocation matrices.

2. BASIC NOTATIONS

We need the following additional notations for $N \in \mathbb{N}$,

$$\omega^N := \{k \in \mathbb{Z}^2 \mid x_k^N \in \overline{\Omega}\} = \{k \in \mathbb{Z}^2 \mid |k_j| \leq N \,,\, j = 1,2\}\,. \tag{2.1}$$

A nodal basis for the space $S^d(\square^N)$ of tensor product splines is constructed as follows:
Let $\phi^0(t) := 1$ for $t \in \mathbb{R}$, $-1/2 < t < 1/2$, $\phi^0(t) := 0$ for $|t| \geq 1/2$, $\phi^1(t) := \max\{0, 1 - |t|\}$ for $t \in \mathbb{R}$ and, more generally, for $d \in \mathbb{N}$,

$$\phi^d := \phi^0 * \cdots * \phi^0 \quad \text{(convolution of } d+1 \text{ copies of } \phi^0).$$

For $x \in \mathbb{R}^2$ we define

$$\varphi_0^{1,d}(x) \;:=\; \phi^d(x_1) \cdot \phi^d(x_2)\,, \tag{2.2}$$

$$\varphi_k^{N,d}(x) \;:=\; \varphi_0^{1,d}(N(x - x_k^N)) = \varphi_0^{1,d}(Nx - k)\,. \tag{2.3}$$

Then $S^d(\square^N) = \operatorname{span}\{\varphi_k^{N,d} \mid k \in \mathbb{Z}^2\}$. We need the following $(2N+1)^2$ - dimensional subspace of $S^d(\square^N)$,

$$S^d(\square^N \cap \overline{\Omega}) := \operatorname{span}\{\varphi_k^{N,d} \mid k \in \omega^N\}\,. \tag{2.4}$$

The collocation points ω^N correspond to the break points for the splines in $S^d(\square^N \cap \overline{\Omega})$ if the degree d is odd and to the midpoints between the break points if d is even. Note that the support of the spline functions in $S^d(\square^N \cap \overline{\Omega})$ is slightly larger than Ω, namely the set

$$[-1 - \frac{d+1}{2N}, 1 + \frac{d+1}{2N}] \times [-1 - \frac{d+1}{2N}, 1 + \frac{d+1}{2N}]\,.$$

For any continuous function f on $\mathbb{R}^2$, we define the restriction to $\square^N$ by

$$r^N f := (f(x_k^N))_{k \in \mathbb{Z}^2}\,. \tag{2.5}$$

For any sequence $(f_k)_{k\in\mathbb{Z}^2}$ we define the corresponding spline function by

$$i^{N,d}(f_k) := \sum_{k \in \mathbb{Z}^2} f_k \varphi_k^{N,d}\,. \tag{2.6}$$

Note that for $d = 0, 1$, the f_k are the nodal values of $i^{N,d}(f_k)$.

The system of equations corresponding to the collocation equations (1.3) can be written as

$$r^N V i^{N,d}(u_k) = r^N f \quad \text{on } \omega^N\,, \tag{2.7}$$

where $(u_k)_{k\in\omega^N}$ are the components of $u^N \in S^d(\square^N \cap \overline{\Omega})$,

$$u^N = \sum_{k \in \omega^N} u_k \varphi_k^{N,d}\,. \tag{2.8}$$

The matrix elements of the system (2.7) are given by

$$\left((V\varphi_k^{N,d})(x_{k'}^N)\right)_{k,k'\in\omega^N} . \tag{2.9}$$

From the translation invariance of the operator V (1.1) and the definition (2.3) of the basis functions $\varphi_k^{N,d}$ we see that this is a Toeplitz matrix.

By $H^s(\mathbb{R}^2)$ ($s\in\mathbb{R}$) we denote the usual Sobolev spaces [7]. For a bounded domain $\mathcal{O}$,

$$p_{\mathcal{O}} : u \mapsto u|_{\mathcal{O}} \tag{2.10}$$

is the operator of restriction to $\mathcal{O}$. Then

$$H^s(\mathcal{O}) := \{u|_{\mathcal{O}} \mid u \in H^s(\mathbb{R}^2)\} = p_{\mathcal{O}} H^s(\mathbb{R}^2) \tag{2.11}$$

and

$$\tilde{H}^s(\mathcal{O}) := \{u \in H^s(\mathbb{R}^2) \mid \operatorname{supp} u \subset \overline{\mathcal{O}}\} . \tag{2.12}$$

Then with respect to the natural extension of the L^2 duality

$$(f, g) := \int_{\mathbb{R}^2} f(x)\, \overline{g(x)}\, dx , \tag{2.13}$$

$\tilde{H}^s(\mathcal{O})$ is the dual space of $H^{-s}(\mathcal{O})$.

The basic mapping properties of the operator V in the energy spaces are well known (see e.g. [8], [16]):

$p_{\mathcal{O}}V : \tilde{H}^{-1/2}(\mathcal{O}) \to H^{1/2}(\mathcal{O})$ is a positive definite self-adjoint isomorphism.

3. STABILITY AND CONVERGENCE OF THE COLLOCATION METHOD

Let $\Pi^N := i^{N,1} r^N$ be the interpolation projector onto the space $S^1(\square^N)$ of piecewise bilinear splines. We can write the collocation equations (1.3) in the form

$$p_\Omega \Pi^N V u^N = p_\Omega \Pi^N f . \tag{3.1}$$

We prove the stability and uniform boundedness of the sequence $(p_\Omega \Pi^N V)_{N\in\mathbb{N}}$. Note that the projection Π^N is not bounded on $H^{1/2}$ so that even the uniform boundedness is a nontrivial result.

Theorem 3.1 (Stability) *There is a constant $\gamma > 0$ such that for all $N \in \mathbb{N}$ and all $w \in S^d(\square^N \cap \overline{\Omega})$ there holds*

$$\|p_\Omega \Pi^N V w\|_{H^{1/2}(\Omega)} \geq \gamma \|w\|_{H^{-1/2}(\mathbb{R}^2)} . \tag{3.2}$$

Theorem 3.2 (Uniform boundedness) *There is a constant $C \in \mathbb{R}$ such that for all $N \in \mathbb{N}$ and all $w \in S^d(\square^N \cap \overline{\Omega})$ there holds*

$$\|p_\Omega \Pi^N V w\|_{H^{1/2}(\Omega)} \leq C\|w\|_{H^{-1/2}(\mathbf{R}^2)} . \tag{3.3}$$

For the proof of these theorems one needs besides the well known approximation property and inverse inequality for our spline spaces (see [2], [1], [13]) some results about discrete Sobolev spaces and discrete versions of the $H^{\pm 1/2}$ Sobolev norms. Our spaces are related to those studied by Frank [5] and Stephan [15], see also [12], [11].

In the following we fix the degree d of the splines and write i^N for $i^{N,d}$ and φ_k^N for $\varphi_k^{N,d}$. We define a norm in the space of sequences by

$$|(u_k)_{k\in\mathbb{Z}^2}|_{s,N} := \|i^N(u_k)\|_{H^s(\mathbf{R}^2)} . \tag{3.4}$$

We will see that these norms defined for different d are equivalent, so the discrete Sobolev space is uniquely defined by

$$h_N^s(\mathbb{Z}^2) := \{(u_k)_{k\in\mathbb{Z}^2} \mid |(u_k)|_{s,N} < \infty\} . \tag{3.5}$$

On the grid ω^N we define the finite dimensional subspace

$$\tilde{h}_N^s(\omega^N) := \{(u_k)_{k\in\mathbb{Z}^2} \in h_N^s(\mathbb{Z}^2) \mid u_k = 0 \quad \text{for } k \notin \omega^N\} . \tag{3.6}$$

This is $\mathbb{C}^M$, $M = (2N+1)^2$ with norm $|\cdot|_{s,N}$.

It is easy to see that for $s = 0$ we have the norm equivalence

$$|(u_k)|_{0,N} \approx N^{-1}\Big(\sum_{k\in\mathbb{Z}^2} |u_k|^2\Big)^{1/2} . \tag{3.7}$$

By "$\approx$" we denote the equivalence of norms with constants independent of N. Taking (3.7) into account we define the scalar products in ℓ^2,

$$((u_k),(v_k))_N := N^{-2} \sum_{k\in\mathbb{Z}^2} u_k \overline{v_k} . \tag{3.8}$$

For Fourier series on the square $Q := (-\pi,\pi)\times(-\pi,\pi)$ we use the following notation

$$\tilde{u}(\xi) := \sum_{k\in\mathbb{Z}^2} e^{-\imath k\cdot\xi} u_k =: F(u_k)(\xi) , \tag{3.9}$$

so we have

$$u_k = (2\pi)^{-2} \int_Q \tilde{u}(\xi) e^{\imath k\cdot\xi} d\xi . \tag{3.10}$$

The integral operator V defines two discrete sesquilinear forms:

$$(i^N(u_k), V i^N(v_k)) = \sum_{k,k'} u_k (\varphi_k^N, V\varphi_{k'}^N) \overline{v_{k'}} \tag{3.11}$$

and

$$((u_k), r^N V i^N(v_k))_N = N^{-2} \sum_{k,k'} u_k (V\varphi_{k'}^N)(x_k^N)\overline{v_{k'}}. \tag{3.12}$$

The form (3.11) defines a matrix which corresponds to the Galerkin method with test and trial functions in $S^d(\square^N \cap \overline{\Omega})$ for the equation $Vu = f$. The form (3.12) corresponds to our collocation method (1.3). More precisely, the collocation scheme (1.3) can be considered as a finite section method for the infinite Toeplitz matrix defined by (3.12). The diagonalization of the infinite Toeplitz matrices via Fourier series leads to the following representations of the bilinear forms (3.11), (3.12).

Lemma 3.3 *Let* $\tilde{u} = F(u_k)$, $\tilde{v} = F(v_k)$ *with* $(u_k), (v_k) \in \ell^1(\mathbb{Z}^2)$.

(i) $$(i^N(u_k), V i^N(v_k)) = N^{-3} \int_Q \lambda^G(\xi)\, \tilde{u}(\xi)\, \overline{\tilde{v}(\xi)}\, d\xi\,;$$

(ii) $$((u_k), r^N V i^N(v_k))_N = N^{-3} \int_Q \lambda^C(\xi)\, \tilde{u}(\xi)\, \overline{\tilde{v}(\xi)}\, d\xi\,.$$

Here the functions λ^G *and* λ^C *are defined by*

$$\lambda^G(\xi) = \frac{1}{4\pi^2} \sum_{r\in\mathbb{Z}^2} |\xi + 2\pi r|^{-1} \prod_{j=1}^{2} \left(\frac{2\sin\frac{\xi_j}{2}}{\xi_j + 2\pi r_j}\right)^{2(d+1)} \tag{3.13.a}$$

$$\lambda^C(\xi) = \frac{1}{4\pi^2} \sum_{r\in\mathbb{Z}^2} |\xi + 2\pi r|^{-1} \prod_{j=1}^{2} \left(\frac{2(-1)^{l_j}\sin\frac{\xi_j}{2}}{\xi_j + 2\pi r_j}\right)^{d+1}. \tag{3.13.b}$$

The proof uses the Fourier representation of V,

$$Vu(x) = \frac{1}{2}(2\pi)^{-2} \int_{\mathbf{R}^2} e^{i\xi\cdot x} |\xi|^{-1}\, \hat{u}(\xi) d\xi \tag{3.14}$$

and the Fourier transform of the basis function

$$\widehat{\varphi_0^{1,d}}(\xi) = \prod_{j=1}^{2} \left(\frac{2\sin\frac{\xi_j}{2}}{\xi_j}\right)^{d+1}. \tag{3.15}$$

From the definitions (3.13.a), (3.13.b), we see immediately that the functions λ^G and λ^C are C^∞ functions on $\overline{Q}\backslash\{0\}$ and that λ^G and λ^C for odd d are positive and behave like $\mathcal{O}(|\xi|^{-1})$ at $\xi = 0$. In [3] it is shown that also for even d, λ^C is positive and has this behavior at $\xi = 0$. The proof of the positivity of this alternating series uses the following four inequalities for $\sigma(\xi) = |\xi|^{-1}$, valid for $\xi_1 \neq 0$, $\xi_2 \neq 0$:

$$\begin{aligned}
\sigma(\xi_1, \xi_2) &> 0,\\
\frac{\partial}{\partial\xi_1}(\sigma(\xi_1,\xi_2)\xi_1^{-(d+1)}) &< 0,\\
\frac{\partial}{\partial\xi_2}(\sigma(\xi_1,\xi_2)\xi_2^{-(d+1)}) &< 0,\\
\frac{\partial^2}{\partial\xi_1\partial\xi_2}(\sigma(\xi_1,\xi_2)(\xi_1\xi_2)^{-(d+1)}) &> 0.
\end{aligned}$$

We see that the functions λ^G and λ^C (for any $d \in \mathbb{N}$) define equivalent weights on Q. This implies the following conclusions.

Corollary 3.4 *For any $N \in \mathbb{N}$, the collocation matrix in* (1.3) *is positive definite and symmetric. The collocation equations are always uniquely solvable.*

Corollary 3.5 *(i) Both quadratic forms* (3.11) *and* (3.12) *define norms on $\tilde{h}_N^{-1/2}(\omega^N)$ which are equivalent to the $|\cdot|_{-\frac{1}{2},N}$ norm* (3.4) *with constants not depending on N.*

(ii) Let $\mathcal{O} \subset \mathbb{R}^2$ be a bounded domain. For sequences (v_k) with $\operatorname{supp} i^N(v_k) \subset \mathcal{O}$ *we have the norm equivalence*

$$|(v_k)|^2_{\frac{1}{2},N} \approx N^{-1} \int_Q |\tilde{v}(\xi)|^2 \left(\lambda^G(\xi)\right)^{-1} d\xi \,. \tag{3.16}$$

The proof of Theorems 3.1 and 3.2 uses two further properties of the discrete Sobolev spaces. The first one is the existence of a uniformly bounded extension operator and the second one a discrete analogue of the Sobolev space $H^s(\Omega)$ as dual space of $\tilde{H}^{-s}(\Omega)$.

Let $\hat{\Omega} = 2\Omega = (-2,2) \times (-2,2)$. We use an extension operator L from $H^s(\Omega)$ to $\tilde{H}^s(\hat{\Omega})$. It is well known [14] that such an operator exists as a continuous linear operator

$$L : H^s(\Omega) \to \tilde{H}^s(\hat{\Omega}) \quad \text{for all } s \in \mathbb{R}\,. \tag{3.17}$$

We can assume that L is given in such a way that $\operatorname{supp} \Pi^N Lv \subset \hat{\Omega}$ for all N and all $v \in H^s(\Omega)$. One can then show that there is a constant C independent of N such that

$$\|\Pi^N L p_\Omega u\|_{\tilde{H}^{1/2}(\hat{\Omega})} \leq C \|p_\Omega u\|_{H^{1/2}(\Omega)} \tag{3.18}$$

holds for all $u \in S^d(\square^N)$. This describes the discrete analogue of the extension operator L.

As a replacement for the definition of a discrete analogue of the Sobolev space $H^{1/2}(\Omega)$, we use the following observation:

We show that there is a constant independent of N such that

$$|((u_k),(v_k))_N| \leq C\,|(u_k)|_{-\frac{1}{2},N} \cdot |(v_k)|_{\frac{1}{2},N} \tag{3.19}$$

for all sequences $(u_k),(v_k)$ with $\operatorname{supp} i^N(u_k) \subset \hat{\Omega}$, $\operatorname{supp} i^N(v_k) \subset \hat{\Omega}$.

The proof of the stability theorems now follows. From the stability theorems one can finally derive the following convergence theorem in a standard way, keeping in mind the unbounded nature of the projection operators Π^N.

Theorem 3.6 (Convergence) *Let $f \in H^{1+\varepsilon}(\Omega)$ with some $\varepsilon > 0$ and $u \in \tilde{H}^{-1/2}(\Omega)$ be the solution of the integral equation* (1.1). *Then for any $N \in \mathbb{N}$, the collocation equations* (1.3) *have a unique solution $u^N \in S^d(\square^N \cap \overline{\Omega})$. For any $\eta > 0$ there is a constant $C \geq 0$ independent of N such that*

$$\|u - u^N\|_{H^{-1/2}(\mathbb{R}^2)} \leq C N^{-\frac{1}{2}+\eta} \|f\|_{H^{1+\varepsilon}(\Omega)}\,.$$

Acknowledgement. The first and third authors were partially supported by the DFG-Forschergruppe Ko 634/32-1.

REFERENCES

[1] J. AUBIN. *Approximation of Elliptic Boundary Value Problems.* Wiley-Interscience, New York 1972.

[2] I. BABUŠKA, K. AZIZ. Survey lectures on the mathematical foundation of the finite element method. In A. K. AZIZ, editor, *The Mathematical Foundation of the Finite Element Method with Applications to Partial Differential Equations*, pages 3–359. Academic Press, New York 1972.

[3] M. COSTABEL, W. MCLEAN. Spline collocation for strongly elliptic equations on the torus. To appear.

[4] M. COSTABEL, F. PENZEL, R. SCHNEIDER. Error analysis of a boundary element collocation method for a screen problem in $\mathbb{R}^3$. THD-Preprint 1284, Technische Hochschule Darmstadt, February 1990.

[5] L. S. FRANK. Spaces of network functions. *Math. USSR Sbornik* **15** (1971) 183–226.

[6] M. A. JASWON, G. T. SYMM. *Integral Equation Methods in Potential Theory and Elastostatics.* Academic Press, London 1977.

[7] J. L. LIONS, E. MAGENES. *Nonhomogeneous Boundary Value Problems and Applications*, volume 1. Springer-Verlag, Berlin 1972.

[8] J.-C. NÉDÉLEC. Equations intégrales associées aux problèmes aux limites elliptiques dans des domaines de $\mathbb{R}^3$,. In R. DAUTRAY, J.-L. LIONS, editors, *Analyse Mathématique et Calcul Numérique pour les Sciences et les Techniques*, chapters XI–XIII. Masson, Paris 1988.

[9] S. PRÖSSDORF. Numerische Behandlung singulärer Integralgleichungen. *Z. angew. Math. Mech.* **69** (1989) T5–T13.

[10] S. PRÖSSDORF, A. RATHSFELD. A spline collocation method for singular integral equations with piecewise continuous coefficients. *Integral Equations Oper. Theory* **7** (1984) 536–560.

[11] G. SCHMIDT. Spline collocation for singular integro-differential equations over $(0,1)$. *Numer. Math.* **50** (1987) 337–352.

[12] G. SCHMIDT, H. STRESE. The convergence of a direct BEM for the plane mixed boundary value problem of the Laplacian. *Numer. Math.* **54** (1988) 145–165.

[13] L. SCHUMAKER. *Spline Functions: Basic Theory.* Wiley, New York 1981.

[14] E. M. STEIN. *Singular Integrals and Differentiability Properties of Functions.* Princeton University Press, Princeton, N. J. 1970.

[15] E. P. STEPHAN. Differenzenapproximationen von Pseudo-Differentialoperatoren. Thesis, Technische Hochschule Darmstadt 1975.

[16] E. P. STEPHAN. Boundary integral equations for screen problems in $\mathbb{R}^3$. *Integral Equations Oper. Theory* **10** (1987) 236–257.

Martin Costabel
Institut für Angewandte Mathematik
Universität Heidelberg
Im Neuenheimer Feld 293
D-6900 Heidelberg

Frank Penzel and Reinhold Schneider
Fachbereich Mathematik
Technische Hochschule Darmstadt
Schloßgartenstraße 7
D-6100 Darmstadt

TO THE STUDY OF THE NEGATIVE SPECTRUM OF AN ELLIPTIC OPERATOR

Yu. V. Egorov, V. A. Kondrat'ev

1 Introduction

This paper is a continuation of our work [1], in which some estimates were obtained for the negative spectrum of the elliptic operator

$$L = L_0 - V(x), \quad L_0 u = \sum_{|\alpha|\leq m, |\beta|\leq m} D^\alpha(a_{\alpha\beta}(x)D^\beta u), \tag{1}$$

with measurable coefficients $a_{\alpha\beta}$ such that $a_{\alpha\beta} = \overline{a_{\beta\alpha}}$ and
$Re\,(L_0 u, u) \equiv Re \int \sum_{|\alpha|\leq m, |\beta|\leq m} a_{\alpha\beta}(x)D^\alpha u(x)\overline{D^\beta u(x)}dx \geq c_0|u|_m^2$, where c_0 is a positive constant and

$$|u|_m^2 = \int \sum_{|\alpha|=m} |D^\alpha u|^2 dx.$$

The operator L is considered in the space $H_m^0(\Omega)$, where Ω is a domain in R^n, the potential V is assumed to be real-valued. Our aim is to estimate the number N of negative eigen-values of L.

Of the new works on this theme note these of M. S. Birman-M. Z. Solomyak [2] and of M. Schechter [3].

Here we state some new results: the estimates of N from below, the estimates for the number N_μ of eigen-values of L, which are less than μ.

2 Main results

Let $x = (x', x'')$, $x' = (x_1, \ldots, x_k)$, $x'' = (x_{k+1}, \ldots, x_n)$, $1 \leq k \leq n$. Let $\Omega \subseteq Q \times \mathbf{R}^{n-k}$, where Q is a bounded domain in $\mathbf{R}^k$, and the operator L is such that

$$Re \int \sum a_{\alpha\beta}(-D)^\alpha u \overline{D^\beta u} dx \geq Re \int L_1 u\, \overline{u} dx + Re \int L_2 u\, \overline{u} dx$$

for all functions from $C_0^\infty(\Omega)$, where

$$L_1 u = \sum_{\substack{|\alpha|\leq m, |\beta|\leq m \\ \alpha''=0, \beta''=0}} D^\alpha(a_{\alpha\beta}(x')D^\beta u), \quad L_2 u = \sum_{\substack{|\alpha|\leq m, |\beta|\leq m \\ \alpha''=0, \beta''=0}} D^\alpha(a_{\alpha\beta}(x'')D^\beta u),$$

and μ is the least eigen-value of the operator L_1 in the domain Q with the zero Dirichlet conditions.

Corresponding to Theorems 3 – 5 in [1] depending on the oddness of n and k we state the following results.

Theorem 1. *If $n > 2m + k$, then the number of the eigen-values of the operator L, which are less than μ, is estimated as follows:*

$$N_\mu \le C_{n,m,q} \int V_+(x)^q [|x|^{2mq-n} + |x''|^{2mq-n+k}] dx\,,$$

where q is any number greater than $n/2m$.

Theorem 2. *If $2m + k \ge n > 2m$ and $n - k$ is odd, then for $q \ge n/2m$*

$$N_\mu \le C_{n,m,q}(1 + \int V_+(x)^q [|x|^{2mq-n} + |x''|^{2mq-n+k}] dx)\,.$$

Theorem 3. *If $2m + k \ge n > 2m$ and $n - k$ is even, then for $q \ge n/2m$ we have*

$$N_\mu \le C_{n,m,q} \int V_+(x)^q [|x|^{2mq-n} + |x''|^{2mq-n+k}(1 + |\ln|x''||)^{2q-1}] dx\,.$$

Theorem 4. *If $n \le 2m$ and n is odd, k is even, then for $q \ge 1$*

$$N_\mu \le C_{n,m,q}(1 + \int V_+(x)^q [|x|^{2mq-n} + |x''|^{2mq-n+k}] dx)\,.$$

Theorem 5. *If $n \le 2m$ and n is odd, k is odd, then for $q \ge 1$*

$$N_\mu \le C_{n,m,q}(1 + \int V_+(x)^q [|x|^{2mq-n} + |x''|^{2mq-n+k}(1 + |\ln|x''||^{2q-1})] dx)\,.$$

Theorem 6. *If $n \le 2m$ and n is even, k is odd, then for $q \ge 1$*

$$N_\mu \le C_{n,m,q}(1 + \int V_+(x)^q [|x|^{2mq-n}(1 + |\ln|x''||)^{2q-1} + |x''|^{2mq-n+k}] dx)\,.$$

Theorem 7. *If $n \le 2m$, and n, k are even, then for $q \ge 1$*

$$N_\mu \le C_{n,m,q}(1 + \int V_+(x)^q [|x|^{2mq-n}(1 + |\ln|x||)|^{2q-1} + |x''|^{2mq-n+k}(1 + |\ln|x''||)^{2q-1}] dx)\,.$$

Let us state a variant of Theorem 1 for the Schrödinger operator:

Theorem 8. *Let $Q = (0,1) \times \mathbf{R}^{n-1}$ and $\Omega \subseteq Q$. Let $m = 1$ and $n \ge 4$. Denote as N_1 the number of eigen-values of the operator*

$$L = -\Delta - V(x)\,,$$

not exceeding π^2. Then for each $q \ge (n-1)/2$ the inequality

$$N_1 \le C_{q,n} \int_\Omega V_+(x)^q (|x|^{2q-n} + |x'|^{2q-n+1}) dx\,,$$

holds, where $x' = (x_2, \ldots, x_n)$.

The following theorem allows to estimate the number N from below.

Theorem 9. *Let $V(x) \ge 0$, R and ε be positive numbers. Assume that either*

$$\int_{|x|<R} V(x)|x|^{2m-n+\varepsilon} dx = \infty\,,$$

or

$$\int_{|x|>R} V(x)|x|^{2m-n-\varepsilon} dx = \infty$$

and

$$L_0 u = \sum_{|\alpha|=|\beta|=m} D^\alpha(a_{\alpha\beta}(x)D^\beta u(x))\,.$$

Then the essential spectrum of the operator L contains an infinite number of negative numbers.

Theorem 10. *Let $V(x) \geq 0$ and $2\gamma m < 1$. There exists such a positive number b_0, that if either*

$$\int_{|x|<1} V(x)^\gamma |x|^{2m\gamma-n+\varepsilon}dx \geq b_0 k$$

or

$$\int_{|x|>1} V(x)^\gamma |x|^{2m\gamma-n-\varepsilon}dx \geq b_0 k$$

and

$$L_0 u = \sum_{|\alpha|=|\beta|=m} D^\alpha(a_{\alpha\beta}(x)D^\beta u(x))$$

for some positive ε and natural k, then $N \geq k$. In Theorems 9 and 10 it is supposed that $|a_{\alpha\beta}(x)| \leq C$.

3 Proofs of Theorems

Proof of Theorem 1. Our aim is to find a subspace in $C_0^\infty(\Omega)$ of codimension N, for functions u from which the inequality

$$\mu \int u^2 dx + \int V u^2 dx \leq \int \sum a_{\alpha\beta} D^\alpha u \overline{D^\beta u} dx \tag{2}$$

holds. Represent u as follows

$$u = \phi(x')u_1(x'') + u_2(x)\,,$$

where $x' = (x_1, \ldots, x_k)$ and $x'' = (x_{k+1}, \ldots, x_n)$, ϕ is a first eigen-function of the operator L_1, $\int |\phi|^2 dx' = 1$ and

$$\int u_2(x)\overline{\phi(x')}dx' = 0 \qquad \forall x'' \in \mathbf{R}^{n-k} \tag{3}$$

Let $L' = L_1 + L_2$. In order to prove the inequality (2) it suffices to show that

$$\mu \int |u_1(x'')|^2 dx'' + \mu \int |u_2(x)|^2 dx + \int V_+(x)|u(x)|^2 dx \leq$$
$$\leq \mu \int |u_1(x'')|^2 dx'' + \int L_2 u_1 \overline{u_1} dx'' + \int L' u_2 \overline{u_2} dx\,,$$

since $L'(\phi u_1) = \mu\phi u_1 + \phi L_2 u_1$, and from (3) it follows that

$$\int L'(\phi u_1)\overline{u_2} dx = \mu \int u_1(x'')dx'' \int \overline{\phi u_2} dx' + \int L_2 u_1(x'')dx'' \int \overline{\phi u_2} dx' = 0$$

Since

$$\int V_+(x)|u(x')|^2 dx' \leq 2\int V_+(x)|u_1|^2 dx + 2\int V_+(x)|u_2|^2 dx\,,$$

and

$$\int L_1 u_2 \overline{u_2} dx' \geq \mu_1 \int |u_2|^2 dx' \quad \text{where} \quad \mu_1 > \mu\,,$$

we have to test two inequalities:

$$2\int V_+''(x'')|u_1(x'')|^2 dx'' \leq \int L_2 u_1(x'')\overline{u_1(x'')}dx''\,, \tag{4}$$

$$2\int V_+(x)|u_2(x)|^2 dx \leq (\mu_1 - \mu)/\mu_1 \int L' u_2 \overline{u_2} dx \tag{5}$$

Here $V''(x'') = \int V(x)dx'$. In virtue of Theorem 3 of [1] the inequality (4) is valid if $q \geq (n-k)/2m$, $n > k + 2m$ and u_1 belongs to the subspace in $C_0^\infty(\mathbf{R^{n-k}})$ of codimension

$$N_2 \leq C_{n,q} \int (\int V_+(x)dx')^q |x''|^{mq-n+k} dx''.$$

The inequality (5) is fulfilled if u_2 belongs to a subspace in $C_0^\infty(\Omega)$ of codimension

$$N_3 \leq C_{n,q} \int V_+(x)^q |x|^{mq-n} dx.$$

Since $N_\mu \leq N_2 + N_3$, the Theorem is proved for $n > k + 2m$.

The proofs of Theorems 2 - 7 are very similar. We have only to use instead of Theorem 3 of [1] Theorems 4 and 5.

Proof of Theorem 9. Let the second condition be fulfilled. Let us show that there exists an infinite number of functions $\phi_j(x) \in C_0^\infty(\mathbf{R^n})$ with non-intersecting supports, for which $(L\phi_j, \phi_j) < 0$. Remark that for any positive a_0 there exists an infinite sequence of cubes Q_j, mutually non-intersecting and such that

$$\int_{Q_j} V(x)|x|^{2m-n-\epsilon} dx \geq a_0$$

and we can assume that the centers of the cubes Q_j and the lengths of their edges go to infinity as $j \longrightarrow \infty$. Moreover, if M_j are the maximal values of $|x|$ in Q_j, then it is possible to assume that the M_j–neighborhoods of the cubes Q_j do also not intersect.

Let Q be one of these cubes and $h_j \in C_0^\infty(\mathbf{R^n})$, $h_j(x) = 0$ outside of an M_j–neighborhood of the cube Q, $h_j(x) = 1$ in Q and

$$|D^\alpha h_j(x)| \leq C M_j^{-\alpha} \quad \text{for} \quad |\alpha| \leq m.$$

Let us set $\varphi_j(x) = h_j(x)|x|^{m-n/2-\epsilon/2}$. Then

$$(L_0\varphi_j, \varphi_j) \leq C_1 \int_Q \sum_{|\alpha|=m} |D^\alpha \varphi_j|^2 dy \leq c,$$

where the constant c does not depend of j and of the size of Q.

On the other hand

$$\int V\phi_j^2(y)dy \geq \int V(x)\varphi_j^2(x)dx \geq a_0 > 0.$$

The relation $(L\phi_j, \phi_j) < 0$ is true if $a_0 > c$. Therefore, if a_0 is chosen so that this inequality is fulfilled we find the needed sequence of the functions ϕ_j.

In the case when the first inequality is fulfilled, the proof is similar.

Proof of Theorem 10. Denote as $Q(R, d)$ the domain in $\mathbf{R^n}$, consisting of the points x, for which $R < |x| < R + d$. Let us show at first that if

$$\int_{Q(R,d)} V(x)^\gamma dx \geq b_1 d^{1-2\gamma m}(R+d)^{n-1},$$

then there exists a function $u(x)$ from $W_{2,0}^m(Q(R,d))$, for which $(Lu, u) < 0$, whence $N \geq 1$. Let

$$u(x) = ((r-R)(R+d-r))^{m-1/2+\epsilon} \quad \text{for} \quad R < r < R + d,$$

where $r = |x|$, and $u(x) = 0$ for $r < R$ and for $r > R + d$. It is clear that

$$\int \sum_{|\alpha|=m} |D^\alpha u(x)|^2 dx \leq C(R+d)^{n-1} d^{2m-1+4\epsilon}.$$

On the other hand by the Hölder inequality

$$\int V(x)^\gamma dx \le \left(\int V(x)u(x)^2dx\right)^\gamma \left(\int u(x)^{-2\gamma/(1-\gamma)}dx\right)^{1-\gamma},$$

and so

$$\int V(x)u(x)^2dx \ge \left(\int V(x)^\gamma dx\right)^{1/\gamma}\left(\int u(x)^{-2\gamma/(1-\gamma)}dx\right)^{1-1/\gamma}.$$

Remark that

$$\int u(x)^{-2\gamma/(1-\gamma)}dx \le c_1(R+d)^{n-1}d^{1-2\gamma(2m-1+2\varepsilon)/(1-\gamma)}.$$

Therefore,

$$\int V(x)u(x)^2dx \ge c_2\left(\int V(x)^\gamma dx\right)^{1/\gamma}(R+d)^{(n-1)(1-1/\gamma)}d^{1-1/\gamma+2(2m-1+2\varepsilon)},$$

and $(Lu,u)<0$, if

$$c_2\left(\int V(x)^\gamma dx\right)^{1/\gamma}(R+d)^{(n-1)(1-1/\gamma)}d^{1-1/\gamma+2(2m-1+2\varepsilon)} >$$
$$> c_0C(R+d)^{n-1}d^{2m-1+4\varepsilon},$$

i. e. for

$$\int_{Q(R,d)} V(x)^\gamma dx_1 > b(R+d)^{n-1}d^{1-2\gamma m}. \tag{6}$$

Show now that if

$$\int_{Q(1,1)} V(x)^\gamma dx_1 > bk2^{n-1},$$

then in $Q(1,1)$ there exist k functions $u_1,\dots,u_k$ from $W^m_{2,0}(Q(1,1))$ with non–intersecting supports, for which $(Lu_j,u_j)<0$. For this we find on the segment $(1,2)$ $k-1$ points $r_1,\dots,r_{k-1}$ such that

$$1=r_0<r_1<\dots<r_{k-1}<r_k=2$$

and

$$\int_{Q(r_j,r_{j+1}-r_j)} V(x)^\gamma dx > b_12^{n-1} \quad \text{for} \quad j=0,1,\dots,k-1.$$

Then in each layer $Q(r_j,r_{j+1}-r_j)$ the condition (6) is fulfilled and therefore there exists the function u_j.

The similar statement for the layer $Q(2^j,2^j)$ is true if

$$\int_{Q(2^j,2^j)} V(x)^\gamma dx > b_1k2^{n-1+j(n-2m\gamma)}, \tag{7}$$

and for the layer $Q(2^{-j},2^{-j})$ if

$$\int_{Q(2^{-j},2^{-j})} V(x)^\gamma dx > b_1k2^{n-1-j(n-2m\gamma)}. \tag{8}$$

It is easy to see using the homothety

$$x=x'2^{\pm j}.$$

Let the first condition of the Theorem be fulfilled. Then

$$\int_{|x|<1} V(x)^\gamma |x|^{2m\gamma-n+\epsilon} dx = \sum_{j=0}^{\infty} \int_{Q(2^{-j},2^{-j})} V(x)^\gamma |x|^{2m\gamma-n+\epsilon} dx \leq$$
$$\leq \sum_{j=0}^{\infty} 2^{-j(2m\gamma-n+\epsilon)} \int_{Q(2^{-j},2^{-j})} V(x)^\gamma dx .$$

Therefore there exists such a number j_0, that

$$2^{-j_0(2m\gamma-n)} \int_{Q(2^{-j_0},2^{-j_0})} V(x)^\gamma dx > b_2 k ,$$

where $b_2 = b_0/\sum 2^{-j\epsilon}$. Hence if

$$b_1 = b_0 2^{1-n}/\sum 2^{-j\epsilon} ,$$

then for the layer $Q(2^{-j_0}, 2^{-j_0})$ the condition (8) is true and the proof is complete. By the second condition the proof is very similar. We have only to divide the domain $|x| > 1$ in the layers $Q(2^j, 2^j)$ and to use (7). The proof is complete.

REFERENCES

[1] EGOROV YU. V., KONDRATJEV V. A.: *On the negative spectrum of an elliptic operator.* Symposium "Partial Differential Equations", Holzhau, 1988, Teubner–Texte zur Mathematik, pp. 63 – 72.

[2] BIRMAN M. S., SOLOMYAK M. Z.: *Interpolation estimates for the number of negative eigenvalues of a Schrödinger operator.* Schrödinger operators, standard and non–standard. Dubna, 6 – 10 September 1988, World Scientific, Singapore, 1989.

[3] SCHECHTER M.: *The spectrum of the Schrödinger operator.* Trans. of the AMS, Vol. **312**, No. 1, 1989, pp. 115 – 128.

Yu.V. Egorov, V.A. Kondrat'ev
Moscow State University
Department of Mathematics
Moscow V-234, USSR

On the Double Layer Potential Operator over Polyhedral Domains: Solvability in Weighted Sobolev Spaces and Spline Approximation

J. Elschner
Karl-Weierstraß-Institut für Mathematik
Mohrenstraße 39,O-1086 Berlin, Germany

1 Introduction

Let Γ be the boundary of a simply connected bounded polyhedron Ω in $\mathbb{R}^3$. The harmonic double layer potential operator on Γ is defined by

$$Ku(x) := \frac{1}{2\pi}\int_\Gamma u(y)\frac{\partial}{\partial n_y}\frac{1}{|x-y|}do(y) = \frac{1}{2\pi}\int_\Gamma \frac{(x-y)\cdot n_y}{|x-y|^3}u(y)do(y), \; x \in \Gamma \tag{1}$$

where do is the surface measure on Γ and n the outward pointing normal vector to Γ. We consider the second kind integral equation on Γ,

$$Au(x) := (\lambda - K)u(x) = f(x), \; x \in \Gamma, \tag{2}$$

where λ is a complex constant satisfying $|\lambda| \geq 1$. For $\lambda = 1$, (2) occurs in the indirect boundary integral equation method for the Dirichlet problem, whereas the case $\lambda \neq 1$ corresponds to certain transmission problems for the Laplace equation.

Relying on the theory of elliptic boundary value problems in domains with vertices and edges, Maz'ya [13](cf. also the survey [14]) obtained solvability results in weighted Hölder spaces for the more general boundary integral equation of elasticity on polyhedral boundaries. In this note we present solvability results for Eq. (2) in weighted Sobolev spaces which are based on a direct application of Mellin transformation techniques to this integral equation; see also Rempel [18] who studied the adjoint of the double layer potential in certain classical Sobolev spaces, and Schmitz [19] who considered a class of integral operators containing the single layer potential over a polyhedral domain.

Furthermore, we examine the numerical approximation of Eq. (2) by Galerkin methods with splines. While spline approximation methods for boundary integral equations on polygonal domains have been studied by many other authors (see e.g. Atkinson and de Hoog [3], Chandler [4], Costabel and Stephan [6], [7] and the references there), the analysis of boundary element methods for three-dimensional domains with corners and edges is in a rather incomplete stage. There are some very special results by Wendland [20], Kral and Wendland [12], and Atkinson [2] on the stability in the uniform norm of collocation methods with spline trial functions of degree ≤ 2 applied to Eq. (2), which rely on the fact that the essential norm of the double layer potential K in $L^\infty(\Gamma)$ is strictly smaller than one for certain classes of polyhedral boundaries Γ (cp. [10], [11], [14], [20]). Recently Rathsfeld [17] obtained new results on the stability of the panel method (i.e. piecewise constant collocation) for Eq. (2) without any restriction on the essential norm of K.

Moreover, optimal error estimates for Galerkin methods with splines on graded meshes have recently been proved by von Petersdorff [16] who studied the direct boundary element method for elasticity problems in polyhedra. In this case, however, the stability of the method is a consequence of the strong ellipticity of the corresponding integral operator. For the approximate solution of Eq. (2) on a polyhedral boundary Γ, we here present stability results and optimal error estimates in the L^2 norm for Galerkin methods with piecewise polynomials of arbitrary degree on graded meshes. Unfortunately, no result concerning the essential norm of the double layer potential on $L^2(\Gamma)$ seems to be available. To ensure the stability of the Galerkin method, we therefore utilize spline spaces which are modified in that the trial functions vanish on some of the closest triangles to the corners and edges of Γ. This approach is similar to that used in the spline approximation of one-dimensional Mellin convolution operators; see Chandler and Graham [5], Elschner [8].

For more details and the proofs of the theorems stated below, we refer to [9].

2 Localization at a corner

We first localize the operator (1), near each corner, to an integral operator over the corresponding (infinite) tangent cone. In this section, let Γ denote the boundary of a simply connected infinite polyhedral cone with vertex 0. The faces of Γ are (open) plane sectors, say Γ_j $(j = 1, \dots, J)$. The intersection γ of Γ with the two-dimensional unit sphere S^2 is then a spherical polygon consiting of the arcs $\gamma_j = S^2 \cap \Gamma_j$ of great circles. After introducing spherical coordinates $x = r\omega$, $y = r'\omega'$, $r = dist(x, 0)$, $\omega \in S^2$, the double layer potential on Γ can be written

$$\begin{aligned} Ku(x) &= \frac{1}{2\pi} \int_{\mathbb{R}^+ \times \gamma} \frac{n_{\omega'} \cdot \omega r}{|r\omega - r'\omega'|^3} u(r'\omega') d\omega' r' dr' \\ &= \frac{1}{2\pi} \int_{\mathbb{R}^+ \times \gamma} \frac{n_{\omega'} \cdot \omega r/r'}{|(r/r')\omega - \omega'|^3} u(r'\omega') d\omega' \frac{dr'}{r'}, \quad r\omega \in \Gamma, \end{aligned} \tag{3}$$

where $d\omega'$ denotes the arc element on γ. Thus (3) can be regarded as a Mellin convolution operator (with respect to the radial variables) which has the operator kernel

$$k(t)\cdot = \frac{1}{2\pi} \int_\gamma \frac{n_{\omega'} \cdot \omega t}{|t\omega - \omega'|^3} \cdot d\omega'$$

acting on functions defined on γ. The operator-valued Mellin symbol of K is then defined by the Mellin transform of the kernel

$$\mathcal{A}(z)\cdot = \int_0^\infty t^{z-1} k(t) \cdot dt. \tag{4}$$

We now introduce weighted Sobolev spaces on γ and Γ. Let s denote the arc length on γ. The corner points of γ are parametrized by $s_j (j = 1, \dots, J)$. Choose a function $q(s)$ on γ which coincides with the distance function $|s - s_j|$ near each corner s_j and which is C^∞ and vanishes nowhere on each arc γ_j. Then the norm in $X_\rho^k(\gamma)$, $k \in \mathbb{N}_0 = \mathbb{N} \cup \{0\}$, $\rho \geq 0$, is defined by

$$\begin{aligned} \|u; X_\rho^0(\gamma)\| &= \|q^{-\rho} u; L^2(\gamma)\|; \\ \|u; X_\rho^k(\gamma)\| &= \|u; L^2(\gamma)\| + \sum_{1 \leq j \leq J} \sum_{1 \leq i \leq k} \|q^{i-\rho} (d/ds)^i u; L^2(\gamma_i)\|, \, k \geq 1. \end{aligned}$$

Furthermore, for $k \in \mathbb{N}_0$ and $\rho, \mu \geq 0$, $X_{\rho,\mu}^k(\Gamma)$ denotes the space of all functions $u \in L^2(\Gamma)$ with the finite norm

$$\begin{aligned} \|u; X_{\rho,\mu}^0(\Gamma)\| &= \|r^{-\mu}(1+r)^\mu q^{-\rho} u; L^2(\Gamma)\|; \\ \|u; X_{\rho,\mu}^k(\Gamma)\| &= \|u; L^2(\Gamma)\| \\ &\quad + \sum_{1 \leq j \leq J} \sum_{1 \leq i+l \leq k} \|r^{-\mu}(1+r)^\mu (qr)^l (\partial/\partial r)^l q^{i-\rho} (\partial/\partial s)^i u; L^2(\Gamma_j)\|, \; k \geq 1. \end{aligned}$$

Note that $X^k_{\rho,\mu}(\Gamma)$ may be considered as the space of traces of functions in certain weighted Sobolev spaces on domains with singularities, which were introduced by Maz'ya and Plamenevskij [15]. In order to obtain optimal error estimates for boundary element methods, we need another scale of weighted spaces on Γ. The norm in $\Upsilon^k_{\rho,\mu}(\Gamma)$, $k \in \mathbb{N}, \rho, \mu \geq 0$, is given by

$$\begin{aligned} \|u; \Upsilon^k_{\rho,\mu}(\Gamma)\| &= \|u; L^2(\Gamma)\| + \sum_{1\leq j\leq J} \left\{ \sum_{1\leq l\leq k} \|r^{-\mu}(1+r)^{\mu} r^l (\partial/\partial r)^l u; L^2(\Gamma_j)\| \right. \\ &\quad \left. + \sum_{1\leq i, i+l\leq k} \|r^{-\mu}(1+r)^{\mu} r^l (\partial/\partial s)^l q^{i-\rho} (\partial/\partial r)^i u; L^2(\Gamma_j)\| \right\}, \end{aligned}$$

and we set $\Upsilon^0_{\rho,\mu}(\Gamma) = X^0_{\rho,\mu}(\Gamma)$.

Let $(1-\alpha_j)\pi$, $\alpha_j \in (-1,1)$, be the interior angle of the spherical polygon γ at the corner point s_j, and set $\alpha = max\{|\alpha_j|, j = 1, \ldots, J\}$ and $\rho^* = 1/2 + 1/(1+\alpha)$. Note that $\rho^* < 3/2$. The operator norm on a normed space X is denoted by $\|\bullet; \mathcal{L}(X)\|$. The main properties of the operator function $\mathcal{A}(z)$ defined by (4) can now be summarized in the following

Theorem 1 *Let $k \in \mathbb{N}_0$, $\rho \in [0, \rho^*)$, and let $\varepsilon > 0$ be sufficiently small. For $k = 0$, assume in addition that $\rho < 1/2$.*

(i) $\mathcal{A}(z) \in \mathcal{L}(X^k_\rho(\gamma))$ is an analytic operator function in the strip $-1 < Re\ z < 2$.

(ii) On each closed strip $-1+\varepsilon \leq Re\ z \leq 2-\varepsilon$, the representation $\mathcal{A}(z) = \mathcal{A}_0(z) + \mathcal{A}_1(z)$ holds, where $\|\mathcal{A}_0(z); \mathcal{L}(X^k_\rho(\gamma))\| \leq 1-\delta$ for some $\delta \in (0,1)$ and all z, $\mathcal{A}_1(z)$ is compact on $X^k_\rho(\gamma)$ for any z and $\|\mathcal{A}_1(z); \mathcal{L}(X^k_\rho(\gamma))\| \to 0$ as $|Im\ z| \to \infty$.

(iii) The kernel of $\lambda - \mathcal{A}(z)$ in $L^2(\gamma)$ is trivial if $|\lambda| \geq 1$ and $0 \leq Re\ z \leq 1$.

For a proof of (iii), see Rempel [18] who reduced this to the investigation of an eigenvalue problem for the Laplace-Beltrami operator on S^2, and also Rathsfeld [17] for the case $Re\ z = 0$. Another proof of (iii), which relies entirely on integral equation methods, can be found in [9]. The main ingredient of that proof is the a priori estimate

$$\int_\Gamma \overline{u}(1 \pm K)V u do \geq c\|u\|^2, \tag{5}$$

where V is the single layer potential on Γ and $\|\bullet\|$ denotes the norm in the (classical) Sobolev space $H^{-1/2}(\Gamma)$. A version of (5) can be used to obtain another proof of Verchota's theorem [21] on the invertibility of the operator $1-K$ on the boundary of a Lipschitz domain; see [9].

As a simple consequence of (ii), we have

Corollary 1 *Let $|\lambda| \geq 1$ and $-1 < Re\ z < 2$. Then, under the assumption of Theorem 1, $\lambda - \mathcal{A}(z)$ is a Fredholm operator on $X^k_\rho(\gamma)$ with index 0. Moreover, on each closed strip $-1+\varepsilon \leq Re\ z \leq 2-\varepsilon$, $(\lambda - \mathcal{A}(z))^{-1} \in \mathcal{L}(X^k_\rho(\gamma))$ exists and is uniformly bounded whenever $|Im\ z|$ is large enough, and there is only a finite number of z's lying in that strip such that the kernel of $\lambda - \mathcal{A}(z)$ is non-trivial.*

Finally, we remark that, for any fixed λ of absolute value ≥ 1, (iii) implies the existence of a number $\mu^* \in (1,2]$ such that the kernel of $\lambda - \mathcal{A}(z)$ (in $L^2(\gamma)$) is trivial for all z belonging to the strip $1-\mu^* < Re\ z \leq 1$.

3 Solvability results

We first retain the notation of the preceding section and state a theorem on the invertibility of the operator $\lambda - K$ over an infinite cone.

Theorem 2 *Let Γ be the boundary of an infinite polyhedral cone as in Section 2, and assume that $|\lambda| \geq 1$, $0 \leq \mu < \mu^*$, $0 \leq \rho < \rho^*$ if $k \in \mathbb{N}$ and $0 \leq \rho < 1/2$ if $k = 0$. Then the operator $\lambda - K$ is invertible on $X^k_{\rho,\mu}(\Gamma)$ and $\Upsilon^k_{\rho,\mu}(\Gamma)$.*

This follows from Corollary 1, the remark at the end of Section 2 and the characterization of the corresponding weighted Sobolev spaces by means of the Mellin transform.

In case of a bounded polyhedral boundary Γ, let us introduce the weighted spaces $X^k_{\rho,\mu}(\Gamma)$ and $\Upsilon^k_{\rho,\mu}(\Gamma)$ via a partition of unity, where near each corner the corresponding norm is locally defined as in Section 2. (Note that the factor $(1+r)^\mu$ can then be dropped.) Setting $X^k_{\rho,\mu} = X^k_\rho$ etc., we are now in a position to state our main solvability result.

Theorem 3 *Let Γ be the boundary of a bounded polyhedron Ω as in Section 2, and suppose that $|\lambda| \geq 1$.*

(i) The operators $\lambda - K \in \mathcal{L}(X^k_\rho(\Gamma))$ and $\lambda - K \in \mathcal{L}(\Upsilon^k_\rho(\Gamma))$ are Fredholm with index 0 if either $k \in \mathbb{N}$ and $0 \leq \rho < \rho_0$ for some $\rho_0 \in (1, 3/2)$, or $k = 0$ and $0 \leq \rho < 1/2$.

(ii) If, in addition, $\lambda \neq -1$, then the above operators are invertible.

Assertion (i) is a consequence of Theorem 2 and a standard localization procedure. To prove (ii), it is sufficient to verify that the kernel of $\lambda - K$ in $L^2(\Gamma)$ is trivial if $\lambda \neq -1$. This can be done as in the case of a domain with smooth boundary; see [14]. Theorem 3 covers, in particular, the solvability of Eq. (2) in the classical Sobolev space $H^1(\Gamma)$ which corresponds to the choice $k = \rho = 1$. Finally, we remark that this theorem obviously has the following

Corollary 2 *(Smoothness of solutions). Under the assumptions of Theorem 3(i), $u \in L^2(\Gamma)$ and $(\lambda - K)u \in X^k_\rho(\Gamma)$ imply that $u \in X^k_\rho(\Gamma)$. The same results are valid for the Υ-spaces.*

4 Spline Galerkin methods

Throughout this section Γ denotes the boundary of a bounded polyhedron Ω as in Section 1, i.e. Γ consists of a finite number of (plane) polygons. With any $q \geq 1$ we associate a sequence of graded triangulations Δ_n, $n \in \mathbb{N}$, of Γ. Let F be one of the faces of Γ. Using parallels at distance 1, say, to the sides, F is first cut into parallelograms at the convex corners, L-shaped polygons at the non-convex corners, quadrangles near the edges, and a dilate F' of F; see Figure 1. Taking further straight lines at distances $(i/n)^q$, $i = 1, \ldots, n-1$, to the boundary of F, we obtain graded partitions of the polygons near the corner points (cf. Figures 2 and 3), whereas the subdivision of the quadrangles near the edges is completed by choosing a quasi-uniform partition of size $\sim 1/n$ by parallels pointing to the normal direction (cp. Figure 4).

All polygons obtained so far in the subdivision of the boundary strip of unit width are now cut up into triangles, say by non-intersecting diagonals. For the remaining polygon F', we choose a regular family of partitions into triangles of diameter $\sim 1/n$.

Triangulating each face of Γ in this way, we obtain our sequence $\{\Delta_n\}$ of graded meshes which is said to be of class $\mathcal{M}_q$ in the sequel. Note that the total number of triangles of the partition Δ_n is proportional to n^2. For a sequence of triangulations Δ_n of class $\mathcal{M}_q$ we now introduce the modified

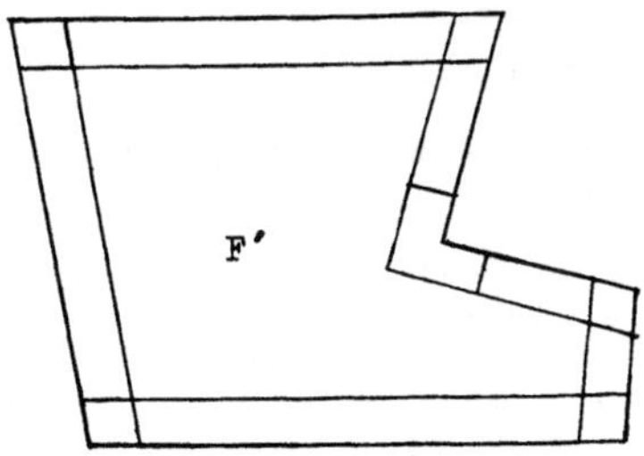

Figure 1

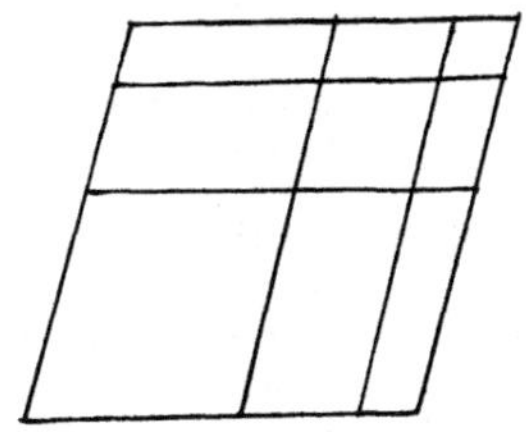

Figure 2

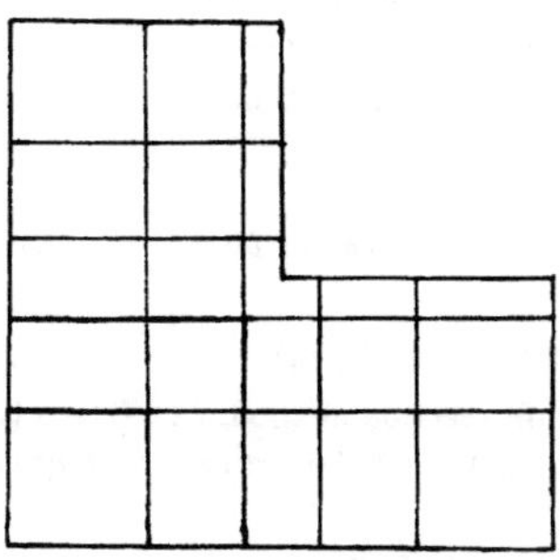

Figure 3

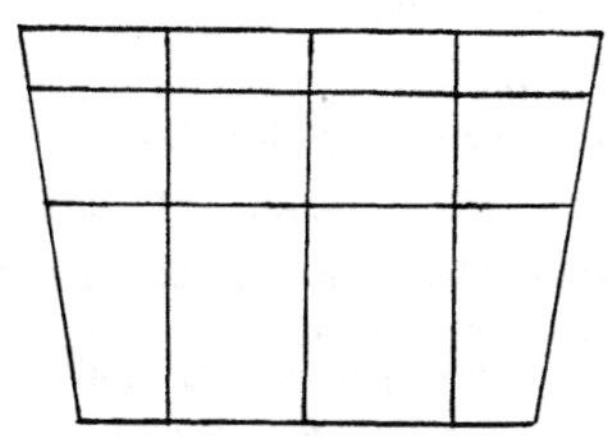

Figure 4

spline spaces $S_d(\Delta_n)$ consisting of all piecewise polynomials of (total) degree $\leq d$ subordinate to Δ_n which vanish on the boundary strip of width $(i_0/n)^q$ on each face of Γ, where the index i_0 does not depend on n. The Galerkin method based on piecewise polynomials for the approximate solution of Eq. (2) is the following: Find $u_n \in S_d(\Delta_n)$ such that

$$P_n A u_n = \lambda u_n - P_n K u_n = P_n f, \tag{6}$$

where P_n is the orthogonal projection of $L^2(\Gamma)$ onto $S_d(\Delta_n)$.

A first step in the convergence analysis of (6) is to prove stability for the following finite section method. For $u \in L^2(\Gamma)$, consider the truncation operators

$$\pi_n u(x) = \begin{cases} 0 & \text{on the boundary strips of width } (i_0/n)^q \\ u(x) & \text{elsewhere.} \end{cases}$$

Stability of the corresponding projection method means that the a priori estimates

$$\|\pi_n A \pi_n u; L^2(\Gamma)\| \geq c \|\pi_n u; L^2(\Gamma)\|, \quad u \in L^2(\Gamma), n \geq n_0 \tag{7}$$

hold, where n_0 is large enough. The study of (7) can now be reduced to the invertibility of certain integral operators. Let Γ^j $(j = 1, \ldots, N)$ be the (infinite) tangent cones at the corner points of Γ, and let Γ^j_* denote the surface obtained after deleting a boundary strip of unit width on each face of Γ^j. Consider the integral operators A^j_* defined by

$$A^j_* u(x) = \lambda u(x) - \frac{1}{2\pi} \int_{\Gamma^j_*} \frac{(x-y) \cdot n_y}{|x-y|^3} u(y) do(y), \quad x \in \Gamma^j_*.$$

Proposition 1 *For each $j = 1, \ldots, N$, the operator A^j_* is Fredholm with index 0 in $L^2(\Gamma^j_*)$.*

Proposition 2 *The validity of (7) is equivalent to the following condition:*

$$\textit{For each } j = 1, \ldots, N \textit{ the kernel of } A^j_* \textit{ in } L^2(\Gamma^j_*) \textit{ is trivial.} \tag{8}$$

Remark 1 *If near each vertex the polyhedron Ω is either convex, or $\mathbb{R}^3 \setminus \Omega$ is convex, or Ω is rectangular there in the sense of [11], then the kernel condition (8) is satisfied.*

Indeed, in those cases the uniform operator norm of the double layer potential on Γ^j is always strictly smaller than one; see [1], [10], [11], [14], [20]. (If Γ^j is the boundary of a rectangular domain, then one has to introduce another norm on $L^\infty(\Gamma^j)$ which is equivalent to the maximum norm; cf. [1], [11].) Moreover, it is not difficult to check that each element of the kernel of A^j_* in $L^2(\Gamma^j_*)$ belongs to $L^\infty(\Gamma^j_*)$, hence condition (8) is fulfilled; cf. Proposition 1.

We are now ready to state our result on the stability of the Galerkin method.

Theorem 4 *Assume $|\lambda| \geq 1$, $\lambda \neq -1$, and suppose that condition (8) is satisfied. If $\{\Delta_n\}$ is of class $\mathcal{M}_q$ and the index i_0 is sufficiently large, then the method (6) is stable in $L^2(\Gamma)$, i.e.*

$$||P_n A u_n; L^2(\Gamma)|| \geq c||u_n; L^2(\Gamma)||, \quad u_n \in S_d(\Delta_n), n \geq n_0 \tag{9}$$

where n_0 is large enough.

We sketch the stability proof in the case $q = 1$, i.e. when a regular family of partitions into triangles of diameter $\sim 1/n$ is utilized. Notice that

$$\pi_n A u_n = P_n A u_n - \pi_n(1 - P_n) K u_n, \quad u_n \in S_d(\Delta_n). \tag{10}$$

We now regard (6) as a small perturbation of the above finite section method. Let T be a triangle of the grid Δ_n not intersecting the boundary strips of width i_0/n. Then the well known approximation property of piecewise polynomials implies that

$$\begin{aligned} ||(1 - P_n) K u_n; L^2(T)|| &\leq c \, diamT ||\nabla K u_n; L^2(T)|| \\ &\leq c \left\{ diamT \inf_{x \in T} \kappa(x) \right\} ||\kappa \nabla K u_n; L^2(T)||, \end{aligned} \tag{11}$$

where $\kappa(x)$ denotes the distance of x to the set of all vertices and edges of Γ and the constant c does not depend on n and T. Note that the expression in braces tends to zero as $i_0 \to \infty$. Since (cf. [9])

$$||\kappa(x) \nabla K u_n; L^2(\Gamma)|| \leq c||u_n; L^2(\Gamma)||,$$

the estimates (11) therefore imply that $||\pi_n(1 - P_n) K P_n; \mathcal{L}(L^2(\Gamma))||$ becomes as small as desired if i_0 and n are sufficiently large. Together with Proposition 2 and relation (10), this yields estimate (9).

For $q > 1$, the proof of (9) is more complicated; see [9]. The next result shows that one obtains the optimal order of convergence when suitably graded meshes are used. Let u be the exact solution of Eq. (2).

Theorem 5 *Suppose the assumption of the previous theorem are satisfied. If, in addition, $f \in \Upsilon^{d+1}_\rho(\Gamma)$, $0 < \rho < 1/2$, and the mesh grading $q \geq (d + 1)/\rho$ is given, then the Galerkin method (6) converges with rate*

$$||u - u_n; L^2(\Gamma)|| = O(n^{-d-1}) \text{ as } n \to \infty. \tag{12}$$

This is a consequence of Theorem 4, Corollary 2 and the approximation properties of the spline spaces. In particular, the optimal rate (12) is obtained if $f \in L^2(\Gamma)$ belongs piecewise to the classical Sobolev space of order $d + 1$ on each face of Γ and if $q > 2(d + 1)$. For a smoother right-hand side, this result cannot be improved (with respect to the mesh grading), since the splines from $S_d(\Delta_n)$ do not belong to the spaces $\Upsilon^{d+1}_\rho(\Gamma)$, $\rho \geq 1/2$, in general.

References

[1] T.S. Angell, R.E. Kleinman, J. Kral, Layer potentials on boundaries with corners and edges, Časopic Pro Pěst. Mathematiky 113 (1968) 387-402.

[2] K.E. Atkinson, Piecewise polynomial collocation for integral equations on surfaces in three dimensions, J. Integral Equations 9 (1985) (Suppl.) 25-48.

[3] K.E. Atkinson, F. de Hoog, The numerical solution of Laplace's equation on a wedge, IMA J. Numer. Anal. 4(1984) 19-41.

[4] G.A. Chandler, Galerkin's method for boundary integral equations on polygonal domains, J. Austral. Math. Soc. B 26 (1984) 1-13.

[5] G.A. Chandler, I.G. Graham, Product integration-collocation methods for non-compact integral operator equations, Math. Comp. 50 (1988) 125-138.

[6] M. Costabel, E.P. Stephan, Boundary integral equations for mixed boundary value problems in polygonal domains and Galerkin approximation, In: Banach Center Publications, Vol. 15, PWN, Warsaw 1985, pp. 175-251.

[7] M. Costabel, E.P. Stephan, On the convergence of collocation methods for boundary integral equations on polygons, Math. Comp. 49 (1987) 461-478.

[8] J. Elschner, On spline approximation for a class of non-compact integral equations, Math. Nachr. 146 (1990) 271-321.

[9] J. Elschner, The double layer potential operator over polyhedral domains I: Solvability in weighted Sobolev spaces, Applicable Analysis (to appear), II: Spline Galerkin methods, Math. Meth. in the Appl. Sci. (to appear).

[10] J. Kral, Integral operators in potential theory, Lecture Notes Math. 823, Springer, Berlin etc. 1980.

[11] J. Kral, W. Wendland, Some examples concerning applicability of the Fredholm-Radon method in potential theory, Aplikace Matematiky 31 (1986) 293-308.

[12] J. Kral, W. Wendland, On the applicability of the Fredholm-Radon method in potential theory and the panel method, In: Notes on Numerical Fluid Mechanics, Vol. 21, Vieweg, 1988, pp. 120-136.

[13] V.G. Maz'ya, Boundary integral equations of elasticity in domains with piecewise smooth boundaries, In: Equadiff 6 Proceedings, Brno University 1985, pp. 235-242.

[14] V.G. Maz'ya, Boundary integral equations, In:Sovrem. Problemy Matem. Fundam. Napravlenia, Vol. 27, Viniti, Moscow 1988, pp. 131-228.

[15] V.G. Maz'ya, B.A. Plamenevskij, Elliptic boundary value problems on manifolds with singularities, In: Problemy Mat. Anal., vyp. 6, Izdat. Leningrad Univ., Leningrad 1977, pp. 85-142.

[16] T. von Petersdorff, Randwertprobleme der Elastizitätstheorie für Polyeder. Singularitäten und Approximationen mit Randelementmethoden, Dissertation, TH Darmstadt 1989.

[17] A. Rathsfeld, The invertibility of the double layer potential operator in the space of continuous functions defined on a polyhedron. The panel method, Complex Variables (to appear).

[18] S. Rempel, Corner singularity for transmission problems in three dimensions, Integral Equations Operator Th. 12 (1989) 835-854.

[19] H. Schmitz, Über das singuläre Verhalten der Lösungen von Integralgleichungen auf Flächen mit Ecken, Dissertation, Univ. Stuttgart 1989.

[20] W. Wendland, Die Behandlung von Randwertaufgaben im $\mathbb{R}_3$ mit Hilfe von Einfach- und Doppelschichtpotentialen, Numer. Math. 11 (1968) 380-404.

[21] G.C. Verchota, Layer potentials and regularity for the Dirichlet problem for Laplace's equation in Lipschitz domains, J. Funct. Anal. 59 (1984) 572-611.

The Bisingular Problems

A.M. Il'in
Sverdlovsk, USSR

The boundary value problem $L(x, u, \varepsilon) = 0$, where L is an operator, $u(x, \varepsilon)$- the solution we look for, $x \in D \subset \mathbb{R}^n$, is said to be singular if the asymptotic expansion

$$u(x, \varepsilon) = \sum_{k=0}^{\infty} \varepsilon^k u(x) \tag{1}$$

is not valid everywhere in $\overline{D}$. We shall consider only problems the solutions of which are slowly varying anywhere in $\overline{D}$ except some narrow subdomains if ε is small. These subdomains are called boundary layers. Outside the boundary layers the solution may be represented by the asymptotic series (1). It is called the outer expansion analogously to the problem arising in hydrodynamics.

Let Γ_ε be a boundary layer of the problem and $\Gamma_0 = \cap \Gamma_\varepsilon$, $(\varepsilon > 0)$. It usually appears that Γ_0 is a subset of the boundary ∂D or of any subset of lower dimension situated inwards of D where the solutions $u(x, \varepsilon)$ undergo rapid changes as $\varepsilon \to 0$ (for example a fluid flow with small viscosity near the boundary or diffused shock wave). The most efficient way of investigating this problem seems to be in stretching of coordinates near Γ_0, in other words an introduction of interior coordinates $\xi(x, \varepsilon)$. The asymptotic expansion of $u(x, \varepsilon)$ near Γ_0 has the form

$$u(x, \varepsilon) = \sum \varepsilon^k \nu_k(\xi). \tag{2}$$

It is called the inner expansion. A recurrent system of boundary value problems arises for the coefficients $\nu_k(\xi)$, which are called boundary layer functions.

This idea was first expressed by Prandtl in the famous report to the III Mathematical Congress in 1904 in connection with the problem of the fluid flow with small viscosity near the boundary. Such methods have been intensively developed in application to a wide range of problems.

Concerning the boundary value problems we study here there are two possibilities. In the first case $u_k(x)$ are smooth functions everywhere in the closure of D (if Γ_0 lies inside of D then $u_k(x)$ may be discontinuous of the first kind). The singularity here arises due to our impossibility to satisfy some of the boundary conditions. Usually the boundary layer functions in such case vanish exponentially away from Γ_0. This boundary layer is sometimes called an exponential one. The correct asymptotic expansion of solution might be obtained as a sum of two series (1) and (2). A similar approach is known as Vishik-Liusternik method.

In the second case the functions $u_k(x)$ have singularities on Γ_0. As a rule the degrees of these singularities increase together with k. Thus the problems for $u_k(x)$ appear to be singular in themselves. We call the original problem with small parameter a bisingular one. As a rule the function $v_k(x)$ grow at infinity and the degree of their growth increases together with k.

One of the variants of the matching method is following. Let $A_{N,x}U$ be a partial sum $\sum_{k=0}^{N} \varepsilon^k u_k(x)$. Changing the variables $x = x(\xi, \varepsilon)$ we obtain

$$A_{N,x}U = \sum_{k=0}^{N} \varepsilon^k u_k(x) = \sum_{k=0}^{N} \varepsilon^k u_k(x(\xi, \varepsilon)) = \sum_{i=0}^{\infty} \mu_i(\varepsilon) v_{i,N}(\xi)$$

where $\mu_i(\varepsilon)$ is a callibrating sequence. We perform the similar operation with the inner asymptotic expansion and obtain

$$A_{N,\xi}V := \sum_{i:\mu(\varepsilon)\neq o(\varepsilon^N)} \mu_i(\varepsilon)v_i(\xi)$$

Matching condition consists of the equality

$$A_{K,\xi}A_{N,x}U = A_{N,x}A_{K,\xi}V, \quad \forall K, N \tag{3}$$

These conditions help us to resolve many boundary value problems for partial differential equations with a small parameter or in the case where the small parameter is contained in a boundary condition. For a very simple example we consider the case where Γ is the origin, L is a second order elliptic differential equation. Let $u(x,\varepsilon)$ be a solution of the first problem in the domain $D_\varepsilon \subset \mathbb{R}^3$ (see fig.) and O is not a spectrum of the first boundary problem for the operator L in the domain D.

Here the domain $D_\varepsilon = D \setminus \omega_\varepsilon$, $x \in \omega_\varepsilon \Leftrightarrow \varepsilon^{-1}x \in \omega$, where ω is a bounded domain.

The following outer and inner asymptotic expansions can be proved

$$\begin{aligned} u(x,\varepsilon) &= \sum_{i=0}^{\infty} \varepsilon^i u_i(x)\ ,\ \varepsilon \to 0\ ,\ \text{for } x >> \varepsilon, \\ u(x,\varepsilon) &= \sum_{i=0}^{\infty} \varepsilon^i v_i(\xi)\ ,\ \varepsilon \to 0\ ,\ \text{for } x = \varepsilon\xi << 1. \end{aligned}$$

The asymptotic expansions of the functions $u_k(x)$ and $v_i(\xi)$ have the forms

$$u_i(x) = \sum_{j=-i}^{\infty} \Phi_{j,i}(x)\ ,\ x \to 0\ ,\ v_i(\xi) = \sum_{j=-i}^{\infty} \Psi_{-j,i}(\xi)\ ,\ \xi \to 0,$$

where $\Phi_{m,i}(x)$ and $\Psi_{m,i}(\xi)$ are the homogeneous functions of order m. At Table 1 the members of the $A_{2,x}A_{2,\xi}V$ and $A_{2,\xi}A_{2,x}U$ are represented.

The matching condition (3) means that in every cell the upper functions are equal to the lower ones.

The main aim of this lecture is to show that this method may be applied to some nonlinear problems. For example we consider the flame wave in the condensed media. It is described by the system of two ordinary differential equations. After some transformations we get the following boundary value problem:

$$\nu\frac{du}{d\xi}\frac{u^\beta}{u-\varepsilon x}\exp\left(\frac{x}{1-b\varepsilon x}\right)\chi\left(x-\frac{z}{\varepsilon}\right), \quad \varepsilon << 1 \tag{4}$$

$U \backslash V$	$v_0(\xi)$	$\varepsilon v_1(\xi)$	$\varepsilon^2 v_2(\xi)$	$\dots$
$u_0(x)$	$\Psi_{0,0}$ $\Phi_{0,0}$	$\varepsilon\Psi_{1,1}(\xi)$ $\Phi_{1,0}(x)$	$\varepsilon^2\Psi_{2,2}(\xi)$ $\Phi_{2,0}(x)$	$\dots$
$\varepsilon u_1(x)$	$\Psi_{-1,0}(\xi)$ $\varepsilon\Phi_{-1,1}(x)$	$\varepsilon\Psi_{0,1}(\xi)$ $\varepsilon\Phi_{0,1}(x)$	$\varepsilon^2\Psi_{1,2}(\xi)$ $\varepsilon\Phi_{1,1}(x)$	$\dots$
$\varepsilon^2 u_1(x)$	$\Psi_{-2,0}(\xi)$ $\varepsilon^2\Phi_{-2,2}(x)$	$\varepsilon\Psi_{0,1}(\xi)$ $\varepsilon^2\Phi_{-1,2}(x)$	$\varepsilon^2\Psi_{1,2}(\xi)$ $\varepsilon^2\Phi_{0,2}(x)$	$\dots$
$\dots$	$\dots$	$\dots$	$\dots$	

Table 1:

$$u(0,\varepsilon) = 0 \ , \ u(\infty,\varepsilon) = 1.$$

Here $\nu = \nu(\varepsilon)$ is the unknown parameter, $\chi(t)$ is a cut-off function ($\chi(t) \equiv 0$ if $t \geq 0$), $b > 0$, $1 < \beta < 2$. (The asymptotics do not depend on z).

W shall find the asymptotic expansion in the form:

$$\nu(\varepsilon) = \sum \nu_k \varepsilon^k \ , \ \varepsilon \longrightarrow 0 \tag{5}$$

$$u(x,\varepsilon) = \sum \varepsilon^k u_k(x) \ , \ \varepsilon \longrightarrow 0. \tag{6}$$

After substituting (5) and (6) to the equation (4) we obtain the recurrent system of the equations:

$$\nu_0 u_0' = u_0^{\beta-1}\exp(-x) \ , \ u_0(0) = 0 \ , \ u_0(\infty) = 1 \ , \tag{7}$$

$$\begin{aligned} \imath u_1 &\equiv \nu_0 u_1' - (\beta-1)u_0^{\beta-2}\exp(-x)u_1 \\ &= -\nu_1 u_0' + u^{\beta-2}(x - bu_0x^2)\exp(-x) \ , \\ \imath u_k &= -\nu_k u_0' - \sum_{i=1}^{k-1} \nu_i u_{k-i}' + F_k(x, u_0, \dots, u_{k-1}) \ , \ k \geq 2, \\ & u_k(0) = u_k(\infty) = 0, k \geq 1. \end{aligned}$$

The equation (7) has an explicit solution $u_0(x) = (1 - e^{-x})^{1/\nu_0}$ where $\nu_0 = 2 - \beta$. The functuion $u_1(x)$ also can be found explicitly. It occurs that on this way it is impossible to determine the coefficient ν_1 because the boundary conditions for u_1 are satisfied for every ν_1. It is a more difficult problem to find $u_2, u_3, \dots$ because the orders of singularities of u_k at zero are increasing with k.

Let us find the other scale and the other asymptotic expansion near the point 0. Naturally the members u and εx are of the same order . Consequently

$$x = \varepsilon^{\alpha}\eta \ , \ u = \varepsilon^{1+\alpha}v(\eta, \varepsilon).$$

Because the left and the right parts of (4) are also of the same order we get $\alpha = \frac{2-\beta}{\beta-1}$.

In the variable η we have

$$\nu\frac{dv}{d\eta} = \frac{v^{\beta}}{v-\eta}\exp(-\varepsilon^{\alpha}\frac{\eta}{1-b\varepsilon^{1+\alpha}\eta}) \ , \ v(0,\varepsilon) = 0.$$

The matching condition (3) shows that the form of the series 5 and 6 is wrong. It is clear that the right form is the following:

$$\nu = \sum_{k\geq 0, i\geq 0} \nu_{k,i}\varepsilon^{k+\alpha i},$$
$$u = \sum_{k\geq 0, i\geq 0} \varepsilon^{k+\alpha i}u_{k,i}(x),$$
$$V = \sum_{k\geq 0, i\geq 0} \varepsilon^{k+\alpha i}v_{k,i}(\eta).$$

The matching condition 3 allows us to determine $\nu_{k,i}$, $u_{k,i}(x)$ and $v_{k,i}(\eta)$. It is interesting to mention that this method permits us to obtain asymptotic expansions for $\beta = 3/2$. The formulas of the previous papers have no sense for such β.

A.M. Il'in
Moscow State University
Department of Mathematics
Moscow V-234, USSR

On a Bimetal Problem for Heat Conductions

Lothar Jentsch
TU Chemnitz, Germany

A bimetal problem is called a boundary contact problem where the contact surface between the different materials and the boundary of the entire body have common points. The following picture shows the typical situation:

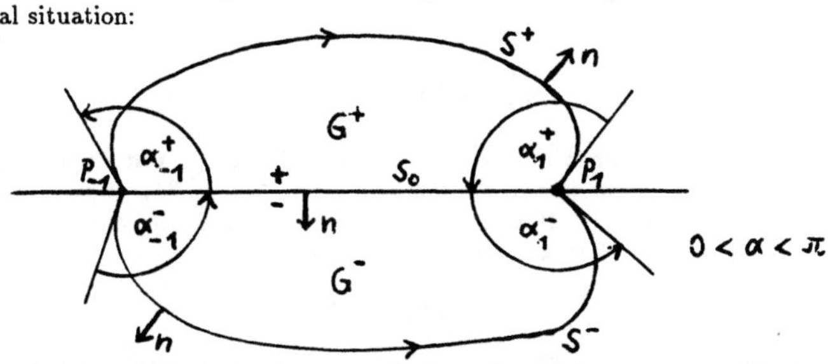

In the domains G^+, G^- different materials are situated so that in G^+ and G^- different differential equations are to be satisfied. On $S = S^+ \cup S^-$ boundary conditions and on S_0 contact conditions are given. The plan for solving such problems consists of several steps.

1. Construction of Green's contact function or contact tensor $G(x,y)$ for two half-spaces or half-planes of different material. The contact tensor $G(x,y)$ is the sum of the corresponding fundamental solution and of a compensatrix such that the contact conditions on the line or plane between the different materials are satisfied. Contact tensors of elastostatics are calculated in several papers [6].

2. Solution of the problem in form of a potential over S with $G(x,y)$ in the kernel of the potential. Such a potential a priori satisfies the differential equation and the contact conditions. It is necessary to investigate these potentials: jump relations, behaviour at infinity and on the contact surface are of interest. The compensatrix has influence on the jump [7]. From the boundary condition and the jump relation we obtain the boundary integral equation $A\varphi = f$ of the bimetal problem where A is a singular matrix integral operator with fixed singularities.

3. Investigation of A: Fredholm property, index of A, solvability conditions of $A\varphi = f$.

4. Asymptotics of the solution φ at the critical points P_1, P_{-1}.

Bimetal problems of elastostatics for the bounded and unbounded plane domain with three different contact conditions are considered in several papers which are cited in [6], [8]. In those papers $\alpha_1^+, \alpha_1^-, \alpha_{-1}^+, \alpha_{-1}^-$ are right angles. In this case results of Duduchava on singular integral equations with fixed singularities [5] are applicable.

In the following the bimetal problem for heat conduction with arbitrary angles is investigated. The Mellin transform technique [3],[4],[11] used is also suited to solve the bimetal problems of elastostatics with arbitrary angles. Of special interest is the fact that the discussion in the case of non-ideal heat contact is easier than in the case of ideal heat contact, although the contact function is more complicated in the former one. As to the derivation of the contact functions and other details we refer to [9].

Let us consider the following boundary contact problem for the stationary heat equation: The temperature $u = u(x) = u(x_1, x_2)$ has to satisfy

a) the differential equation

$$\Delta u = 0 \text{ in } G^+ \cup G^-, \tag{1}$$

b) the contact conditions on S_0

$$\left\{\lambda_1 \frac{\partial u}{\partial n}\right\}^+ = \left\{\lambda_0 \frac{\partial u}{\partial n}\right\}^-, \tag{2}$$

$$\beta\{u\}^+ - \beta\{u\}^- = -\left\{\lambda_1 \frac{\partial u}{\partial n}\right\}^+, \tag{3}$$

c) the boundary condition II

$$\lambda_1 \frac{\partial u}{\partial n} = f \text{ on } S^+ , \; \lambda_0 \frac{\partial u}{\partial n} = f \text{ on } S^-. \tag{4}$$

Here $\{.\}^+$, $\{.\}^-$ denote the limit to a point of S_0 from G^+ and G^-, respectively, $\lambda_1 > 0$, $\lambda_0 > 0$ are the coefficients of heat conduction in the domains G^+, G^-, $0 < \beta < \infty$ is the coefficient of heat transition between G^+ and G^-. If we have the situation of ideal heat contact ($\beta = \infty$) instead of non-ideal heat contact ($0 < \beta < \infty$), then (3) becomes

$$\{u\}^+ = \{u\}^-. \tag{5}$$

The Green's contact function $G(x, y)$ for ideal heat contact can easily be written

$$\begin{aligned}
\pi G(x,y) &= -\ln|x-y| - \frac{\lambda_1 - \lambda_0}{\lambda_0 + \lambda_1} \ln|x - \overline{y}| \text{ for } x_2 > 0, y_2 > 0,\\
\pi G(x,y) &= -\frac{2\lambda_1}{\lambda_0 + \lambda_1} \ln|x-y| \text{ for } x_2 < 0, y_2 > 0,\\
\pi G(x,y) &= -\frac{2\lambda_0}{\lambda_0 + \lambda_1} \ln|x-y| \text{ for } x_2 > 0, y_2 < 0,\\
\pi G(x,y) &= -\ln|x-y| - \frac{\lambda_0 - \lambda_1}{\lambda_0 + \lambda_1} \ln|x - \overline{y}| \text{ for } x_2 < 0, y_2 < 0,
\end{aligned} \tag{6}$$

where $|x-y|^2 = (x_1 - y_1)^2 + (x_2 - y_2)^2$, $|x - \overline{y}|^2 = (x_1 - y_1)^2 + (x_2 + y_2)^2$. It satisfies (2), (5) on the line $x_2 = 0$ for any fixed y.

The Green's contact function $G(x, y)$ for non-ideal heat contact was explicitely calculated for two half-spaces and two half-planes in [9]. In the plane case the result is

$$\pi G(x,y) = \pi G_\infty(x,y) + V(x,y), \tag{7}$$

where $G_\infty(x, y)$ is the contact function (6) at ideal heat contact,

$$\begin{aligned}
V(x,y) &= \frac{\lambda_0}{\lambda_0 + \lambda_1} 2Re(e^{\overline{z}_a} E_1(\overline{z}_a)) \text{ for } x_2 > 0, y_2 > 0,\\
V(x,y) &= -\frac{\lambda_1}{\lambda_0 + \lambda_1} 2Re(e^{z_a} E_1(z_a)) \text{ for } x_2 < 0, y_2 > 0,\\
V(x,y) &= -\frac{\lambda_0}{\lambda_0 + \lambda_1} 2Re(e^{\overline{z}_a} E_1(\overline{z}_a)) \text{ for } x_2 > 0, y_2 < 0,\\
V(x,y) &= \frac{\lambda_1}{\lambda_0 + \lambda_1} 2Re(e^{z_a} E_1(z_a)) \text{ for } x_2 < 0, y_2 < 0,\\
z_a &= a(|x_2| + |y_2| + i(x_1 - y_1)), \; a = \beta \frac{\lambda_0 + \lambda_1}{\lambda_0 \lambda_1},
\end{aligned} \tag{8}$$

and

$$E_1(z) = \int_1^\infty \frac{e^{-zt}}{t} dt = \int_z^\infty \frac{e^{-t}}{t} dt \quad ([1], 5.1).$$

The compensatrix $V(x, y)$ has a fixed logarithmical singularity for $x_1 = y_1$, $x_2 = y_2 = 0$ and

$$V(x,y) = O(1/|z_a|), \ \frac{\partial V(x,y)}{\partial x_k} = O(1/|z_a|^2) \text{ for } z_a \to \infty. \tag{9}$$

We look for a solution of our problem in the form

$$u(x) = \int_S G(x,y)\varphi(y) ds_y. \tag{10}$$

This potential of a single layer type a priori satisfies the Laplace equation and the contact condition on S_0. If $\varphi \in L_2(S)$, then $u(x)$ is bounded in $G^+ \cup G^-$ and in the neighbourhood of P_1, P_{-1} holds

$$\partial u/\partial x_k = O(1/\sqrt{\rho_x}), \tag{11}$$

where ρ_x is the distance of x from S parallel to the x_1-axis [10].

For a solution u of the homogeneous boundary contact problem in this class of regularity, from the Green formula follows

$$\int_{G^+} \lambda_1 |\text{ grad } u|^2 dx + \int_{G^-} \lambda_0 |\text{ grad } u|^2 dx = -\int_{S_0} \beta^{-1} \left(\lambda_1 \left\{ \frac{\partial u}{\partial n} \right\}^+ \right)^2 ds \leq 0.$$

From this we get the uniqueness result that $u = const$. From the Green formula we obtain the necessary condition of solvability

$$\int_S f ds = 0. \tag{12}$$

The jump relation of the potential of the single layer leads to the boundary integral equation of our problem

$$\varphi(x_0) + \int_S \frac{\partial G(x_0, y)}{\partial n_{x_0}} \varphi(y) ds_y = F(x_0), x_0 \in S, \tag{13}$$

where

$$F(x_0) = \begin{cases} \lambda_1^{-1} f(x_0) \text{ for } x_o \in S^+ \\ \lambda_0^{-1} f(x_0) \text{ for } x_o \in S^- \end{cases}.$$

We study the equation (13) in the space $L_2(S)$ and will show, that the operator on the left hand side of (13) is a Φ-operator with index zero. Then we can prove, that the adjoint homogeneous integral equation

$$\psi(x_0) + \int_S \frac{\partial G(y, x_0)}{\partial n_y} \psi(y) ds_y = 0 \tag{14}$$

has exactly one linear independent solution

$$\psi(x_0) = \begin{cases} \lambda_1 \text{ for } x_o \in S^+ \\ \lambda_0 \text{ for } x_o \in S^- \end{cases}.$$

Therefore, the condition (12) is also sufficient for the existence of a solution. For all arguments we suppose $S^+ \in C^{1,\mu}$, $S^- \in C^{1,\mu}$. We choose a parameter representation of

$$S^+ : x = y^+(t) \text{ and } S^- : x = y^-(t), \ -1 \leq t \leq 1$$

such that in the neighbourhood of $t = +1$ and $t = -1$ the parameter t is the arc length of S^+ and S^-.

Following [10], if we introduce the vectors

$$\Phi(t) = \begin{pmatrix} \Phi^+(t) \\ \Phi^-(t) \end{pmatrix} = \begin{pmatrix} \varphi(y^+(t)) \\ \varphi(y^-(t)) \end{pmatrix},$$
$$F(t) = \begin{pmatrix} F^+(t) \\ F^-(t) \end{pmatrix} = \begin{pmatrix} F(y^+(t)) \\ F(y^-(t)) \end{pmatrix},$$

then (13) can be written as a system

$$\Phi(t) + \frac{1}{\pi}\int_{-1}^{+1} F(t,\tau)\Phi(\tau)d\tau + (K\Phi)(t) = F(t), \tag{15}$$

where K is a compact operator in the space $(L_2[-1,1])^2$.

The elements of the matrix $F(t,\tau)$ are in the case of non-ideal heat contact $(0 < \beta < \infty)$

$$\begin{aligned} f_{11}(t,\tau) &= -\frac{2a_1^+(1-\tau)}{(a_1^+)^2(t-\tau)^2+(2-t-\tau)^2} - \frac{2a_{-1}^+(1+\tau)}{(a_{-1}^+)^2(t-\tau)^2+(2+t+\tau)^2}, \\ f_{12}(t,\tau) &= 0,\ f_{21}(t,\tau)=0, \\ f_{22}(t,\tau) &= -\frac{2a_1^-(1-\tau)}{(a_1^-)^2(t-\tau)^2+(2-t-\tau)^2} - \frac{2a_{-1}^-(1+\tau)}{(a_{-1}^-)^2(t-\tau)^2+(2+t+\tau)^2}, \end{aligned} \tag{16}$$

and in the case of ideal heat contact $(\beta = \infty)$

$$\begin{aligned} f_{11}(t,\tau) &= -2\frac{\lambda_1-\lambda_0}{\lambda_0+\lambda_1}\frac{a_1^+(1-\tau)}{(a_1^+)^2(t-\tau)^2+(2-t-\tau)^2} \\ &\quad -2\frac{\lambda_1-\lambda_0}{\lambda_0+\lambda_1}\frac{a_{-1}^+(1+\tau)}{(a_{-1}^+)^2(t-\tau)^2+(2+t+\tau)^2}, \\ f_{12}(t,\tau) &= -2\frac{\lambda_0}{\lambda_0+\lambda_1}\frac{\sin(\alpha_1^++\alpha_1^-)(1-\tau)}{(t-1)^2+(\tau-1)^2-2(t-1)(\tau-1)\cos(\alpha_1^++\alpha_1^-)} \\ &\quad -2\frac{\lambda_0}{\lambda_0+\lambda_1}\frac{\sin(\alpha_{-1}^++\alpha_{-1}^-)(1+\tau)}{(t+1)^2+(\tau+1)^2-2(t+1)(\tau+1)\cos(\alpha_{-1}^++\alpha_{-1}^-)}, \\ f_{21}(t,\tau) &= -2\frac{\lambda_1}{\lambda_0+\lambda_1}\frac{\sin(\alpha_1^++\alpha_1^-)(1-\tau)}{(t-1)^2+(\tau-1)^2-2(t-1)(\tau-1)\cos(\alpha_1^++\alpha_1^-)} \\ &\quad -2\frac{\lambda_1}{\lambda_0+\lambda_1}\frac{\sin(\alpha_{-1}^++\alpha_{-1}^-)(1+\tau)}{(t+1)^2+(\tau+1)^2-2(t+1)(\tau+1)\cos(\alpha_{-1}^++\alpha_{-1}^-)} \\ f_{22}(t,\tau) &= -2\frac{\lambda_0-\lambda_1}{\lambda_0+\lambda_1}\frac{a_1^-(1-\tau)}{(a_1^-)^2(t-\tau)^2+(2-t-\tau)^2} \\ &\quad -2\frac{\lambda_0-\lambda_1}{\lambda_0+\lambda_1}\frac{a_{-1}^-(1+\tau)}{(a_{-1}^-)^2(t-\tau)^2+(2+t+\tau)^2}. \end{aligned} \tag{17}$$

Here are $a_1^+ = \cot\alpha_1^+$, $a_{-1}^+ = \cot\alpha_{-1}^+$, $a_1^- = \cot\alpha_1^-$, $a_{-1}^- = \cot\alpha_{-1}^-$.

The equation (15) shows the structure of the boundary operator of bimetal problems

$$A = I + S + K^+ + K^- + T,$$

where I is the identity matrix operator , K^+ is a matrix integral operator with fixed singularity at $t = \tau = 1$, K^- is a matrix integral operator with fixed singularity at $t = \tau = -1$, S is a singular matrix operator

$$(Su)(t) = c\int_{-1}^{1}\frac{u(\tau)}{\tau - t}d\tau,$$

and T is a compact operator. In our special problem the operator S is absent. In [9] the following two theorems are proved :

Theorem 1 *Let $I+S+K^+$, $I+S+K^-$ be Φ-operators, then $A = I+S+K^+ + K^- + T$ is a Φ-operator.*

Theorem 2 *Let $I+K^+$, $I+K^-$ be Φ-operators with index zero, then $A = I+K^+ + K^- + T$ is a Φ-operator with index zero.*

According to that, we have to prove the Fredholm property of the operator $I+K^+$ only. In order to get the index of the operator $I+K^+$, we introduce a parameter $\gamma \in [0,1]$ and study the operator $I+\gamma K^+$. If $I+\gamma K^+$ is a Φ-operator for all $\gamma \in [0,1]$, then $I+K^+$ is a Φ-operator with index zero.

The operator

$$((I+\gamma K^+)\Phi)(t) = \Phi(t) + \gamma\frac{1}{\pi}\int_{-1}^{+1} F^+(t,\tau)\Phi(\tau)d\tau \tag{18}$$

can be written as a Mellin convolution operator

$$\Psi(x) + \frac{2\gamma}{\pi}\int_0^1 g(\frac{x}{y})\frac{1}{y}\Psi(y)dy, \tag{19}$$

where

$$\begin{aligned} x = -\frac{1}{2}t + \frac{1}{2},\ y &= -\frac{1}{2}\tau + \frac{1}{2},\ \Psi(x) = \Phi(-2x+1), \\ g(\frac{x}{y})\frac{1}{y} &= F^+(-2x+1, -2y+1). \end{aligned} \tag{20}$$

We apply the theorem ([11],[9]) : If

$$\det(I + \frac{2\gamma}{\pi}Mg(z))_{z=\frac{1}{2}+iy} \neq 0 \quad \text{for all } y \in (-\infty, +\infty), \gamma \in [0,1],$$

then $I+K^+$ is a Φ-operator with index zero. Hereby, M denotes the Mellin transform

$$\tilde{u}(z) = Mu(z) = \int_0^\infty u(t)t^{z-1}dt.$$

The kernels, which appear are of the form

$$\frac{1}{a^2(t-1)^2 + (t+1)^2} = \frac{1}{a^2+1}\frac{1}{t^2 + 2t\cos(2\alpha - \pi) + 1}, \tag{21}$$

where $a = \cot\alpha$. From the table of Bateman ([2],6.2(12)) it follows that

$$M(\left(\frac{1}{a^2(t-1)^2 + (t+1)^2}\right) = -\pi\frac{1}{a^2+1}\frac{\sin[(z-1)\Theta]}{\sin\Theta\sin(\pi z)},$$

where $\Theta = 2\alpha - \pi$, $-\pi < \Theta < \pi$.

For non-ideal heat contact $(0 < \beta < \infty)$ we have

$$\det(I + \frac{2\gamma}{\pi}Mg(z)) = \left[1 - \gamma\frac{\sin[(z-1)\Theta_1^+]}{\sin\pi z}\right]\left[1 - \gamma\frac{\sin[(z-1)\Theta_1^-]}{\sin\pi z}\right], \tag{22}$$

where $\Theta_1^+ = 2\alpha_1^+ - \pi$, $\Theta_1^- = 2\alpha_1^- - \pi$.

By elementary calculations we obtain

$$f(y) := \frac{\sin[(z-1)\Theta]}{\sin\pi z}|_{z=\frac{1}{2}+iy} = -\sin\frac{\Theta}{2}\frac{\cosh\Theta y}{\cosh\pi y} + i\cos\frac{\Theta}{2}\frac{\sinh\Theta y}{\cosh\pi y} \tag{23}$$

and

$$|Re\ f(y)| \le \left|\sin\frac{\Theta}{2}\right| < 1. \tag{24}$$

Therefore the symbol curve

$$w = \det(I + \frac{2\gamma}{\pi}Mg(z))|_{z=\frac{1}{2}+iy}$$

does not cut $w = 0$ for any $\gamma \in [0,1]$. Hence, $I + K^+$ is a Φ-operator with index zero.

In the case of ideal heat contact ($\beta = \infty$) we have

$$I + \frac{2\gamma}{\pi}Mg(z) =: A(z)$$

with

$$\begin{aligned}
A_{11}(z) &= 1 - \gamma\frac{\lambda_1 - \lambda_0}{\lambda_0 + \lambda_1}\frac{\sin[(z-1)\Theta_1^+]}{\sin\pi z},\\
A_{12}(z) &= -\gamma\frac{2\lambda_0}{\lambda_0 + \lambda_1}\frac{\sin[(z-1)\Theta_A]}{\sin\pi z},\\
A_{21}(z) &= -\gamma\frac{2\lambda_1}{\lambda_0 + \lambda_1}\frac{\sin[(z-1)\Theta_A]}{\sin\pi z},\\
A_{22}(z) &= 1 - \gamma\frac{\lambda_0 - \lambda_1}{\lambda_0 + \lambda_1}\frac{\sin[(z-1)\Theta_1^-]}{\sin\pi z},\\
\Theta_A &= \alpha_1^+ + \alpha_1^- - \pi.
\end{aligned} \tag{25}$$

The symbol curve $\det A(\frac{1}{2} + iy)$ remains unchanged, if we interchange Θ_1^+, Θ_1^- and λ_0, λ_1 or if we replace Θ_1^+ by $-\Theta_1^+$, Θ_1^- by $-\Theta_1^-$ and interchange λ_0, λ_1. Therefore, without loss of generality, we can assume that $\Theta_1^+ > 0$, $|\Theta_1^-| \le \Theta_1^+$. It holds

$$\det A(\frac{1}{2} + iy) = \overline{\det A(\frac{1}{2} - iy)}, \quad \lim_{|y|\to\infty} \det A(\frac{1}{2} + iy) = 1.$$

The point $\det A(\frac{1}{2})$ lies on the real positive half-axis, because

$$\begin{aligned}
\det A(\frac{1}{2}) &= \frac{1}{(\lambda_0 + \lambda_1)^2}\left[\lambda_0^2(1 - \gamma\sin\frac{\Theta_1^+}{2})(1 + \gamma\sin\frac{\Theta_1^-}{2})\right.\\
&+ \left.\lambda_1^2(1 + \gamma\sin\frac{\Theta_1^+}{2})(1 - \gamma\sin\frac{\Theta_1^-}{2}) + 2\lambda_0\lambda_1(1 - \gamma^2 + \gamma^2\cos\frac{\Theta_1^+}{2}\cos\frac{\Theta_1^-}{2})\right] > 0.
\end{aligned}$$

The real part of $\det A(\frac{1}{2} + iy)$ is

$$\begin{aligned}
Re\det A(\frac{1}{2} + iy) &= \frac{1}{(\lambda_0 + \lambda_1)^2}\left\{\lambda_0^2\left[(1 - \gamma\sin\frac{\Theta_1^+}{2}\frac{\cosh\Theta_1^+ y}{\cosh\pi y})(1 + \gamma\sin\frac{\Theta_1^-}{2}\frac{\cosh\Theta_1^- y}{\cosh\pi y})\right.\right.\\
&+ \left.\gamma^2\cos\frac{\Theta_1^+}{2}\cos\frac{\Theta_1^-}{2}\frac{\sinh\Theta_1^+ y}{\cosh\pi y}\frac{\sinh\Theta_1^- y}{\cosh\pi y}\right]\\
&+ \lambda_1^2\left[(1 + \gamma\sin\frac{\Theta_1^+}{2}\frac{\cosh\Theta_1^+ y}{\cosh\pi y})(1 - \gamma\sin\frac{\Theta_1^-}{2}\frac{\cosh\Theta_1^- y}{\cosh\pi y})\right.\\
&+ \left.\gamma^2\cos\frac{\Theta_1^+}{2}\cos\frac{\Theta_1^-}{2}\frac{\sinh\Theta_1^+ y}{\cosh\pi y}\frac{\sinh\Theta_1^- y}{\cosh\pi y}\right]\\
&+ 2\lambda_0\lambda_1\left[1 - \gamma^2\frac{1}{\cosh^2\pi y} + \gamma^2\cos\frac{\Theta_1^+}{2}\cos\frac{\Theta_1^-}{2}\frac{\cosh\Theta_1^+ y}{\cosh\pi y}\frac{\cosh\Theta_1^- y}{\cosh\pi y}\right.\\
&- \left.\left.\gamma^2\sin\frac{\Theta_1^+}{2}\sin\frac{\Theta_1^-}{2}\frac{\sinh\Theta_1^+ y}{\cosh\pi y}\frac{\sinh\Theta_1^- y}{\cosh\pi y}\right]\right\}.
\end{aligned}$$

Now we can estimate that

$$Re\det A(\frac{1}{2}+iy)>0 \text{ when } \Theta_1^+\geq 0 \text{ or } \Theta_1^-<0,\ \lambda_0\leq\lambda_1.$$

The imaginary part is

$$\begin{aligned} Im\det A(\frac{1}{2}+iy) &= \gamma\frac{1}{(\lambda_0+\lambda_1)^2}\left\{\lambda_0^2\left[\cos\frac{\Theta_1^+}{2}\frac{\sinh\Theta_1^+y}{\cosh\pi y}(1+\gamma\sin\frac{\Theta_1^-}{2}\frac{\cosh\Theta_1^-y}{\cosh\pi y})\right.\right.\\ &- \left.\cos\frac{\Theta_1^-}{2}\frac{\sinh\Theta_1^-y}{\cosh\pi y}(1-\gamma\sin\frac{\Theta_1^+}{2}\frac{\cosh\Theta_1^+y}{\cosh\pi y})\right]\\ &+ \lambda_1^2\left[-\cos\frac{\Theta_1^+}{2}\frac{\sinh\Theta_1^+y}{\cosh\pi y}(1-\gamma\sin\frac{\Theta_1^-}{2}\frac{\cosh\Theta_1^-y}{\cosh\pi y})\right.\\ &+ \left.\cos\frac{\Theta_1^-}{2}\frac{\sinh\Theta_1^-y}{\cosh\pi y}(1+\gamma\sin\frac{\Theta_1^+}{2}\frac{\cosh\Theta_1^+y}{\cosh\pi y})\right]\\ &+ 2\lambda_0\lambda_1\gamma\left[\sin\frac{\Theta_1^+}{2}\cos\frac{\Theta_1^-}{2}\frac{\sinh\Theta_1^+y}{\cosh\pi y}\frac{\cosh\Theta_1^-y}{\cosh\pi y}\right.\\ &+ \left.\left.\cos\frac{\Theta_1^+}{2}\sin\frac{\Theta_1^-}{2}\frac{\cosh\Theta_1^+y}{\cosh\pi y}\frac{\sinh\Theta_1^-y}{\cosh\pi y}\right]\right\}. \end{aligned}$$

We easily obtain that

$$Im\det A(\frac{1}{2}+iy)\neq 0 \text{ for } y\neq 0 \text{ when } \Theta_1^-<0,\ \lambda_0>\lambda_1.$$

Therefore, we have proved, that in the case $\beta=\infty$ the operator $I+K^+$ is a Φ-operator with index zero.

In order to determine the asymptotics of the solution of the integral equation at the point P_1 we write (15) in the form

$$\Phi(t)\chi_1(t)+\frac{1}{\pi}\int_{-1}^{+1}F^+(t,\tau)\Phi(\tau)\chi_1(\tau)d\tau=U(t), \tag{26}$$

where $\chi_1(t)$ is a cut-off function equal to unity in a neighbourhood of $t=1$. If we put (cf. (20))

$$\Psi(x)=\Phi(-2x+1)\chi_1(-2x+1),\ H_0(x)=U(-2x+1),$$

and multiply by a cut-off function $\chi(x)$ such that $\chi(x)\Psi(x)=\Psi(x)$, then (26) becomes

$$\Psi(x)+\frac{2}{\pi}\int_0^1 g(\frac{x}{y})\frac{1}{y}\Psi(y)dy=H(x), 0\leq x<\infty. \tag{27}$$

In the case $0<\beta<\infty$ we have for (27)

$$\begin{aligned} \Psi^+(x)-\frac{1}{\pi}\int_0^1\frac{2a_1^+}{(a_1^+)^2(\frac{x}{y}-1)^2+(\frac{x}{y}+1)^2}\frac{1}{y}\Psi^+(y)dy &= H^+(x),\\ \Psi^-(x)-\frac{1}{\pi}\int_0^1\frac{2a_1^-}{(a_1^-)^2(\frac{x}{y}-1)^2+(\frac{x}{y}+1)^2}\frac{1}{y}\Psi^-(y)dy &= H^-(x). \end{aligned} \tag{28}$$

For the Mellin transformed system we obtain

$$\begin{aligned} \tilde{\Psi}^+(z)-\frac{\sin[(z-1)\Theta_1^+]}{\sin\pi z}\tilde{\Psi}^+(z) &= \tilde{H}^+(z),\\ \tilde{\Psi}^-(z)-\frac{\sin[(z-1)\Theta_1^-]}{\sin\pi z}\tilde{\Psi}^-(z) &= \tilde{H}^-(z). \end{aligned}$$

Hence we have

$$\tilde{\Psi}^{\pm}(z) = \tilde{H}^{\pm}(z)\frac{\sin \pi z}{\sin \pi z - \sin[(z-1)\Theta_1^{\pm}]}.$$

The asymptotics of $\Psi^+(x)$ only depends on the angle α_1^+. From

$$\int_0^1 x^\nu x^{z-1} dx = (z+\nu)^{-1}, \quad Re z > - Re \nu \ ([2], 6.2(18))$$

it follows, that the leading term in the development of $\Psi^+(x)$ and therefore the exponent ν is determined by that pole in the half-plane $Re z < 1/2$, which lies nearest to the line $Re z = 1/2$. The equation $\sin \pi z - \sin[(z-1)\Theta_1^+] = 0$ has roots only for $Re z = 0$. The desired zero and exponent for $0 < \alpha_1^+ < 2\pi/3$ are

$$z_0 = \frac{\pi - 2\alpha_1^+}{2(\pi - \alpha_1^+)}, \ \nu_0 = -z_0 = -1 + \frac{1}{2}\frac{\pi}{\pi - \alpha_1^+}.$$

Therefore, we have

$$\tilde{\Psi}^+(z) = \tilde{H}^+(z_0)\frac{1}{2(\pi - \alpha_1^+)} \tan \pi z_0 \frac{1}{z - z_0} + \cdots$$

In the case $\beta = \infty$ we have

$$\begin{aligned} \tilde{\Psi}^+(z) &= \tilde{H}^+(z)\frac{A_{22}(z)}{\Delta(z)} - \tilde{H}^-(z)\frac{A_{12}(z)}{\Delta(z)}, \\ \tilde{\Psi}^-(z) &= -\tilde{H}^+(z)\frac{A_{21}(z)}{\Delta(z)} + \tilde{H}^-(z)\frac{A_{11}(z)}{\Delta(z)}, \end{aligned}$$

where $A_{ij}(z)$ are given by (25) for $\gamma = 1$ and $\Delta(z) = det A(z)$.

Let us consider some special cases.

1. $\lambda_0 = \lambda_1$, $\alpha_1^+ + \alpha_1^- = \omega$, $\Theta_A = \omega - \pi$. Here we have

$$\Delta(z) = 1 - \frac{\sin^2[(z-1)\Theta_A]}{\sin^2 \pi z}.$$

The poles being of interest are

$$z_1 = 1 - \frac{\pi}{2\pi - \omega}, \ z_2 = 1 - \frac{\pi}{\omega},$$

and the corresponding exponents are

$$\nu_1 = \frac{\pi}{2\pi - \omega} - 1, \ \nu_2 = \frac{\pi}{\omega} - 1.$$

For $\omega > \pi$ is $\nu_2 < \nu_1$. Thus for <u>$\omega > \pi$</u> we have the exponent $\nu_2 = \frac{\pi}{\omega} - 1$ and

$$\begin{pmatrix} \tilde{\Psi}^+(z) \\ \tilde{\Psi}^-(z) \end{pmatrix} = \begin{pmatrix} 1 & -1 \\ -1 & 1 \end{pmatrix} \begin{pmatrix} \tilde{H}^+(z_2) \\ \tilde{H}^-(z_2) \end{pmatrix} \frac{1}{2\omega} \tan \pi z_2 \frac{1}{z - z_2} + \cdots$$

and for <u>$\omega < \pi$</u> we have the exponent $\nu_1 = \pi/(2\pi - \omega) - 1$ and

$$\begin{pmatrix} \tilde{\Psi}^+(z) \\ \tilde{\Psi}^-(z) \end{pmatrix} = \begin{pmatrix} 1 & 1 \\ 1 & 1 \end{pmatrix} \begin{pmatrix} \tilde{H}^+(z_1) \\ \tilde{H}^-(z_1) \end{pmatrix} \frac{1}{2(2\pi - \omega)} \tan \pi z_1 \frac{1}{z - z_1} + \cdots$$

2. $\alpha_1^+ = \alpha_1^- = \alpha$, $\Theta_1^+ = \Theta_1^- = \Theta = 2\alpha - \pi$. Here the determinant

$$\Delta(z) = 1 - \frac{\sin^2[(z-1)\Theta]}{\sin^2 \pi z}$$

is independent of λ_0 and λ_1. For $\underline{\alpha > \pi/2}$ we have

$$\nu_2 = \frac{1}{2}\frac{\pi}{\alpha} - 1, \; z_2 = 1 - \frac{1}{2}\frac{\pi}{\alpha},$$

$$\begin{pmatrix} \tilde{\Psi}^+(z) \\ \tilde{\Psi}^-(z) \end{pmatrix} = \begin{pmatrix} \frac{2\lambda_0}{\lambda_0+\lambda_1} & -\frac{2\lambda_0}{\lambda_0+\lambda_1} \\ -\frac{2\lambda_1}{\lambda_0+\lambda_1} & \frac{2\lambda_1}{\lambda_0+\lambda_1} \end{pmatrix} \begin{pmatrix} \tilde{H}^+(z_2) \\ \tilde{H}^-(z_2) \end{pmatrix} \frac{1}{4\alpha} \tan \pi z_2 \frac{1}{z - z_2} + \cdots$$

and for $\underline{\alpha < \pi/2}$ we have

$$\nu_1 = \frac{1}{2}\frac{\pi}{\pi - \alpha} - 1, \; z_1 = 1 - \frac{1}{2}\frac{\pi}{\pi - \alpha}$$

$$\begin{pmatrix} \tilde{\Psi}^+(z) \\ \tilde{\Psi}^-(z) \end{pmatrix} = \begin{pmatrix} \frac{2\lambda_1}{\lambda_0+\lambda_1} & \frac{2\lambda_0}{\lambda_0+\lambda_1} \\ \frac{2\lambda_1}{\lambda_0+\lambda_1} & \frac{2\lambda_0}{\lambda_0+\lambda_1} \end{pmatrix} \begin{pmatrix} \tilde{H}^+(z_1) \\ \tilde{H}^-(z_1) \end{pmatrix} \frac{1}{4(\pi - \alpha)} \tan \pi z_1 \frac{1}{z - z_1} + \cdots$$

3. $\alpha_1^+ + \alpha_1^- = \pi$, $\Theta_A = 0$, $\Theta_1^+ = -\Theta_1^- = \Theta = 2\alpha_1^+ - \pi$. The determinant becomes $\Delta(z) = (\Delta_0(z))^2$, where

$$\Delta_0(z) = 1 - \delta \frac{\sin[(z-1)\Theta]}{\sin \pi z}, \; \delta = \frac{\lambda_1 - \lambda_0}{\lambda_0 + \lambda_1}.$$

The equation $\sin \pi z = \delta \sin[(z-1)\Theta]$ has only real solutions. In the situation $0 < \alpha_1^+ < \pi/2$, $\lambda_1 > \lambda_0$ or $\pi/2 < \alpha_1^+ < \pi$, $\lambda_1 < \lambda_0$ one solution x_0 with $0 < x_0 < 1/2$ exists. The corresponding exponent is $\nu_0 = -x_0$, $-1/2 < \nu_0 < 0$.

In the case $0 < \alpha_1^+ < \pi/2$, $\lambda_1 < \lambda_0$ or $\pi/2 < \alpha_1^+ < \pi$, $\lambda_1 > \lambda_0$ there is of interest the first negative root x_0 of the equation $\sin \pi x = \delta \sin[(x-1)\Theta]$ hence the exponent $\nu_0 = -x_0 > 0$. Further we get

$$\begin{aligned} \tilde{\Psi}^+(z) &= \tilde{H}^+(x_0)\Delta_0'(x_0)^{-1}\frac{1}{z - x_0} + \cdots, \\ \tilde{\Psi}^-(z) &= \tilde{H}^-(x_0)\Delta_0'(x_0)^{-1}\frac{1}{z - x_0} + \cdots. \end{aligned}$$

References

[1] Abramowitz, M., Stegun, I.:Handbook of mathematical functions (Russian). Nauka, Moskau 1979.

[2] Bateman, H.:Tables of integral transforms (Russian). Nauka, Moskau 1969.

[3] Costabel, M., Stephan, E.: Boundary integral equations for mixed boundary value problems in polygonal domains and Galerkin approximation, TH Darmstadt, Preprint 593, März 1981.

[4] Costabel, M.,Stephan, E.: Curvature terms in the asymptotic expansions for solutions of boundary integral equations on curved polygons, TH Darmstadt, Preprint 673, Juni 1982.

[5] Duduchava, R.: Integral equations with fixed singularities. Teubner Texte zur Mathematik, Leipzig 1979.

[6] Jentsch, L.: Über Randintegralmethoden bei Randkontaktaufgaben. 9.TMP (1988) Proceedings, Teubner-Texte zur Mathematik, Bd.111, S.110-119.

[7] Jentsch, L.: On a thermal stress problem for contacting half-spaces with inclusions of other material involving a new method for computing potentials and singular integrals. Mathematical Methods in the Applied Sciences, Vol.10 (1988), 225-243.

[8] Jentsch, L.: Bimetallprobleme mit verschiedenen Kontaktbedingungen. 8.TMP (1983) Proceedings, Teubner-Texte zur Mathematik, Bd. 63, S.111-121.

[9] Jentsch, L.: The Green's contact functions for heat conduction of two half-spaces and half-planes with application to several boundary contact problems. To appear in: Mathematical Methods in the Applied Sciences, Vol 14 (1991).

[10] Jentsch, L.: Über ein Bimetallproblem in der Ebene. Zeitschr. für Analysis und ihre Anwendungen, Bd. 1(5)(1982), S.67-92.

[11] Rempel, S.,Schulze,B.-W.:Asymptotics for elliptic mixed boundary problems. Mathem. Research, Vol.50, Akademie-Verlag Berlin 1989.

L. Jentsch
FB Mathematik
Technische Universität Chemnitz
Reichenhainer Str. 39-41
O-9022 Chemnitz, Deutschland

Spectral properties of the operator pencils, associated with elliptic boundary problems in a cone

V.A.Kozlov
Academy of Siences of the USSR
Leningrad Branch of Mechanical Engineering Research Institute
Bolshoy 61, V.O., Leningrad, 199178, USSR

It is known, that behaviour of solutions of elliptic boundary value problems near conic points of the boundary is determined by the eigenvalues of the boundary value problems polynomially depending on a spectral parameter, in regions of the unit sphere. If the positions of these eigenvalues are known, then one can draw conclusions about singularities, continuity, or smoothness of the solutions. Therefore this information is of interest for applications.

The present paper deals with the recent investigations of operator pencils generated by elliptic boundary value problems in a cone.

The Dirichlet problem

Consider a strongly elliptic operator in $\mathbb{R}^n$

$$\mathcal{A}(\partial_x) = \sum_{|\alpha|=2m} A_\alpha \partial_x^\alpha$$

where $\{A_\alpha\}$ are constant selfadjoint $i \times i$ matrices.

Let K be the cone in $\mathbb{R}^n$ determined by the inequality

$$x_n > \varphi(x'), \quad x' = (x_1, \dots, x_{n-1}); \tag{1}$$

the function φ is of the class $C^\infty(\mathbb{R}^{n-1} \setminus \{0\})$.

We are interested in solutions of the problem

$$\mathcal{A}(\partial_x) u = 0 \text{ on } K \tag{2}$$

which have a form

$$u(x) = |x|^{\lambda_0} \sum_{0 \le k \le N} \frac{(\log |x|)^k}{k!} u_{N-k}(x/|x|) \tag{3}$$

where $u_k \in [\overset{0}{H}{}^m(\Omega)]^i$, $\Omega = K \cap S^{n-1}$.

In spherical coordinates (r, ω), the operator $\mathcal{A}$ may be written as

$$r^{2m} \mathcal{A}(\partial_x) = A(r\partial_r, \omega, \partial_\omega).$$

For $\lambda \in \mathbb{C}$, $P(\lambda)$ denotes $\mathcal{A}(\lambda, \omega, \partial_\omega)$ acting from $[\overset{0}{H}{}^m(\Omega)]^i$ to $[H^{-m}(\Omega)]^i$.

It is easy to show that the function (3) is a solution of the problem (2) iff λ_0 is the eigenvalue of the operator pencil P, u_0 is eigenfunction and $u_1, \dots, u_N$ are associated functions.

If λ_0 is eigenvalue of the operator pencil P, then $-\overline{\lambda}_0 + 2m - n$ is also eigenvalue and their geometric and algebraic multiplicities coincide.

Theorem 1 *[1]. There are no eigenvalues of P in the strip $|Re\lambda - m + n/2| \leq 1/2$.*

Remarks:

1^0. The estimate of the strip width cannot be improved in the case of $2m \geq n - 1$.

2^0. Theorem 1 fails without the assumption that the coefficients of operator $\mathcal{A}$ are real. In order to show this consider the strongly elliptic operator

$$\mathcal{A}(\partial_{x_1}, \partial_{x_2}) = \partial^2_{x_2} + (t-1)i\partial_{x_1}\partial_{x_2} + t\partial^2_{x_1}$$

in the angle $K_\varphi = \{x \in \mathbb{R}^2 : 0 < r < \infty,\ \Theta \in (0, \varphi)\}$, $\varphi \in (0, 2\pi)$. Here (r, Θ) are the polar coordinates of point $x \in \mathbb{R}^2$ and $t > 0$. The set $\{\lambda_k : k = \mp 1, \mp 2, \dots\}$ where

$$\begin{aligned} Re\ \lambda_k &= 2\pi k[\rho + (\varphi - \alpha)^2/\rho]^{-1}, \\ Im\ \lambda_k &= -Re\ \lambda_k(\varphi - \alpha)/\rho \end{aligned}$$

and

$$\begin{aligned} \rho &= \log(\cos^2\varphi + t\sin\varphi), \\ \alpha &= \arg(\cos\varphi - it\sin\varphi) \in [-2\pi, 0], \end{aligned}$$

is the spectrum of the corresponding pencil. If t is sufficiently large and $\varphi \neq \pi, 2\pi$, then $Re\ \lambda_1$ is close to 0. Hence Theorem 1 is not true.

Now consider the case $n = 2$, $K = K_\varphi$, $\varphi \in (0, 2\pi]$. Suppose, $\mathcal{A}$ is a scalar operator, i.e. the coefficients A_α are real numbers. The corresponding operator pencil will be denoted by P_φ.

Theorem 2 *[2]. Let $\varphi \in (0, \pi)$. Then the strip $m - 1 \leq Re\ \lambda \leq m$ does not contain any eigenvalue of the operator pencil P_φ. Besides, the eigenvalues of this pencil do not coincide with $m + k$, $k = 0, 1, \dots, m - 1$.*

Denote the roots with the positive imaginary part of the equation $\mathcal{A}(1, z) = 0$ by $\alpha_1, \dots, \alpha_m$.

Theorem 3 *[3]. There exist real analytic functions $\varphi_k : (m - 1/2, m] \to [\pi, 2\pi)$, $k = 1, \dots, m$, with the following properties*

(i) $\varphi'_k(s) < 0$ *for* $s \in (m - 1/2, m)$, $k = 1, \dots, m$;

(ii) $\varphi_1(s) < \varphi_2(s) < \dots < \varphi_m(s)$, $s \in (m - 1/2, m]$;

(iii) $\varphi_1(m) = \pi$, $\varphi'_1(m) = -\pi/Im\ (\alpha_1 + \dots + \alpha_m)$, $\lim_{s \to m-1/2} \varphi_k(s) = 2\pi$ $(1 \leq k \leq m)$.

(iv) Let $\varphi \in [\pi, 2\pi)$. Then the eigenvalues of the operator pencil P_φ belonging to the strip $m - 1 \leq Re\ \lambda \leq m$ coincide with one of the roots of the equation $\varphi_k(s) = \varphi$ for some k, $k = 1, \dots, m$. The eigenvalues belonging to the strip $m - 1 \leq Re\ \lambda < m$ are simple. If m is an eigenvalue, then its algebraic multiplicity is not larger than 2.

From Theorem 3 we deduce the existence of exactly m simple eigenvalues $\mu_1(\varphi), \ldots, \mu_m(\varphi)$, $\varphi \in (\varphi_m(m), 2\pi)$, belonging to the strip $m-1 \le Re\ \lambda \le m$. Moreover, the following theorem is true.

Theorem 4 *[3] As $\varphi \to 2\pi$*

$$\mu_k(\varphi) = m - 1/2 + \gamma_k(2\pi - \varphi)^{2k-1} + 0(|2\pi - \varphi|^{2k}),\ k = 1, \ldots, m,$$

where

$$\gamma_k = \frac{-1}{4\pi}\left[\frac{(k-1)!\Gamma(k-1/2)}{(2k-2)!\Gamma(1/2)}\right]^2 \frac{\det D_{2k}}{\det D_{2k-2}}$$

Here D_{2k} is a $2k \times 2k$ matrix determined in the following way.

Let

$$\sum_{0\le s\le m} (\alpha_s + i\beta_s) z^{m-s} = \prod_{1\le k\le m} (z - \alpha_k)$$

then $D_0 = 1$ and

$$D_{2k} = \begin{pmatrix} \alpha_0 & \alpha_1 & \alpha_2 & \ldots & \alpha_{2k-1} \\ \beta_0 & \beta_1 & \beta_2 & \ldots & \beta_{2k-1} \\ 0 & \alpha_0 & \alpha_1 & \ldots & \alpha_{2k-2} \\ 0 & \beta_0 & \beta_1 & \ldots & \beta_{2k-2} \\ . & . & . & \ldots & . \\ . & . & . & \ldots & . \end{pmatrix}$$

where $\alpha_s = \beta_s = 0$ for $s > m$.

Biharmonic equation, [4]. Let K be an arbitrary cone in $\mathbb{R}^n$ and $\Omega = K \cap S^{n-1}$. The operator pencil generated by the Dirichlet problem for the biharmonic equation in the cone K has the form

$$L(\lambda)u = \delta^2 u + 2[\lambda(\lambda+n-4) - (n-4)]\delta u + \lambda(\lambda+n-4)(\lambda-2)(\lambda+n-2)u$$

$$u \in \overset{0}{W}{}_2^2(\Omega),$$

where δ is the Laplace operator on S^{n-1}.

Denote by $\mu_0(\Omega)$ the first eigenvalue of the Dirichlet problem for the operator $-\delta$ in Ω.

Let us introduce the number

$$\sigma_0 = \sigma_0(\Omega) = 2 - n/2 + \sqrt{(1-n/2)^2 + 1 + \mu_0(\Omega)}$$

It monotonically depends on Ω and

$$\begin{aligned} \sigma_0(S^{n-1}) &= 2 - n/2 + \sqrt{(1-n/2)^2 + 1} > 1, \\ \sigma_0(S_+^{n-1}) &= 2 - n/2 + \sqrt{1 - n^2/4} > 2 \end{aligned}$$

Theorem 5 *[4]. The following propositions are valid in the strip $2 - n/2 \le Re\ \lambda \le \sigma_0(\Omega)$:*

1) All eigenvalues of the pencil L are real and non-degenerate.

2) The nearest to the line $Re\ \lambda = 2 - n/2$ eigenvalue of the pencil L is given by the formula

$$\sigma_1(\Omega) = inf\ \{R(u),\ \ u \in \overset{0}{W}{}_2^2(\Omega),\ u \ne 0\}$$

where $R(u)$ is the lowest root of the equation $(L(\sigma)u, u)_{L_2(\Omega)} = 0$ belonging to the strip under consideration.

Corollary If $\Omega \subset S_+^{n-1}$ then the strip $2 - 2/n \le Re\ \lambda \le 2$ does not contain eigenvalues of the pencil L. If $\Omega \supset S_+^{n-1}$ then the strip $2 - 2/n \le Re\ \lambda \le 2$ contains at least one eigenvalue of L.

Note, that the Theorem 5 and the corollary is not true for the operator pencil corresponding to the Dirichlet problem for the operator $\Delta^2 + \alpha\partial_{x_1}^4$, $\alpha > 0$, in a cone $\mathbb{R}^4$ (see [4]).

Lame system, [5]. Consider the Dirichlet problem for Lamé system

$$\Delta u + \gamma^{-1} \text{ grad div} u = o \text{ on } K$$

$$u = 0 \text{ on } \partial K \setminus \{o\}$$

where $K = (0,\infty) \times \Omega$ is an arbitrary cone, $\gamma > 0$, $u = (u_1, u_2, u_3)$ is a displacement vector. We denote the corresponding operator pencil by $L(\lambda)$.

Introduce the number

$$F_\gamma(\Omega) = \sqrt{\Gamma(\Omega)/2 + 3/4 + (\gamma+1)(2\gamma+1)}$$

where

$$\begin{aligned}\Gamma(\Omega) &= \inf \left\{ Q(u_\omega) + (1+\gamma)^{-1} \int_\Omega |\nabla_\omega u_\omega|^2 d\omega \right\} / \int_\Omega |u_\omega|^2 d\omega \\ Q(u_\omega) &= \int_\Omega \left\{ |\partial_\Theta u_\Theta|^2 + |\partial_\Theta u_\varphi|^2 + \left| \frac{1}{\sin\Theta}\partial_\varphi u_\Theta - \cot\Theta u_\varphi \right|^2 + \left| \frac{1}{\sin\Theta}\partial_\varphi u_\varphi - \cot\Theta u_\Theta \right|^2 \right\} d\omega.\end{aligned}$$

Here $(u_r, u_\Theta, u_\varphi)$ are the spherical components of a vector $u = u(r,\Theta,\varphi)$, $u_\omega = (u_\Theta, u_\varphi)$ and

$$\nabla_\omega u_\omega = \frac{1}{\sin\Theta}\partial_\Theta(\sin\Theta u_\Theta) + \frac{1}{\sin\Theta}\partial_\varphi u_\varphi$$

It can be shown that $\Gamma(\Omega) > 1$ and hence $F_\gamma(\Omega) > 3/2$.

Theorem 6 *[5]. In the strip $-1/2 \le Re\ \lambda < F_\gamma(\Omega) - 1/2$ there holds for $\gamma > 0$:*

1) All eigenvalues of the pencil $L(\lambda)$ are real.

2) The eigenfunctions have no associated functions.

3) The eigenvalues in the interval $[-2/3, 1+2\gamma)$ can be found by means of a varitional principle: Let

$$\begin{aligned}S(\lambda; u, u) &= (2+\lambda+2\gamma)\left\{ \int_\Omega |\nabla_\omega u_r|^2 d\omega - (\lambda+2)(\lambda-1)\frac{\gamma+1}{\gamma} \int_\Omega |u_r|^2 d\omega \right\} \\ &+ (1-\lambda+2\gamma)\left\{ \gamma^{-1} \int_\Omega |\nabla_\omega u_\omega|^2 d\omega - (\lambda(\lambda+1)-1) \int_\Omega |u_\omega|^2 d\omega \right. \\ &\left. + Q(u) + 2\gamma^{-1}(\lambda+2+2\gamma) Re \int_\Omega u_r \nabla_\omega \overline{u}_\omega d\omega \right\}\end{aligned}$$

and

$$(\mathcal{L}(\lambda)u, u)_{(L_2(\Omega))^3} = S(\lambda; u, u)$$

then $\mathcal{L}(\lambda)$ is a selfadjoint semibounded operator with discrete spectrum . Denote by $\mu_1(\lambda) \le \mu_2(\lambda) \le \ldots$ the corresponding eigenvalues. The functions $\mu_k : (-1/2, 1+2\gamma) \to \mathbb{R}^1$ are continuous and each of them has at most one root on the interval $(-1/2, \min\{1+2\gamma, F_\gamma(\Omega) - 1/2\})$. The eigenvalues of the operator pencil L belonging to the strip $-1/2 \le Re\lambda < \min\{1 + 2\gamma, F_\gamma(\Omega) - 1/2\}$ coincide with one of the roots of the functions μ_k

Corollary

a) All eigenvalues of $L(\lambda)$ in $(-1/2, \min\{1+2\gamma, F_\gamma(\Omega)-1/2\})$ depend monotonically on Ω.

b) If $\Omega \subseteq S^2_+$ then the interval $[-1/2, 1]$ is free of eigenvalues;

c) if $S^2_+ \subseteq \Omega \subseteq S^2$ then there are exactly three real eigenvalues $\lambda_k \in [0,1)$.

The Neumann Problem

Consider a differential operator

$$\mathcal{A}(\partial_x) = (-1)^m \sum_{|\alpha|=|\alpha|=m} A_{\alpha\beta}\partial_x^{\alpha+\beta},$$

$A_{\alpha\beta}$ are constant $i \times i$ matrices, $A^*_{\alpha\beta} = A_{\alpha\beta}$. Let K be the cone in $\mathbb{R}^n$ determined by the inequality (1), where φ is a smooth function on $\mathbb{R}^n \setminus \{0\}$.

We are interested in the solution u of the problem

$$\int_K \sum_{|\alpha|=|\beta|=m} [A_{\alpha\beta}\partial_x^\alpha u, \partial_x^\beta v]\, dx = 0, \ \forall v \in [W_2^m(K)]^i, v = 0 \text{ near } 0, \infty. \tag{4}$$

Let

$$Q_\alpha(\lambda, \omega, \partial_\omega)u(\omega) = r^{-\lambda+|\alpha|}\partial_x^\alpha(r^\lambda u(\omega))$$

and

$$P(u,v;\lambda) = \sum_{|\alpha|=|\beta|=m} \int_\Omega (A_{\alpha\beta}Q_\beta(\lambda)u, Q_\alpha(-\overline{\lambda}+2m-n)v)d\omega$$

It is easy to verify that

$$P(u,v;\lambda) = \frac{1}{2|\log\varepsilon|}\int_{\varepsilon<|x|<1/\varepsilon, x\in K} \sum_{|\alpha|=|\beta|=m} (A_{\alpha\beta}\partial_x^\beta u, \partial_x^\alpha v)dx$$

where $u(x) = r^\lambda u(\omega)$, $v(x) = r^{-\overline{\lambda}+2m-n}v(\omega)$, $\varepsilon \in (0,1)$.

For every $\lambda \in \mathbb{C}$ the form P induces a continuous operator

$$\mathcal{P}(\lambda) : (W_2^m(\Omega)) \longrightarrow (W_2^m(\Omega))^{*i}$$

The function

$$u(x) = r^{\lambda_0} \sum_{0\le k\le N} \frac{(\log r)^k}{k!} u_{N-k}(\omega),$$

$u_k \in (W_2^m(\Omega))^i$, is a solution of the problem (4) iff λ_0 is the eigenvalue of the operator pencil $\mathcal{P}$, u_0 is the eigenfunction and $u_1, \dots, u_N$ are associated functions.

Theorem 7 *[1],[6]. Let*

$$\sum_{|\alpha|=|\beta|=m} (A_{\alpha\beta}f_\beta, f_\alpha) \ge c \sum_{|\alpha|=m} |f_\alpha|^2, c > 0$$

for all vectors $f_\alpha \in \mathbb{C}^i$.

Then the following propositions are valid.

(i) *If* $2m < n-1$ *then the strip* $|Re\lambda - m + n/2| \leq 1/2$ *does not contain any eigenvalue of the operator pencil* $\mathcal{P}$.

(ii) *If* $2m \geq n-1$ *and* n *is even then the strip* $|Re\lambda - m + n/2| \leq 1/2$ *contains exactly one eigenvalue* $\lambda_0 = m - n/2$. *The vector function* u *is a eigenfunction iff* $u = p|_\Omega$, *where* p *is homogeneous vector polonomial in* $\mathbb{R}^n$ *of degree* $m - n/2$. *Every eigenfunction has exactly one associated function.*

(iii) *If* $2m \geq n-1$ *an* n *is odd then the strip* $|Re\lambda - m + n/2| \leq 1/2$ *contains exactly two eigenvalues* $\lambda_\pm = m - n/2 \pm 1/2$, *which the geometric and algebraic multiplicities of which are equal to* $iC^{n-1}_{m+(n-1)/2}$ *and* $i[C^{n-1}_{m+(n-1)/2} + C^{n-1}_{m+(n-3)/2}]$, *respectively. The vector-function* u *is a eigenfunction of* $\mathcal{P}$ *corresponding to an eigenvalue* λ_+ *iff* $u = p|_\Omega$, *where* p *is an homogeneous vector polynomial in* $\mathbb{R}^n$ *of degree* $m - n/2 + 1/2$. *Every eigenfunction has at most one associated function.*

References

[1] Kozlov V.A. ,Mazya V.G. Spectral properties of operator pencils generated by elliptic boundary value problems in a cone, Funct. Anal. and its Application 1988, vol.22 No2, pp.38-46.

[2] Kozlov v.A. Singularities of solutions to the Dirichlet problem for elliptic equations in vicinity of angle points. Algebra and Analysis, No 4,1989, pp.161-177.

[3] Kozlov V.A. On the spectrum of operator pencils arising in Dirichlet problem for elliptic equations in an angle. Math.Zamjetki, 45, No 2, 1989, pp 117-118.

[4] Kozlov V.A. On the Dirichlet problem for elliptic equations on domains with conical points. Dif. Uravnenia, v.26, No6, 1990, pp1014-1023.

[5] Kozlov V.A. ,Mazya V.G. , Schwab C. On the first boundary value problem of 3-d elasticity on conical domains, Preprint TR 89-46, Department of Mathematics, Univ. of Maryland at college park MD 20742, USA, August 1989.

[6] Kozlov V.A. Mazya V.G. On the spectrum of an operator pencil generated by the Neumann problem in a cone. Algebra and Analysis, to appear.

CRACK PHENOMENA IN HETEROGENEOUS MEDIA

D. LEGUILLON
and
E. SANCHEZ-PALENCIA
(Paris)

1. INTRODUCTION

This paper is devoted to a survey of recent work of the authors on applications of singularity theory to the study of the propagation of cracks in elastic-brittle materials. In particular we consider singular situations (crack arriving to an interface between two media, for example) appearing in composite materials. Generally speaking, we only give results and some rapid indications on the proofs. Nevertheless, new results are presented in sections 4 and 6, which are a little more explicit.

Section 2 is devoted to a rapid description of singularities for the system of elasticity in the general framework of [5], [3], [8], [9]. We mainly describe the numerical methods for computing singularities developed in [12]. In Section 3 we describe a method of studying perturbation of the domain in the vicinity of a singularity. This method, based on the procedure of the matched asymptotic expansions allows us to study the energy release (perturbation of the elastic energy) associated with a small increase of a crack. This study is mainly based on [12] lecture 7, and [11] (see also [15]). Two very different kinds of singularities (called "strong" and "weak" hereafter) appear when a crack touches an interface between two materials according to the behavior of the energy release ("large" or "small") with respect to the corresponding energy release in a crack in a homogeneous material. Both kinds of singularities appear in composite materials.

In Section 5 we introduce classical theory of fracture (see [2], [17] as well as recent developments in [10], [16], [18]). Equating the energy release obtained in Section 3 to the energy necessary to generate a new element of crack, it is possible to give a criterion to obtain the direction of propagaiton of a crack in somewhat general situations. We use here some results of dependence on parameters previously proved in Section 4.

Finally, Section 6 is devoted to some remarks on the propagation of smooth cracks in anisotropic materials (in the sense that the energy necessary to produce a crack depends on its orientation). It appears that, in some cases, the smooth cracks may only propagate along certain directions of the material.

2. GENERALITIES ON NUMERICAL COMPUTATION OF SINGULARITIES

In this section we give a very rapid account of our method of computing the exponents, eigenvectors and intensity factors of singularities ([12], lectures 5 and 6) as well as some examples.

We consider 2-dimensional problems of elasticity in a sector $\theta \in (0, \omega)$ of the plane x_1, x_2, where r, θ are the polar coordinates. Let $\mathbf{U}(x_1, x_2)$ be the displacement vector (with two components U_i, i=1,2 in the direction of the axes x_1, x_2 (We note that a variant of the method allows us to consider 3-dimensional vectors depending on x_1, x_2 for elasticity problems in edges). Let $a_{ijkh}(\theta)$ be the elasticity coefficients satisfying the classical symmetry and positivity conditions of an elastic (generally anisotropic) material. They are piecewise smooth functions of θ, in particular, piecewise constant, describing the case of a composite material with interfaces in the direction of some $\theta = const$. The equations are

$$(2.1) \qquad -\frac{\partial \sigma_{ij}}{\partial x_j} = 0 \text{ (equilibrium)}$$

$$(2.2) \qquad \sigma_{ij} = a_{ijkh}\, e_{kh}(\mathbf{U}) \quad \text{(strain} - \text{stress)}$$

$$(2.3) \qquad e_{kh} = \frac{1}{2}\left(\frac{\partial U_k}{\partial x_h} + \frac{\partial U_h}{\partial x_k}\right) \quad \text{(strain)}$$

We of course add boundary conditions on $\theta = 0$ and $\theta = \omega$ corresponding to free- stress (= Neumann condition $\sigma_{ij}n_j = 0$) for instance, and classical transmission conditions at the interfaces (continuity of $\mathbf{U}$ and $\sigma_{ij}n_j$).
Taking advantage of the variational formulation in the x_1, x_2 variables and with an appropriate choice of the test functions, searching solutions of the form

$$(2.4) \qquad \mathbf{U}_\alpha = r^\alpha \mathbf{u}_\alpha(\theta)$$

reduces to an implicit eigenvalue problem of the form

$$(2.5) \qquad \left|\begin{array}{l} \text{Find non zero } \mathbf{u}_\alpha(\theta) \in \mathbf{H}^1(0,\omega) \text{ and } \alpha \in \mathbf{C} \text{ such that :} \\ A(\alpha)\mathbf{u}_\alpha \equiv -\alpha^2 A_1 \mathbf{u}_\alpha + \alpha A_2 \mathbf{u}_\alpha + A_3 \mathbf{u}_\alpha = 0 \end{array}\right.$$

where A_1 and A_3 (resp.A_2) are hermitian (resp. skew-hermitian) operators continuous from $H^1(0,\omega)$ into its dual. The numerical implementation of this 1-dimensional eigenvalue problem is easily derived from the coefficients of the classical sesquilinear form in x_1, x_2 and incorporates implicitly the interface and boundary conditions.

Moreover, the problem may also be written as a classical (explicit) eigenvalue problem considering $\mathbf{U}(x_1, x_2)$ as a function of r with values in $\mathbf{H}^1(0,\omega)$. Then, (2.1)- (2.3) becomes

$$(2.6) \qquad r\frac{\partial}{\partial r}\vec{\mathbf{U}} = \mathcal{A}\vec{\mathbf{U}} \text{ with}$$

$$(2.7) \qquad \vec{\mathbf{U}} = \begin{bmatrix} \mathbf{U} \\ r\dfrac{\partial}{\partial r}\mathbf{U} \end{bmatrix} ; \mathcal{A} = \begin{bmatrix} 0 & I \\ A_1^{-1}A_3 & A_1^{-1}A_2 \end{bmatrix}$$

and (2.5) becomes

$$(2.8) \qquad \mathcal{A}\vec{\mathbf{u}}_\alpha = \alpha\vec{\mathbf{u}}_\alpha ; \ \vec{\mathbf{u}}_\alpha = \begin{bmatrix} \mathbf{u}_\alpha \\ \alpha\mathbf{u}_\alpha \end{bmatrix}$$

Of course, at defective eigenvalues of the problem (2.8) we have also solutions of the form

$$U = r^{\alpha}(\log r\ \mathbf{u}_\alpha + \mathbf{v}_\alpha) \tag{2.9}$$

where $\mathbf{u}_\alpha, \mathbf{v}_\alpha$ are the eigenvector and the associated vector. In the case of the (defective) eigenvalue $\alpha = 0$, these solutions describe point forces applied to the vertex of the sector (see [13] for numerical computations in the composite case).

Numerically, after discretization of the problem (2.5) with $P1$ 1-dimensional finite elements it is possible to solve the eigenvalue problem (2.8) using standard codes for (complex or not) eigenvalues and eigenvectors. Nevertheless, it is easier solving (2.5) under the form

$$\mathrm{Det}(A^h(\alpha)) = 0 \tag{2.10}$$

where A^h denote the discretized matrix A, using a bissection method, and then computing the eigenvectors (and even the associated vectors) by the Gauss method or by inverse iterations (when only the eigenvectors are searched). This method is easier than the first one and gives the same precision.

Examples :

Let us give some examples of numerical results. The results given hereafter are concerned with a composite made of successive plies of unidirectional carbon-epoxy composite (each ply is considered "homogenized"). The elastic coefficients of the ply (with fibers in the direction 1) are :

$$a_{1111} = 145000\ ,\quad a_{1122} = a_{1133} = 6720\ ,\quad a_{2233} = 6220\ ,$$
$$a_{2222} = a_{3333} = 12700\ ,\quad a_{1212} = a_{1313} = 4850\ ,\quad a_{2323} = 3240\ .$$

In the two first cases (fig. 2.1 and 2.3) there are singularities of boundary free of stresses : **1)** between a ply 0 and a ply 90°, giving $\alpha = 0.892$, **2)** between a ply 0 and a ply 45°, giving $\alpha = 0.963$.

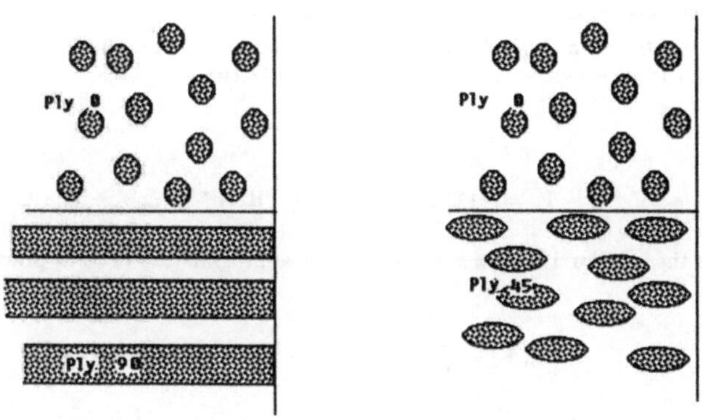

Fig.2.1and2.2 : Stress free boundaries (0/90 and 0/45)

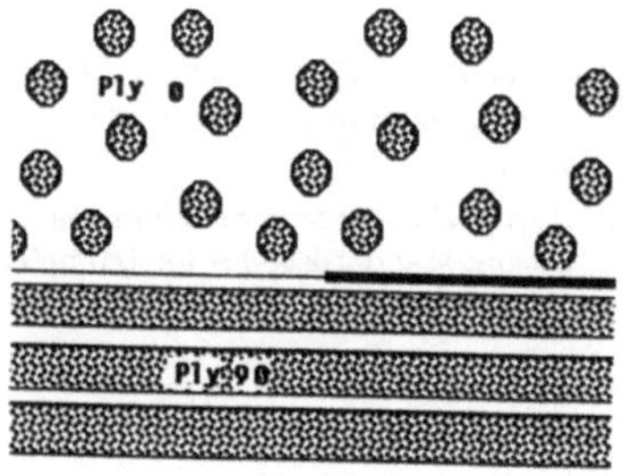

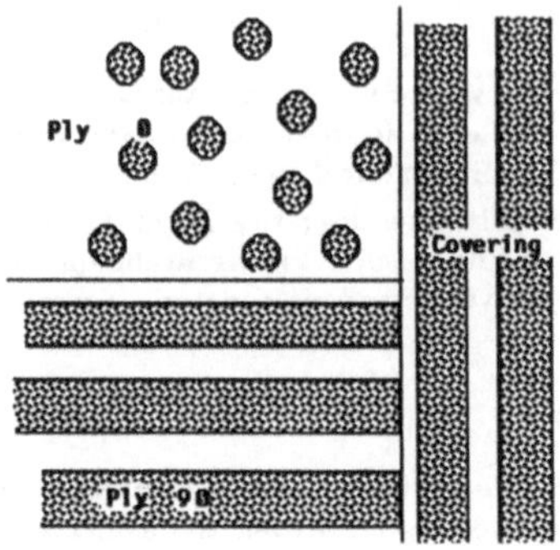

Fig. 2.3 *and* 2.4 : Other singularities (de lamination and covering)

The third example (Fig. 2.3) is a crack of delamination between a ply 0 and a ply 90°, giving $\alpha_1 = 0.5$ and $\alpha_2 = 0.5 \pm i\ 0.035$. The complex exponents with real part 0.5 are classical in interfacial cracks (cf. [18] and Section 4 hereafter). The fourth example (Fig. 2.4) is concerned with a reinforcement of the situation of the first example by a covering made of the some material with fibers normal to the interface. Surprisingly, the singularity of example 1 does not disappear, and becomes even stronger : $\alpha = 0.852$. This result must be pondered by the stress intensity factors.

Computing the stress intensity factors :

Let $\mathbf{U}(x_1, x_2)$ be a solution of the elasticity system in a domain containing an angular point at the origin, in the preceding framework. Then it has an expansion in terms of the functions (2.4) (or (2.9)...) for all the eigenvalues α_i, with coefficient k_{α_i}. If $\mathbf{U}$ is of class H^1 in a neighbourhood of the origin, only the functions with Re $\alpha_i \geq 0$ appear. The terms with Re $\alpha_i < 1$, if there are, are called the singularities at the origin. The coefficients kx_i (intensity factors) which depend on the loading, may be computed using an extraction technique [1] and the following property.

LEMMA 1.1 : *If α is an eigenvalue, then $-\alpha$, (as well as $\bar{\alpha}$ and $-\bar{\alpha}$ in the complex case) is too, with the same geometric and algebric multiplicities.*

The result is :

$$(2.11) \qquad k_\alpha = \psi\,(\mathbf{U}, r^{-\alpha}\mathbf{u}_{-\alpha}) / \psi\,(r^\alpha \mathbf{u}_\alpha, r^{-\alpha}\mathbf{u}_{-\alpha})$$

where $\mathbf{U}$ is the solution itself (or a finite element approximation of it, in practice), and

$$(2.12) \qquad \psi(\mathbf{U}, \mathbf{V}) \equiv \int_\Gamma [\sigma_{ij}(\mathbf{U})n_j V_i - \sigma_{ij}(\mathbf{V})n_j \mathbf{U}_i]\, ds.$$

Here Γ denote any path around the origin joining the two boundaries of the sector, and $\mathbf{n}$ is the unit normal towards the origin.

Remark 1.2 : Formula (2.12) works for simple eigenvalues. It may be adapted to multiple and defective eigenvalues.■

Remark 1.3 : The expansion in eigenvalues and eigenvectors, as well as (2.12) also hold in the case when **U** is a solution in a domain containg a sector in a neighbourhood of infinity. If **grad u** is of class $\mathbf{L}^2$ there, only the terms with $Re\alpha \leq 0$ appear in the expansion.■

3. PERTURBATION OF THE DOMAIN. ASYMPTOTIC BEHAVIOR OF ENERGY. STRONG AND WEAK SINGULARITIES

Perturbation of the domain near singularity plays an important role in the sequel (choice of the direction of propagation of a crack, for instance). We consider here a "perturbed domain" Ω^ε defined from an unperturbed one Ω^0 in the following way. Ω^ε coincides with Ω^0 unless in the vicinity of the origin (fig. 3.1) where it is homothetic with with ratio ε (ε = small parameter) of a fixed domain in Ω'^n of the auxiliary plane $y = x/\varepsilon$. The unperturbed domain Ω^0 is merely Ω^ε for $\varepsilon = 0$. ε may be interpreted as the ratio ℓ/L, where L and ℓ are characteristic lengths of Ω_0 and the perturbation, respectively.

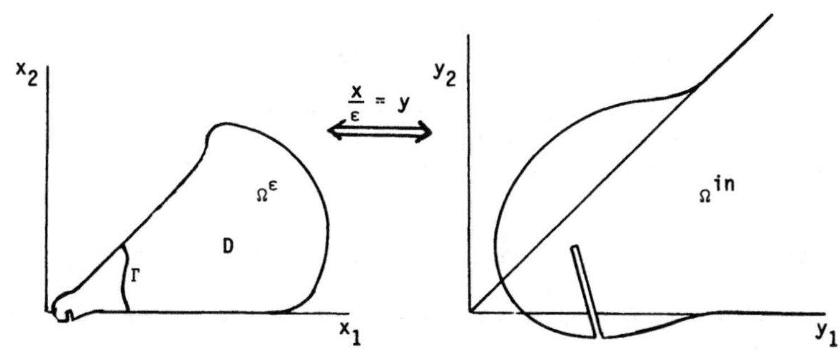

Fig. 3.1

We prescribe appropriate boundary conditions, which are of the Neumann type $\sigma_{ij}n_j = 0$ in a neighbourhood of the origin, and a loading (given volume or surface forces independent of ε) vanishing in a neighbourhood of the origin. The asymptotics of the solution for small ε may be obtained by the inner and outer asymptotic expansion procedure (cf. [12], Lecture 8, and [15]). Let $\mathbf{U}^\varepsilon$ be solution of the perturbed problem, defined on Ω^ε. We admit that there exists an "outer expansion"

$$\text{(3.1)} \qquad \mathbf{U}^\varepsilon(x) = \mathbf{U}^0(x) + f_1(\varepsilon)\,\mathbf{U}^1(x) + \ldots \textit{ with } \lim_{\varepsilon \longrightarrow 0} f_1(\varepsilon) \longrightarrow 0$$

for $x \in \Omega^0$. The leading term $\mathbf{U}^0$ is the solution of the unperturbed problem. Moreover, in the vicinity of the origin we consider the dilated variable $y = x/\varepsilon$. The corresponding domain $\Omega^\varepsilon/\varepsilon$ becomes the semi-infinite domain Ω'^n as $\varepsilon \rightarrow 0$. We admit in Ω'^n the "inner expansion"

$$\text{(3.2)} \qquad \mathbf{U}^\varepsilon(\varepsilon y) = F_0(\varepsilon)\,\mathbf{V}^0(u) + F_1(\varepsilon)\,\mathbf{V}^1(y) + \ldots\text{with } \lim_{\varepsilon \longrightarrow 0} \frac{F_1(\varepsilon)}{F_0(\varepsilon)} = 0$$

By substituting (3.2) into the equations (2.1) we see that the terms $\mathbf{V}^1(y)$ are solutions of the elasticity problem in Ω'^n with Neumann boundary conditions ; in addition we must prescribe conditions of behavior at infinity, which are in fact matching conditions between (3.1) and (3.2) in an intermediate region (with small x and large y) defined by

$$(3.3) \qquad z = \varepsilon^{-\gamma} x = \varepsilon^{1-\gamma} y \text{ with } \quad 0 < \gamma < 1$$

The matching is obtained by identification of the asymptotic behavior of (3.1) and (3.2) in the region (3.3) (see [19], chap. 6 for a more precise description of the matching). Using the expansions of the preceding section for small x in (3.1) and for large y in (3.2), we obtain

$$(3.4) \qquad \begin{cases} \mathbf{U}^0(x) &= kr^\alpha \mathbf{u}_\alpha(\theta) + \ldots & as\ r \longrightarrow 0 \\ F_0(\varepsilon)\mathbf{V}^0(y) &= F_0(\varepsilon)\left[k'\rho^\beta \mathbf{u}_\beta(\theta) + \ldots\right] & as\ \rho \longrightarrow \infty \end{cases}$$

where $r = \mid x \mid$ and $\rho = \mid y \mid = r/\varepsilon$. Here r^α is the singularity of $\mathbf{U}^0(x)$ at the origin supposed to be simple, and ρ^β the leading term,, to be obtained of $\mathbf{V}^0(y)$ at infinity. The identification of the two expressions of (3.4) in the region z gives

$$(3.5) \qquad \beta = \alpha \ ; \ k' = 1 \ ; \ F_0(\varepsilon) = k\varepsilon^\alpha$$

The leading term $\mathbf{V}^0(y)$ of the inner expansion is then the solution of the well- posed problem

$$(3.6) \qquad -\frac{\partial \sigma_{ij}(\mathbf{V}^0)}{\partial y_j} = 0 \ \ in \ \Omega'^n$$

$$(3.7) \qquad \sigma_{ij} n_j = Q \ \ on \ \partial\Omega'^n$$

$$(3.8) \qquad \mathbf{V}^0 = k \mid y \mid^\alpha \mathbf{u}_\alpha(\theta) + \text{smaller terms as } \mid y \mid \longrightarrow \infty$$

<u>Remark 3.1</u> : Additive constants or rigid motions may be added in the preceding expansions according to the precise definition of the problems. The well-posedness of (3.6)-(3.8) should be understood in this way.■

When $\mathbf{V}^0(y)$ is known (by an approximate finite element method, in practice), we may consider its asymptotic behavior as $\mid y \mid \longrightarrow \infty$;

$$(3.9) \qquad \mathbf{V}^0(y) = \rho^\alpha \mathbf{u}_\alpha(\theta) + K\rho^{-\alpha}\mathbf{u}_{-\alpha}(\theta) + \ldots$$

because according to Lemma 1.1, $\rho^{-\alpha}$ is the leading term at infinity (after ρ^α of course, and discarding constant terms). The coefficient K (i.e. the intensity factor of $\rho^{-\alpha}$ at infinity of V^0) plays an important role in the sequel. It is computed in practice by formula (2.11).

At the present level we may perform the matching at a higher order. It is easily seen that the term $f^1\mathbf{U}^1)$ in (3.1) is such that

$$f^1(\varepsilon) = kK\varepsilon^{2\alpha} \tag{3.10}$$

$$\mathbf{U}^1(x) = r^{-\alpha}\mathbf{u}_{-\alpha}(\theta) + ... \tag{3.11}$$

Remark 3.2 : According to (3.11), $\mathbf{U}^1(x)$ is very singular at the origin, and it is not of class H^1 there. This is a classical feature of these expansions (cf. [12] sect. 8.3 in particular, where all the terms are written down). As the singularity in $r^{-\alpha}$ of $\mathbf{U}^1$ is known, $\mathbf{U}^1$ may be calculated by solving a well-posed problem. But this will not be used in the sequel.■

From (3.10) and (3.11) it appears that the leading terms containing $r^{\alpha}\mathbf{u}_{\alpha}$ and $r^{-\alpha}\mathbf{u}_{-\alpha}$ in $\mathbf{U}^{\varepsilon}$ given by (3.1) are :

$$\mathbf{U}^{\varepsilon} = kr^{\alpha}\mathbf{u}_{\alpha}\theta + \varepsilon^{2\alpha}kKr^{-\alpha}\mathbf{u}_{-\alpha} + ... \tag{3.12}$$

The preceding developments may be used for calculating the asymptotic behavior as $\varepsilon \longrightarrow 0$ of the energy of the body [11]. To fix ideas, we consider the body Ω^0 (and the perturbed one Ω^{ε}) submited to given surface forces **H**independent of ε, acting on the boundary of the body, but vanishing in a neighbourhood of the origin. Let $\partial\Omega^2$ be the part of the boundary where they do not vanish. The energy stored in the body Ω^{ε} is

$$W^{\varepsilon} = \frac{1}{2}\int_{\Omega^{\varepsilon}} \sigma_{ij}^{\varepsilon} e_{ij}^{\varepsilon} dx \equiv \frac{1}{2}\int_{\partial\Omega^2} H_i U_i^{\varepsilon} dS \tag{3.13}$$

and an analogous expression for the energy W^0 instead of $\mathbf{U}^{\varepsilon}$. Here σ^{ε}, e^{ε} as well as σ^0, e^0 are the expressions defined in (2.2), (2.3). The Green formula for the elasticity system and the functions $\mathbf{U}^0, \mathbf{U}^{\varepsilon}$ in the domain D delimited by a curve Γ as in fig. 3.1 gives :

$$\int_D \sigma_{ij}^0 e_{ij}^{\varepsilon} dx = \int_{\partial D} \sigma_{ij}^0 n_j U_i^{\varepsilon} ds = \int_{\partial\Omega^2} H_i U_i^{\varepsilon} ds + \int_{\Gamma} \sigma_{ij}^0 n_j U_i^{\varepsilon} ds \tag{3.14}$$

We now write an analogous expression exchanging the roles of $\mathbf{U}^0$ and $\mathbf{U}^{\varepsilon}$; taking the difference of both expressions and using the symmetry of the elasticity system and the definitions of W^{ε}, W^0 given in (3.13), we have

$$0 = 2W^{\varepsilon} - 2W^0 + \int_{\Gamma} (\sigma_{ij}^0 n_j U_i^{\varepsilon} - \sigma_{ij}^{\varepsilon} n_j U_i^0)\, ds$$

or, using the definition (2.12) of ψ :

$$\delta W^{\varepsilon} \equiv W^{\varepsilon} - W^0 = \frac{1}{2}\psi(\mathbf{U}^{\varepsilon}, \mathbf{U}^0) \tag{3.15}$$

This expression is clearly independent of the curve Γ of fig. 3.1, provided the perturbation of the domain is on one side of it and the region $\partial\Omega^2$ where the forces are given, on the other side. Using the expressions (3.4) and (3.12) to compute (3.15), it becomes

$$\delta W^\varepsilon \equiv W^\varepsilon - W^0 = \frac{1}{2} k^2 K \varepsilon^{2\alpha} \tag{3.16}$$

provided $\mathbf{u}_{-\alpha}$ is normalized according to

$$\psi\left(r^{-\alpha}\mathbf{u}_{-\alpha}\,,\, r^{\alpha}\mathbf{u}_{\alpha}\right) = 1\,. \tag{3.17}$$

Remark 3.3. : Formula (2.16) gives the leading term of the perturbation of energy as a function of the stress intensity factor k of r^α at the origin for $\mathbf{U}^0$ defined in (3.4), which depend on the loeading, but not on the perturbation itself, and of the stress intensity factor K of $r^{-\alpha}$ at infinity for $\mathbf{V}^0$, defined in (3.9), which depend on the perturbation, but not on the loading.■

We see that δW^ε is proportional to ε only if $\alpha = 1/2$. It will prove useful to define *strong and weak singularities* as those for which $\alpha < 1/2$ and $\alpha > 1/2$ respectively. According to (3.16), *the perturbation of the energy is infinitely large or small with respect to the characteristic length of it according to the fact that the singularity is strong or weak.* The singularities with $\alpha = 1/2$ (or $Re(\alpha) = 1/2$ in the complex case) constitute the limit between the strong and weak singularities. The cracks in a homogeneous medium ($\alpha = 1/2$) or at the interface between two media ($Re\,\alpha = 1/2$) belong to this limit.

Analogous properties may be exhibited by defining the energy release ratio :

$$G \equiv -\frac{\partial W}{\partial \ell} = -\frac{\alpha}{L} k^2 K \varepsilon^{2\alpha-1} \tag{3.18}$$

in the vicinity of the singularity. It tends to either infinity or zero as $\varepsilon \longrightarrow 0$ in a strong or weak singularity, respectively.

In an analogous way [11] in the case of a singularity α real but double, we have

$$G = -\frac{\alpha}{L}\left[k_I^2 K_I' + k_{II}^2 K_{II}'' + k_I k_{II}\left(K_I'' + K_{II}'\right)\right]\varepsilon^{2\alpha-1} \tag{3.19}$$

and in the case of singularity with complex α

$$G = -\frac{4}{L}\left[Re(\alpha)K' k\bar{k} + Re\left(\alpha K k^2 \varepsilon^{2iIm(\alpha)}\right)\right]\varepsilon^{2Re(\alpha)-1} \tag{3.20}$$

where the different constants K are independent of the loading.

A typical example of strong (resp. weak) singularity is a crack arriving to an interface between two isotropic media from the rigid (resp. soft) one, fig. 3.2 (resp. fig. 3.3) giving $\alpha_1 = 0.195$ and $\alpha_2 = 0.245$ (resp. $\alpha_1 = 0.667$ and $\alpha_2 = 0.806$)

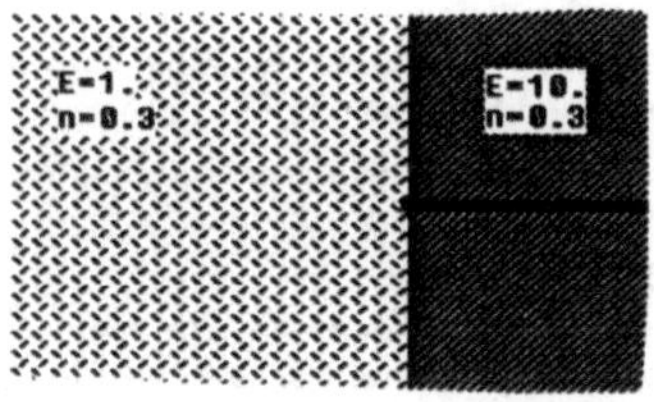

Fig. 3.2.

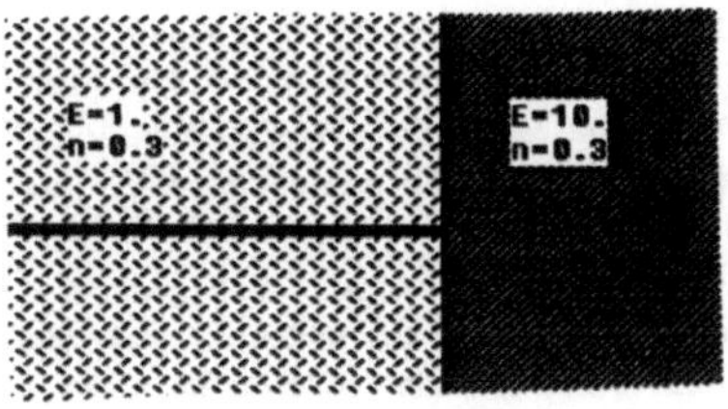

Fig. 3.3.

4. FURTHER STUDY OF δW^ϵ.DEPENDENCE ON PARAMETERS

We shall consider in the next section a criterion to obtain the direction of progress of a crack in somethat general situations. The criterion involves the values of δW^ϵ for "virtual elements of crack" in different directions. We begin here studying the continuity and analyticity properties of δW^ϵ with respect to the direction of the element of crack.

To this end, we consider the case of Ω^{in} formed by a sector with an element of crack of unit length in the direction φ, with φ belonging to the open set $(0,\omega)$ (see Fig. 4.1).

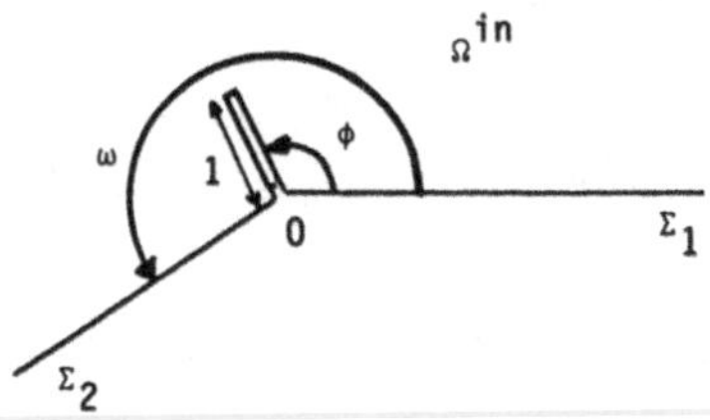

Fig. 4.1.

We then consider problem (3.6)-(3.8) which we write under a slightly different form. Let $\mathcal{V}$ be the Hilbert space of the vector fields defined in Ω^{in} up to a rigid motion and having finite energy

$$\int_{\Omega^{in}} a_{ijkh} e_{kh}(\mathbf{U}) e_{ij}(\mathbf{U}) dy \ < \infty \tag{4.1}$$

which can be taken as the square of the norm in $\mathcal{V}$. Let also $\mathbf{b}(y)$ be an auxiliary vector field equal to $| \ y \ |^\alpha \ \mathbf{u}_\alpha(\theta)$ (resp. to zero) in a neighbourhood of infinity (resp. of the origin), for $| \ y \ |> 5$ (resp. $| \ y \ |< 4$), say. Then, (3.6)-(3.8), with $k = 1$, and $\mathbf{V}^0 = \mathbf{b} + \mathbf{u}$, amounts to

$$\begin{cases} \text{Find } \mathbf{u} \in \mathcal{V} \quad \text{such that,} \\ \displaystyle\int_{\Omega^{in}} a_{ijkh} \ e_{kh}(\mathbf{u}) \ e_{ij}(\mathbf{v}) \ dy = \int_{\Omega^{in}} \frac{\partial}{\partial x_j} \ [a_{ijkh} e_{kh}(b)] \ v_i dy \quad \forall \ \mathbf{v} \in \mathcal{V} \end{cases} \tag{4.2}$$

where the right side is taken in the usual distribution sense if interfaces are involved. We note that the corresponding support is contained in $4 <| \ y \ |< 5$. This is a well-posed problem.

The coefficients $a_{ijkh}(\theta)$ being given, we consider the dependence of $K(\varphi)$ on the angle φ. K is of course the intensity factor of $| y |^{-\alpha}$ at infinity. In this section the singularities α and $-\alpha$ are considered to be simple, but obvious modifications apply to other cases.

<u>THEOREM 4.1</u> - *Let the coefficients a_{ijkh} be constant (or even analytic functions of θ) in $[0,\omega]$. Then $K(\varphi)$ is analytic for $\varphi \in (0,\omega)$.*

We now give an outline of the proof of this theorem. Details are omitted, in particular those arising from the fact that **u** is defined up to a rigid motion. The main difficulty to prove this theorem is that the space $\mathcal{V} = \mathcal{V}(\varphi)$ depends on the paramter φ. Then to prove analyticity in a neighbourhood of $\varphi = \varphi_0 \in (0,\omega)$ we perform an analytic family of diffeomorphisms sending to a fixed space.

To this end, we construct a smooth function $f(\theta)$ of the shape indicated in fig. 4.2, equal to 0 (resp. to 1) for θ near 0 and ω (resp. φ_0). Then, we have :

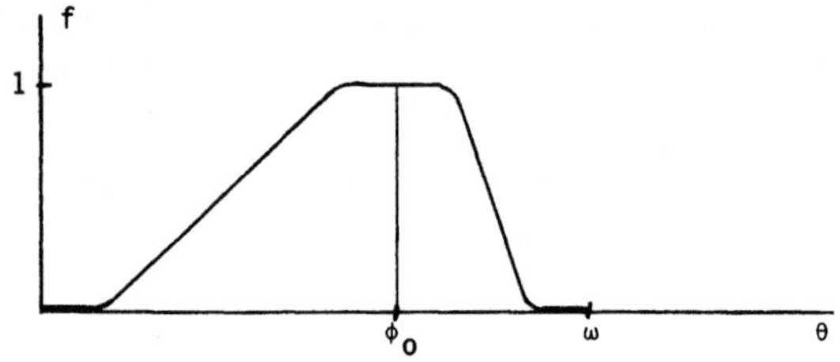

Fig. 4.2.

<u>LEMMA 4.2</u> - *Let $\varphi_0 \in (0,\omega)$ be given. Then, the transformation*

$$\theta' = \theta - \left(\varphi - \frac{\omega}{2}\right) f(\theta) \tag{4.3}$$

defines for φ in a neighbourhood of φ_0, a C^∞-diffeomorphism of $(0,\omega)$ on itself. Moreover, there are neighbourhoods of 0 and ω that remain fixed, and a neighbourhood of $\theta = \varphi$ which undergoes the translation $(\omega/2) - \varphi$.

Now, let $F(\rho)$ be a out off function as indicated in fig. 4.3. We then have :

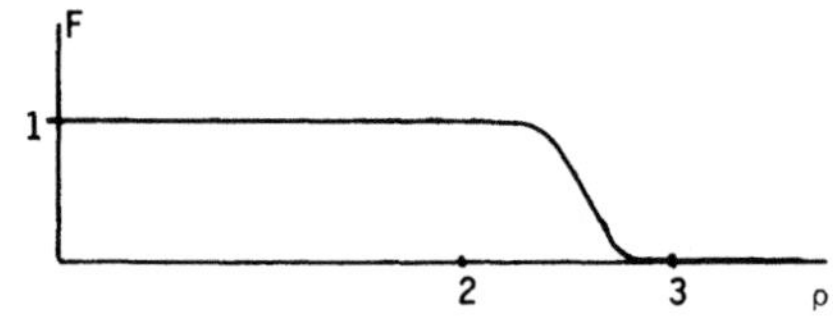

Fig. 4.3.

<u>LEMMA 4.3</u> - *Let $\varphi_0 \in (0,\omega)$ be given. Then, the transformation*

$$\begin{cases} \rho' &= \rho \\ \theta' &= \theta - \left(\varphi - \frac{\omega}{2}\right) f(\theta) F(\rho) \end{cases} \tag{4.4}$$

defines for φ in a neighbourhood of φ_0, a C^∞-diffeomorphism of the infinite half- strip $0 < \theta < \omega$, $0 < \rho < +\infty$ on itself. Moreover, there are neighbourhoods of the edges $\theta = 0$, $\theta = \omega$ which remain fixed and of the segment $\theta = \varphi$, $0 < \rho < 1$ which undergoes a translation $(\omega/2) - \varphi$ in the direction of the axis θ.

Now, writing (4.2) in polar coordinates ρ, θ and performing the family of diffeomorphisms of Lemma 4.3 with parameter φ, this problem reduces to another problem in the fixed space with fissure at the angle φ_0. Moreover, from (4.4) we see that the family of diffeomorphisms depends holomorphically on φ, and of course it preserves the fact of belonging to L^2 and H^1. Then, the problem (4.2) becomes an analogous one with modified coefficients, which depend holomorphically on φ, but in the space $\mathcal{V}(\varphi_0)$. Using standard holomorphic dependence theory for operators (cf. [7] or [19], chap. V), we see that problem (4.2) is equivalent to

$$A(\varphi)\, \mathbf{u}(\varphi) = F(\varphi) \Leftrightarrow \mathbf{u}(\varphi) = A(\varphi)^{-1}\, F(\varphi) \tag{4.5}$$

where $A(\varphi)$ is a holomorphic function of φ with values in the space of operators $\mathcal{L}(\mathcal{V}(\varphi_0), \mathcal{V}(\varphi_0)')$ and $F(\varphi)$ is a holomorphic function with values in $\mathcal{V}'(\varphi_0)$ (representing the right side of (4.2)). As A^{-1} is also holomorphic, we see that $\mathbf{u}$ depends holomorphically on φ in the vicinity of φ_0. Moreover, for $\rho \geq 3$ the diffeomorphism is the identity, and then the very solution depends holomorphically on φ with values in $H^1(3 <\mid y \mid< 4)$. Using standard regularity theory for the elasticity system up to the boundaries Σ_1 and Σ_2 (see perhaps [19], sect. II.9), the convergence is also holomorphic with values in $H^2(3 <\mid y \mid< 4)$, and this suffices for proving that the intensity factor of $\rho^{-\alpha}\mathbf{u}.(\theta)$ depends holomorphically on φ. Theorem 4.1. is proved.

The same proof is valid in the case of discontinuous coefficients at the interfaces. In that case the diffeomorphism is performed in the sector between two consecutive interfaces, proving there the holomorphy. At the end of the proof, regularity theory for the transmission problem is used, giving holomorphy in a space of finitions belonging piecewise to H^2 :

<u>THEOREM 4.4</u> - *Let the coefficient $a_{ijkh}(\theta)$ be constant (or even analytic functions of θ) in the intervals between the interfaces (where they undergo first kind discontinuities). Then, $K(\varphi)$ is analytic in the intervals between interfaces.*

Moreover, it is possible to use exactly the family of diffeomorphism of the proof of theorem 4.1 even in the case when the element of fissure crosses an interface (i.e. $\theta = \varphi_0$ is an interface). But in that case the coefficients after the diffeomorphism are no longer holomorphic in φ, as the lines of discontinuity move smoothly with φ. In that case we obtain only continuity instead of holomorphy, as follows easily from [19] sect. V.9, in particular Proposition V.9.4 and theorem V.9.1. The solution after the family of diffeomorphisms is a continuous function of φ with values in $\mathcal{V}(\varphi_0)$. Using as before regularity for the transmission elasticity problem, we see that the solutions are continuous functions with values in a space of functions belonging piecewise to H^2 (but only to H^1 in the whole). In addition, for $\varphi = 0$ and $\varphi = \omega$, the element of fissure does not modify the domain and consequently the corresponding δW^ε vanishes, as well as K. We have :

<u>THEOREM 4.5</u> - *Under the hypotheses of Theorem 4.4, $K(\varphi)$ is a continuous function of $\varphi \in [0, \omega]$, which vanishes at $\varphi = 0$ and $\varphi = \omega$.*

As a typical numerical example we give in fig. 4.4 the shape of $K(\varphi)$ for a crack arriving with an angle of 36° to an interface Γ between two elastic media with E and ν given in the figure. The corresponding value of the eigenvalue is $\alpha = 0.207$, (strong singularity) :

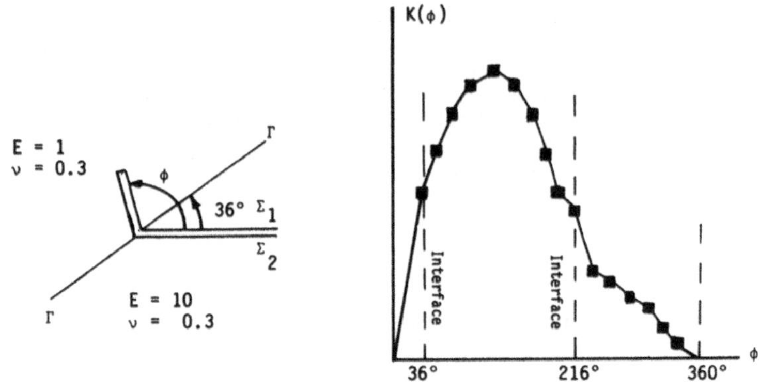

Fig. 4.4.

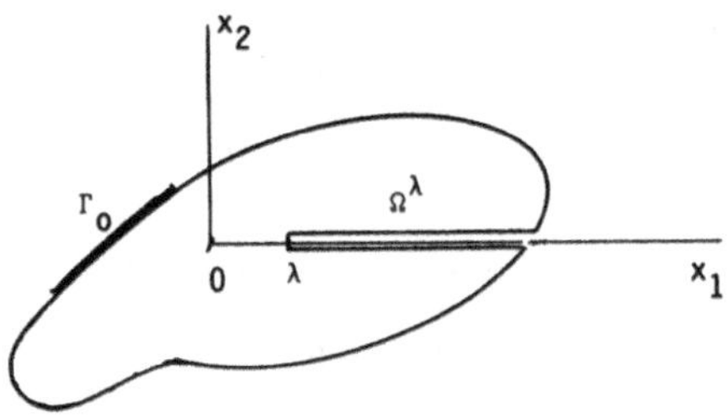

Fig. 4.5

Let us now consider another problem of smoothness with respect to parameters. In two-dimensional elasticity we consider a crack propagating along a smooth interface between 2 materials, which we take straight, for the sake of simplicity. If the loading is given, we prove that the stored energy is an indefinitely differentiable function of the abscisse of the extremity of the crack. Analogous results in the case of a homogeneous body were proved in [4] and [15]. As a consequence, in the context of formula (3.16) we have necessarily $\alpha = 1/2$. In fact, it is known [18] that in that case there are two complex conjugate singularities,$\alpha = 1/2 \pm i\beta$. On the basis of (3.16) it appears that the oscillating term $\varepsilon^{2\beta i}$ must vanish i.e. the coefficient K' vanishes in that case.

Let us consider the body Ω^λ of fig. 4.5 with a crack along the axis $x_2 = 0$, occuping the region $x_1 > \lambda$, where λ denotes a parameter, either positive or negative. The body is in elastic equilibrium under the action of given body forces (for instance) f, vanishing in a neighbourhood of the origin. Moreover, the body is fixed by part Γ_0 of its boundary. Denoting

(4.6) $$V^\lambda = \{\mathbf{v} \in \mathbf{H}^1(\Omega^\lambda)\ ,\ \mathbf{v}\ =\ 0 \text{ on } \Gamma_0\}\,,$$

the solution $\mathbf{u}^\lambda$ is defined by

(4.7) $$\begin{cases} \text{Find } \mathbf{u}^\lambda \in V^\lambda \text{ sucli that} \\ \displaystyle\int_{\Omega^\lambda} a_{ijkh}(x)\frac{\partial u_k^\lambda}{\partial x_h}\frac{\partial v_i}{\partial x_j}dx = \int_{\Omega^\lambda} f_i v_i dx \qquad \forall\, \mathbf{v} \in V^\lambda \end{cases}$$

and the corresponding energy is

(4.8) $$W^\lambda\ = \frac{1}{2}\int_{\Omega^\lambda} a_{ijkh}\frac{\partial u_k^h}{\partial x_h}\frac{\partial v_i}{\partial x_j}dx.$$

<u>THEOREM 4.6.</u> - *Let the coefficients be piecewise constant, taking the values $a^{\pm}_{ijkh}$ for $x_2 \gtrless 0$. Then, the energy W^λ is an analytic function of λ.*

This theorem is proved by methods somewhat analogous to those of theorem 4.1. The only important difference is in the definition of the appropriate family of diffeomorphisms, depending analytically on λ.

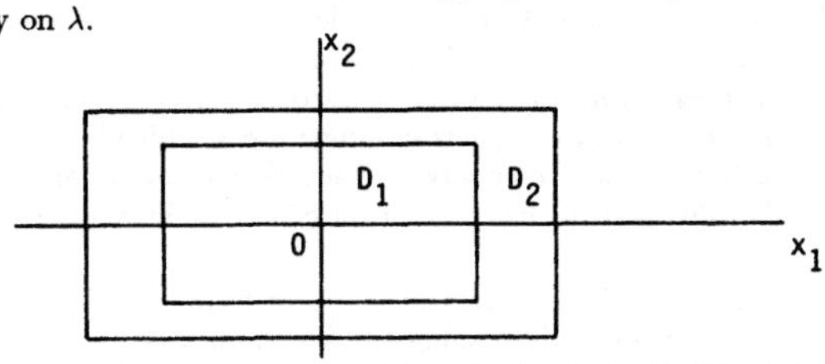

Fig. 4.6.

To this end, let us consider in the plane x_1x_2 two rectangles centered at the origin, disposed as shown in fig. 4.6., and sufficiently small to be contained in the solid Ω (the fissure is not taken into account here). Let $\theta(x)$ be a function of $\mathcal{D}(\mathbb{R}^2)$ taking the value 1 in D_1, vanishing out of D_2 and taking values comprised between 0 and 1 in $D_2\backslash D_1$. Let $\mathbf{e}_1, \mathbf{e}_2$ be the unit vectors of the axes. Then we define the transformation $x \longrightarrow z$:

(4.9) $$z\ = x - \lambda\theta(x)\mathbf{e}_1$$

which is a translation $-\lambda\mathbf{e}_1$ for $x \in D_1$ and keeps invariant the points x out of D_2. Moreover, any point is submitted to a translation parallel to the axis $x_2 = 0$. It is easily seen that, for $\mid \lambda \mid$ sufficiently small it defines a diffeomorphism between $x \in \Omega^\lambda$ and $z \in \Omega^0$, which depends analytically on the parameter λ. Then, we may continue the proof as in the case of Theorem 4.1. In particular, the fact that in the diffeomorphism (4.9) each point is translated along a parallel to the interface $x_2 = 0$ ensures that the coefficients of the equation after the diffeomorphism depend analytically on λ, as the interface does not depend on λ (note the difference with Theorem 4.5).

<u>Remark 4.7.</u> : Obviously, the coefficient a_{ijkh} may be taken analytic in $x_2 > 0$ and $x_2 < 0$. Moreover, the result also hold true in the case when this hypothesis is only satisfied in

a neighbourhood of the origin : the coefficients may be arbitrary elsewhere. It suffices to take D_1 and D_2 sufficiently small. We also note that the case of a (smooth) curve fissure may also be considered performing a previous diffeomorphism to transform it into a straight one.■

5. THE CRITERION OF FRACTURE AND EXAMPLES

We now present the thermodynamic criterion of fracture (see [2], for instance). It stipulates that each material has some specific energy γ per unit surface which is transformed into heat in the irreversible process of generating by fracture a surface.

It is to be noted that ancient thermodynamic theories considered γ as a reversible energy associated with some kind of surface tension. Modern interpretation consider γ as an energy which is dissipated in the irreversible process of "fracture of constitutive links". In any case, these interpretations are not useful for our purposes here. In order to have a heuristic interpretation of the fracture phenomenon, we may think of an elastic medium as an agregate of "molecules" linked to each other by elastic strings, which broke away when the corresponding lengths attend some specific threshold. The energy to break one string is equal to one half the product of the final force by the corresponding lengthening. By multiplying it by the number of links to be broken by unit surface we obtain γ.

Generally speaking, γ depends on the material and also on the orientation of the surface generated. This amounts to *anisotropy of the fracture behavior*, which may be easily interpreted with our previous heuristic model. Generally speaking, materials which are anisotropic from the elastic point of view are also anisotropic in fracture. Nevertheless, according to th microscopic structure, materials which are elastically isotropic may be anisotropic in fracture. An example is laminate steel, which is roughly isotropic as for the elastic behavior but its specific energy of fracture γ is lower (sometimes 100 times lower) for cracks in the direction of the orientation of the perlitic phase than in the orthogonal direction [6].

The energy γ may also depend on the mode of fracture in the case of multiple modes (think of the different ways of breaking the strings in our heuristic model). But we *shall consider only simple modes in the sequel* (or cases where the ratio of the different modes is given, which reduces to the precedent one). The corresponding energy of fracture will be $\gamma(\psi)$ where ψ denote the angle of the fissure with some characteristic direction of the material(if this one is fracture-anisotropic ; otherwise, γ is a constant).

In the case of an interface between two elastic media a and b, obviously the energies γ^a and γ^b are involved in the cracks in each medium, but in addition, we must consider the energy γ^I corresponding to the interface. According to the nature of the pasting, it may be very different from γ^and γ^b. We shall see later an example of this.

With this criterion, it is easy to obtain the *direction φ of progress* of an element of crack in a somewhat general case. Let us consider, always in the framework of two-dimensional elasticity a body in the situation of sect. 4, in particular, fig. 4.1. We have an "unperturbed state" with the body bounded (in the vicinity of the point O what we consider) by the free boundaries Σ_1 and Σ_2, without the new element of crack of angle φ. For instance, in the case when the unperturbed state is a crack we have $\omega = 2\pi$. Now, we submit the body to an external loading, which implies some intensity factor k for the singularity at the origin. We consider an incremental loading, and we increase it until the value of k which produces a small new crack of length ε (which is considered as knwon for the time being and small in order to use the asymptotic theory of sect. 4. We consider the process to be quasistatic, i.e. inertia effects are neglected. Then, if an element of crack of direction φ is produced, the energy of fracture $\varepsilon\gamma(\varphi)$ must be equal to the energy released. According to (3.16) we have

$$\gamma(\varphi)\varepsilon = (1/2)k^2 K(\varphi)\varepsilon^{2\alpha} \tag{5.1}$$

Then, in the incremental process, the φ for which (5.1) holds with the smallest k is the φ for

$$\inf_{\delta \in [0,\omega]} \frac{\gamma(\varphi)}{K(\varphi)} \tag{5.2}$$

is attained. Thus, the problem of the *direction* of the progress of a crack in a quasi-static process is solved by (5.2).

Let us give an example. We consider the case of fig. 4.4. (see rather fig. 5.1.). We give specific energies $\gamma^a = 4$

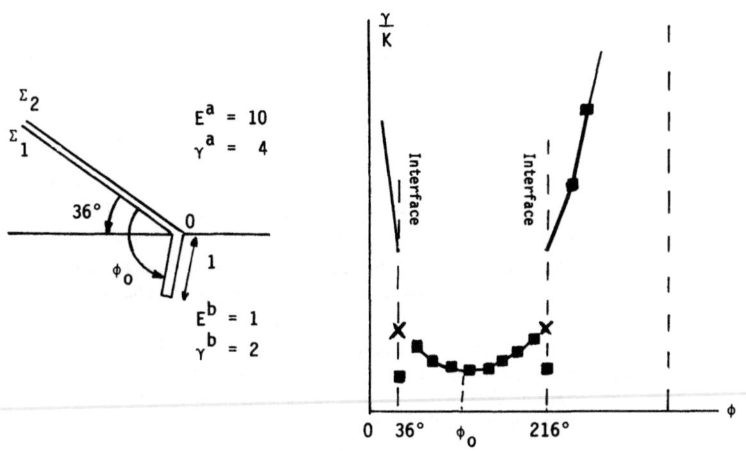

Fig. 5.1.

and $\gamma^b = 2$ (note that only the mutual ratios have an influence on φ). Moreover, we consider two cases. In the first one the specific energy of the interface is taken to be $\gamma^I = 2$. The curve γ/K is plotted in fig. 5.1. (compare with fig. 4.4.). At the interface, the corresponding points are marked with a cross. The inf. is obtained for φ_0, and the crack is in the medium b. As a second example, we take now $\gamma^I = 1$ ("bad pasted" with respect to the previous one). Then in the curve γ/K the points at the itnerfaces are marked with a ■. The inf. is attainded at the interfaces, and in this case, the crack propagates along the interface.

It should be noticed that the (small) length ε of the new element of fissure was considered above to be known. This point is only important for $\alpha \neq 1/2$. In fact, when φ is known by the preceding considerations, the corresponding value of the intensity factor k which produces the element of fissure is

$$(1/2)k^2 = \operatorname{Inf}\frac{\gamma}{K}\,\varepsilon^{2\alpha-1}. \tag{5.3}$$

So, in the case of a strong (resp. weak) singularity, $2\alpha - 1 < 0$ (resp. > 0) and strictly speaking, k is equal to zero (resp. to infinity). In practice, things are a little different, as lengths ε arbitrarily small do not make sense (they would be smaller than the characteristic length of the microstructure (micro-crystals in a metal, etc). In fact, *for each material there is a characteristic length ε_0 which may* be considered as the smallest one making sense for a fissure, to be replaced in (5.3) [6]. The corresponding values of k are then "small or large", but not "zero or infinity" in a strong or weak singularity.

But there is another important property of (5.3). If a k is given satisfying (5.3) for $\varepsilon = \varepsilon_0$, then it will not be (resp. be also) satisfyied for $\varepsilon > \varepsilon_0$ provided the singularity is strong (resp. weak). Consequently, a *crack starting from a weak singularity tends to propagate provided it starts.* Oppositely, a *crack starting from a strong singularity needs more and more loading to propagate.* Of course these considerations are only valid for small length ε. Otherwise, a new study must be performed at the extremity of the fissure.

Remark 5.1. : We adopted here the strict point of view of an element of crack εstarting from a singularity. Cracks starting "near" a singularity may also be considered, as well as the corresponding energy release ratio [14].

6. COMMENTS ON CRACKS PROPAGATING IN ANISOTROPIC MATERIALS

We consider in this section "smooth cracks" propagating along curves of class C^1(i.e. having a continuously varying tangent) in a homogeneous medium. We have of course $\alpha = 1/2$ at any point. We consider, as in the preceding sections, a simple singularity (or pherhaps a multiple one with given ratios for the corresponding intensity factors). We shall see that the anisotropy properties of the fracture energy γ defines in most cases the directions of the smooth cracks. The isotropic case appears as non-generic in some sense.

At any point of the cracks, the progress of the crack is in the prolongation of the previous one. In the context of section 5 we have $\omega = 2\pi$, and the inf. in (5.2) is attained for $\varphi = \varphi_0 = \pi$.

Let us consider, as a first example the case of an anisotropic material with fracture energy $\gamma = \gamma(\varphi - \psi)$ where ψ is the angle of the previous crack with some characteristic direction of the material. Let us admit that *$\gamma(\varphi - \psi)$ is a sooth function.* The function K in (5.2) depends of course on the orientation of the anisotropic material.

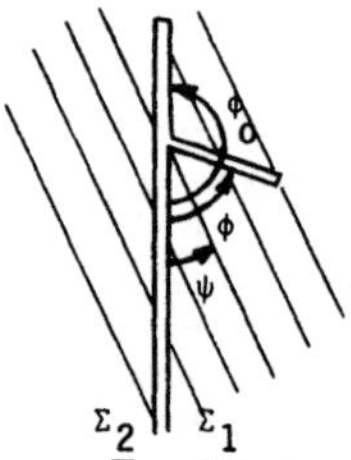

Fig. 6.1.

(6.1)
$$\operatorname*{Inf}_{\varphi \in [0, 2\pi]} \frac{\gamma(\varphi - \psi)}{K(\varphi, \psi)}$$

is attained for $\varphi = \pi$, and this gives

(6.2)
$$\frac{\gamma'(\pi - \psi)}{\gamma(\pi - \psi)} = \frac{K'_{\varphi}(\pi, \psi)}{K(\pi, \psi)}$$

which is, generically, an equation giving (one of several) ψ. Consequently, the smooth cracks are necessarily oriented by a given angle ψ with respect to the characteristic direction of the medium. For instance, if the elastic properties are isotropic K is independent of ψ, and (6.2) shows that the smooth fissures are necessarily oriented along the direction of smallest energy of fracture.

We now consider an example of a material with a *non-smooth energy of fracture* $\gamma = \gamma(\hat{\varphi})$, $\hat{\varphi} = \varphi - \psi$ given by

$$(6.3) \qquad \gamma(\hat{\varphi}) = a \mid cos\hat{\varphi} \mid + b \mid sin\hat{\varphi} \mid , \; a < b$$

This is a model for some kind of "fibered material" such that a "macroscopic crack" is microscopically "step-wise" (fig. 6.2). The direction $\varphi = \psi$ has an energy less than its orthogonal.

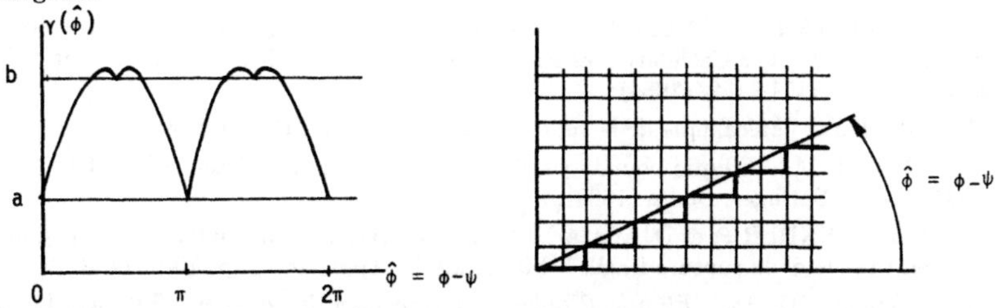

Fig. 6.2.

We consider the material to be elastic-isotropic, K is independent of ψ and we shall take $K(\varphi)$ to be symmetric with respect to $\varphi = \pi$ (for instance a fissure submitted to the classical mode I, fig. 6.3.). It is then easily seen that in that case,

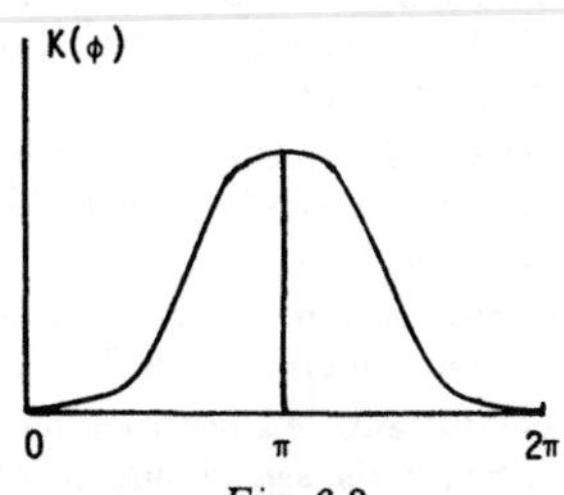

Fig. 6.3.

$$(6.4) \qquad \text{Inf} \frac{\gamma(\varphi - \psi)}{K(\varphi)}$$

is attained for $\varphi = \pi$ when $\psi = 0$, or $\psi = \pi$, but not for ψ near (but different) from 0 and π. This means that the smooth crack may propagate along the direction of smallest energy, but not in an angular neighbourhood of it (this direction is an "attractor" for the crack). In addition, and depending on the function $K(\varphi)$ and on b/a, the crack may or not propagate along the orthogonal direction $\varphi = \psi \pm \pi/2$.

As another example, we consider instead of (6.3) the new function (6.5) where a and b are two constants. This situation appears physically for a material cut and pasted ; the corresponding direction has an energy $\gamma^1 = a$ which is less or more than the energy of the body $\gamma = b$ according to the quality of the pasting (bad or good). the function $K(\varphi)$ is taken as in the preceding example (fig. 6.3.).

$$\gamma(\hat{\varphi}) = \begin{cases} a & \text{for } \hat{\varphi} = 0, \pi, 2\pi \\ b & \text{otherwise} \end{cases} \tag{6.5}$$

It is then easily seen that, *in the case $a < b$(bad pasted)*, the inf. in (6.4) is attained for $\varphi = \pi$ when $\psi = 0$ or $\psi = \pi$, but not for ψ near this values (the direction of smallest energy is an "attractor"). The inf. is also attained for $\varphi = \pi$ when φ is near $\pi/2$. This means that *the only directions of propagation of smooth fissures are the direction of smallest energy and the directions making with it an angle sufficiently large. In the case $a > b$ (weel pasted) the directions of smooth cracks are any one unless the direction of largest energy.*

REFERENCES

[1] BABUSKA I. and MILLER A., "The post-processing approach in the finite element method. Part 2 : The Calculation of stress intensity factors". Intern. Journ. Num. Meth. Engin., 20, p. 1111-1129, (1984).

[2] BUI H.D., "*Mécanique de la rupture fragile*", Masson, Paris (1978).

[3] DAUGE M., "Elliptic boundary value problems in corner domains" (= Lect. Notes Math. 1341), Springer, Berlin (1988).

[4] DESTUYNDER P. et DJAOUA M., "Intepretatation mathématique de l'intégrale de Rice en théorie de la rupture fragile", Math. Meth. Appl. Sc., 3, p. 70-87 (1981).

[5] GRISVARD D., "*Elliptic Problems in Non-smooth domains*", Pitman, London, (1985).

[6] GOLDSTEIN R.V., Personnal communication, april 1990.

[7] KATO T., "Perturbation theory for linear operators", Springer, Berlin (1966).

[8] KONDRATIEV V.A., "Boundary value problems for elliptic equations in domains with conical or angular points", Trudy Moskovs. Mat. Obs., 16 p. 209- 292 (1967) (= Transac. Moscow Math. Soc. 16 p. 227-313).

[9] KONDRATIEV V.A. and OLEINIK O.A., "Boundary value problems for partial differential equations in non-smooth domains", Uspekhi Mat. Nauk, 38, 2, p. 3-76 (= Russian Math. Surveys, 38, 2, p. 1-86, 1983).

[10] LEBLOND J.B., "Crack paths in plane situations. I General form of the expansions of the stress intensity factors". Intern. Journ. Solids Structures, 25, p. 1311-1325 (1989).

[11] LEGUILLON D., "Calcul du taux de restitution d'énergie au voisinage d'une singularité", Compt. Rend. Acad. Sc., Paris, sér. II, 309, p. 945-950 (1989).

[12] LEGUILLON D. et SANCHEZ-PALENCIA E., "*Computation of singular solutions in elliptic problems and elasticity*", Paris-New York, Masson-John Wiley, (1987).

[13] LEGUILLON D. et SANCHEZ-PALENCIA E., "Problems in singularity theory and related topics", in Contemporary Problems in Mathematical Physics (in memorian of I.N. VEKUA), Acad. Sci. Georgia (1987), p. 64-77.

[14] LEGUILLON D. et SANCHEZ-PALENCIA E., "Solutions locales en élasticité. Effets de concentration de contraintes". To appear in "Calcul des structures et intelligence articielle, vol. 4, ed. J.M. Fouet, P. Ladevèze et R. Ohayon, ed. Pluralis, Paris (1990).

[15] MAZ'YA V.G. and NAZAROV S.A., "The asymptotic behavior of energy integrals under small perturbations of the boundary near corner points and conical points", Trud. Moskov. Matem. Obsch., 50 (1987), (= Trans. Moscow Math. Soc. 1988, p. 77-127).

[16] NEEDLEMAN A., "Computational micromechanics" in Theoretical and applied mechanics, IUTAM Symposium 1989, North-Holland, p. 217-240.

[17] NGUYEN Q.S., "Méthodes énergétiques en mécanique de la rupture", Journ. Mécanique, 19, p. 363-386 (1980).

[18] RICE J.R., "Elastic fracture mechanics concepts for interfacial cracks", Journ. Appl. Mech., 55, p. 98-103 (1988).

[19] SANCHEZ-HUBERT J. and SANCHEZ-PALENCIA E., "*Vibration and coupling of continuous systems. Asymptotic theory*", Springer, Berlin (1979).

Laboratoire de Modélisation en Mécanique
associé au CNRS
Université Pierre et Marie Curie
4 place Jussieu
75252 PARIS CEDEX 05.

Propagation of Branching Conormal Asymptotics for Hyperbolic Cauchy Problems

Michael Lorenz
Technical University Chemnitz
Bert-Wolfgang Schulze
Karl-Weierstraß-Institute Berlin

Contents

1 Introduction

This exposition deals with the conormal asymptotics transversal to the jumps of solutions of the Cauchy problem for strongly hyperbolic equations. The monograph [3] of Courant, Hilbert contains classical statements for second order equations: If the Cauchy data have a jump the solution of the Cauchy problem also jumps in the normal direction to the characteristic manifold. Many boundary value problems lead to solutions with asymptotics (cf. e.g. Schulze [13]). For elliptic problems the discrete asymptotics in terms of the distance y to the vertex of a cone have the form

$$u(y) \sim \sum_{j=0}^{\infty} \sum_{k=0}^{m_j} \zeta_{jk}\, y^{-p_j} \ln^k y \quad \text{as } y \to 0 \tag{1}$$

(cf. [7], [14]), where $p_j \in \mathbb{C}$ is a sequence with $Re\ p_j \to -\infty$ as $j \to \infty$, $m_j \in \mathbb{N} = \{0, 1, 2, \ldots\}$. (The other parameters such as angle or tangent variables are dropped.) Along an edge we have to expect variable discrete asymptotics, where the p_j, m_j and ζ_{jk} depend on the parameter x of the edge. In general the functions p_j, m_j may have jumps. The branching conormal asymptotics was introduced by Rempel and Schulze [11] $K \subset\subset \Omega$, [15], [16]. Instead of $y^{-p_j} \ln^k y$ in (1) there are used analytic functionals carried by single points, of finite orders, applied to the holomorphic function y^{-z}. For the case of branching asymptotics along an edge we have smooth functions of analytic functionals with a special pattern of compact sets as carriers, where for each parameter x of the edge the asymptotics are discrete.

We investigate the Cauchy problem for the strongly hyperbolic operator A of order m

$$Au = f\ ,\ 0 < t < T \tag{2}$$

$$D_t^j u|_{t=0} = g_j \quad j = 0, \dots, m-1. \tag{3}$$

The functions g_j are assumed to have branching, constant or variable smooth discrete asymptotics transversal to closed compact smooth manifolds Γ_j of codimension 1.

The phenomenon of propagation of singularities is well understood in terms of wave front sets for large classes of operators (cf. Hörmander [5], and many other authors). The discontinuities of (1), (2) propagate along $\Gamma_j(t)$. The main result of our paper is that the solution of the Cauchy problem also has branching or constant discrete asymptotics transversal to $\Gamma_j(t)$ for times t, $0 < t < T$, where T is chosen so small that caustics do not arise. The variable smooth discrete asymptotics remain preserved only under some additional (transmission-) condition.

As a byproduct we obtain the simpler result that the branching and constant discrete asymptotics are preserved (modulo shifts of exponents and differentiations) under the action of pseudodifferential operators. Moreover the asymptotics are invariant under applying a diffeomorphism. It is then remarkable that the result in general, i.e. the propagation of asymptotics may be reduced to compositions of pseudodifferential actions and diffeomorphisms. The adequate transformation of the variable discrete asymptotics under pseudodifferential operators itself also is not evident, since at a first glance one might expect that the asymptotics are smeared over the tangential space. For the constant asymptotics, i.e. when the exponents do not depend on the tangent variable, the problem of evaluating the asymptotics under the action of a pseudodifferential operator is equivalent to the study of the transmission property (or violated transmission property) with respect to a surface, cf. [12]. Our result on the propagation of asymptotics for Fourier integral operators extend a result of Hirschowitz-Piriou [4] on the transmission property which states the propagation of the Taylor asymptotics with other methods. The approach in our paper als appplies to hyperbolic operators with multiple characteristics in the sense of Chazarain [2].

Note that the conormal asymptotics imply the conormality in the sense of the regularity of derivatives along totally characteristic vector fields with respect to the jumps. The asymptotics, of course, are a stronger property. The conormality of solutuions of non-linear hyperbolic equations along jumps is rather typical and was studied by many authors, in particular by Bony [1], Rauch, Reed [9], [10], Metivier [8].

2 Spaces of functions with discrete and continuous conormal asymptotics

First remind of some definitions and elementary results from Jeanquartier [6], Rempel, Schulze [12].

Let $\mathbb{R}_+ = \{y \in \mathbb{R}, y > 0\}$. The Mellin transform of $u \in C_0^\infty(\mathbb{R}_+)$ is defined by

$$\tilde{u}(z) = Mu(z) := \int_0^\infty u(y) y^{z-1} dy.$$

There exists an extension to functions $u \in L^2(\mathbb{R}_+)$, then the Mellin transform is an isomorphism between $L^2(\mathbb{R}_+)$ and $L^2(Re\ z = 1/2)$. The following conditions are equivalent:

(i) $\tilde{u}(z)$ is holomorphic in $Re\ z > 1/2$, the function $\tilde{u}_{Re\ z}(Im\ z) : Im\ z \to \tilde{u}(z)$ belongs to $L^2(\mathbb{R}_{Im\ z})$ for all $Re\ z > 1/2$, and there are constants $a, c > 0$ such that

$$||\tilde{u}_{Re\ z}||_{L^2(\mathbb{R})} \leq c a^{(Re\ z - 1/2)} \quad Re\ z > 1/2.$$

(ii) $\tilde{u}(z)$ is the Mellin transform of some function $u \in L^2(\mathbb{R}_+)$, $supp\ u \subset [0, a]$.

If (i) is satisfied $\tilde{u}_{Re\ z}$ tends to some limit $\tilde{u}_{1/2}$ in $L^2(\mathbb{R})$ as $Re\ z \to 1/2$ and then

$$||\tilde{u}_{1/2}||_{L^2(\mathbb{R})} = \sqrt{2\pi}||u||_{L^2(\mathbb{R}_+)}.$$

Now we define the space of functions with discrete conormal asymptotics.

Let $P = \{(p_j, m_j); j \in \mathbb{N}\}$ be a sequence of pairs $(p_j, m_j) \in \mathbb{C} \times \mathbb{N}$ with $Re\ p_j \to -\infty$ for $j \to \infty$, $Re\ p_{j+1} \le Re\ p_j < 1/2$, $j \in \mathbb{N}$.

Denote by $\mathcal{P}$ the set of all such P. We call $\omega \in C_0^\infty(\overline{\mathbb{R}}_+)$ a cut-off function if $\omega(y)$ is identically equal 1 in a neighbourhood of the origin.

Definition 2.1 *Let $\delta \ge 0$ and $N = N(\delta, P) \in \mathbb{N}$ be the smallest number for which $Re\ p_j < 1/2 - \delta$ for all $j > N(\delta, P)$. By $S_P(\mathbb{R}_+)$ we denote the subspace of functions $u \in C^\infty(\mathbb{R}_+)$ with the following properties. For arbitrary $\delta, \eta \in \overline{\mathbb{R}}_+$, $l \in \mathbb{N}$ there exists a sequence $\{\zeta_{jk} = \zeta_{jk}(u); 0 \le k \le m_j, j \in \mathbb{N}\}$ such that for $N = N(\delta, P)$*

$$\pi_P^\delta u(y) := u(y) - \omega(y) \sum_{j=0}^{N} \sum_{k=0}^{m_j} \zeta_{jk} y^{-p_j} \ln^k y$$

satisfies the condition

$$y^{\eta-\delta}(y\frac{d}{dy})^l \pi_P^\delta u \in L^2(\mathbb{R}_+).$$

S_P is a nuclear Fréchet space.

Example 2.2 *A function $u \in S(\mathbb{R})|_{\mathbb{R}_+}$ has Taylor asymptotics at $y = 0$, this means $u \in S_{P_0}(\mathbb{R}_+)$, where $P_0 := \{(-j, 0), j \in \mathbb{N}\}$.*

Example 2.3 *There are oscillating functions, which have discrete asymptotics:*

$$\omega(y)\cos(\ln y) = 1/2(y^i + y^{-i})\omega(y) \in S_P(\mathbb{R}_+),$$

where $P = \{(\pm i, 0)\}$.

Now we describe the image of S_P under Mellin transform. $\chi \in C^\infty(\mathbb{C})$ is called a P excision function if $|\chi(z)|$ is bounded in the complex plane and for open neighbourhoods U, V of $\bigcup_{j=0}^\infty \{p_j\}$ with $\overline{U} \subset V$ we have $\chi(z) = 1$ for $z \in \mathbb{C} \setminus V$ and $\chi(z) = 0$ for $z \in U$. Using the fact that the Mellin transform of $\omega(y)y^{-p}\ln^k y$, $P \in \mathbb{C}$, $Re\ p < 1/2$, $k \in \mathbb{N}$ is a meromorphic function with exactly one pole at $z = p$ of order $k + 1$ we obtain

Proposition 2.4 *$u \in L^2(\mathbb{R}_+)$ belongs to S_P if and only if $\tilde{u}(z)$ is meromorphic with poles at $z = p_j$ of multiplicities $m_j + 1$ and for any P excision function χ*

$$\int_{Re\ z=c} |z^l \tilde{u}(z)\chi(z)|^2 |dz| < \infty \qquad l \in \mathbb{N},$$

uniformly for $\delta \le c \le \delta'$ for all $-\infty < \delta < \delta' \le 1/2$.

Now we interprete the discrete conormal asymptotics in terms of analytic functionals. By $\mathcal{A}(\mathbb{C})$ we denote the space of holomorphic functions in $\mathbb{C}$, equipped with the topology of uniform convergence on compact sets. The topological dual $\mathcal{A}'(\mathbb{C})$ of $\mathcal{A}(\mathbb{C})$ is called the space of analytic functionals on $\mathbb{C}$. Let $K \subset \mathbb{C}$ be a compact set. Then $\lambda \in \mathcal{A}'(\mathbb{C})$ is said to be carried by K if for every open neighbourhood U of K there is a constant C_U such that

$$|<\lambda, h>| \le C_U \sup_{w \in U} |h(w)| \qquad h \in \mathcal{A}(\mathbb{C}).$$

By $\mathcal{A}'(K)$ we denote the subspace of all $\lambda \in \mathcal{A}'(\mathbb{C})$ carried by K. For fixed $P \in \mathbb{C}$ the functional $\frac{d^k}{dw^k}\delta_{w-p}$ defined by

$$\left\langle \frac{d^k}{dw^k}\delta_{w-p}, h \right\rangle := (-1)^k(\frac{d^k}{dw^k}h)(p)$$

is an element in $\mathcal{A}'(\mathbb{C})$ carried by p. We may represent the singular functions in the definition of discrete asymptotics as the value of an analytic functional on the special holomorphic function y^{-w} :

$$y^{-p}\ln^k y = \left\langle (-1)^k\frac{d^k}{dw^k}\delta_{w-p}, y^{-w} \right\rangle = \frac{k!}{2\pi i}\int_C \frac{y^{-w}}{(w-p)^{k-1}}dw,$$

C being a curve clockwise surrounding p.

This formula may be used to extend the notion of asymptotics by inserting more general analytic functionals, carried by a countable number of disjoint compact sets. By B_d we denote the set of all closed subsets $B = \bigcup_{j\in\mathbb{N}} B_j$ with the following properties: There is a sequence $d = \{\delta_j\}_{j\in\mathbb{N}}$, $\delta_0 = 0$, $\delta_j < \delta_{j+1}$, $\delta_j \to \infty$ as $j \to \infty$ such that

$$\sup_{z\in B_{j+1}} Re\ z < 1/2 - \delta_{j+1} < \inf_{z\in B_j} Re\ z < 1/2, \quad j \in \mathbb{N}.$$

Furthermore, we assume that $B = B^c$, where B^c denotes the complement of the union of the unbounded connected components of $\mathbb{C} \setminus \overline{B}$.

By $C_{\delta,\varepsilon}$ we denote a piecewise smooth curve surrounding the set $U_{\delta,\varepsilon}$ consisting of a finite number of horizontals and verticals which is contained in the set $\{z \in \mathbb{C} \setminus B; dist(z, B) \leq \frac{1}{1+\varepsilon}\} \cap \{z; Re\ z > 1/2 - \delta\}$.

Definition 2.5 *By $S_B(\mathbb{R}_+)$ we denote the subspace of all $u \in C^\infty(\mathbb{R}_+)$ such that for every $\delta \in d$ there exists some $\zeta_\delta \in \mathcal{A}(\mathbb{C})$ carried by $B_\delta := B \cap \{Re\ z > 1/2 - \delta\}$ and*

$$y^{\eta-\delta}(y\frac{d}{dy})^l(u - \omega(y) < \zeta_\delta, y^{-w} >) \in L^2(\mathbb{R}_+)\ , \quad (\eta, \delta, l) \in \overline{\mathbb{R}}_+ \times d \times \mathbb{N}.$$

We define on S_B a system of norms

$$\pi_\gamma(u) := ||u||_{L^2(\mathbb{R}_+)} + ||y^{\eta-\delta}(y\frac{d}{dy})^l(u - \omega(y) < \zeta_\delta, y^{-w} >)||_{L^2(\mathbb{R}_+)} + || < \zeta_\delta, M(\omega(y)y^{-w}) > ||_{L^2(C_{\delta,\varepsilon})}$$

where $\gamma = (\eta, \delta, \varepsilon, l) \in \mathbb{R}_+ \times d \times \mathbb{R}_+ \times \mathbb{N}$.

Denote by $S_B^{(\delta)}(\mathbb{R}_+)$ the closure of $S_B(\mathbb{R}_+)$ with respect to the norms π_γ, $\gamma = (\eta, \delta, \varepsilon, l) \in \mathbb{R}_+ \times d \times \mathbb{R}_+ \times \mathbb{N}$.

Example 2.6 *To guarantee the L^2 decrease at infinity we multiply the function $e^{1/z}$ by Mv, the Mellin transform of a $v \in C_0^\infty(\mathbb{R}_+)$ and see that $M^{-1}(e^{1/z}Mv)$ has not discrete asymptotics but belongs to $S_B(\mathbb{R}_+)$, $B = \{0\}$.*

There is a well defined linear map $\zeta_\delta : S_B(\mathbb{R}_+) \to \mathcal{A}'(B_\delta)$. We may represent the analytic functional $\zeta_\delta(u)$ by the values of the analytic extension of the Mellin transform of u. We have

$$< \zeta_\delta(u), h > = \frac{1}{2\pi i}\int_{C_{\delta,\varepsilon}} u(z)h(z)dz\ , \quad h \in \mathcal{A}(V_{\delta,\varepsilon})$$

for some open neighbourhood $V_{\delta,\varepsilon} \supset \overline{U}_{\delta,\varepsilon}$, where $\varepsilon \in \mathbb{R}_+$ is sufficiently large. This equation can easily be verified by using the fact that $M(y^{-w}\chi(y)) = (z - w)^{-1}$, χ being the characteristic function of the interval (1).

The solutions of the Cauchy problem will be represented as a finite sum of functions, which belongs to S_{B^k}, $k = 1,\ldots,N$; $B^k = \sum_{j=1}^{\infty} B_j^k \in \mathcal{B}_{d_k}$; the discrete sets d_k may be different of each other. For instance, the set $\bigcup_{k=1}^{N} B^k$ may be in imaginary direction a bounded neighbourhood of the negative real axis.

Definition 2.7 *Denote by $\mathcal{B}$ the set of all $B = \sum_{k=1}^{N} B^k := (\bigcup_{k=1}^{N} B^k)^c$, $B^k \in \mathcal{B}_{d_k}$. Using the systems of norms π_{γ_k} on S_{B^k} we introduce the system on S_B*

$$\pi_\gamma(u) = \inf \sum_{k=1}^{N} \pi_{\gamma_k}(u_k),\ u_k \in S_{B^k}\ ,\quad u = \sum_{k=1}^{N} u_k.$$

It can be shown that S_B is a nuclear Frechet space and that S_B does not depend on the representation of the decomposition of B in $\sum B^k$.

Example 2.8 *The function $\omega(y)(\ln y)^{-1}$ belongs to $S_{\mathbb{R}_-}(\mathbb{R}_+)$, $\mathbb{R}_- = (-\infty, 0)$ which follows from*

$$\frac{d}{dz} M\omega(y)(\ln y)^{-1} = M\omega(y) = z^{-1}M(-y\frac{d}{dy})\omega(y) = z^{-1}h(z),$$

where $h(z) = \sum_{j=0}^{\infty} h_j z^j \in \mathcal{A}(\mathbb{C})$. $M\omega(y)(ln y)^{-1} = h_0 \int_1^z z^{-1}dz + k(z) = h_0 ln z + k(z)$ for some $k(z) \in \mathcal{A}(\mathbb{C})$.

3 Spaces of functions with constant discrete, smooth discrete and branching asymptotics

Let $\Omega \subset \mathbb{R}^{n-1}$ be an open set. Then the topological vector spaces $C^\infty(\Omega, S_B(\mathbb{R}_+))$ and $C_0^\infty(\Omega, S_B(\mathbb{R}_+))$ are defined in a standard way. In more general cases the set B also depends on the variable $x \in \Omega$. We assume this dependence to be continuous in the following sense. Set

$$\begin{aligned} \Pi P(x) &:= \left\{ \bigcup_{j=0}^{\infty} \{p_j\}, (p_j, m_j) \in P, j \in \mathbb{N} \right\} \\ \Pi_\delta P(x) &:= \Pi P(x) \cap \{Re\ z > 1/2 - \delta\}. \end{aligned}$$

An element $U \in \mathcal{B}$ is called a closed neighbourhood of $\Pi_\delta P(x_0)$ if $\Pi_\delta P(x_0) \subset int\ U$. Then $P : \omega \to \mathcal{P}$ is called continuous at $x_0 \in \Omega$ if for each $\delta > 0$ and each sufficiently small closed neighbourhood U of $\Pi_\delta P(x_0)$ there is a closed set $K \subset \Omega$, $x_0 \in int\ K$ such that $\Pi_\delta P(x_0) \subset int\ U$ for all $x \in K$. Denote by $N(x, \delta, P)$, $x \in \Omega$, the number of points of $\Pi_\delta P(x_0)$ counted with their multiplicities. Set $N(K, \delta, P) := \sup_{x \in K} N(x, \delta, P)$ for $K \subset\subset \Omega$. Then we define $\mathcal{C}(\Omega, \mathcal{P})$ as the class of all mappings $P : \Omega \to \mathcal{P}$, which are continuous at each $x \in \Omega$ and satisfy $N(K, \delta P) < \infty$ for each $K \subset\subset \Omega$, $\delta \in \mathbb{R}_+$. For each fixed $x \in \Omega$ we choose a neighbourhood $U(x) \in \mathcal{B}$ of $\Pi_\delta P(x)$. The corresponding compact sets $K(x)$, for which $\Pi_\delta P(\tilde{x}) \subset\subset int\ U(x)$ for all $\tilde{x} \in K(x)$, cover Ω. It is sufficient to consider a locally finite covering $\{K_j\}_{j \in J}$ of Ω. Denote by $\mathcal{K}_\delta(P)$ the system of all sequences $\{K_j, U_j\}$ with these properties, associated with $P \in \mathcal{C}(\Omega, \mathcal{P})$. Denote by $\varrho_{\{x_0\}}$ and ϱ_{K_j} the operator of restriction to $\{x_0\}$ and $int\ K_j$, respectively.

Definition 3.9 *Let $P \in \mathcal{C}(\Omega, \mathcal{P})$ be given and denote by $C^\infty(\Omega, S_P)$ the space of all functions $u(x,y) \in C^\infty(\Omega \times \mathbb{R}_+)$ with the properties*

(i) $u(x_0, .) \in S_{P(x_0)}(\mathbb{R}_+)$ for each $x_0 \in \Omega$

(ii) for all $\delta > 0$, $\{K_j, U_j\} \in \mathcal{K}_\delta(P)$, $l \in \mathbb{N}$ we have for every j

$$\varrho_{K_j} u(x,y) \in C^\infty(int\ K_j, S^{(\delta)}_{U_j}(\mathbb{R}_+)).$$

$C^\infty(\Omega, S_P)$ is a nuclear Fréchet space in the projective limit topology with respect to the mappings $\varrho_{\{x\}}$ and ϱ_{K_j}. A function $u \in C^\infty(\Omega, S_P)$ has "variable discrete conormal asymptotics".

Example 3.10 *We consider the simplest case where the variable asymptotics arises.*

Let $a(x)$, $b(x) \in C^\infty(\Omega)$ be bounded functions. There exists a compact set $K \subset \mathbb{C}$ such that $(a(x)+z)(b(x)+z) \neq 0$ $\forall x \in \Omega$, $z \notin K$. C is a positively oriented connected curve surrounding K, then we define

$$u(x,y) = (2\pi i)^{-1} \int_C \frac{y^{-z}\omega(y)}{(a(x)+z)(b(x)+z)} dz.$$

This function has asymptotics of the form

$$\zeta_1(x)y^{a(x)} + \zeta_2(x)y^{b(x)} \text{ for } x \notin N := \{x \in \Omega : a(x) = b(x)\}$$
$$\zeta_3(x)y^{a(x)} + \zeta_4(x)y^{a(x)} \text{ for } x \in N.$$

However, in general it is not possible to represent the variable asymptotics in terms of C^∞-families of rational functions. The adequate represantation may be given in terms of functions $g(x,z) \in C^\infty(\Omega, \mathcal{A}(\mathbb{C} \setminus K))$, where $g(x,z)$ extends to a meromorphic function for every fixed x, and K is a compact set which is so large that it contains the poles for all x. If C is a curve in $\mathbb{C} \setminus K$ surrounding K clockwise then

$$u(x,y) = \frac{1}{2\pi i} \int_C g(x,z) y^{-z} dz \tag{4}$$

represents the corresponding variable asymptotics. Since an infinite asymptotic expansion (i.e. for real part of the poles tending to $-\infty$) may always be understood as a finite expansion modulo a remainder which is flat of some order, tending to infinity, this picture applies also for the asymptotics expansion. Below we use the representation (4) whenever a finite part of the asymptotics is to be expressed in the Mellin image.

Example 3.10 shows that the differentiation with respect to x and passing from a function to its asymptotics do not commute, because the asymptotics do not depend continuously on x.

Theorem 3.11 *[11], [15]. For every $P \in \mathcal{C}(\Omega, \mathcal{P})$ and any $\alpha \in \mathbb{N}^{n-1}$ there is a $Q_\alpha = Q_\alpha(P) \in \mathcal{C}(\Omega, \mathcal{P})$ such that the operator $D^\alpha_x : C^\infty(\Omega, S_P) \to C^\infty(\Omega, S_{Q_\alpha})$ is continuous and*

$$\begin{aligned} \Pi P(x) &\supset \Pi Q_\alpha(x), \quad x \in \Omega \\ N(x,\delta,Q_\alpha) &\leq 2^{|\alpha|} N(x,\delta,P), \quad x \in \Omega,\ \delta \in \mathbb{R}_+. \end{aligned}$$

Now we transfer the concept to the following situation. Γ is a compact C^∞-manifold of codimension 1, which is embedded in $\mathbb{R}^n$, u is a function, which has variable asymptotics in local coordinates in each chart L in normal direction to Γ. This means u satisfies Definition 3.9, where $\tilde{x} \in \Gamma \cap L = \Gamma_L$ and the $\tilde{y}$-coordinate represents one normal direction.

In an analogous manner there are defined the spaces $\mathcal{C}(\Gamma_L, \mathcal{P})$ and $C^\infty(\Gamma_L, S_P)$. Let us choose a diffeomorphism

$$\tilde{x} = \Psi(x,y), \quad \tilde{y} = \Phi(x,y), \tag{5}$$

which transforms Γ_L in a subset of $\mathbb{R}^{n-1}$. Define $(\Omega, 0) = \Psi^{-1}(\Gamma_L, 0)$. Now we show the invariance (modulo "shadow") of variable asymptotics under diffeomorphisms. Define the translation of $P = \{(p_j, m_j), j \in \mathbb{N}\}$ by

$$T^k P := \{(p_j + k, m_j), j \in \mathbb{N}\}.$$

For future references we call the $T^{-k}P$ the shadows of P (under the translation to the left).

Theorem 3.12 *If u has variable asymptotics, i.e.*

$$u \in C^\infty(\Gamma_L, S_P)$$

then

$$u(\Psi(x,y), \Phi(x,y)) \in C^\infty(\Omega, S_Q), \quad Q = \bigcup_{k=0}^{\infty} T^{-k}P.$$

Proof. *We have to consider integrals of the type*

$$I = \int_C g(\tilde{x}, z)\tilde{y}^{-z}dz = \int_C g(\Psi(x,y), z)(\Phi(x,y)/y)^{-z}y^{-z}dz.$$

As $\Phi(x,0) = 0$ we have

$$\begin{aligned} I &= \int_C g(\Psi(x,0), z)\Phi'_y(x,0)^{-z}y^{-z}dz \\ &+ \sum_{k=1}^{\infty} \frac{1}{k!}y^k \int_C \left(\frac{\partial}{\partial y}\right)^k g(\Psi(x,y), z)(\Phi(x,y)/y)^{-z}|_{y=0}y^{-z}dz + R, \end{aligned}$$

where R does not generate asymptotics.

Since $\Phi'_y(x,0)$ is a non-vanishing smooth function the first integral belongs to $C^\infty(\Omega, S_P)$. Theorem 3.11 and the fact that $\Phi(x,y)/y$ is smooth up to the boundary imply that each item of the second term belongs to $C^\infty(\Omega, S_{T^{-k}P})$.

Now we define special classes of functions with variable asymptotics.

Remark 3.13 *Let us give an example, where the shadow arises. The function $u(\tilde{x}, \tilde{y}) = \tilde{y}$ belongs to $C^\infty(\Gamma, S_P)$, $P = \{(-1,0)\}$. Using the diffeomorphism $\tilde{x} = x$, $\tilde{y} = \sin y$ for $0 < y < \pi/2$ we see that $u = \sin y \in C^\infty(\Omega, S_Q)$, $Q = \{(k,0), k \in -2\mathbb{N} - 1\}$.*

Definition 3.14 *If $(p_j(x), m_j(x))$ does not depend on x for all $j \in \mathbb{N}$ the function $u \in C^\infty(\Gamma_L, S_P)$ has "constant conormal asymptotics".*

Denote this class by $C_c^\infty(\Gamma_L, S_P)$. If for each x the function u can be represented by a finite sum $u = \sum_{l=1}^N u_l$ in a neighbourhood of x and $(p_j(x), m_j(x))$ depends smoothly on x for each $j \in \mathbb{N}$ with respect to each u_l, the function $u \in C^\infty(\Gamma_L, S_P)$ has "variable smooth discrete conormal asymptotics", we write $u \in C_v^\infty(\Gamma_L, S_P)$. A function u has "branching conormal asymptotics" if it belongs to $C_b^\infty(\Gamma_L, S_P) := C^\infty(\Gamma_L, S_P) \setminus C_v^\infty(\Gamma_L, S_P)$.

The Theorems 3.11 and 3.12 extend to the general case.

Theorem 3.15 *For every $P \in \mathcal{C}(\Gamma_L, \mathcal{P})$ and any $\alpha \in \mathbb{N}^{n-1}$ there is a $Q_\alpha = Q_\alpha(P) \in \mathcal{C}(\Gamma_L, \mathcal{P})$ such that the operators*

$$\begin{aligned} D_x^\alpha &: C_c^\infty(\Gamma_L, S_P) \longrightarrow C_c^\infty(\Gamma_L, S_{Q_\alpha}) \\ D_x^\alpha &: C_v^\infty(\Gamma_L, S_P) \longrightarrow C_v^\infty(\Gamma_L, S_{Q_\alpha}) \\ D_x^\alpha &: C_b^\infty(\Gamma_L, S_P) \longrightarrow C_b^\infty(\Gamma_L, S_{Q_\alpha}) \end{aligned}$$

are continuous and $\Pi P(x) \supset \Pi Q_\alpha(x)$, $x \in \Gamma_L$, $N(x, \delta, Q(P)) \le 2^{|\alpha|}N(x, \delta, P)$, $x \in \Gamma_L$, $\delta \in \mathbb{R}_+$.

Theorem 3.16 *The pull-back with respect to the diffeomorphism (5) preserves the classes $C_c^\infty(\Gamma_L, S_P)$, $C_v^\infty(\Gamma_L, S_P)$, $C_b^\infty(\Gamma_L, S_P)$ (modulo shadow).*

4 The Cauchy problem for strictly hyperbolic equations

We shall study the Cauchy problem in $\mathbb{R}^n \times [0,T]$ for the strictly hyperbolic operator A of order m

$$Au = f \tag{6}$$

$$D_t^j u|_{t=0} = g_j, \quad j = 0,\ldots,m-1. \tag{7}$$

Γ_j are closed compact C^∞-manifolds of codimension 1, embedded in $\mathbb{R}^n$. We assume f to be smooth, the functions g_j have constant variable smooth or branching asymptotics in the normal direction to Γ_j; outside Γ_j the functions g_j are smooth. For simplicity assume that the Cauchy problem has a unique solution (cf. [5]). The diffeomorphism (5) transforms the problem (6), (7) locally to

$$\begin{aligned} Au &= \sum_{j=0}^{m} A_j(t,x,y,D_x,D_y)D_t^j u = f \qquad (8)\\ &\quad 0<t<T,\ x\in\mathbb{R}^{n-1},\ y\in(-y_0,y_0),\\ &\quad \text{for some small } y_o\in\mathbb{R}\\ &\quad D_j^t u|_{t=0} = g_j,\ \ j=0,\ldots,m-1. \qquad (9)\end{aligned}$$

We may assume $A_m = 1$ and $A_j(t,x,y,\xi,\eta) \in S_{cl}^{m-j}$ if $j < m$. A_j has the principal symbol $a_j(t,x,y,\xi,\eta)$, which is positively homogeneous in ξ,η of degree $m-j$. The zeros $\tau = \lambda_j$, $j = 0,\ldots,m$ of the principal symbol of A

$$a(t,x,y,\xi,\eta) = \sum_{j=0}^{m} a_j(t,x,y,\xi,\eta)\tau^j = 0$$

are real and uniformly simple.

We may factorize microlocally the operator A such that for all j

$$A - (D_t - \Lambda_j(t,x,y,D_x,D_y))Q_j$$

is smoothing, Λ_j has the principal symbol λ_j and Q_j is an elliptic operator.

The solution u of the Cauchy problem (8), (9) may be expressed in the following form

$$u = \sum_{r=0}^{m-1} E_r g_r \tag{10}$$

where the operator E_r is a solution of

$$\begin{aligned} AE_r &= 0\\ \frac{\partial^k}{\partial t^k}E_r|_{t=0} &= \delta_{rk}I,\ k=0,\ldots,m-1;\\ &\quad \delta_{rk} \textit{ the Kronecker symbol.}\end{aligned}$$

E_r is a finite sum over j of Fourier integral operators with kernels

$$\int_{\mathbb{R}^n} e^{i[\Psi_j(t,x,y,\xi,\eta)-v\xi-w\eta]}k^{(j,r)}(t,x,y,\xi,\eta)d\xi d\eta \tag{11}$$

where Ψ_j is the solution of the Hamilton-Jacobi equation

$$\begin{aligned}\frac{\partial}{\partial t}\Psi_j(t,x,y,\xi,\eta) &= \lambda_j(t,x,y,\frac{\partial}{\partial x}\Psi_j,\frac{\partial}{\partial y}\Psi_j)\\ \Psi|_{t=0} &= x\xi + y\eta\end{aligned}$$

and $k^{(j)}$ can be obtained as solution of the transport equations [17]. The above factorization and equation (10) are satisfied microlocally. Therefore choose the support of the functions $K^{(j)}$ in a neighbourhood of the bicharacteristics such that the equation (10) holds modulo a smoothing operator. In order to obtain the singularities of the solution u we have to calculate the Hamilton forward flow out with respect to $\tau - \lambda - j$ starting at the wave front set of g_j. Denote by $H(y)$ the Heaviside function.

Lemma 4.17 *The wave front set of $g_j \in C^\infty(\Gamma_L, S_P)$ is contained in the conormal bundle of Γ_L.*

Proof. We have to evaluate for example

$$\xi^\alpha F(H(y)\omega(y)\varphi(x)g_j(x,y)) = \int\int e^{i(x\xi+y\eta)}H(y)\omega(y)\varphi(x)\int D_x^\alpha g(x,z)y^{-z}dzdxdy,$$

where $\varphi \in C_0^\infty(\mathbb{R}^{n-1})$, and $g(x,z)$ is associated with the asymptotics of g_j in the sense of the formula (4).

The Fourier transform of an integrable function with compact support is bounded, hence, the lemma is proved.

Remark 4.18 *The following example shows that the wave front set of similar functions (which have not variable discrete asymptotics) are not always contained in the conormal bundle.*

$$WF(H(y)/x^2+y^2)) \not\subset N^*(\mathbb{R}^{n-1}).$$

In the sequel we need the projection to the configuration space of the forward flow out starting at $\{(x,0,0,\eta),\ x \in \mathbb{R}^{n-1}, \eta \neq 0\}$, which is the conormal bundle to $\mathbb{R}^{n-1}$.

$x_j^+(t,x,\eta)$, $y_j^+(t,x,\eta)$ (and $x_j^-(t,x,\eta)$, $y_j^-(t,x,\eta)$, respectively) is the solution of the Hamilton system

$$\begin{aligned}
\dot{x}_j^+ &= -\frac{\partial}{\partial\zeta^+}\lambda_j(t,x_j^+,y_j^+,\zeta^+,\Theta^+),\ x_j^+(0,x,\eta) = x \\
\dot{y}_j^+ &= -\frac{\partial}{\partial\Theta^+}\lambda_j(t,x_j^+,y_j^+,\zeta^+,\Theta^+),\ y_j^+(0,x,\eta) = 0 \\
\dot{\zeta}_j^+ &= \frac{\partial}{\partial x_j^+}\lambda_j(t,x_j^+,y_j^+,\zeta^+,\Theta^+),\ \zeta_j^+(0,x,\eta) = 0 \\
\dot{\Theta}_j^+ &= -\frac{\partial}{\partial y_j^+}\lambda_j(t,x_j^+,y_j^+,\zeta^+,\Theta^+),\ \Theta_j^+(0,x,\eta) = \eta > 0.
\end{aligned}$$

If $\eta < 0$ the solution of the analogous Hamilton system is denoted by x_j^-, y_j^-, ζ_j^- and Θ_j^-.

The aim of this section is to show that for small time $0 < t < T$ the solution u of the Cauchy problem has the same quality of asymptotics at the manifolds generated by $x_j^\pm$ as the Cauchy data in $\mathbb{R}^{n-1}$. Therefore we have to shift the solution u by the operator

$$S_j^\pm u(t,x,y) = u(t, x_j^\pm(t,x,\pm1), y + y_j^\pm(t,x,\pm1)).$$

If a function depends furthermore on ξ and η we set

$$S_j^\pm f(t,x,y,\xi,\eta) = f(t, x_j^\pm(t,x,\pm1), y + y_j^\pm(t,x,\pm1), \xi, \eta).$$

To reduce the action of a Fourier integral operator to the composition of a diffeomorphism with the action of a pseudodifferential operator we split up the phase functions into two parts

$$\begin{aligned}
\Psi_j(t,x,y,\xi,\eta) &= \frac{\partial}{\partial\xi}\Psi_j(t,x,y,0,\pm1)\xi + \frac{\partial}{\partial\eta}\Psi_j(t,x,y,0,\pm1)\eta \\
&+ [\Psi_j(t,x,y,\xi,\eta) - \frac{\partial}{\partial\xi}\Psi_j(t,x,0,\pm1)\xi - \frac{\partial}{\partial\eta}\Psi_j(t,x,y,0,\pm1)\eta].
\end{aligned}$$

The amplitude $k^{(j,r)}$ of (11) may be represented by an asymptotic sum

$$k^{(j,r)}(t,x,y,\xi,\eta) \sim \sum_{l=0}^{\infty} k_{-l}^{(j,r)}(t,x,y,\xi,\eta),$$

where the functions $k_{-l}^{(j,r)}$ are positively homogeneous in (ξ,η) of order $-l$.

Using the Taylor formula

$$\begin{aligned} h_{-1}^{(j,r)}(t,x,y,\xi,\eta) &- \sum_{|\alpha|\leq N} \frac{1}{\alpha!}\xi^\alpha (\frac{\partial}{\partial\xi})^\alpha h_{-l}^{(j,r)}(t,x,y,0,\eta) \\ &= \sum_{|\alpha|=N+1} \frac{1}{\alpha!}\xi^\alpha (\frac{\partial}{\partial\xi})^\alpha h_{-l}^{(j,r)}(t,x,y,\delta\xi,\eta),\ 0<\delta<1, \end{aligned}$$

with

$$\begin{aligned} h_{-l}^{(j,r)} &= S_j^{\pm}(\exp i\left\{\Psi_j(t,x,y,\xi,\eta) - \frac{\partial}{\partial\xi}\Psi_j(t,x,y,0,\pm1)\xi\right. \\ &\left. - \frac{\partial}{\partial\eta}\Psi_j(t,x,y,0,\pm1)\eta\right\}) \cdot k_{-l}^{(j,r)}(t,x,y,\xi,\eta) \end{aligned} \tag{12}$$

we have to compute $(\frac{\partial}{\partial\xi})^\alpha h_{-l}^{(j,r)}$.

For instance, we get the following terms

$$S_j^{\pm}(\frac{\partial}{\partial\xi}\Psi_j(t,x,y,\delta\xi,\eta) - \frac{\partial}{\partial\xi}\Psi_j(t,x,y,0,\pm1))^\alpha h_{-l}^{(j,r)}$$

$$S_j^{\pm}(\frac{\partial}{\partial\xi})^\alpha\Psi \cdot k_{-l}^{(j,r)},\quad S_j^{\pm}(\frac{\partial}{\partial\xi})^\alpha k_{-l}^{(j,r)}.$$

We obtain the estimate

$$|\frac{\partial}{\partial\xi}\Psi_j(\dots,\delta\xi,\eta) - \frac{\partial}{\partial\xi}\Psi_j(\dots,0,\pm1)| \leq C|\xi|(1+|\xi|+|\eta|)^{-1} \leq C'(1+|\xi|)(1+|\eta|)^{-1}$$

and finally for $h^{(j,r)} = \sum_{l=0}^{\infty} h_{-l}^{(j,r)}$

$$|h^{(j,r)}(t,x,y,\xi,\eta) - \sum_{|\alpha|\leq N}\frac{1}{\alpha!}\xi^\alpha(\frac{\partial}{\partial\xi})^\alpha h_{-l}^{(j,r)}(t,x,y,\xi,\eta)|_{\xi=0} \leq C(1+|\xi|)^{2N+2}(1+|\eta|)^{-N-M-2}.$$

On the other hand we have

$$|(\frac{\partial}{\partial\xi})^\alpha h_{-l}^{(j,r)}(t,x,y,0,\eta)| \leq C(1+|\eta|)^{1-l-|\alpha|}.$$

Denote by $Op(h)$ the pseudodifferential operator with symbol h.

Lemma 4.19 *For every $m \in \mathbb{N}$ there exists an $N = N(l,m) \in \mathbb{N}$ such that*

$$[\sum_{|\alpha|=N+1}\frac{1}{\alpha!}Op(\xi^\alpha(\frac{\partial}{\partial\xi})^\alpha h_{-l}^{(j,r)}(t,x,y,\delta\xi,\eta))]\varphi(x)\omega(|y|)g_r(x,y)\ \in\ C^m(\mathbb{R}^n),$$

where the symbol $h_{-l}^{(j,r)}$ is defined by (12), $0<\delta<1$, $\varphi(x) \in C_0^\infty(\Omega)$.

Proof. First observe that $u_r := \varphi(x)\omega(|y|)g_r(x,y) \in C_0^\infty(\mathbb{R}^{n-1}, L^2(\mathbb{R}^1))$. Then it suffices to show that for every $s \in \mathbb{R}^1$ there is an $N = N(s)$ such that for $|\alpha| = N+1$ $Op(\xi^\alpha(\frac{\partial}{\partial\xi})^\alpha h_{-l}^{(j,r)}$: $C_0^\infty(\mathbb{R}^{n-1}, L^2(\mathbb{R}^1)) \to H^s(\mathbb{R}^n)$ for every $0 < t < T$. The Fourier transform of every $u_r \in C_0^\infty(\mathbb{R}^{n-1}, L^2(\mathbb{R}^1))$ satisfies an estimate of the form $(1+|\xi|)^M (Fu_r)(\xi,\eta) \in L^2(\mathbb{R}^n)$ for arbitrary $M \in \mathbb{N}$.

We may interprete $Op(\xi^\alpha(\frac{\partial}{\partial\xi})^\alpha h_{-l}^{(j,r)})$ as pseudodifferential operator with operator-valued symbol, namely

$$Op(\xi^\alpha(\frac{\partial}{\partial\xi})^\alpha h_{-l}^{(j,r)})u_r = \int e^{i(x-v)\xi} op_{(y)}(\xi^\alpha(\frac{\partial}{\partial\xi})^\alpha h_{-l}^{(j,r)})(t,x,\xi)u_r(v,y)dvd\xi$$

$op_{(y)}$ being the pseudodifferential action with respect to y depending on the parameters t,x,ξ. Then for given s and $|\alpha| \geq -l+s$ we get for fixed t

$$||D_x^\beta D_\xi^\gamma op_{(y)}(\xi^\alpha(\frac{\partial}{\partial\xi})^\alpha h_{-l}^{(j,r)})u_r(v,y)||_{H^s(\mathbb{R}_y^1)} \leq C(1+|\xi|)^{N_0}||u_r||_{L^2(\mathbb{R}_y^1)}$$

for all $x \in supp\varphi$ and all multi-indices β,γ with a constant $C = C(\beta,\gamma,\varphi)$.

This implies $Op(\xi^\alpha(\frac{\partial}{\partial\xi})^\alpha h_{-l}^{(j,r)}) : C_0^\infty(\mathbb{R}^{n-1}, L^2(\mathbb{R}^1)) \to C^\infty(\mathbb{R}^{n-1}, H^s(\mathbb{R}^1))$. Since $C^\infty(\mathbb{R}^{n-1}, H^s(\mathbb{R}^1) \subset C^m(\mathbb{R}^n)$ for s large enough, we get the assertion.

Let φ_k be a partition of unity on $\mathbb{R}^{n-1}$. Choosing a constant c such that $\omega \equiv 1$ on $[0,c]$, the shifted solution (10) may be represented by

$$\begin{aligned}
(S_j^\pm u)(t,x_j^\pm,y+y_j^\pm) &= \sum_k \sum_{r=0}^{m-1}(-1)^{\pm1}\int_0^{\pm\infty}\int_{R^{2n-1}}(\exp iS_j^\pm[(\frac{\partial}{\partial\xi}\Psi_j(t,x,y,0,\pm1)-v)\xi \\
&+ (\frac{\partial}{\partial\eta}\Psi_j(t,x,y,0,\pm1)-w)\eta])\sum_{l=0}^{\infty}h_{-l}^{(j,r)}(t,x,y,\xi,\eta) \\
&\varphi_k(v)\omega(|w|)g_r(v,w)dvdwd\xi d\eta. \qquad (13)
\end{aligned}$$

The Cauchy data g_r belong to the classes $C_c^\infty(\mathbb{R}^{n-1}, S_P)$, $C_v^\infty(\mathbb{R}^{n-1}, S_P)$ or $C_b^\infty(\mathbb{R}^{n-1}, S_P)$ in each normal direction to $\mathbb{R}^{n-1}$; therefore

$$g_r(v,w) = H(w)\sum_f\int_{C_f} g_{1,f}^{(r)}(v,z)|w|^{-z}dz \;+\; H(-w)\sum_f\int_{D_f} g_{2,f}^{(r)}(v,z)|w|^{-z}dz.$$

The non-trivial part of the asymptotics for $y \to 0$ is determined by the terms

$$\begin{aligned}
I &= \sum_{f,|\alpha|\leq N, l\leq M, s=0}\frac{1}{\alpha!}(-1)^{\pm1}\int_0^{\pm\infty}\int_{R^{2n-1}}(\exp iS_j^\pm[(\frac{\partial}{\partial\xi}\Psi_j - v)\xi \\
&+ (\frac{\partial}{\partial\eta}\Psi_j - w\eta])\cdot\xi^\alpha\cdot h_{-l,\alpha,s}^{(j,r)}\cdot|\eta|^{-l-|\alpha|+s}(H(w)\int_{C_f} g_{1,f}^{(r)}(v,z)|w|^{-z}dz \\
&+ H(-w)\int_{D_f} g_{2,f}^{(r)}(v,z)|w|^{-z}dz)\cdot dvdwd\xi d\eta \qquad (14)
\end{aligned}$$

where $\sum_{s=0,1} h_{-l,\alpha,s}^{(j,r)}|\eta|^{-l-|\alpha|+s} = (\frac{\partial}{\partial\xi})^\alpha h_{-l}^{(j,r)}|_{\xi=0}$ and $h_{-l,\alpha,s}^{(j,r)}$ are positively homogeneous functions of degree zero with respect to η.

Integrating with respect to v and ξ we have to substitute $g^{(r)}(v,z)$ by $D_v^\alpha g^{(r)}(S_j^\pm\frac{\partial}{\partial\xi})\Psi_j(t,x,y,0,\pm1),z)$.

Now we have to evaluate the integral

$$(-1)^{\pm1}\int_0^{\pm\infty}\int_{\mathbb{R}^1} e^{i(y-w)\eta}\eta^{-1}|w|^{-z}dwd\eta. \qquad (15)$$

A direct computation shows that (15) is equal to

$$(\Theta_1(l,z,\pm y)+\Theta_2(l,z,\pm y)\ln|y|)|y|^{-z+1}$$

where Θ_1 and Θ_2 are positively homogeneous functions of degree zero with respect to y, Θ_1 is an analytic function with respect to z if $-z \notin \mathbb{N}$ and $\Theta_2 \equiv 0$ if $-z \notin \mathbb{N}$. Using the Mellin representation of the pseudodifferential operator on the half axis (cf. [12]) we obtain that operators without the transmission property generate an additional *ln* term. We obtain

Theorem 4.20 *For each $i \in \mathbb{N}$ and small t, $0 < t < T$, there exists a number $N \in \mathbb{N}$ such that the solution u of the Cauchy problem (8), (9) satisfies*

$$\begin{aligned} & u(t,x,y)-\sum_{r,j=0}^{m-1}\sum_{f,|\alpha|,l<N,s=0,1}\sum_{\pm}\frac{1}{\alpha}h_{-l,\alpha,s}^{(j,r)}(t,x,y,0,\pm 1) \\ \cdot\ & [\int_{C_f}(\Theta_1(l+|\alpha|-s,z,w)+\Theta_2(l+|\alpha|-s,z,w)\ln|w|)D_v^\alpha g_{1,f}^{(j,r)}(v,z) \\ \cdot\ & |w|^{-z+l+|\alpha|-s}dz \\ +\ & \int_{D_f}(\Theta_1(l+|\alpha|-s,z,-w) \\ +\ & \Theta_2(l+|\alpha|-s,z,-w)\ln|w|)D_v^\alpha g_{2,f}^{(j,r)}(v,z)|w|^{-z+l+|\alpha|-s}dz]\ \in\ C^i, \end{aligned} \tag{16}$$

where $w = S_j^{\pm}\frac{\partial}{\partial\eta}\Psi_j(t,x,y,0,\pm1)$, $v = S_j^{\pm}\frac{\partial}{\partial\xi}\Psi_j(t,x,y,0,\pm1)$.

If every point of the set $-N$ does not belong to the carrier of the analytic functionals, which are generated by $g_{1,f}^{(j,r)}$ and $g_{2,f}^{(j,r)}$, the function Θ_2 is identically zero.

Analogously it can be shown

Theorem 4.21 *The action of a pseudodifferential operator Q of order zero with classical symbol*

$$q \sim \sum_{l=0}^{\infty} q_{-l}(t,x,y,\tau,\xi,\eta)$$

on functions

$$H(y)\int_C g_1(t,x,z)|y|^{-z}dz\ +\ H(-y)\int_D g_2(t,x,z)|y|^{-z}dz,$$

with variable discrete asymptotics, have asymptotics of the form

$$\begin{aligned} & \sum_{\alpha,\beta,\pm}\frac{1}{\alpha!\beta!}(\frac{\partial}{\partial\xi})^\alpha(\frac{\partial}{\partial\tau})^\beta q_{-l}(t,x,y,0,0,\pm1) \\ \cdot\ & [\int_C(\Theta_1(l+|\alpha|+|\beta|,z,y)+\Theta_2(l+|\alpha|+|\beta|,z,y)\ln y) \\ \cdot\ & D_x^\alpha D_t^\beta g_1(t,x,z)|y|^{-z+l+|\alpha|+|\beta|}dz \\ +\ & \int_D(\Theta_1(l+|\alpha|+|\beta|,z,-y)+\Theta_2(l+|\alpha|+|\beta|,z,-y)\ln|y|) \\ \cdot\ & D_x^\alpha D_t^\beta g_2(t,x,z)|y|^{-z+l+|\alpha|+|\beta|}dz]. \end{aligned}$$

The composition of formulas for the asymptotics (16), the substitution rule under a diffeomorphism, derivation and action of a pseudodifferential operator implies the following theorem.

Denote by $\Gamma_j^{\pm}$ the manifolds which are generated by the Hamilton forward flow out with respect to $\tau - \lambda_j$ starting at the conormal bundle of Γ_j.

Theorem 4.22 *For small t, $0 < t < T$, the solution u of the Cauchy problem (6), (7) has constant discrete or branching discrete asymptotics in each normal direction to $\Gamma_j^{\pm}(t)$ $j = 0, \ldots, m-1$ if this is the case for all Cauchy data g_j in each normal direction to Γ_j. If every point of the set $-\mathbb{N}$ does not belong to the carrier of the analytic functionals generated by g_j the variable smooth discrete asymptotics are preserved.*

The solution u has branching discrete asymptotics in the other case.

The explicit form of the asymptotics follows easily by composing the asymptotics in the above calculations.

References

[1] J.-M. Bony. Second microlocization and propagation of singularities for semilinear hyperbolic equations, Preprint Univ. de Paris Sud, Orsay, 1985, 11.

[2] J. Charazain. Opérateurs hyperboliques a caracteristiques de multiplicité constante, Ann. Inst. Fourier, Grenoble 24, 1 (1974) 173 -202.

[3] R. Courant, D. Hilbert. Methoden der Mathematischen Physik II, Springer-Verlag, Berlin-Heidelberg-New-York 1968.

[4] A. Hirschowitz, A. Piriou. Properties de transmission pour les distributions integrales de Fourier, Comm. P.D.E. 4, 2 (1979) 113-217.

[5] L. Hörmander. The Analysis of Linear Partial Differential Operators III, Springer-Verlag, Berlin-Heidelberg-New-York-Tokyo, 1985.

[6] P. Jeanquartier. Transformation de Mellin et developpements asymptotiques. L'Enseignement mathématique 25, 3-4 (1979) 285-308.

[7] V.A. Kondrat'ev. Boundary value problems for elliptic equations in domains with conical points, Trudy Mosk. Mat. Obsc. 16 (1967), 209-292.

[8] G. Metivier. The Cauchy problem for semilinear hyperbolic systems with with discontinuous date Preprint.

[9] J. Rauch, M. Reed. Jump discontinouities of semilinear strictly hyperbolic systems in two variables: creation and propagation, Comm. Math. Phys. 81 (1981) 203-227.

[10] J. Rauch, M. Reed. Discontinuous progressing waves for semilinear systems, Comm. P.D.E. 10 (1985), 1033-1075

[11] S. Rempel, B.-W Schulze. Branching asymptotics for elliptic operators on manifolds with edges, Partial. Diff. Equ. 1984, Banach Center Publ. Vol. 19 pp. 231-236, Warzaw 1987.

[12] S. Rempel, B.-W Schulze. Pseudo-differential and Mellin operators in spaces with conormal singularity, Berlin, Akad. d. Wiss. d. DDR, Inst. f. Math. 1984, Report R-MATH-01/84.

[13] B.-W. Schulze. Pseudo-Differential Operators on Manifolds with singularities, North' Holland, Amsterdam, 1991.

[14] B.-W. Schulze. Ellipticity and continuous conormal asymptotics on manifolds with conical singularities, Preprint Univ. de Paris Sud, Orsay, 1986, TO6, Math. Nachr. 136 (1986) 7-57.

[15] B.-W. Schulze. Regularity with continuous and branching asymptotics on manifolds with edges, Preprint Univ. Bonn, SFB 72, 1986, no 777. Integral Equ. and Operator Theory 1 (1988) 557 - 602.

[16] B.-W. Schulze. Opérateurs pseudo-différentiels et asymptotique sur des variétés a singularités, Seminaire Bony-Meyer, Equ. Deriv. Part., Ecole Polytechnique 1985/86.

[17] F. Treves. Introduction to Pseudodifferential and Fourier Integral Operators vol. 2,Plenum Press, New York and London, 1980.

M. Lorenz, B.-W. Schulze
FB Mathematik
Technische Universität Chemnitz
Reichenhainer Str. 39-41
O-9022 Chemnitz, Deutschland

Reflection of Branching Asymptotics for Hyperbolic Problems

Michael Lorenz and Yasser Al Smaiel
Technical University Chemnitz

1 Introduction

We consider first order systems of strictly hyberbolic equations in $\mathbb{R}^1_+ \times \mathbb{R}^n_+$ with initial and boundary conditions as well as first order pseudodifferential equations obtained by reducing strictly hyperbolic equations of higher order.

We assume that bicharacteristic rays hit the boundary transversally. Nirenberg [1], Taylor [2], Majda and Osher [3] treted the problem of reflection of singularities. Results on reflection of jumps were obtained by Hirschowitz and Piriou [4].

For $n = 1$ we assume that the Cauchy data have singularities only on a finite set of points and there are on both sides discrete asymptotics. The right hand sides of equations and the boundary conditions are chosen smoothly. Furthermore we assume that no caustics may arise. If several singularities carrying bicharacteristics cross transversally the boundary we showed (cf. Lorenz, Smaiel [5]) using some microlocal elliptic boundary conditions that the equality of discrete asymptotics (that means the exponents) remains valid after reflection modulo shadow or the singularities are swallowed at the boundary. We show in the present paper the quality of branching asymptotics remains valid or the singularities are not reflected.

2 Branching asymptotics

$$\mathcal{S}_{\mathfrak{p}}(R_+) \subset C^\infty(R_+) \cap L^2(R_+)$$

is a space of functions with discrete asymptotics (cf. Schulze [6]). These functions can be identified with elements of the Schwartz space $\mathcal{S}(R_+)$ in a neighbourhood of infinity and have in a neighbourhood of the origin the asymptotics expansion

$$\sum_{j=0}^{\infty} \sum_{k=0}^{m_j} \zeta_{jk} y^{p_j} (\ln y)^k$$

where $\mathfrak{p} = \{(p_j, m_j) \in \mathbb{C} \times \mathbb{Z}_+\}$, $Re\ p_j \to -\infty$ for $j \to \infty$

$$Re\ p_{j+1} \leq Re\ p_j < \frac{1}{2} \text{ and } \zeta_{jk} \in \mathbb{C}.$$

Definition 1 *A function $u(x, y)$, $x \in \mathbb{R}^n$, $y \in \mathbb{R}^1_+$, has branching asymptotics at $y = 0$ if for every x there are discrete asymptotics with respect to y which is characterized by a discrete number of pairs $\{(p_j(x), m_j(x))\}, p_j(x) \in \mathbb{C}, m_j(x) \in \mathbb{Z}_+, j \in \mathbb{Z}_+$ and coefficients $\zeta_{jk}(x)$.*

$p_j(x)$, $m_j(x)$ and $\zeta_{jk}(x)$ may depend continuously on x; then one can write the asymptotics as follows:

$$\sum_{j=0}^{\infty}\sum_{k=0}^{m_j(x)} \zeta_{jk} y^{-p_j(x)} (\ln y)^k.$$

If two curves $p_j(x)$ and $p_i(x)$ intersect each other on the complex plane then one deals with sums of asymptotic terms of the following form:

$$\oint \frac{y^{-z} dz}{(p_j(x)-z)(p_i(x)-z)}.$$

More generally we consider asymptotic terms of the form

$$\oint g(x,z) y^{-z} dz$$

where the function g is smooth in the variable x and analytic with respect to z outside a certain union of compact subsets. In this case we have variable asymptotics (cf. Schulze [6]).

Now we define a space of functions over $\mathbb{R}_+^{n-1} \times \mathbb{R}_+^1$, which are infinitely smooth outside a finite set of smooth manifolds of codimension one and have branching asymptotics in normal direction to these manifolds.

Definition 2 *The function g belongs to the space*

$$C^\infty((M_1, \mathcal{S}_{p_1}), \ldots, (M_L, \mathcal{S}_{p_L})),$$

if in local coordinates near the smooth manifold M of the codimension one it holds

$$g(x,y) = H(y) \sum_j \int_{C_j} g_1(x,z)|y|^{-z} dz + H(-y) \sum_j \int_{D_j} g_2(x,z)|y|^{-z} dz,$$

where g_i, $i = 1,2$ are functions belonging to $C^\infty(M, A(\mathbb{C}\setminus K))$ (cf. [6]). $H(y)$ is the Heaviside function $H(y) = 0$, $y \leq 0$, $H(y) = 1$, $y > 0$. If we assume on both sides of the manifolds M_j different asymptotics types we have an overlapping of the types for $t > 0$.

3 Reflection of singularities

We shall examine reflection of singularities of solutions of first order equations of the form

$$\frac{\partial u}{\partial y} = Gu$$

in the domain $(y,t,x) \in \mathbb{R}_+^1 \times \mathbb{R}_+^1 \times \mathbb{R}_+^{n-1}$ with a boundary given by $y = 0$, where $G = G(y) = G(y,t,x,D_t,D_x)$ is a smooth one parameter valued function, G is an $N \times N$ matrix of operators with the principal part $G_1(y,t,x,\tau,\xi)$, which is homogeneous of degree one in (τ,ξ).

At the boundary $y = 0$ we pose conditions

$$\beta u(0) = 0$$

where $\beta \in OPS^0$ is a matrix pseudodifferential operator of order zero.

We assume j bicharacteristic curves pass over some point $(t_0, x_0, \tau_0, \xi_0) \in T^*(\mathbb{R}_+^1 \times \mathbb{R}_+^{n-1}) \setminus \{0\}$.

This means there are j real solutions $\eta_1, \ldots, \eta_j$ of

$$\mathcal{A}(0, t_0, x_0, \eta_\nu, \tau_0, \xi_0) = 0, \ \nu = 1, \ldots, j.$$

The associated bicharacteristics $\gamma_\nu(s) = (y(s), t(s), x(s), \eta(s), \tau(s), \xi(s))$ solve the equations

$$\begin{aligned} \dot{x} &= \frac{\partial \mathcal{A}}{\partial \xi}, \ x(0) = x_0 \quad ; \quad \dot{\xi} = -\frac{\partial \mathcal{A}}{\partial x}, \xi(0) = \xi_0 \\ \dot{t} &= \frac{\partial \mathcal{A}}{\partial \tau}, \ y(0) = t_0 \quad ; \quad \dot{\tau} = -\frac{\partial \mathcal{A}}{\partial t}, \ \tau(0) = \tau_0 \\ \dot{y} &= \frac{\partial \mathcal{A}}{\partial \eta}, \ y(0) = 0 \quad ; \quad \dot{\eta} = -\frac{\partial \mathcal{A}}{\partial y}, \ \eta(0) = \eta_j. \end{aligned}$$

We assume that the bicharacteristic curves intersect $y = 0$ transversally. This means that $\dot{y} = \frac{\partial \mathcal{A}}{\partial \eta} \neq 0$ at $(0, t_0, x_0, \eta_\nu, \tau_0, \xi_0)$, which implies the real zeros $\eta_1, \ldots, \eta_j$ are simple.

$\mathcal{A}(y, t, x, \eta, \tau, \xi)$ has exactly j simple roots $\eta_\nu(t, x, \tau, \xi)$, $\nu = 1, \ldots, j$ for (t, x, τ, ξ) in a small conic neighbourhood of $(t_0, x_0, \tau_0, \xi_0)$ for small $y \geq 0$.

Hence, $G_1(y, t, x, \tau, \xi)$ has j simple purly imaginary eigenvalues $-i\lambda_1, \ldots, -i\lambda_j$ in a conic neighbourhood of $(t_0, x_0, \tau_0, \xi_0)$ for small $y > 0$.

We may define a smooth invertible matrix $U(y, t, x, \tau, \xi)$, which is homogeneous of degree zero in (τ, ξ) such that

$$UG_1U^{-1} = \begin{pmatrix} -i\lambda_1 & & & & \\ & \ddots & & & \\ & & -i\lambda_j & & \\ & & & E_+ & \\ & & & & E_- \end{pmatrix} =: \tilde{G}_1$$

in a conic neighbourhood of $(t_0, x_0, \tau_0, \xi_0)$ for small $y > 0$. $E_+(E_-)$ are square matrices, whose eigenvalues have positive (negative) real part. The corresponding operator $U(y)$ of $U(y, t, x, \tau, \xi)$ is elliptic and has a parametrix $U^{-1}(y)$. In such a way we can decompose this system completely (cf. Taylor [2]) and get finally the system in the form

$$\frac{\partial}{\partial y} w = \begin{pmatrix} -i\lambda_1 & & & & \\ & \ddots & & & \\ & & -i\lambda_j & & \\ & & & E_+ & \\ & & & & E_- \end{pmatrix} w + \begin{pmatrix} a_1 & & & & \\ & \ddots & & & \\ & & a_j & & \\ & & & \alpha & \\ & & & & \beta \end{pmatrix} w + Rw \tag{1}$$

with boundary conditions

$$\mathcal{M}w(0) = h \tag{2}$$

where a_ν, α, β are pseudodifferential operators of order zero and R is a smoothing operator.

We denote the first l components of the vector w by w^I, the next $j-1$ by w^{II}. Thus we define analogously w^{III}, w^{IV} acting on E_+, E_- respectively.

Taylor [2] proved the following statement: The solution w is smooth along the bicharacteristics associated with $-\lambda_1, \ldots, -\lambda_l$, for these it holds: The $t-$ component increases, when the $y-$ component starting from $y = 0$ increases and hence $w^I(0)$ is smooth in a conic neighbourhod of $(t_0, x_0, \tau_0, \xi_0)$ If (2) is microlocally an elliptic or at last hypoelliptic equation for $w^{II}(0)$ and $w^{IV}(0)$, it immediately implies the smoothness of w in a conic neighbourhood.

The same fact holds for the function u since it is connected with w through a smooth family of elliptic pseudodifferential operators of order zero.

4 The wave front set of functions with asymptotics

We give some examples of wave front sets of functions which have constant discrete asymptotics if x varies (cf. [5]). (x,y) are the coordinates in $\mathbb{R}^2$. $H(y)$ is the Heaviside function.

$$\begin{aligned}
WF(H(y)) &= \{(x,0,0,\eta),\eta\neq 0\}\\
WF(\mathcal{F}^{-1}H(\pm\eta)\mathcal{F}H(y)) &= \{(x,0,0,\eta),\eta \gtrless 0\}\\
WF(e^{-1/x^2}H(x)\mathcal{F}^{-1}H(\eta)\mathcal{F}H(y)) &= \{(x,0,0,\eta),\eta>0,x\geq 0\}\\
WF(e^{-1/x^2}[H(x)\mathcal{F}^{-1}H(\eta)\mathcal{F}(y) &+ H(-x)\mathcal{F}^{-1}H(-\eta)\mathcal{F}H(-y)])\\
&= \{(x,0,0,\eta),x\cdot\eta>0 \text{ or } x=0\}.
\end{aligned}$$

The third function has for fixed $x>0$ and $|y|\to 0$ a logarithmic singularity.

The projection on the configuration space may be a non simply connected set after propagation of the singularities of the fourth example.

5 Theorem about the turn of branching asymptotics

We consider the following pseudodifferential boundary value problem (BVP)

$$\mathcal{A}u = f \tag{3}$$

in a region $\mathbb{R}^1_+\times\mathbb{R}^{n_1}\times\mathbb{R}^1_+$ with boundary conditions

$$\gamma u|_{y=0} = h \tag{4}$$

and initial data

$$u|_{t=0} = g \tag{5}$$

$\mathcal{A}$ is required to be a strictly hyperbolic $N\times N$-system of pseudodifferential operators of order one, is a $k\times N$-matrix of pseudodifferential operators of order zero.

In addition we assume the boundary $t=0$ and $y=0$ is noncharacteristically and the problem is well posed. Every component of the Cauchy data $g=(g_1,\ldots,g_N)$ may have branching asymptotics in normal direction to a finite set of smooth manifolds (cf. Def.2)

We write

$$g_j\in C^\infty((M_1^{(j)},\mathcal{S}_{p_1}(j)),\ldots,(M_{L_{(j)}}^{(j)},\mathcal{S}_{p_{L_{(j)}}}(j))),\ j=1,\ldots,J.$$

Let f and h be smooth functions satisfying compatibility conditions at $t=y=0$. First we solve a Cauchy problem in a neighbourhood of $t=0$. We find the following representation

$$\mathcal{A}=B(t,x,y,D_t,D_x,D_y)C^{-1}(t,x,y,D_x,D_y)[\frac{\partial I}{\partial t}-A(t,x,y,D_x,D_y)]\cdot C(t,x,y,D_x,D_y),$$

where I is the unit matrix of order N. A is a diagonal matrix of smooth families of pseudodifferential operators of order one, B and C are elliptic operators of order zero and C^{-1} is the parametrix of C (cf. Nirenberg [1]).

The solution can be written microlocally as composition of elliptic pseudodifferential and Fourier integral operators, whose phase function is the solution of the Hamilton-Jacobi equations for

$$\tau - a(t,x,y,\xi,\eta).$$

We diagonalize the system as in (1) and put

$$w=\tilde{C}u,\ \text{with}\ \tilde{C}=(I+K_1)U$$

If bicharacteristics carrying singularities cross the boundary $y = 0$ the components w^{II} and w^{IV} are transformed to Cauchy data for w^{I} and w^{III} microlocally by means of the elliptic boundary conditions.

The following result on propagation of asymptotics were proved by Lorenz and Schulze [7]: For a fixed and small time t the solution vector u belongs to a space type defined in point 1 . For fixed t the type of branching asymptotics remains valid modulo shadow.

If several bicharacteristics carrying singularities cross the boundary at $(t_0, x_0, 0, \tau_0, \xi_0, \eta_0)$ the following question is posed: Which type of asymptotics have the single components of u in t-direction after reflection ?

A diffeomorphism transforms M locally to $y = t = 0$.

The type of branching asymptotics is invariant under this action. We need a representation of u in a microlocal neighbourhood of $(t_0, x_0, 0, \tau_0, \xi_0, \eta_0)$. It is sufficient to consider one component of the diagonal matrix $\frac{\partial}{\partial t} I - A(t, x, y, D_x, D_y)$.

For such a component the solution is given by

$$v(t,x,y) = \int e^{i(\psi(t,x,y,\xi,\eta) - \tilde{v}\xi - \tilde{w}\eta)} K(t,x,y,\xi,\eta) g(\tilde{v},\tilde{w}) d\tilde{v} d\tilde{w} d\xi d\eta \qquad (6)$$

where $v = Cu$.

We consider the projection of the bicharacteristics starting at M, which represents some $M^{(j)}$, on the configuration space (t, x, y-coordinates). We want to cut this manifold with the plane $y = y_0$. The solution of the Hamilton system is given by in the form

$$\begin{array}{rcl} x^{\pm} = x(t,x,\eta) & , & x^{\pm}(0,x,\eta) = x_0 \\ y^{\pm} = y(t,x,\eta) & , & y^{\pm}(0,x,\eta) = 0 \\ \xi^{\pm} = \xi(t,x,\eta) & , & \xi^{\pm}(0,x,\eta) = 0 \\ \Theta^{\pm} = \Theta(t,x,\eta) & , & \Theta^{\pm}(0,x,\eta) = \eta \end{array}$$

$\pm$ is the sign of η, $\eta \in \mathbb{R}^1 \setminus \{0\}$.

We fix $y^{\pm}$ at y_0 and solve the equation $y_0 = y(t, x, \eta)$ with respect to t. In general there exists no unique solution. On the other hand there is an unique intersection of each bicharacteristic curve with the contangent bundle of the boundary because of the transversality condition. Hence, for each $x_0 \in M$ and η_0 and sufficiently small y_0 one can find one t which satisfies the equation $y_0 = y(t, x_0, \eta_0)$.

We denote the solution by $t^{\pm} = t^{\pm}(x, \eta)$ and define the manifold $\Gamma(y_0)$ by the equations

$$X^{\pm}(x,\eta) = x^{\pm}(t(x,\eta), x, \eta)\ , t = t(x,\eta).$$

Now we assume the Cauchy data have branching asymptotics in the normal direction to the smooth manifold M of codimension $n - 1$. Consider the image which arises as a result of the propagation of the singularities.

We set $\zeta = (x, \eta)$, where $x \in M$. For $t > 0$ the singularities are concentrated on the projections $x(t, \zeta), y(t, \zeta)$ of bicharacteristics on the configuration space.

For each ζ we fix $t = t(\zeta)$ satisfying $y(t(\zeta), \zeta) = 0$

In general the obtained manifold $\Gamma(0)$ is not smooth. We assume $\Gamma(y_0), 0 < y_0, y_0$ sufficiently small, is a smooth manifold and the rank of the matrix $\{t'_\zeta(\zeta), x'_\zeta(t(\zeta), \zeta)\}$ is equal to $n - 1$.

We have a diffeomorphism between M and Γ. Our aim is to show that for fixed $y = y_0$ the solution of the (BVP) has branching asymptotics in the normal direction to $\Gamma(y_0)$ if this is supposed for the Cauchy data at M.

Now we shift the solution v (6) using the operator

$$S^{\pm}v(t,x,y) = v(t+t^{\pm}(x,\pm 1), X^{\pm}(x,\pm 1), y_0).$$

Applying $S^{\pm}$ on the microlocal solution (6) we obtain

$$\begin{aligned} v(t+t^{\pm}(x,\pm 1), X^{\pm}(x,\pm 1), y_0) &= \int e^{i(\psi(t+t^{\pm}(x,\pm\eta)),X^{\pm}(x,\pm 1),y_0,\xi,\eta)-\tilde{v}\xi-\tilde{w}\eta)} \\ & K(t+t^{\pm}(x,\pm 1), X^{\pm}(x,\pm 1), y_0,\xi,\eta)\cdot g(\tilde{v},\tilde{w})d\tilde{v}d\tilde{w}d\xi d\eta \quad (7)\end{aligned}$$

Theorem 3 *Let $0 < y = y_0$ be fixed. Then we obtain:*

$$v(t+t^{\pm}(x,\pm 1), X^{\pm}(x,\pm 1), y_0) \in C^{\infty}(\Gamma(y_0), \mathcal{S}_Q),\ Q = \bigcup_{k=0}^{\infty} T^k \mathfrak{p}$$

if $g \in C^{\infty}(\{(x,y), x \in \mathbb{R}^n, y = 0\}, \mathcal{S}_{\mathfrak{p}(x)})$.

Proof. Assume that $g \in C^{\infty}(\{(x,y), x \in \mathbb{R}^n, y = 0\}, \mathcal{S}_{\mathfrak{p}(x)})$ Then g has the form

$$g(\tilde{v},\tilde{w}) = H(w)\sum_{f}\int_{C_f} g_1(v,z)|w|^{-z}dz + H(-w)\sum_{f}\int_{D_f} g_2(v,z)|w|^{-z}dz. \quad (8)$$

The phase function ψ is a composition of a diffeomorphism and a pseudodifferential phase function. Further, we write ψ in the form

$$\begin{aligned}\psi(t+t^{\pm}(x,\pm 1), X^{\pm}(x,\pm 1), y_0,\xi,\eta) &= \frac{\partial}{\partial\xi}\psi(t+t^{\pm}(x,\pm 1), X^{\pm}(x,\pm 1), y_0, 0, \pm 1)\cdot\xi \\ &+ \frac{\partial}{\partial\eta}\psi(t+t^{\pm}(x,\pm 1), X^{\pm}(x,\pm 1), y_0, 0, \pm 1)\cdot\eta \\ &+ [\psi(t+t^{\pm}(x,\pm 1), X^{\pm}(x,\pm 1), y_0,\xi,\eta) \\ &- \frac{\partial}{\partial\xi}\psi(t+t^{\pm}(x,\pm 1), X^{\pm}(x,\pm 1), y_0, 0, \pm 1)\cdot\xi \\ &- \frac{\partial}{\partial\eta}\psi(t+t^{\pm}(x,\pm 1), X^{\pm}(x,\pm 1), y_0, 0, \pm 1)\cdot\eta]. \quad (9)\end{aligned}$$

The first both items of (9) present a composition of a diffeomorphism and a pseudodifferential phase function. The remaining part of ψ given in (9), too, will be added to which occurs in the solution of (3)-(5).

We apply the Euler formula to ψ:

$$\psi = \frac{\partial}{\partial\xi}\psi\cdot\xi + \frac{\partial}{\partial\eta}\psi\cdot\eta \quad (10)$$

Replacing (10) in (9) yields ψ in the form

$$\begin{aligned}\psi &= \frac{\partial}{\partial\xi}\psi(t+t^{\pm}(x,\pm 1), X^{\pm}(x,\pm 1), y_0, 0, \pm 1)\xi \\ &+ \frac{\partial}{\partial\eta}\psi(t+t^{\pm}(x,\pm 1), X^{\pm}(x,\pm 1), y_0, 0, \pm 1)\eta \\ &+ [\frac{\partial}{\partial\xi}\psi(t+t^{\pm}(x,\pm 1), X^{\pm}(x,\pm 1), y_0,\xi,\eta) \\ &- \frac{\partial}{\partial\xi}\psi(t+t^{\pm}(x,\pm 1), X^{\pm}(x,\pm 1), y_0, 0, \pm 1)\xi \\ &+ (\frac{\partial}{\partial\eta}\psi(t+t^{\pm}(x,\pm 1), X^{\pm}(x,\pm 1), y_0,\xi,\eta) \\ &- \frac{\partial}{\partial\eta}\psi(t+t^{\pm}(x,\pm 1), X^{\pm}(x,\pm 1), y_0, 0, \pm 1))\eta]. \quad (11)\end{aligned}$$

Both differences in (11) can be evaluated using Taylors formula. This yields

$$\begin{aligned}|\frac{\partial}{\partial\xi}\psi(t+t^{\pm}(x,\pm1),X^{\pm}(x,\pm1),y_0,\xi,\eta) & - \\ \frac{\partial}{\partial\xi}\psi(t+t^{\pm}(x,\pm1),X^{\pm}(x,\pm1),y_0,0,\pm1)\xi| & < C_1\frac{1+|\xi|}{1+|\eta|}, \\ |\frac{\partial}{\partial\eta}\psi(t+t^{\pm}(x,\pm1),X^{\pm}(x,\pm1),y_0,\xi,\eta) & - \\ \frac{\partial}{\partial\eta}\psi(t+t^{\pm}(x,\pm1),X^{\pm}(x,\pm1),y_0,0,\pm1)\eta| & < C_2\frac{1+|\xi|}{1+|\eta|}.\end{aligned}$$

Further we consider the amplitude $K(t+t^{\pm}(x,\pm1),X^{\pm}(x,\pm1),y_0,\xi,\eta)$ and look at the Taylor expansion of K in a neighbourhood of $\xi=0$

$$K=\sum_{\alpha}\frac{1}{\alpha!}\xi^{\alpha}(\frac{\partial}{\partial\xi})^{\alpha}k(t+t^{\pm}(x,\pm1),X^{\pm}(x,\pm1),y_0,0,\eta), \tag{12}$$

wher $k(t+t^{\pm}(..),X^{\pm}(..),y_0,0,\eta)$ is a sum of homogeneous functions in η. Then we write K in the form

$$K=\sum_{l\geq0}\sum_{\beta}c_{\beta-l}\xi^{\beta}\eta^{-l} \tag{13}$$

where $c_{\beta l}$ are smooth functions in t,x,y_0.

Now we put (12), (13), (11) and (7) in (7) and get the following form

$$\begin{aligned}v(t+t^{\pm}(x,\pm1),X^{\pm}(x,,\pm1),y_0) = \\ \int e^{i[\frac{\partial}{\partial\xi}\psi(t+t^{\pm}(x,\pm1),X^{\pm}(x,\pm1),y_0,0,\pm1)\xi+\frac{\partial}{\partial\eta}\psi(t+t^{\pm}(x,\pm1),X^{\pm}(x,\pm1),y_0,0,\pm1)\eta-\tilde{v}\xi-\tilde{w}\eta]} \\ \cdot\sum c'_{\beta l}\xi^{\beta}\eta^{-l}H(\pm\tilde{w})\int g(\tilde{v},z)|\tilde{w}|^{-z}dzd\tilde{v}d\tilde{w}d\xi d\eta.\end{aligned}$$

We show the solution v given in (13) belongs to the space introduced in Definition 2 of branching asymptotics with respect to t and varying x, that is, $v\in C^{\infty}(\Gamma,\mathcal{S}_Q)$.

We pull ξ^{β} in (13) under the integral sign.

$$\begin{aligned}v(t+t^{\pm}(..),X^{\pm}(..),y_0) = \\ \int e^{i[(\frac{\partial}{\partial\xi}\psi)\xi+(\frac{\partial}{\partial\eta}\psi)\eta-\tilde{v}\xi-\tilde{w}\eta]}\eta^{-l}H(\pm w)\int D^{\beta}_{\tilde{v}}g(\tilde{v},z)|\tilde{w}|^{-z}dzd\tilde{v}d\tilde{w}d\xi d\eta.\end{aligned} \tag{14}$$

We compute the integral $\int e^{i(\alpha-\tilde{w})}\eta^{-l}|\tilde{w}|^{-z}d\tilde{w}d\eta$ and obtain

$$\int e^{i(\alpha\eta-\tilde{w}\eta)}\eta^{-l}H(\pm\tilde{w})|w|^{-z}d\tilde{w}d\eta=[\Theta_1(l,z,\pm\alpha)+\Theta_2(l,z,\pm\alpha)\ln|\alpha|]|\alpha|^{-z+1}$$

where

$$\alpha=\frac{\partial}{\partial\eta}\psi(t+t^{\pm}(x,\pm1),y_0,0,\pm1). \tag{15}$$

Put (15) in (14) and apply the Fourier transform then we obtain

$$\begin{aligned}v(t+t^{\pm}(..),X^{\pm}(..),y_0) = \\ \int D^{\beta}_{\tilde{\tilde{v}}}g(\tilde{\tilde{v}},z)[\Theta_1(l,z,\pm\alpha)+\Theta_2(l,z,\pm\alpha)\ln|\alpha|]|\alpha|^{-z+l}dz\end{aligned} \tag{16}$$

where

$$\tilde{\tilde{v}}=\frac{\partial}{\partial\xi}\psi(t+t^{\pm}(x,\pm1),X^{\pm}(x,\pm1),y_0,0,\pm1). \tag{17}$$

Now we may apply the theorem on the diffeomorphism (cf. [7]) because $\frac{\partial}{\partial\xi}\psi$ and $\frac{\partial}{\partial\eta}\psi$ vanish identically on the bicharacteristics.

Using Taylor expansion we obtain

$$\begin{aligned} v &= \int D_v^\beta g(\psi(x,0),z)[\Theta_1 + \Theta_2 \ln|\Phi(x,0)|].|\Phi_t'(x,0)|^{-z+l}|t|^{-z+l}dz \\ &+ \sum_{k=1} \frac{1}{k!}t^k \int (\frac{\partial}{\partial t})^k D_v^\beta g(\psi(x,t),z)[\Theta_1 + \Theta_2 \ln|\Phi_t'(x,0)|]. \\ & \left|\frac{\Phi(x,t)}{t}\right| |_{t=0}^{-z+1}.|t|^{-z+l}dz + R. \end{aligned}$$

since $\frac{\Phi(x,t)}{t}$ is smooth up to the boundary. The solution v has discrete asymptotics with respect to t for fixed x,

$$v \in \mathcal{S}((t(x_0, y_0),\ \cup_{k=0}^{\infty} T^{-k}\mathfrak{p})).$$

If x runs over $(\Gamma(y_0)$ the solution v belongs to the space

$$C^\infty(\Gamma(y_0), \mathcal{S}_Q)\ ,\ Q = \bigcup_{k=0}^{\infty} T^k\mathfrak{p}.$$

6 Reflection of asymptotics at the boundary

We consider the solution $w(y,t,x)$ of (BVP) given in (1) with the boundary condition

$$\mathcal{M}w(0) = h(t,x) \tag{18}$$

This can be written in the following form

$$w(0) = S^I w^I + s^{II} w^{II} + s^{III} w^{III} + S^{IV} w^{IV} = h(t,x), \tag{19}$$

Further, if (S^I, S^{III}) is microlocally an elliptic operator we can write (19) in the form

$$(w^I, w^{III}) = ((S^I, S^{III})^{-1}(h(t,x) - (S^{II}w^{II} + S^{IV}w^{IV})) \tag{20}$$

where $w^{II} \in C^\infty(\Gamma(y_0), \mathcal{S}_{\mathfrak{p}_0})$ at the boundary, h is a regular function $(S^{II}w^{II}, S^{IV}w^{IV})$ are smooth functions in (x,y,t). Then we have

$$(S^{II}, S^{IV})(w^{II}, w^{II}) \in C^\infty(\Gamma, \mathcal{S}_{\mathfrak{p}})$$

and since (S^I, S^{III}) is a pseudodifferential operator of order zero it follows

$$(w^{II}, w^{III}) \in C^\infty(\Gamma, \mathcal{S}_Q),\ Q = \bigcup_{k=0}^{\infty} T^{-k}\mathfrak{p}.$$

Finally we get the following

Theorem 4 *Let $0 < t = t_0$ be fixed. If the Cauchy data of a strictly hyperbolic boundary value problem (3)-(5) have branching asymptotics at a manifold of codimension one*

$$g_1 \in C^\infty((M_1^{(j)}.\mathcal{S}_{\mathfrak{p}_1^{(j)}}), \dots, (M_{L(j)}^{(j)}, \mathcal{S}_{\mathfrak{p}_{L(j)}^{(j)}})),\ j = 1, \dots, J$$

then the solution u of the problem has under the ellipticity conditions (20) modulo shadow the same type of asymptotics as g after reflection or it is smooth.

References

[1] L. Nirenberg. Reflections of bicharacteristics for differential equations and singularities near a boundary, Seminar Goulaouic-Lions-Schwartz 1974-1975.

[2] M. Taylor. Pseudodifferential operators, Princeton, 1981.

[3] A. Mazja and S. Osher. Reflection of singularities at the boundary. Comm. Pure Appl. Math. 28 (1975), 479-499.

[4] A. Hirschowitz, A. Piriou. Propriétés de transmission pour les distributions intégrales de Fourier, Comm. P.D.E. 4, 2 (1979) 113-217.

[5] M. Lorenz and Y. Smaiel. Reflexion von diskreter Asymptotik für streng hyperbolische Randwertaufgaben. WZ TU K.-M.-Stadt 32, 1 (1990), 33-40.

[6] B.-W. Schulze. Pseudo-differential operators on manifolds with singularities, North Holland 1991.

[7] M. Lorenz and B.-W. Schulze. Propagation of branching conormal asymptotics for hyperbolic Cauchy problems. Proceedings of the Intern. Workshop Analysis in Domains and on Manifolds with Singularities, Breitenbrunn 1990, this volume.

[8] B.-W. Schulze. Regularity with continuous and branching asymptotics for elliptic ooerators on manifolds with edges. Int. Equ. and Operator Theory 11 (1988) 557 - 602.

M. Lorenz, Y.A. Smaiel
FB Mathematik
Technische Universität Chemnitz
Reichenhainer Str. 39-41
O-9022 Chemnitz, Deutschland

Edge Operators in Geometry

RAFE MAZZEO

Stanford University

§*1. Introduction* Differential operators with degeneracies of various sorts have been studied by a large number of mathematicians, with results ranging from those concerning very special operators to those concerning analytic and algebraic properties of general classes of operators. Much of the analysis has focused on elliptic operators, and the main concern has been to show that many properties usually associated with elliptic operators or elliptic boundary problems have analogues in a more general context. In this note we describe how a certain class of operators with uniformly controlled degeneracies may be analyzed. These are called edge operators because they arise when nondegenerate elliptic operators are written in polar coordinates around an edge of a domain, or indeed around any distinguished submanifold; however as we shall indicate, they also arise in many other geometrically natural situations. Their analysis is undertaken by constructing an algebra of pseudodifferential operators sharing the same type of degeneracy which is large enough to contain parametrices for many of the elliptic elements of the original ring of edge differential operators. These parametrices are constructed explicitly, and this allows us to conclude a great deal of detailed information about solutions of the original equations. Operators in a somewhat restricted subclass of the class of elliptic operators have very good analytic properties on a scale of weighted Sobolev or Hölder spaces for all but a discrete set of values of the weight parameter – for many values of the weight parameter the operators are semi-Fredholm, with infinite dimensional kernel or cokernel, although occasionally it will have neither hence be Fredholm. On the other hand for certain values of the weight parameter it may have both, hence the most we can conclude is that the range is closed. Simultaneously we obtain a structure theory for the generalized inverses for these operators as well as the 'Bergman' projectors onto the kernel or cokernel.

We shall also indicate how this analysis may be used in the study of geometric problems. We discuss briefly two such problems: the first concerns the study of harmonic differential forms, and more broadly, the differential form spectrum for the Hodge Laplacian on geometrically finite quotients of hyperbolic space as well as on a class of Riemannian manifolds for which the metric shares the same sort of degeneracy at infinity. This is previous work of the author **[M1]**, **[M2]**, **[M-P]** (the last paper is joint work with Ralph Phillips). The second problem concerns the elliptic equation arising in the Yamabe problem in which one seeks metrics of constant scalar curvature conformal to a given metric. In the situation described here the background metric is the standard metric on the sphere and one is searching for a complete metric of constant positive scalar curvature on a given subdomain of the sphere, or equivalently a solution of the relevant PDE with prescribed singular set. This is joint work with Nat Smale **[M-S]**, and a part of the general analysis described here was worked out in that paper. The more complete analysis is contained in **[M3]**. It fits into a framework initiated some time ago by Richard Melrose to study many different types of degenerate problems. Parts of this program are expounded in **[Me2]**. In addition many particulars of the

The author was partially supported by an NSF Postdoctoral Fellowship

theory here were discovered by Melrose and Mendoza in 1983. We are very grateful to Melrose for many discussions and suggestions on various aspects of this work.

The analytic results here bear close resemblance to the work of many other authors, of which we most particularly want to mention B.-W. Schulze and S. Rempel, cf. Schulze's article in this volume and **[R-S]**, **[S]** and the references therein. The results there are very complete and more general in many respects. Schulze's analysis of elliptic edge operators covers also operators with variable indicial roots (see below) which we do not undertake, and very general results on asymptotics are obtained. The strategy there is to complete the operator to a system of the type introduced by Boutet de Monvel with supplementary trace and Poisson operators and to prove Fredholm properties for this system. However, it appears that our results are not a consequence of Schulze's when the operator has both infinite dimensional kernel or cokernel, although they do follow if at most one of these spaces has infinite dimension. Nonetheless, it is probable that the full extent of our results could be deduced by those methods.

§2. Edge Operators We now introduce in detail the classes of differential and pseudodifferential operators of edge type and describe the main theorems. Throughout the rest of this paper X will be a compact manifold with boundary, with the property that its boundary $\partial X = M$ is itself the total space of a fibration

$$F \longrightarrow M \qquad \Big\downarrow \pi \qquad Y \tag{2.1}$$

with fibre F and base space Y. For example, if $\overline{X}$ is a compact manifold without boundary and Y a submanifold of arbitrary dimension, then the process of blowing up $\overline{X}$ around Y, cf. [Me2], which may be regarded as the introduction of polar coordinates around Y, results in a new manifold X with boundary M which is fibred by spheres, which correspond to the angular directions in the polar coordinates, and with base space Y. It is frequently the case that the fibration (2.1) is trivial and $M = F \times Y$, so this may be assumed by the reader without significant loss of generality. We shall frequently use coordinates (x, y, z) on X near ∂X adapted to this picture as follows. The coordinate x will be a defining function for ∂X, i.e. a function which vanishes on the boundary with nonvanishing gradient there; the coordinate y will be a local coordinate on Y lifted up to M and then extended locally to a neighbourhood of the boundary; finally, z will denote a local coordinate along the fibre. We shall assume that $\dim Y = p$ $\dim F = q$ and $\dim X = n$.

The fundamental object in our point of view is the space of all smooth vector fields on X which are completely unrestricted in the interior but which are required to lie tangent to the fibres of π at ∂X. We shall call this space $\mathcal{V}_e$. In local coordinates as described above, it is clear that $\mathcal{V}_e$ is spanned over $\mathcal{C}^\infty(X)$ by

$$x\partial_x, x\partial_{y^\alpha}, \partial_{z^\mu}. \tag{2.2}$$

For convenience, here and later we shall always assume that $1 \leq \alpha, \beta, \cdots \leq p$, $1 \leq \mu, \nu, \cdots \leq q$.

(2.3) Definition: The ring $Diff_e^*(X)$ of differential operators of edge type is the subset of the space $\mathrm{Diff}^*(X)$ of all differential operators on X consisting of those which

locally can be expressed as sums of products of elements of $\mathcal{V}_e$. The subspace $\mathrm{Diff}_e^m(X)$ consists of those for which each summand in any such decomposition contains at most m factors.

Thus in local coordinates, any $L \in \mathrm{Diff}_e^m(X)$ may be expressed as

$$L = \sum_{j+\alpha+\beta\leq m} a_{j,\alpha,\beta}(x,y,z)(x\partial_x)^j(x\partial_y)^\alpha\partial_z^\beta. \tag{2.4}$$

Note that $\mathcal{V}_e$ is closed under Lie bracket of vector fields, hence may be considered a Lie algebra, and this guarantees that $Diff_e^*(X)$ is well defined.

There is an invariant symbol mapping for $Diff_e^*(X)$. Formally, for any L which has been expressed as in (2.4) we set

$$^e\sigma_L(x,y,z;\xi,\eta,\zeta) = \sum_{j+\alpha+\beta=m} a_{j,\alpha,\beta}(x,y,z)\xi^j\eta^\alpha\zeta^\beta. \tag{2.5}$$

This is a function not on the ordinary cotangent bundle T^*X but rather on the 'edge' cotangent bundle $^eT^*X$. This is a bundle naturally associated to the edge structure and is dual to the edge tangent bundle ^eTX, of which $\mathcal{V}_e$ is the full space of smooth sections. The exact definitions of these bundle are given in **[M3]** and need not concern us here, but we note that $^eT^*X$ is canonically isomorphic to T^*X in the interior of X, and that the symbol above is a natural object. As usual, it is a homogeneous polynomial on the fibres of $^eT^*X$. L is called elliptic if $^e\sigma_L(x,y,z,\xi,\eta,\zeta) \neq 0$ for $(\xi,\eta,\zeta) \neq 0$.

As in the theory of elliptic boundary problems, (symbol) ellipticity of L is not sufficient to guarantee that it has reasonable analytic properties. Thus we need to introduce further constraints, such as boundary conditions, at the boundary. Although it is possible to develop a full theory of 'elliptic boundary problems' for edge operators **[S]**, we shall simply consider the mapping properties of L in weighted spaces, which may be regarded as imposing generalized Dirichlet conditions. Even here only certain weights are allowable for a good theory, and this might be thought of as some sort of Lopatinski-Schapiro condition. This requires the following concept. Any $L \in Diff_e^*(X)$ possesses two simpler model operators at ∂X. These are analogues of Schulze's notion of boundary symbol. Rather than describe these models invariantly, which is done in **[M3]**, we define them in local coordinates. The first, and most primitive, model for L is the indicial operator I_L. For L as in (2.4)

$$I_L = \sum_{j+\beta\leq m} a_{j,0,\beta}(0,y,z)(s\partial_s)^j\partial_z^\beta \tag{2.6}$$

In this operator the variable y has been reduced to a parameter and s is a new auxiliary variable in $\mathbf{R}^+$. I_L is an operator on $\mathbf{R}^+ \times F$ of 'totally characteristic type' as in **[Me1]**, **[Me-Me]** depending on the parameter y. Rempel and Schulze have also undertaken a thorough study of such operators, and these are used as building blocks for edge operators also in Schulze's theory. The second model for L is called the normal operator N_L. It is the model on the space $\mathbf{R}_s^+ \times \mathbf{R}_u^p \times F_z$, where the subscripts denote the variables on each factor, given by

$$N_L = \underset{j+\alpha+\beta\quad m}{\qquad} a_{j,\alpha,\beta}(0,y,z)(s\partial_s)^j(s\partial_u)^\alpha\partial_z^\beta. \tag{2.7}$$

Note that once again y is a parameter. Both I_L and N_L are invariant with respect to a group, hence their solvability properties become tractable using Mellin and Fourier transform techniques. For each y and z I_L is $\mathbf{R}^+$-invariant in s and N_L is invariant in (s,u) with respect to the group which is the semi-direct product of the dilation group $\mathbf{R}^+$ and the translation group $\mathbf{R}^p$ in u.

(2.8) Definition: For each $y \in Y$ the boundary spectrum $\mathrm{spec}_b(L,y)$ is defined to be the set of values of $\zeta \in \mathbf{C}$ for which the operator

$$I_L(y,\zeta;z,\partial_z) = \sum_{j+\beta \leq m} a_{j,0,\beta}(0,y,z)(i\zeta)^j \partial_z^\beta \tag{2.9}$$

fails to be invertible on $L^2(F_y)$, $F_y = \pi^{-1}(y)$. The points of $\mathrm{spec}_b(L,y)$ are called indicial roots of I_L or of L.

If L is elliptic in the sense of edge operators then $I_L(y,z,s\partial_s,\partial_z)$ is elliptic in the sense of totally characteristic operators and $I_L(y,\zeta;z,\partial_z)$ is elliptic in the ordinary sense. In this case, $\mathrm{spec}_b(L,y)$ is a discrete set in $\mathbf{C}$ for each y. It is shown in **[Me-Me]** that this set has the property that each horizontal strip $a \leq \Im\zeta \leq b$ in the complex plane contains only a finite number of points of $\mathrm{spec}_b(L,y)$ for fixed y.

The boundary spectrum of L will determine exactly which values of the weight parameter in the weighted Sobolev spaces below will be omitted. At these weight values L turns out not to have closed range. This sort of behaviour is familiar from the many studies of conic operators, cf. **[Me-Me]**, **[R-S]**, **[Mc]**. In order to simplify the theory we shall always make the assumption that the set $\mathrm{spec}_b(L,y)$ is independent of y and for simplicity we shall denote this set by Λ. As Schulze has shown, this is not necessary to build up a satisfactory theory, but it simplifies matters immensely. The methods here immediately imply many of the simpler properties below when the indicial roots do vary with y, and it is probable that the more delicate properties may be extended using these techniques. However, we have chosen to work in this (lack of) generality because the edge operators which arise in geometric problems seem to always have this property.

Before stating the main theorem concerning elliptic edge operators we must introduce the function spaces on which they will act. We shall use scales of weighted degenerate Sobolev and Hölder spaces. These spaces incorporate the approximate dilation invariance properties of the edge calculus.

(2.10) Definition: For $\delta \in \mathbf{R}$ $k \in \mathbf{N}$ and $0 < \alpha < 1$ let

$$x^\delta H_e^k(X) = \{u = x^\delta v : (x\partial_x)^j (x\partial_y)^\alpha \partial_z^\beta v \in L^2(X; dx\,dy\,dz/x)\}$$
$$\mathcal{C}^{k,\alpha,\delta}(X) = \{u = x^\delta v : (x\partial_x)^j (x\partial_y)^\alpha \partial_z^\beta v \in \mathcal{C}^{0,\alpha,0}\}$$

where

$$\mathcal{C}^{0,\alpha,0} = \left\{ v : \sup|v| + \sup \frac{|(x+\tilde{x})^\alpha (v(x,y,z) - v(\tilde{x},\tilde{y},\tilde{z}))|}{|(|x-\tilde{x}| + |y-\tilde{y}| + |z-\tilde{z}|)|^\alpha} < \infty \right\}.$$

The Sobolev spaces have only been defined when k is a nonnegative integer, but by duality and interpolation there exist spaces $x^\delta H_e^k$ for k any real number. The pseudodifferential calculus described below provides an interpolation method to realize these spaces concretely.

(2.11) THEOREM. *Let $L \in \mathit{Diff}_e^m(X)$ be elliptic, and suppose as above that the set Λ of indicial roots for L is independent of y. Then, for $\delta \notin \Lambda$ the maps*

$$L : x^\delta H_e^{k+m}(X) \longrightarrow x^\delta H_e^k(X)$$
$$L : \mathcal{C}^{k+m,\alpha,\delta}(X) \longrightarrow \mathcal{C}^{k,\alpha,\delta}(X) \tag{2.12}$$

have closed range, while for $\delta \in \Lambda$ their range is not closed. If L has the 'unique continuation property' (UCP), i.e. if there exist no solutions of $Lu = 0$ with $u \in \dot{\mathcal{C}}^\infty(X)$, then there exist two numbers δ' and δ'' such that if $\delta < \delta'$ then L is surjective and if $\delta > \delta''$ then L is injective in (2.12). These properties are understood to hold for all $k \in \mathbf{R}$ in the Sobolev framework, but only for $k \in \mathbf{N}$ in the Hölder framework.

The proof of this Theorem involves the introduction of a class of pseudodifferential operators of edge type. We describe these now very briefly. Any putative parametrix Q for an elliptic differential edge operator L has a Schwartz kernel κ_Q which is a distribution on X^2. Of course, since L is degenerate at ∂X, we expect that κ_Q must have some sort of extraordinary singularity at the corner $\partial X \times \partial X$ (as well as at the rest of the boundary of X^2). The basic idea, due to Melrose **[Mel]** in his study of conic operators, and applicable in a wide variety of degenerate problems, is to regard κ_Q as a simpler distribution on a somewhat more complicated manifold. This new manifold is the product X^2 blown up around the 'fibre diagonal' S in its corner. In local coordinates, S is given as $S = \{x = \tilde{x} = 0,\ y = \tilde{y}\}$, and invariantly it is given as the set of points in $M \times M$ which project down onto the diagonal in $Y \times Y$ under the map $\pi \times \pi$. The product blown up around S will be denoted X_e^2, the edge double product. Its interior is naturally diffeomorphic to that of X^2, as is most of its boundary. Indeed the only difference is that X_e^2 has a new codimension one boundary face, the 'front face', which projects down onto S via the blow-down map $\beta : X_e^2 \to X^2$. It should be observed that the diagonal in X^2 lifts to a smooth submanifold with boundary in X_e^2 which intersects the boundary only in the interior of the front face, and that intersection is transversal.

It is easy to check that any $L \in \mathit{Diff}_e^*(X)$, elliptic or not, when regarded as acting on, say, the left factor of X in X^2, lifts smoothly to X_e^2. This may be checked by noting that any $V \in \mathcal{V}_e$ lifts in the same way to a smooth vector field in X_e^2 which in fact is tangent to all boundaries. Furthermore, the lift of L acts tangentially to the front face of X_e^2, hence its restriction to that boundary face is well defined. It may be shown that this restriction is canonically isomorphic to the normal operator N_L defined above. There are two calculi of edge pseudodifferential operators: the small calculus, $\Psi_e^*(X)$ consists of operators A with Schwartz kernels κ_A on X^2 which are pushforwards of distributions on X_e^2 which are singular only along the lifted diagonal in X_e^2, with classical conormal singularities of the usual type, and which vanish to infinite order at the 'side boundaries' (all boundary faces except the front face). These Schwartz kernels are also required to be smooth up to the front face, and their conormal singularities along the diagonal are required to be extendible across this face. This class of operators is called the small calculus because it takes account only of the degeneracies of L but ignores completely more refined boundary behaviour. $\Psi_e^*(X)$ is graded by the subspaces $\Psi_e^\ell(X)$, where the ℓ is the conormal order of the (pre-)Schwartz kernel on the lifted diagonal. Given an edge operator L of order m it is possible to find $Q_i \in \Psi_e^{-m}(X)$ such that both $QL - I$ and $LQ - I$ are elements of $\Psi_e^{-\infty}(X)$. Unfortunately elements of this residual space are not compact operators, hence the need for a larger calculus. The large calculus $\Psi_e^{*,E,F}(X)$ then consists of operators A with Schwartz kernels κ_A again pushforwards from X_e^2 with the same behaviour near the lifted diagonal, but which are now allowed to be classical conormal at the side boundary faces. The 'index sets' E and F describe the classical conormal behaviour precisely at each of the two side boundary components. (It is also necessary to add residual terms to the calculus; these are kernels on X^2 smooth in the interior and conormal at both boundaries. However, we suppress these in the interests of simplicity.) Suffice it to say that with this extended definition it is possible to prove

(2.13) THEOREM. *For $L \in \mathrm{Diff}_e^m(X)$, elliptic, with constant indicial roots in the set Λ and for $\delta \notin \Lambda$ there exist operators $Q_i \in \Psi_e^{-m,E,F}(X)$ and $P_i \in \Psi_e^{-\infty,E,F}(X)$, $i = 1,2$, such that*

$$LQ_1 = I - P_1, \quad Q_2 L = I - P_2. \tag{2.14}$$

Here the Q_i are bounded between $x^\delta H_e^k$ and $x^\delta H_e^{k+m}$, with similar behaviour between the Hölder spaces. The P_i are projectors onto the kernel and cokernel of L and the Q_i are the generalized inverses.

Theorem (2.11) is an immediate corollary.

We state two results concerning asymptotics which follow from Theorem (2.13).

(2.15) PROPOSITION. *Suppose L satisfies the hypotheses of (2.13) and $u \in x^\delta H_e^k(X)$ for some δ satisfies $Lu = 0$. Then u has an expansion as $x \to 0$ of the form*

$$u \sim \sum_{\Re(s_j)\to\infty} \sum_{\ell=0}^{\infty} \sum_{r=0}^{N_{j,\ell}} x^{s_j+\ell}(\log x)^r u_{j,\ell,r}(y,z) \tag{2.16}$$

The s_j here are all elements of Λ, and the coefficient functions $u_{j,\ell,r}$ are smooth as functions of z but in general have the property that $u_{j,\ell,r} \in H^{p_j-\ell}$ as functions of y.

Although expansions of this type seem very weak, this assertion is frequently sharp. For example, the function $u = x/(x^2+y^2)$ on the half-plane $\{(x,y) : x > 0\}$, which satisfies $x^2(\partial_x^2 + \partial_y^2)u = 0$ may be shown to have the expansion

$$u \sim \sum (-1)^j \pi \frac{\partial_y^{2j}\delta_0}{(2j)!} x^{2j} + \sum (-1)^{j+1} \frac{\partial_y^{2j+2}(\log|y|)}{(2j+1)!} x^{2j+1}.$$

2.17 PROPOSITION. *Suppose L satisfies the hypotheses of (2.13) and $\delta \notin \Lambda$. If $\ker L \cap x^\delta H_e^k$ is finite dimensional then any u in this space admits an expansion of the form (2.16) with all $u_{j,\ell,r}$ smooth in both (y,z). On the other hand, if this space is infinite dimensional and if P is the orthogonal projection onto this space, then its restriction along the diagonal, $P(x,y,z,x,y,z)$ admits an expansion of the form (2.16) with all coefficient functions smooth.*

§3. *Harmonic Forms* We now describe an application of the ideas in §2 to the problem of developing a Hodge theory on infinite volume geometrically finite quotients of hyperbolic space and other manifolds with the same asymptotic structure at infinity. For simplicity we shall limit ourselves to manifolds modelled on convex cocompact quotients $\mathbf{H}^n/\Gamma$. Hodge theory on arbitrary geometrically finite hyperbolic quotients is the subject of **[M-P]**. This involves the analysis of the Laplacian on intermediate rank cusps. The more general Riemannian analogue is the motivation for the work in progress **[L-M-M]**. The term convex cocompact means that the discrete group Γ has no parabolic elements and the quotient has infinite volume. In this case it may be compactified to a manifold with boundary X by adding on the quotient of the domain of discontinuity in the sphere at infinity for hyperbolic space. Guided by the structure of the hyperbolic metric at this ideal boundary one is led to introduce the class of conformally compact manifolds.

(3.1) Definition: The Riemannian manifold (X, g) is called conformally compact if X admits a compactification as a smooth manifold with boundary in such a way that the metric g may be written as $g = \rho^{-2}h$ where h is a smooth metric nondegenerate up to the boundary of the compactification and ρ is a defining function for ∂X, i.e. $\rho^{-1}(0) = \partial X$ and $\nabla\rho \neq 0$ there.

This definition is not intrinsic, but this class of manifolds has proved quite useful. Besides **[M1]**, where they were introduced, see also **[G-L]** where the edge calculus is used to construct complete Einstein metrics with prescribed conformal infinity on the ball. It may be shown that the Riemann curvature tensor of the metric g, although arbitrary in the interior, becomes isotropic with all sectional curvatures converging to the negative of the square of the normal derivative of ρ with respect to h along any geodesic diverging to infinity. In particular, these metrics are all asymptotically hyperbolic in some sense. As expected, any convex cocompact quotient $\mathbf{H}^n/\Gamma$ is in this class. For simplicity we shall assume that this limiting curvature is identically -1 along all of ∂X. All results below have suitable modifications in the more general case.

For any degree k, $0 \leq k \leq n = \dim X$, the Hodge Laplacian is defined on sections of the bundle of k-forms by the formula

$$\Delta = d\delta + \delta d : \Omega^k(X) \to \Omega^k(X). \tag{3.2}$$

Here δ is the adjoint of d with respect to the metric g. A simple calculation gives

(3.3) LEMMA. *$\Delta \in \mathit{Diff}^2_e(X; \bigwedge^k, \bigwedge^k)$, where the edge structure $\mathcal{V}_e$ on X is given by the trivial fibration on ∂X where the fibres are points, so that $\mathcal{V}_e$ consists of all smooth vector fields on X which vanish at ∂X.*

It is also simple to compute that the $\mathrm{spec}_b(\Delta) = \Lambda = \{-1, 0, n-2k-1, n-2k\}$. The fact that Λ is a finite set is a consequence of the fact that the fibres of the edge structure are 0-dimensional.

We are interested in Δ as an operator on $L^2\Omega^k(X)$. The fact that it is essentially self-adjoint on compactly supported forms is a classical fact which uses only the completeness of the metric g. Hence it has a unique closure, and we would like to determine as much as possible about its spectrum. Of particular interest is the space of L^2 harmonic forms $L^2\mathcal{H}^k$ in each degree. Here the main questions are whether this space is finite dimensional and, if so, whether its dimension can be computed in topological terms. Proofs of all parts of the following theorems are to be found in **[M1]**, **[M2]**, **[M-P]**.

(3.4) THEOREM. *Let (X, g) be conformally compact with limiting curvature -1. Then for $k < n/2$ the essential spectrum occupies the whole half-line $[(n-2k-1)^2/4, \infty)$. There are isolated eigenvalues of finite multiplicity below the continuous spectrum* *and in the interval* $[(n-2k-1)^2/4,(n-2k+1)^2/4)$*, but the half-line* $[(n-2k+1)^2/4,\infty)$ *is devoid of embedded point spectrum. In particular, the null-space* $L^2\mathcal{H}^k$ *is finite dimensional and may be canonically identified with the relative cohomology space* $H^k(X,\partial X)$*. When* $k>n/2$ *the same result is true if the statement above is modified by interchanging the roles of the numbers* $(n-2k-1)^2/4$ *and* $(n-2k+1)^2/4$*. In addition,* $L^2\mathcal{H}^k$ *is now identified with the absolute cohomology space* $H^k(X)$*. When* $k=n/2$ *the half-line* $[1/4,\infty)$ *is completely filled out by essential spectrum, and there are no embedded eigenvalues. There a finite number of finite multiplicity eigenvalues in the interval* $(0,1/4)$*, and* 0 *is an isolated eigenvalue of infinite multiplicity, i.e.* $L^2\mathcal{H}^{n/2}$ *has infinite dimension.*

The spectrum below $(n-2k\pm1)^2/4$ is analyzed by finding parametrices in $\Psi_e^{-2,E,F}(X)$ for which the remainder terms are compact. In particular, the first part of Proposition (2.17) may be applied to show that an arbitrary element of $L^2\mathcal{H}^k$ admits an expansion of the form (2.16) with all coefficient functions smooth. Using this the classical arguments of Hodge theory may be used to deduce the dimension of the harmonic space. Embedded eigenvalues in the appropriate intervals of the continuous spectrum are ruled out by first showing that an eigenform corresponding to such an eigenvalue would have to vanish to infinite order at ∂X (this uses a strengthening of Proposition (2.15)) and then applying a unique continuation for second order operators for this edge structure which is proved in **[M2]**. Recent work shows that there is some hope of obtaining an index theorem in the middle dimension using Proposition (2.17) as a starting point. The projector here is a real analogue of the Bergman projector in several complex variables. Similarly, it is possible to characterize the space of boundary values of $L^2\mathcal{H}^{n/2}$ as coinciding, up to a compact error, with the negative eigenspace for a certain Dirac operator on the boundary. In particular, the projector onto this space is an analogue of the classical Szegö projector.

§4. Scalar Curvature We now describe briefly another geometric problem, the analysis of which uses edge operators in a crucial way. The Yamabe theorem on compact manifolds has attracted a great deal of attention recently. The problem may be formulated as fixing a metric g_0 of volume 1 on a compact manifold without boundary and searching for a new metric g in the conformal class determined by g_0 also of volume 1 with constant scalar curvature. This problem was finally solved in 1984 by R. Schoen. It is possible to formulate this problem variationally, and the solution consists of showing that the minimum value for the 'energy' functional for the problem is always obtained by a positive smooth function. It turns out that this minimum is always unique except in the exceptional case of the sphere and the conformal class of the standard 'round' metric. Furthermore, although there may be other metrics of constant scalar curvature in a conformal class, the sign of the constant is an invariant of the conformal class. Actually, the only constant scalar curvature metric with normalized volume in a 'nonpositive' conformal class is the minimum of the functional. However, a positive conformal class may contain many normalized solutions, although R. Schoen has proved that the family of solutions in any conformal class (except the standard one on the sphere) is compact and generically finite.

The scalar curvature problem on noncompact manifolds has a rather different nature. Most attention has focussed on finding complete metrics on open manifolds obtained by omitting a closed subset Λ from a compact manifold N conformal to a given metric g_0 on N. Equivalently, one may think of this as finding a positive solution of the associated partial differential equation with prescribed singular set. This equation takes the following general form. If g_0 has scalar curvature R_0 and the conformally related metric $g = u^{4/(n-2)}g_0$ has scalar curvature R, where both R and R_0 may be variable, then the conformal factor u satisfies the equation

$$(4.1) \qquad \Delta_0 u - \frac{n-2}{4(n-1)}R_0 u = -\frac{n-2}{4(n-1)}Ru^{\frac{n+2}{n-2}}$$

Here the Laplacian is the one with negative spectrum. As stated before, if g_0 is a given metric on N, so that R_0 is some function, we are seeking a positive solution u to (4.1) on $N\backslash\Lambda$ which blows up at a sufficient rate along Λ that g is complete.

If Λ is a smooth submanifold of dimension k, and $n = \dim N$, then following earlier work of Loewner and Nirenberg, Aviles and McOwen **[A-M]** showed that in order that a complete conformal metric g on $N\backslash\Lambda$ exist with constant negative scalar curvature it is necessary and sufficient that $k > (n-2)/2$. Supplementary to this result is a theorem of Schoen and Yau **[S-Y]** which states that if $N = S^n$ and there exists a metric of constant positive scalar curvature on $S^n\backslash\Lambda$ (where now Λ is an arbitrary closed subset) then the Hausdorff dimension of Λ is less than or equal to $(n-2)/2$. The intermediate case $k = (n-2)/2$, presumably corresponding to vanishing scalar curvature, is rather unstable and not much is known about it.

An example in which the rôle of the relative dimensions k and n is clearly seen is when Λ is an equatorial S^k in S^n. Due to the conformally invariant nature of this equation, we may as well start with the flat metric on $\mathbf{R}^n\backslash\mathbf{R}^k$. Let y be the linear variable along $\mathbf{R}^k$ and (x, z) be polar coordinates in the orthogonal $\mathbf{R}^{n-k}$, $x \in \mathbf{R}^+$, $z \in S^{n-k-1}$. Then the flat metric $g_0 = dx^2 + x^2dz^2 + dy^2$ is clearly conformally related to

$$(4.2) \qquad g = \frac{dx^2 + dy^2}{x^2} + dz^2.$$

But this is simply the product metric on $\mathbf{H}^{k+1} \times S^{n-k-1}$. Its scalar curvature is $R_{n,k} = (n-2k-2)(n-1)$. In particular $R_{n,k} < 0$ when $k > (n-2)/2$ and $R_{n,k} > 0$ when $k > (n-2)/2$.

No general results like the Aviles-McOwen result are known in the positive case. However, R. Schoen **[SR]** was able to construct solutions on many subsets of the sphere $S^n\backslash\Lambda$. In particular, there exist solutions of the equation (4.1) with $R = R_0 = R_{n,k}$ whenever Λ is a finite subset of the sphere containing more than one point. The other types of sets Λ for which he was able to construct solutions are all rather similar to limit sets of Kleinian groups. In particular they are unrectifiable. Hence it is natural to ask whether there exist solutions on the complement of a smooth submanifold in the sphere. The only case where solutions had been known to exist is in the example above, where Λ is a 'round' subsphere (it need not be an equator because of the conformal invariance of the problem). Furthermore, nothing was known about uniqueness in the positive case. (In the negative case it is possible to show that there is a unique complete solution to (4.1) in any conformal class.) Although inconclusive in many respects, the following result is proved in **[M-S]**.

(4.3) THEOREM. *Let Λ be any k-dimensional submanifold of S^n with $k < (n-2)/2$ sufficiently close to the equator S^k in the $C^{3,\alpha}$ norm for some $0 < \alpha < 1$. Then there exists an infinite dimensional family of complete metrics on $S^n\backslash\Lambda$ conformal to the standard metric and with scalar curvature $R_{n,k}$.*

In fact, it is shown that the solution space forms an open Banach submanifold in an appropriate ambient Banach manifold consisting of the total deformation space for the metric and Λ. It is also shown that solutions may be found in a small tubular neighbourhood of an arbitrary smooth k-dimensional submanifold of S^n.

This result is proved by an application of the implicit function theorem. We sketch some of the difficulties involved. For notational convenience we shall only indicate how infinitely many solutions of (4.1) may be constructed on $S^n \backslash S^k$, i.e. with fixed singular set. The more general case only involves more notation. One solution of the problem is already known when $\Lambda = S^k$, namely the product solution (4.2). We take this as our background metric g_0. Set

$$H(u) = \Delta_0 u - \frac{n-2}{4(n-1)} R_{n,k} u + \frac{n-2}{4(n-1)} R_{n,k} u^{\frac{n+2}{n-2}}. \tag{4.4}$$

Then by construction $H(1) = 0$ and we seek to find a family of solutions near this one. The linearization of H about $u = 1$ is

$$L\phi = \Delta_0 \phi + (n - 2k - 2)\phi \tag{4.5}$$

An immediate simplification in this special case is that an arbitary positive solution of $H(u) = 0$ on $\mathbf{H}^{k+1} \times S^{n-k-1}$ may be shown to be spherically symmetric with respect to the second factor, hence may be considered as a function on $\mathbf{H}^{k+1}$. This uses the Alexandrov reflection principle. Hence we shall assume that Δ_0 in (4.5) is the Laplacian on hyperbolic space. In order to apply the implicit function theorem we must find spaces V and W such that $L : V \to W$ and such that V admits a decomposition $V = E \oplus F$ such that $E = \ker L \cap V$ and $L : F \to W$ is an isomorphism. Clearly this shows that we may not use classical L^2 theory; if $(n - 2k - 2) > k^2/4$ then it is well known, in particular cf. (3.4) above, that L does not have closed range on $L^2\mathbf{H}^{k+1}$. However, the weighted Hölder spaces of §2 prove to be suitable. In this case the set of indicial roots contains only two points $\{\frac{k}{2} \pm \sqrt{k^2/4 - (n-2k-2)}\}$, hence by Theorem (2.17) the linearization L has closed range as a map $C^{2,\alpha,\delta} \to C^{0,\alpha,\delta}$ whenever δ does not lie in the set of real parts of $\operatorname{spec}_b(L)$. Notice in particular that when $n - 2k - 2 > k^2/4$ this merely restricts δ to not equal $k^2/4$. Furthermore, when $\delta < k^2/4$ then L becomes surjective. Finally the boundedness of the projector P_2 on $C^{2,\alpha,\delta}$ shows that $C^{2,\alpha,\delta}$ decomposes into subspaces E and F with the properties stated above. E is the kernel of L in $C^{2,\alpha,\delta}$, hence is infinite dimensional since it may be parametrized by functions on the boundary S^k of $\mathbf{H}^{k+1}$. In order to ensure that the solutions of $H(u) = 0$ correspond to complete metrics we demand that u be bounded away from 0 and $+\infty$. This requires that $0 < \delta$. Hence we take $0 < \delta < \Re(k^2/4 - \sqrt{k^2/4} - (n-2k-2))$ and apply the implicit function theorem as described.

The full theorem, when Λ varies as well, requires the full linearization L on $\mathbf{H}^{k+1} \times S^{n-k-1}$ for which $\operatorname{spec}_b(L)$ is an infinite set.

It should also be noted that it is possible to prove an analogue of Proposition (2.15) for the nonlinear equation $H(u) = 0$ in certain cases, to conclude that the solution u admits an asymptotic expansion of the form (2.16) with smooth coefficients. It is unknown whether this expansion holds in general.

We conjecture that there is a set of positive solutions u to $H(u) = 0$ on $S^n \backslash S^k$ which are in one to one correspondence with the set of bounded positive measurable functions on S^k.

References

[A-M] P. Aviles and R. McOwen, *Complete Conformal Metrics with Negative Scalar Curvature in Compact Riemannian Manifolds*, Duke Math. Jour. **56 no. 2** (1988), 395-398.

[GL] R. Graham and J. Lee, *Einstein Metrics with Prescribed Conformal Infinity on the Ball*, to appear, Advances in Math..

[L-M-M] B. Livingston, R. Mazzeo and R.B. Melrose, *Fibred Cusps and Harmonic Forms*, in preparation.

[M1] R. Mazzeo, *Hodge Cohomology of a Conformally Compact Metric*, J. Diff. Geom. **28** (1988), 309-339.

[M2] —, *Unique Continuation at Infinity and Embedded Eigenvalues for Asymptotically Hyperbolic Manifolds*, to appear, Amer. Jour. Math..

[M3] —, *Elliptic Theory of Differential Edge Operators*, in preparation.

[M-P] — and R.S. Phillips, *Hodge Theory on Hyperbolic Manifolds*, Duke Math. Jour. **60 no. 2** (1990), 509-559.

[M-S] — and N. Smale, *Conformally Flat Metrics of Constant Positive Scalar Curvature on Subdomains of the Sphere*, preprint, 1989.

[Me1] R.B. Melrose, *Transformation of Boundary Problems*, Acta Math. **147:3-4** (1981), 149-236.

[Me2] —, *Analysis on Manifolds with Corners*, In preparation.

[Me-Me] — and G. Mendoza, *Elliptic Operators of Totally Characteristic Type*, MSRI preprint, 1983.

[Mc] R. McOwen, *The Behaviour of the Laplacian on Weighted Sobolev Spaces*, Comm. Pure Appl. Math. **XXXII** (1979), 783-795.

[R-S] S. Rempel and B.-W. Schulze, *Asymptotics for Mixed Elliptic Boundary Problems*, Vol. 50 Akademie-Verlag Berlin 1989.

[S] B.-W. Schulze, *Pseudodifferential Operators on Manifolds with Edges*, in "Symposium "Partial Differential Equations" Holzau 1988", Teubner-Texte zur Mathematik, Band 112, Leipzig, 1988, pp. 259 - 280.

[SR] R. Schoen, *The Existence of Weak Solutions with Prescribed Singular Behaviour for a Conformally Invariant Scalar Equation*, Comm. Pure and Appl. Math. **XLI** (1988), 317-392.

[S-Y] — and S.T. Yau, *Conformally Flat Manifolds, Kleinian Groups and Scalar Curvature*, Inventiones Math. **92** (1988), 47-71.

Department of Mathematics, Stanford, Ca. 94305

On solvability of systems of vector fields

by

GERARDO A. MENDOZA AND FRANÇOIS TREVES

§1. Introduction. Suppose L is a vector field with smooth coefficients defined in a neighborhood Ω of 0 in $\mathbf{R}^N$, let σ be its principal symbol, and assume $L \neq 0$ everywhere. Then

$$(P)\ \forall\ \zeta \in \mathbb{C}\setminus 0,\ \Im(\zeta\sigma) \text{ does not change sign along null-bicharacteristics of } \Re(\zeta\sigma)$$

is a sufficient condition for solvability of L, namely, if (P) holds, then there exists a neighborhood Ω_0 of 0 in Ω such that $\forall\ f \in L^2(\Omega_0)\ \exists\ u \in L^2(\Omega_0)$ such that $Lu = f$ in Ω_0, in the distributional sense. Conversely, if L is solvable in the above sense, then (P) holds in Ω_0. See [NT] and references therein, and [BF], [H].

If $N = 2$ and there is a C^∞ function Z on Ω with $LZ = 0$ and $dZ(0) \neq 0$, then the solvability of L, or equivalently, condition (P), can be related to a certain homological condition on the sets $Z^{-1}(z_0)$, the "fibers" of Z. In order to clarify this claim, let us assume that $Z = x + i\varphi(x,t)$ with φ real valued, a situation to which one is easily reduced by replacing Z with a suitable function $\tilde{Z} = F \circ Z$ (F holomorphic, defined near Z(0), with $F'(Z(0)) \neq 0$), and choosing an appropriate system of coordinates. The solvability of L near 0 is equivalent to the solvability of aL when a is a smooth nonvanishing function, so we may assume $L = D_t + i(\varphi_x D_t - \varphi_t D_x)$. The symbol of L is $\sigma = \tau + i(\varphi_x\tau - \varphi_t\xi)$, a null-bicharacteristic of $\Re(\sigma)$ is a curve $s \to (s+t_0, x_0, 0, \xi_0)$ with $\xi_0 \neq 0$ and along such a curve we have that $\Im(\sigma)$ does not change sign if and only if for each x, $\varphi_t(t,x)$ does not change sign as t ranges over any interval J such that $\{x\}\times J \subset \Omega$. If we take intervals I and J containing 0, with $I\times J \subset \Omega$, a neighborhood of 0, we conclude that (P) implies that φ is monotone in $t \in J$, for each $x \in I$, or equivalently, that the sets

$$\{(x,t) \in I\times J: x = x_0,\ \varphi(x,t) = y_0\}$$

are connected, for each $z_0 = x_0 + iy_0$. Conversely, if this last satement is true, then φ is monotone and thus condition (P) is satisfied by σ in U.

Connectedness is a homological property and we shall describe in Section 3 sufficient conditions of a similar nature for the solvability of certain systems of vector fields. The latter will be introduced in Section 2, where we also state the problem more precisely. Complete proofs of the theorems stated in Section 3 can be found in [MT].

Related results, concerning either necesary or sufficient conditions, can be found in [T1], [T3], [CH], [T5] and [CT]. In particular, [CT] contains results on necessary conditions for the existence

The first author was partially supported by a grant from Fundación Polar while visiting Rutgers University, New Brunswick, N. J. The second author was partially supported by NSF Grant DMS-8903007

of solutions, for systems of vector fields which are more general than the ones we treat in [MT].

§2. Statement of the problem. The systems of vector fields we are interested in are local expressions of certain complexes

$$L_q: C^\infty(M,E_q) \to C^\infty(M,E_{q+1}),\ q = 0,\ 1,\ \ldots$$

of first order differential operators between bundles over a (C^∞ paracompact N-dimensional) manifold M which are generalizations of the Cauchy-Riemann complex. The basic datum needed for the construction of the complexes, see [T2] or [T4], is a rank n subbundle V of CTM, the complexified tangent bundle of M, having the property that given two vector fields X, Y of the distribution V, the Lie bracket [X,Y] is also a vector field of V. Given such V, we let $E = V^*$ be the dual bundle, let $E_q = \Lambda^q E$, and define $L_q: C^\infty(M,E_q) \to C^\infty(M,E_{q+1})$ as follows. If u is a section of E_0, i. e., a function, then $L_0(u)(v) = du(v)$ for $v \in V$, and if $u \in C^\infty(M,E_q)$, $q > 0$, and $v_1, \ldots v_{q+1} \in V_p$, the fiber of V over p, then with smooth extensions $V_1, \ldots, V_{q+1}$ in V we set

$$(2.1)\quad L_q(u)(v_1,\ldots,v_{q+1}) = (q+1) \sum_{i=0}^{q+1} (-1)^{i+1} V_i\{u(V_1,\ldots,\hat{V}_i,\ldots,V_{q+1})$$
$$+ \sum_{i<j} (-1)^{i+j} u([V_i,V_j],V_1,\ldots,\hat{V}_i,\ldots,\hat{V}_j,\ldots,V_{q+1})$$

with the right hand side computed at p. Using the formal integrability of V one verifies that this definition is independent of the extensions V_i of the v_i, and that $L_{q+1}L_q = 0$. We shall drop the subindex from the notation and write simply L.

Many questions may be asked about the complex (2.1). Here we are interested in the local exactness at a point $p_0 \in M$. Namely, under what conditions is the following true:

$(2.2)_q$ Given a neighborhood Ω of p_0, there exists a neighborhood Ω' of p_0 in Ω such that $\forall\, f \in C^\infty(\Omega,E_q)$ with $Lf = 0\ \exists\, u \in C^\infty(\Omega,E_{q-1})$ such that $Lu = f$ in Ω'.

Our most fundamental hypothesis on L, or V, will be that the structure bundle V be "locally integrable" near p_0. This means that the annihilator $V^\perp$ of V in CT*M is generated, near p_0, by the differentials of $m = N - n$ smooth functions. That is, there are smooth function $Z_1,\ldots,Z_m$ defined in a neighborhood of Ω with $dZ_1,\ldots,dZ_m$ independent at every point, such that $LZ_j = 0$ for $j = 1,\ldots,m$. The existence of such sets of "solutions" is by no means automatic: there are examples (see [N], [JT], [T3]) of structures V where no such sets exist in any neighborhood of a point.

We shall also assume $m = 1$, although there are still a number of general statements one can make concerning the situation $m \geq 1$. We write Z instead of Z_1. Since the sufficient condition for solvability will involve the sets $Z^{-1}(z_0)$, $z_0 \in \mathbf{C}$, it is important to point out that these sets are determined by V itself, in the following sense: any point p in the domain of Z has a neighborhood

U such that if h is a C^1 solution of $Lh = 0$ with the same domain as Z then $\forall\ z_0 \in \mathbf{C}$, h is constant on the sets $Z^{-1}(z_0) \cap U$. This was shown in [BT].

If dZ generates $V^{\perp}$ over Ω, then the same is true of $\zeta Z + \eta$, $\zeta \in \mathbf{C}\backslash 0$, $\eta \in \mathbf{C}$. Thus by choosing ζ and η we may assume $Z(p_0) = 0$ and $d\Re(Z) \neq 0$ at p_0. Let $\tilde{x} = \Re(Z)$, $t_1, \ldots, t_n$ be coordinates centered at p_0, let $\tilde{Z} = Z(\tilde{x},t)/Z_{\tilde{x}}(0,0)$. Then $x = \Re(\tilde{Z})$, $t_1, \ldots, t_n$ is a local chart centered at p_0, $d\tilde{Z}$ spans $V^{\perp}$ over Ω, and $\tilde{Z} = x + i\varphi(x,t)$, with φ real valued, smooth and $\varphi(0,0) = \varphi_x(0,0) = 0$. We shall assume that we have coordinates $x, t_1, \ldots, t_n$ in Ω centered at p_0 with respect to which $Z = x + i\varphi$ with φ as described above. A basis for V over Ω is then given by the commuting vector fields

$$(2.3) \qquad L_i = \partial_{t_i} - \frac{i\varphi_{t_i}(x,t)}{1 + i\varphi_x(x,t)}\partial_x.$$

The forms $dZ, dt_1, \ldots, dt_n$ span $\mathbf{C}T^*\Omega$, so $V^*|_{\Omega}$ is isomorphic to the span of the dt_i, and any element $u \in C^{\infty}(\Omega,E_q)$ can be identified to an element $\tilde{u} = \sum_{|I|=q}\tilde{u}_I(x,t)dt_I$ with $\tilde{u}_I \in C^{\infty}(\Omega)$. With this identification we have

$$(2.4) \qquad L\Big(\sum_{|I|=q} u_I(x,t)dt_I\Big) = \sum_{|I|=q}\ \sum_{i=1}^{n} L_i u_I(x,t)dt_i \wedge dt_I$$

Notice that if $\varphi_t(0,0) \neq 0$ then the complex L is elliptic near p_0 and $(2.2)_q$ holds for all q. Thus we shall always assume $\varphi_t(0,0) = 0$.

§3. Results. Throughout this section Ω will be a neighborhood of 0 in $\mathbf{R}_x \times \mathbf{R}^n_t$, $Z = x + i\varphi(x,t)$ with $\varphi \in C^{\infty}(\Omega)$, real valued and satisfying $\varphi(0,0) = d\varphi(0,0) = 0$. The vector bundle V is spanned by the vector fields (2.3), E is the subbundle of $\mathbf{C}T^*\Omega$ spanned by the differentials dt_i and $L: C^{\infty}(M,E_q) \to C^{\infty}(M,E_{q+1})$, $q = 0, 1, \ldots, n$ is given by (2.4).

It was shown in [T3] that if the coefficients of the vector fields L_i are real-analytic and every point $p \in \Omega$ has a basis N_p of neighborhoods such that $Z^{-1}(z_0) \cap U$ is connected whenever $U \in N_p$, then $(2.2)_1$ holds. As regards to top degree, [CH] showed that if there exists a neighborhood U of 0 such that the sets $Z^{-1}(z_0) \cap U$ have trivial $(n-1)$ homology, and again Z is real-analytic, then $(2.2)_n$ holds. The condition that Z be real analytic in the case of top degree forms was removed in [T5]. In [MT] we consider the condition

$(3.1)_d$ There exist two bases of neighborhoods $\{U_k\}_{k\in N}$, $\{U'_k\}_{k\in N}$ of 0 with $U'_k \subset U_k$, $U_k = I_k \times Q_k$ with Q_k homologically trivial, such that $\forall\ z \in \mathbf{C}$, if $Z^{-1}(z) \cap U'_k \neq \emptyset$, then the d-homology of $Z^{-1}(z) \cap U_k$ is trivial.

We show

THEOREM 3.1. If $(3.1)_0$ holds, then $(2.2)_1$ holds.

THEOREM 3.2. If $(3.1)_{q-1}$ and $(3.1)_{q-2}$ hold, then $(2.2)_q$ holds.

Theorem (3.1) improves the result on sufficiency in [T3] in that the vector fields need not have real-analytic coefficients. On the other hand, Theorem 3.2, in the case $q = n$, is weaker than the results in [CH] and [T5]: the additional condition on the $(n - 2)$ homology of the fibers is not necessary then, and in fact, there are examples of solvable systems in degree q where $(3.1)_{q-1}$ holds but not $(3.1)_{q-2}$.

We shall end with an outline of our proof of theorem 3.1. Assume $(3.1)_0$ holds. There is no loss in assuming that the neighborhoods U_k' have the form $J_k \times B_k$ where the B_k are balls, so we shall assume this from now on. Fix $x_0, y_0 \in \mathbf{R}$, let $z_0 = x + iy_0$. We claim that if v is a C^∞ locally constant function defined in the set $F = \{t \in Q_k: \varphi(x_0,t) < y_0\}$ then v is constant on $F^+(z_0) = \{t \in B_k: \varphi(x_0,t) < y_0\}$ (if this set is nonempty). Indeed, either $\varphi(x_0,t) < y_0$ when $t \in B_k$ and then v is constant on the connected set B_k, or there exists $t_0 \in B_k$ where $\varphi(x_0,t_0) = y_0$ in which case one can show, using the augmented Mayer-Vietoris sequence, Sard's Lemma and the conditions on U_k, U_k' and the fibers of Z expressed by $(3.1)_0$, that F itself is connected. A similar statement holds on the sets $F^-(z_0) = \{t \in B_k: \varphi(x_0,t) > y_0\}$.

Let $i:\Omega \to \mathbf{C}\times\mathbf{R}^n$ be the map $i(x,t) = (Z(x,t),t)$; note that

$$i_*(L_i) = \partial_{t_i} - \frac{2i\varphi_{t_i}}{1 + i\varphi_x(x,t)}\partial_{\bar{z}}.$$

Denote by $\pi:\mathbf{C}\times\mathbf{R}^n \to \mathbf{C}$ the proyection and let $\rho:\mathbf{C}\times\mathbf{R}^n \to \mathbf{R}\times\mathbf{R}^n$ be the map $\rho(z,t) = (x,t)$; we write $z = x + iy$. Let

$$O^+ = \{(z,t): \rho(z,t) \in \Omega,\ y > \varphi(x,t)\}$$
$$O^- = \{(z,t): \rho(z,t) \in \Omega,\ y < \varphi(x,t)\}$$

and let $\Sigma = i(\Omega)$. There exists a neighborhood W of 0 in $\mathbf{C}\times\mathbf{R}^n$ such that for any $f = \sum f_i(x,y)dt_i$, $f_i \in C^\infty(\Omega)$, there exists $f^\pm = \sum f_i^\pm(z,t)dt_i$, $f_i^\pm \in C^\infty(W\cap O^\pm)$ smooth up to $W\cap\Sigma$, holomorphic in z and such that $i^*(f^+|_{\Sigma\cap W} + f^-|_{\Sigma\cap W}) = f|_{i^{-1}(W)}$. Furthermore, on can choose W so that if $Lf = 0$ in Ω then $d_t f^\pm = 0$. Here d_t indicates differentiation in the t variables only, z is a parameter. The construction of f^+ and f^- in [MT] is such that not only is $d_t f^\pm = 0$, but in fact, for each z_0, $f^\pm|_{\pi^{-1}(z_0)}$ is exact. With this information and the validity of $(3.1)_0$ one can find another neighborhood W_1 of 0 in W, independent of f, and $u^\pm \in C^\infty(W_1\cap O^\pm)$ such that

$$d_t u^\pm = f^\pm \text{ in } W_1\cap O^\pm$$
$$|\partial_x^\alpha \partial_y^\beta u(z,t)| \le C|y - \varphi(z,t)|^{-\kappa}\|f\|_{K,N} \text{ if } (z,t) \in W_1\cap O^\pm,$$

with C, κ, N positive, depending on α, β but not on f, and K some fixed compact set in Ω. Of course $u^{\pm}$ need not be holomorphic. However, $v^{\pm} = \partial u^{\pm}/\partial \bar{z}$ is necesarily locally constant on each set $\pi^{-1}(z_0)\cap W_1\cap O^{\pm}$, since its differential in t vanishes. Choose k and s such that $W_2 = \{(z,t): \rho(z,t) \in U_k = J_k\times Q_k, |y| < s\} \subset W_1$ and let $W_3 = \{(z,t): \rho(z,t) \in U_k' = J_k\times B_k, |y| < s\}$. Then $v^{\pm}|_{W_3\cap O^{\pm}}$ is constant on the sets $\pi^{-1}(z_0)\cap W_3\cap O^{\pm}$. Indeed, $v^{+}(z_0,\cdot)$ is locally constant on $\{t \in Q_k: \varphi(x_0,t) < y_0\} \approx \pi^{-1}(z_0)\cap W_2\cap O^{\pm}$ when $z_0 \in \pi(W_2\cap O^{\pm})$, thus constant on $\{t \in B_k: \varphi(x_0,t) < y_0\} \approx \pi^{-1}(z_0)\cap W_3\cap O^{\pm}$ when $z_0 \in \pi(W_3\cap O^{\pm})$, and similarly for v^{-}. Let $D^{\pm} = \pi(W_3\cap O^{\pm})$, $g^{\pm}$ the function on $D^{\pm}$ defined by $v^{\pm}$. If Δ is a sufficiently small neighborhood of 0 in **C**, depending only on W_3, one can find $\psi^{\pm} \in C^{\infty}(\Delta\cap D^{\pm})$ such that $\partial\psi^{\pm}/\partial\bar{z} = g^{\pm}$ in $\Delta\cap D^{\pm}$ and such that

$$|\pi^*\psi^{\pm}| \le C|y - \varphi(z,t)|^{-\kappa}\|f\|_{K,N} \text{ in } W_4\cap O^{\pm} \tag{3.2}$$

where $W_4 = \pi^{-1}(\Delta)\cap W_3$, for certain numbers C, M and N and compact set K in Ω independent of f. Let $w^{\pm} = \pi^*\psi^{\pm}$.

If $\tilde{u}^{\pm} = u^{\pm} - w^{\pm}$, in $W_4\cap O^{\pm}$, then $\tilde{u}^{\pm}$ is smooth, holomorphic in z, and satisfies an estimate of the form (3.2), again with C, κ, N and K independent of f. This implies that $\tilde{u}^{\pm}$ has distributional boundary values $\mu^{\pm}$ on $\Sigma\cap W_4 \approx i^{-1}(W_4) = \Omega_0$ and $L(\mu^{\pm}) = f^{\pm}$ in Ω_0, thus $L(\mu^{+} + \mu^{-}) = f$ in that set. The distribution $\mu = \mu^{+} + \mu^{-}$ has finite order independent of f, and one can use this together with the integrability of the structure V to show that in a smaller neighborhood Ω' of 0 in Ω_0 there is in fact a C^{∞} solution u of Lu = f.

References

[BF] Beals, R. & Fefferman, C. *On local solvability of linear partial differential equations*, Ann. of Math. **97** (1973), 482-498.

[BT] Baouendi, M. S., & Treves, F., *A property of the functions and distributions annihilated by a locally integrable system of complex vector fields*. Ann. of Math. **113** (1981), 387-421.

[CH] Cordaro, P. & Hounie, J., *On local solvability of undetermined systems of vector fields*, Amer. J. of Math. **112** (1990), 243-270.

[CT] Cordaro, P. & Treves, F., *Homology and cohomology in hypoanalytic structures of the hypersurface type*, to appear.

[H] Hörmander, L. *Pseudo-differential operators of principal type*, in "Singularities in Boundary Value Problems", H. G. Garnir, ed., 1981, D. Reidel Pub. Co., pp. 69-96.

[JT] Jacobowitz, H. & Treves, F., *Non-realizable CR structures*, Invent. Math. **66** (1982) 231-249.

[MT] Mendoza, G. A., & Treves, F., *Local solvability in a class of overdetermined systems oflinear PDE*, preprint.

[N] Nirenberg, L., *Lectures on linear partial differential equations*, Reg. Conf. Series in Math. **17** (1973), AMS.

[NT] Nirenberg, L. & Treves, F., *On local solvability of linear partial differential equations, I* and *II*, Comm. on Pure Appl. Math. **23** (1970), 1-38, 459-509.

[T1] Treves, F., *Study of a model in the theory of complexes of pseudo-differential operators*, Ann. of Math. **104** (1976), 269-324.

[T2] Treves, F., *Approximation and representation of functions and distributions annihilated by a system of complex vector fields*, Centre Math. Ecole Polytechnique, Palaiseau, France

(1981).
[T3] Treves, F., *On the local integrability and local solvability of systems of vector fields*, Acta. Math. **151** (1983), 1-48.
[T4] Treves, F., *Hypo-analytic structures*, Cont. Math. **27** (1984), 23-44.
[T5] Treves, F., *On the local solvability for top degree forms in hypo-analytic structures*, Amer. J. of Math **113** (1990), 403-421

G. A. Mendoza: IVIC-Matemáticas, Apartado 21827, Caracas 1020-A, Venezuela, and Department of Mathematics, University of Washington, Seattle, WA 98195, USA
F. Treves: Mathematics Department, Rutgers University, New Brunswick, N.J. 08903, USA

On the Neumann Problem in Domains with Edges

S.A. Nazarov, B.A. Plamenevskij

The report is devoted to the discussion of different statements of the Neumann problem for self-adjoint elliptic systems in domains with edges on the boundary. The aim consists in the determination of such statements that give Fredholm operators. The problem is considered in appropriate functional spaces with weighted norms. In some cases it is natural to submit the solution to the "radiation conditions" on edges.

Different results about the Neumann problem for concrete elliptic equations in the domains with edges have been obtained in [1]-[4]. The class of elliptic systems, which is investigated in the present report, has been extracted by Nazarov [5]. In connection with the radiation conditions we indicate the paper [6]. The Neumann problem in the domains with the intereserting edges is treated in [7]. Certain subjects of this report are discussed in detail in the book [8].

1 Fredholm property of Neuman problem operator

1.1 Statement of the problem

Let G be a domain in $\mathbb{R}^n$ with boundary ∂G and let $M \subset \partial G$ be smooth (of class C^∞) closed d-dimensional submanifold in $\mathbb{R}^n$, $0 < d < n-1$. Suppose that $\partial G \backslash M$ is a smooth $(n-1)$-dimensional submanifold in $\mathbb{R}^n$ and for each point z^0 in "the edge" M there exists a neighbourhood U in $\mathbb{R}^n$ and a diffeomorphism $\kappa : U \to B_\rho^n(0)$ such that $\kappa(U \cap G) = B_\rho^n(0) \cap \mathbb{D}(z^0)$; here $B_\rho^n(0)$ is a ball with the radius $\rho > 0$ and the center at 0, $\mathbb{D}(z^0)$ is the wedge $\mathbb{K}(z^0) \times \mathbb{R}^d$ and $\mathbb{K}(z^0)$ is an open $(n-d)$-dimensional cone, which cuts out on the sphere $S^{n-d-1} = \partial B_1^{n-d}(0)$ the domain $\Omega(z^0)$ with smooth $(n-d-2)$-dimensional boundary $\partial\Omega(z^0)$.

Let $\mathcal{L}(x, D_x)$ be a $k \times k$-matrix differential operator having smooth coefficient in $\overline{G}$, $ord\ \mathcal{L}_{hj} = t_h + t_j$ while $\{t_1, \dots, t_k\}$ is a set of natural numbers, $m = t_1 + \dots + t_k$ and $D_x = -igrad$. We assume there is a GNreen formula

$$(\mathcal{L}u, v)_G + (\mathfrak{N}u, \mathcal{D}v)_{\partial G} = a(u, v; G), \tag{1}$$

where $u = (u_1, \dots, u_k)$, $v = (v_1, \dots, v_k)$ belong to $C_c^\infty(\overline{G} \backslash M)^k$, $(\ ,\ )_G$ and $(\ ,\ )_{\partial G}$ are scalar products in $L_2(G)^k$ and $L_2(\partial G)^m$, respectively, $\mathcal{D}(x, D_x)$ is a Dirichlet system, $\mathfrak{N}(x, D_x)$ is a $(m \times k)$-matrix differential operator and $a(u, v; G)$ is a symmetric form

$$a(u, v; G) = \sum_{h,j=1}^{k} \sum_{|\sigma| \le t_h} \sum_{|\alpha| \le t_j} \int_G a_{\alpha,\sigma}^{j,h}(x) D_x^\alpha u_j(x) \overline{D_x^\sigma v_h(x)}\, dx \tag{2}$$

with $a_{\alpha,\sigma}^{j,h} \in C^\infty(\overline{G})$. We denote elements of matrices $\mathcal{D}, \mathfrak{N}$ by $\mathcal{D}_{qh}, \mathfrak{N}_{qh}$ and put $ord\ \mathfrak{N}_{qj} = \sigma_q + t_j$. It is clear that $ord\ \mathcal{D}_{qh} = t_h - \sigma_q - 1$ and $\sigma_q \ge 0$, $q = 1, \dots m$.

We consider the boundary value problem

$$\mathcal{L}(x, \mathcal{D}_x)u(x) = f(x),\ x \in G;\ \mathfrak{N}(x, \mathcal{D}_x)u(x) = g(x),\ x \in \partial G \setminus M, \tag{3}$$

supposing that the operator $\{\mathcal{L}, \mathfrak{N}\}$ is elliptic on $\overline{G} \setminus M$ and doesn't degenerate on M (see the assumptions about $\{\mathcal{L}^0, \mathfrak{N}^0\}$ in section 2). From (1), (2) we obtain that the problem (3) is formally self-adjoint.

1.2 List of model problems

Fix a point $z^0 \in M$. In a neighbourhood U we write the operator $\mathcal{L}$ in coordinates $(y,z) \equiv \kappa(x) \in \mathbb{K}(z^0) \times \mathbb{R}^d$ and denote by $\mathcal{L}^0(z^0, D_y, D_z)$ the principal part at the point $0 = \kappa(z^0)$ of the operator obtained. In the sequel we shall not indicate the point z^0 in $\mathbb{D}(z^0)$, $\mathbb{K}(z^0)$, etc.

The equality (1) implies the Green formula of the form

$$(\mathcal{L}^0 u, v)_{\mathbb{D}} + (\mathfrak{N}^0 u, \mathcal{D}^0 v)_{\partial\mathbb{D}} = a^0(u, v; \mathbb{D}), \tag{4}$$

where $u, v \in C_c^\infty(\overline{\mathbb{D}} \setminus \Gamma)$, Γ is the edge $0 \times \mathbb{R}^d$ of $\mathbb{D}$. $\mathcal{D}^0$ is a Dirichlet system,

$$a^0(u, v; \mathbb{D}) = \sum_{h,j=1}^{k} \sum_{|\sigma|=t_h} \sum_{|\alpha|=t_j} \int_{\mathbb{D}} a_{\alpha,\sigma}^{0j,h}(x) D_{y,z}^{\alpha}(y,z) \overline{D_{y,z}^{\sigma}(y,z)}\, dydz. \tag{5}$$

Suppose that the operator $\{\mathcal{L}^0, \mathfrak{N}^0\}$ is elliptic on $\overline{\mathbb{D}} \setminus \Gamma$ and the condition

$$a^0(v, v; Q) = 0 \Longrightarrow v \in \Pi|_Q \tag{6}$$

is satisfied; here Q is an arbitrary domain in $\mathbb{R}^n$, Π is some fixed subspace of polynomials with coefficients from $\mathbb{C}^k$.

Remark 1.1 1^0. *if $G \subset \mathbb{R}^3$ and $\{\mathcal{L}, \mathfrak{N}\}$ is the operator of the elasticity theory problem the the lineal Π consists of rigid shifts $v(x) = B + c \times x$, where B, c are constant vectors and the vector product is denoted by $\times$,*

2^0. *By (5) the lineal Π contains each function $p = (p_1, \dots, p_k)$ with polynomials p_j, $\deg p_j \le t_j - 1$, $j = 1, \dots, k$. Since $\mathcal{D}^0$ is a Dirichlet system we obtain from (4), (5) that these polynomials are solutions of the problem $\{\mathcal{L}^0, \mathfrak{N}^0\}u = 0$ in $\mathbb{D}$.*

We denote by $V_\beta^l(\mathbb{D}, \Gamma)$, where $\beta \in \mathbb{R}$, $l = 0, 1, \dots$, the space of functions in $\mathbb{D}$ with norm

$$||v; V_\beta^l(\mathbb{D}, \Gamma)|| = \left(\sum_{|\alpha| \le l} |||y|^{\beta - l + |\alpha|} D_{y,z}^\alpha v; L_2(\mathbb{D})||^2 \right)^{1/2} \tag{7}$$

and denote by $V_\beta^{l-1/2}(\partial\mathbb{D}, \Gamma)$ for $l \ge 1$ the corresponding space of traces on $\partial\mathbb{D} \setminus \Gamma$. The operator $\{\mathcal{L}^0, \mathfrak{N}^0\}$ of the boundary value problem in $\mathbb{D}$ maps continuously

$$\begin{aligned} \mathcal{D}_\beta^l V(\mathbb{D}, \Gamma) &\equiv \prod_{j=1}^{k} V_\beta^{l+t_j}(\mathbb{D}, \Gamma) \ni u \to \{\mathcal{L}^0, \mathfrak{N}^0\}u \in \mathcal{R}_\beta^l V(\mathbb{D}, \Gamma) \\ &\equiv \prod_{h=1}^{k} V_\beta^{l-t_h}(\mathbb{D}, \Gamma) \times \prod_{q=1}^{m} V_\beta^{l-\sigma_q - 1/2}(\partial\mathbb{D}, \Gamma) \end{aligned} \tag{8}$$

for every $\beta \in \mathbb{R}$ and $l > \max\{t_1, \dots, t_k\} \equiv \max t_j$.

If we apply the Fourier transfom $\mathcal{F}_{z\to\xi}$ and change coordinates and functions

$$y = |\xi|^{-1}\eta, \hat{u}_j(y,\xi) = |\xi|^{-t_j}\, U_j(\eta,\xi), \; \hat{f}_h(y,\xi) = |\xi|^{t_h} F_h(\eta,\xi), \; g_q(y,\xi) = |\xi|^{\sigma_q} G_q(\eta,\xi)$$

then we obtain from equation

$$\{\mathcal{L}^0(D_y, D_z), \mathfrak{N}^0(y, D_y, D_z)\}u = \{f, g\} \tag{9}$$

the following problem, depending on the parameter $\Theta = |\xi|^{-1}\,\xi \in S^{d-1}$

$$\overset{0}{A}(\Theta)U \equiv \{\mathcal{L}^0(D_y,\Theta), \mathfrak{N}^0(y,D_y,\Theta)\}U = \{F,G\} \tag{10}$$

in the cone $\mathbb{K}$. The following norms are equivalent:

$$||u;\mathcal{D}^l_\beta V(\mathbb{D},\Gamma)|| \sim \left(\int_{\mathbb{R}^d} |\xi|^{2(l-\beta)-n+d} ||U(\cdot,\xi);\mathcal{D}^l_\beta E(\mathbb{K})||^2\, d\xi\right)^{1/2} \tag{11}$$

Here the vector space $\mathcal{D}^l_\beta E(\mathbb{K})$ is constructed similary to (8) while

$$||v;E^l_\beta(\mathbb{K})|| = \left(\sum_{|\alpha|\leq l} ||\rho^\beta(1+\rho)^{l-|\alpha|}D^\alpha_\eta V;L_2(\mathbb{K})||^2\right)^{1/2} \tag{12}$$

and $\rho = |\eta|$. The obvious alterations in the second line of (8) define the spaces of $\mathcal{R}^l_\beta E(\mathbb{K})$. The operator

$$\overset{0}{A}{}^l_\beta(\Theta) : \mathcal{D}^l_\beta E(\mathbb{K}) \longrightarrow \mathcal{R}^l_\beta E(\mathbb{K}) \tag{13}$$

is continuous. The mapping (8) is Fredholm iff the operator (13) is an isomorphism.

Now we introduce one more model problem. We write the differential operator in the spherical coordinates (ρ,ω),

$$\mathcal{L}^0_{hj}(D_\eta,0) = \rho^{-t_h-t_j}L_{hj}(\omega,D_\omega,\rho D_\rho),\ \mathfrak{N}^0_{qj}(\eta,D_\eta,0) = \rho^{-\sigma_q-t_j}N_{qj}(\omega,D_\omega,\rho D_\rho),$$

and define the operator pencil

$$\mathbb{C} \ni \lambda \to \mathfrak{a}(\lambda) = \left\{||L_{h_j}(\omega,D_\omega,\lambda-it_j)||^k_{h,j=1},\ ||N_{qj}(\omega,D_\omega,\lambda-it_j)||^{m,k}_{q,j=1}\right\} \tag{14}$$

associated with the boundary value problem in the domain $\Omega = \mathbb{K}\cap S^{n-d-1}$. The spectrum of the pencil (14) consists of normal eigenvalues situated inside the set $\{\lambda\in\mathbb{C} : |Im\ \lambda| < s|Re\ \lambda|\}$ $\cup\{\lambda\in\mathbb{C} : |Re\ \lambda| < s\}$, where s is certain positive number. In view of formal self-adjointness of the problem (10) points of the spectrum are disposed symmetrically with respect to the line $\mathbb{R}+i(n-d)/2$.

1.3 Properties of the operator $\overset{0}{A}{}^l_\beta(\Theta)$

The mapping (13) is a Fredholm operator iff the line $\mathbb{R}+i(\beta-l+(n-d)/2)$ doesn't contain eigenvalues of the pencil $\mathfrak{a}$. The assumption (6) implies

Lemma 1.2 1^0. *If $n-d$ is odd or $n-d > 2\max t_j$, then the line $\mathbb{R}+i(n-d)/2$ is free from the spectrum of the pencil (14).*

2^0. *If $n-d$ is even, $n-d \leq 2\max t_j$, then only the eigenvalue $\lambda_0 = i(n-d)/2$ belongs to $\mathbb{R}+i(n-d)/2$. The space of eigenvectors coincides with the lineal of traces on the sphere S^{n-d-1} of the elements of the set $\Pi_0 = \{(p_1,\dots,p_k)\in\Pi : \deg\ p_j = t_j-(n-d)/2\}$. Every eigenvector has an associated vector and there are no Jordan chains with length > 2.*

In the sequel the dimensions n and d from 2^0 and 1^0 will be called critical and non-critical, respectively. We denote by δ the maximal positive number such that the strip $\{\lambda\in\mathbb{C} : Im\ \lambda - (n-d)/2 \in (0,\delta)\}$ does not contain eigenvalues of the pencil $\mathfrak{a}$. Taking into account (6) it is possible to check the mapping (13) is a monomorphism for $\beta = \beta_- \in (l-\delta,l)$ and an epimorphism $\beta = \beta_+ \in (l,l+\delta)$. In view of Lemma 1.2 and formal self-adjointness of the problem (10) the equalities $Ind\ \overset{0}{A}{}^l_{\beta_+}(\Theta) = 2dim\ \Pi_0 + Ind\ \overset{0}{A}{}^l_{\beta_-}(\Theta)$, $Ind\ \overset{0}{A}{}^l_{\beta_+}(\Theta) = -\ \overset{0}{A}{}^l_{\beta_-}(\Theta)$ are valid; here $IndA = dim\ \ kerA - dim\ \ cokerA$ and $\Pi_0 = \{0\}$ in the critical case.

Theorem 1.3 1^0. *For non-critical dimensions the operator (13) is an isomorphism iff* $|\beta - l| < \delta$.

2^0. *For critical dimensions the operator (13) is not an isomorphism for every* β.

1.4 Fredholm property of the original problem.

Let

$$A^l_\beta = \{\mathcal{L}, \mathfrak{N}\} : \mathcal{D}^l_\beta V(G, M) \longrightarrow \mathcal{R}^l_\beta V(G, M) \tag{15}$$

be the operator of the problem (13). Fredholm property of A^l_β is deduced as usually from the assertion 1.3(1^0) in the case $|\beta - l| < \delta_0$, where $\delta_0 = \min\{\delta(z^0) : z^0 \in M\}$ and $\delta(z^0)$ is the number, defined after Lemma 1.2. Functional spaces in (16) are defined similary to (18) while $\mathbb{D}, \Gamma, |y|$ and (y, z) are replaced in (7) by $G, M, dist(x, M)$ and x.

Theorem 1.4 *If the dimension is non critical and* $|\beta - l| < \delta_0$ *then the mapping (9) is an isomorphism and the mapping (16) is a Fredholm operator, while* $Ind A^l_\beta = 0$.

If $n - f$ is even and $n - d \leq 2 \max t_j$ (i.e. we have the critical case) then the mapping is not Fredholm for every $\beta \in \mathbb{R}$. Indeed, in view of (15) we have

$$\begin{aligned} dim\ \ ker\ \overset{0}{A^l_\beta} (\Theta) \ &\geq\ dim\ ker\ \overset{0}{A^l_{\beta_+}} (\Theta) \ = \ dim\ \Pi_0 > 0 \text{ for } \beta > l, \\ dim\ \ coker\ \overset{0}{A^l_\beta} (\Theta) \ &\geq\ dim\ coker\ \overset{0}{A^l_{\beta_-}} (\Theta) \ = \ dim\ \Pi_0 > 0 \text{ for } \beta > l. \end{aligned} \tag{16}$$

Hence $dim\ ker A^l_\beta = \infty$ (for $\beta > l$) or $dim\ coker A^l_\beta = \infty$ (for $\beta < l$); besides $Im A^l_\beta$ is not closed because of presence of the eignevalue λ_0 on the line $\mathbb{R} + i(n - d)/2$. Thus to provide a Fredholm property of problem (13) in the critical case we must change the domain of the operator $A + \{\mathcal{L}, \mathfrak{N}\}$ (and the operator $\overset{0}{A} (\Theta)$).

If $\beta \in (l - \delta_0, l)$ then $\overset{0}{A^l_\beta} (\Theta)$ is a monomorphism and $dim\ coker\ \overset{0}{A^l_\beta} (\Theta) = dim\ \Pi_0$. The function χp (where $p \in \Pi_0$ and χ is a cutoff function from $C^\infty_c(\overline{\mathbb{K}})$, $\chi = 1$ near the vertex of the cone) doesn't belong to $\mathcal{D}^l_\beta E(\mathbb{K})$ but the inclusion $\overset{0}{A} (\Theta)(\chi p) \in \mathcal{R}^l_\beta E(\mathbb{K})$ is correct for $\beta > l - 1$. (Diminishing δ_0 we shall suppose the last inequality is valid). Thus it is convenient to extend the domain of $\overset{0}{A}$ by joining to $\mathcal{D}^l_\beta E(\mathbb{K})$ the set $\{\chi p : p \in \Pi_0\}$ (the formal stimulus : in view of such extension the index $Ind\ \overset{0}{A}$ increases by $dim\ \Pi_0$ and becomes zero). To this end we introduce the space $E^{l,s}_\beta(\mathbb{K})$ with the norm

$$||V; E^{l,s}_\beta(\mathbb{K})|| = \left(\sum_{|\alpha| \leq l} ||\rho^\beta (1 + \rho^{-1})^{|\alpha| - l - \Theta(s - |\alpha|)} D^\alpha_\eta V; L_2(\mathbb{K})||^2 \right)^{1/2} \tag{17}$$

wher $\Theta(t) = 1$ for $t \geq 0$ and $\theta(t) = 0$ for $t < 0$, $s \leq l$. If $s < 0$ then the norms (12) and (17) coincide.

Lemma 1.5 *For each function*

$$U \in \mathcal{D}^{l,(d-n)/2}_\beta E(\mathbb{K}) \equiv \prod_{j=1}^{k} E^{l+t_j, t_j + (d-n)/2}_\beta (\mathbb{K}) \tag{18}$$

the representation of the form $U = \chi p + W$ *is correct; here* $p \in \Pi_0$ *and* $W \in \mathcal{D}^l_\beta E(\mathbb{K})$.

Theorem 1.6 *Let* $n - d$ *be even,* $n - d \leq 2 \max t_j$ *and* $\beta \in (l - \delta_0, l)$ *while* $\delta_0 \leq 1$. *Then the mapping* $\overset{0}{A} (\Theta) : \mathcal{D}^{l,(n-d)/2}_\beta E(\mathbb{K}) \to \mathcal{R}^l_\beta E(\mathbb{K})$ *is an isomorphism.*

Now we are ready to investigate the problem (9) in the wedge $\mathbb{D}$ and the problem (4) in the domain G for the critical case. Introduce the space $V_{\beta}^{l,s}(G,M)$ by taking

$$\|v;V_{\beta}^{l,s}(G,M)\| = \left(\sum_{|\alpha|\le l} \|dist(x,M)^{\beta-l+|\alpha|+\Theta(s-|\alpha|)}D_x^{\alpha}v;L_2(G)\|^2\right)^{1/2};$$

the space $\mathcal{D}_{\beta}^{l,(d-n)/2}V(G,M)$ is constructed similarly to (8), (18).

Theorem 1.7 *Under the condition of Theorem 1.6 the operator*

$$A_{\beta}^{l,(d-n)/2} \equiv \{\mathcal{L},\mathfrak{N}\} : \mathcal{D}_{\beta}^{l,(d-n)/2}V(G,M) \longrightarrow \mathcal{R}_{\beta}^{l}V(G,M)$$

is Fredholm and $IndA_{\beta}^{l,(d-n)/2} = 0$.

2 Generalized solution of Neumann problem

2.1 Generalized statements of the problem.

Let $H^l(G)$ be a Sobolev space in G, $\mathcal{D}^l H(G) = \prod_{j=1}^{k} H^{l+t_j}(G)$. A generalized solution of Neumann problem is the element $u \in \mathcal{D}^0 H(G)$ subject to condition

$$a(u,v;G) = < Y, v > \tag{1}$$

for every $v \in \mathcal{D}^0 H(G)$; here $< \, , \, >$ is the duality for pair $\mathcal{D}^0 H(G)^*$, $\mathcal{D}^0 H(G)$. If righthand side $\{f,g\}$ of 1.(3) belongs to $L_2(G)^k \times L_2(\partial G)^m$ then

$$< Y, v >= (f,v)_G + (g,\mathcal{D}v)_{\partial G}. \tag{2}$$

When $n-d < 2\max t_j$ it is possible to consider the Sobolev problem containing additional conditions on M for the function u. Let h be positive integer, $h \le N \equiv \max t_j + (d-n)/2$, and $\mathcal{D}^{0,h} \overset{0}{H}(G,M)$ be the completion of the set $\{w \in C^{\infty}(\overline{G})^k : \mathcal{D}_x^{\alpha} w_j(x) = 0,\ x \in M,\ |\alpha| \le h + t_j - 1 - \max t_j\}$ with respect to the norm of $\mathcal{D}^l H(G)$. A function $u \in \mathcal{D}^{0,h} \overset{0}{H}(G,M)$ is, by definition a solution of the Sobolev problem (with "number" h) if the equality (1) is satisfied for each $v \in \mathcal{D}^{0,h} \overset{0}{H}(G,M)$ where $Y \in \mathcal{D}^{0,h} \overset{0}{H}(G,M)^*$. Since $\mathcal{D}^0 H(G) = \mathcal{D}^{0,h} \overset{0}{H}(G,M)$ we describe the index $h=0$ to the first statement problem.

The question of existence for generalized solution is reduced to investigation of the operators

$$A_h : \mathcal{D}^{0,h} \overset{0}{H}(G,M) \longrightarrow \mathcal{D}^{0,h} \overset{0}{H}(G,M)^* \tag{3}$$

defined by (1). We consider one special case only. Assume the following condition is satisfied for the form 1.(2):

$$a(u,u;G) = 0 \Longrightarrow u \in \Pi \tag{4}$$

here Π is the finite dimensional lineal of polynomials subject to 1.(3). The Intersection $\Pi \cap \mathcal{D}^{0,h} \overset{0}{H}(G,M)$ is denoted by Π_h; if $h \le 0$ then $\Pi_h = \Pi$.

Theorem 2.8 *If* $< Y, p >= 0$ *for each* $p \in \Pi_h$, *then there exists a generalized solution* $u \in \mathcal{D}^{0,h} \overset{0}{H}(G,M)$ *of Sobolev problem. This solution is defined up to an arbitrary term from* Π_h. *(In other words, kernel and cokernel of the operator 3) coincide with* Π_h*).*

2.2 Connection between the operators (3) and (16), (19).

We suppose at first that the inequality $n-d > 2\max t_j$ holds and $0 < l-\beta < \delta_0$, $\{f,g\} \in \mathcal{R}^l_\beta V(G,M)$. It is possible to check that on formula (2) implies the definition of the functional $Y = Y(f,g) \in \mathcal{D}^0 H(G)^*$ and for $h=0$ the estimate

$$||Y(f,g); \mathcal{D}^{0,h} \overset{0}{H} (G,M)^*|| \le c||\{f,g\}; \mathcal{R}^l_\beta V(G,M)|| \tag{5}$$

is valid. Besides, $\mathcal{D}^l_\beta V(G,M) \subset \mathcal{D}^0 H(G)$ and according to 1.(1) the solution $u \in \mathcal{D}^l_\beta V(G,M)$ of the problem 1.(14) satisfied to (1) and $ker A^l_\beta = \Pi^\beta$ (see (4)). It implies that the restriction of the operator (3) (for $h=0$) to the space $\mathcal{D}^l_\beta V(G,M)$ coincides with the mapping $YA^l_\beta : \mathcal{D}^l_\beta V(G,M) \to \mathcal{D}^0 \underset{\bullet}{H}(G)$. This fact could be considered as smoothness increasing of a generalized solution. Of course, because of possible singularities on the edge M the solution $u \in \mathcal{D}^l_\beta V(G,M)$ is not a classical solution from $\mathcal{D}^{\max t_j} H(G)$. The question of precise description of differential properties of such solutions is clarified by asymptotic formulas near M, see Section 3.

The case $n-d \le 2\max t_j$ is considered similarly. Let $0 < l-\beta < \delta_0$ and $\delta_0 \le 1$. The space $\mathcal{D}^l_\beta V(G,M)$ for odd $n-d$ (or the space $\mathcal{D}^{l,(d-n)/2} V(G,M)$ for even $n-d$) is embedded in $\mathcal{D}^{0,[N]} \overset{0}{H} (G,M)$; the estimate (5) holds for $h=[N]$. Thus we have

Theorem 2.9 *The restriction of $A_{[N]}$ on $\mathcal{D}^l_\beta V(G,M)$ (or on $\mathcal{D}^{l,(d-n)/2}_\beta V(G,M)$) coincide with the operator YA^l_β (or $YA^{l,(d-n)/2}_\beta$).*

2.3 Comparison of generalized solution for different h

Let $n-d$ be odd and let $n-d < 2\max t_j$ (for even $n-d$ there are additional difficulties see below). Suppose the inequalities $0 \le h < N$, $l \ge h$ hold and $u \in \mathcal{D}^{0,h} \overset{0}{H} (G,M)$ is subject to (1), where $Y = Y(f,g) \in \mathcal{D}^{l,N} \overset{0}{H} (G,M)^* \subset \mathcal{D}^{l,h} \overset{0}{H} (G,M)^*$ and $\{f,g\} \in \mathcal{R}^l_l V(G,M)$. The Hardy inequalities imply

$$||dist(x,M)^{-\kappa} D^\alpha_x u_j; L_2(G)|| \le c||u; \mathcal{D}^0 H(G)||, \ |\alpha| = t_j - \kappa, \ j = 1,\dots,k,$$

where $\kappa = (n-d-1)/2$. Now using the subdivision of G, refined near M and known local estimates of elliptic boundary value problem solutions we obtain the inequality

$$\sum_{j=1}^{k} \sum_{|\alpha|=t_j-\kappa} ||D^\alpha_x u_j; V^{l+\kappa}_l(G,M)|| \le c\left(||\{f,g\}; \mathcal{R}^l_l V(G,M)|| + ||u; \mathcal{D}^0 H(G)||\right). \tag{6}$$

This inequality together with the assertion analogous to Lemma 1.2 allows us to deduce asymptotic expansion of the solution with a remainder from $\mathcal{D}^l_l V(G,M)$. At first we note that by localization the matter is reduced to consideration of a solution of the problem in a wedge $\mathbb{D}$ with compact support. By applying the Fourier transform $\mathcal{F}_{z\to\xi}$, by passing to the coordinates $\eta = |\xi| y$, by using the variants of the Hardy inequality (similar to Lemma 1.5) and by returning to the coordinates y, we deduce the formula

$$\hat{u}_j(y,\xi)\chi_0(|\xi|r) \sum_{\mu=0}^{\max t_j-\kappa-1-h} \sum_{q=1}^{dim\, P_\mu} P^{(q,\mu)}_j(y)\hat{\sigma}_{q,\mu}(\xi) + \hat{w}_j(y,\xi), \tag{7}$$

where $r = |y|$, χ_0 is a cutoff function from $C^\infty_c(\mathbb{R})$, $\chi_0 = 1$ near the point $r = 0$,$\{P^{(q,\mu)}\}$ is a basis in the space of polynomials $P_\mu = \{(p_1,\dots,p_k)\} \in \Pi : deg\ p_j = t_j - \kappa - 1 - \mu\}$.

Lemma 2.10 *Let $u \in \mathcal{D}^{0,h} \overset{0}{H} (G, M)$ is a function with a compact support while the expression from the first line of (6) is finite. Then the representation*

$$u(y,z) = \sum_{\mu=0}^{\max t_j - \kappa - 1 - h} \sum_{q=1}^{\dim P_\mu} P^{(q,\mu)}(y)(\mathcal{C}_0\sigma_{q\mu})(|y|, z) + w(y,z) \tag{8}$$

is valid, where $\sigma_{q\mu} \in H^{\mu+1/2}(\mathbb{R}^d)$, $w \in \mathcal{D}_l^l V(G, M)$ and $\mathcal{C}_0$ is an extension operator (from Γ to $\mathbb{D}$), defined by the equality

$$(\mathcal{C}_0 k)(r, z) = \int_{\mathbb{R}^d} \mathcal{X}(t) k(z - rt) dt; \tag{9}$$

at last, $\mathcal{X}$ is the Fourier transform preimage of the function $\xi \to (2\pi)^{-d}\chi_0(|\xi|)$.

The formula (8) shows that the distinction between generalized solutions u for different h but with the same $\{f, g\}$ appears in the principal terms of asymptotics (the different sets of polynomials $P^{(q,\mu)}$ correspond to these solutions). Nevertheless being based on Lemma ?? it doesn't succeed to describe the connection between A_h and $A_{[N]}$. We explain it by consideration of the problem 1.(9) in $\mathbb{D}$. By Theorem 1.4 the mapping 1.(8) for $\beta = l$ is an isomorphism (recall that $n - d$ is odd). It is possible to check that the inclusion $\{\mathcal{L}^0, \mathfrak{N}^0\}(P^{(q,\mu)}\mathcal{C}_0\sigma) \in \mathcal{R}_l^l V(G, M)$ is correct for each function $\sigma \in H^{\mu+1/2}(\mathbb{R}^d)$, i.e. there is the solution $(y, z) \to \zeta^{(q,\mu)}(\sigma; y, z)$ which admits the representation (8). Hence, the statement of the problem in the class of form (8) leads to appearing of infinite dimensional kernel. We emphasize that the annihilation of such kernel at the expense of additional conditions for the set of functions $\{\sigma_{q\mu}\}$ doesn't provide a physically sensible statement of the problem. The correct conclusion consists in choosing the special set of functions $\{\sigma_{q\mu}\}$ and prescribing to the remainder some special properties. To investigate this question we must face lowest terms of the asymptotics of solutions. The following sections are devoted to this theme.

3 Asymptotics of solutions and contribution of singularities to the Green formula

3.1 Spectrum of the pencil 1.(14)

The pencil $\lambda \to \mathfrak{a}(\lambda; z^0)$ has been introduced in section 1.2. Assume that there is such a number $\beta < l$ that the line $\mathbb{R} + i(\beta - l + (n-d)/2)$ doesn't contain eigenvalues of $\mathfrak{a}(\cdot; z^0)$ for every z^0. We fix the point z^0 and shall not indicate the dependece on z^0 in notations. Denote by $\lambda_1, \ldots, \lambda_T$ the set of eigenvalues of $\mathfrak{a}(\cdot)$ situated in the strip $\{\lambda \in \mathbb{C} : |Im\ \lambda - (n-d)/2| < \overline{l - \beta}\}$. In account to the end of the section 1.2. we enumerate eigenvalues in such order that $\lambda_\nu = \lambda_{T-\nu-1} + i(n-d)$, $Im\ \lambda_\nu < (n-d)/2$ for $\nu = 1, \ldots .T_0 \equiv [T/2]$ and $\lambda_{T_0+1} = i(n-d)/2$ for odd T (in case of critical dimensions). Canonic systems of Jordan chains $\{\varphi_{(\nu)}^{(\sigma,\tau)} : \tau = 1, \ldots, J_\nu, \sigma = 0, \ldots, \kappa_{\tau\nu} - 1\}$consists of eigenvectors $\varphi_{(\nu)}^{(0,\tau)}$ and associated vectors $\varphi_{(\nu)}^{(1,\tau)}, \ldots, \varphi_{(\nu)}^{(\kappa_{\tau\nu}-1,\tau)}$ of that pencil $\mathfrak{a}$, corresponding to number λ_ν; here $\kappa_{\tau\nu}$ are partial multiplicities, $J_\nu = dim\ ker\mathfrak{a}(\lambda_\nu)$. In account of Lemma 1.2 we note that the total multiplicity of $\lambda_1, \ldots, \lambda_T$ is an even number 2ρ. The system of chains generates the set of solutions of the problem, $\{\mathcal{L}^0(D_y, 0), \mathfrak{N}^0(D_y, 0)\}$ in the cone $\mathbb{K}$ of the form

$$u_{(\nu)}^{(\sigma,\tau)}(y) = ||r^{i\lambda_\nu + t_j} \sum_{q=0}^{\sigma} \frac{1}{q!}(i \log r)^q (\varphi_{(\nu)}^{(\sigma-q,\tau)})_j(\omega)||_{j=1}^k, \tag{1}$$

where $r = |y|, \omega \in \Omega$.

Theorem 3.11 1^0. *There are such Jordan chains, corresponding to numbers* λ_ν *and* $\lambda_{T-\nu-1}$, *that*

$$q(u^{(\sigma,r)}_{(\nu)}, u^{(\xi,\eta)}_{(\zeta)}) = i\delta_{\nu,T-\nu-1}\delta_{\tau,J-\eta}\delta_{\sigma,\kappa_{\eta\zeta}-\xi-1}, \tag{2}$$

$\nu,\zeta = 1,\ldots,T$, $r = 1,\ldots,J_\nu$, $\eta = 1,\ldots,J_\zeta$, $\sigma = 0,\ldots,\kappa_{\tau\nu}-1$, $\xi = 0,\ldots,\kappa_{\eta\zeta}-1$ *where* $y \to \chi_0(|y|)$ *is the cutoff function from 2.(7),* $q(u,v) = ([\mathcal{L}^0,\chi_0]u,v)_{\mathbb{K}} + ([\mathfrak{N}^0,\chi_0]u,\mathcal{D}^0 v)_{\partial\mathbb{K}}$.

2^0. *If* T *is odd then there are such Jordan chains, corresponding to* $\lambda_{T_0+1} = i(n-d)/2$ *that*

$$q(u^{(\sigma,\tau)}_{(T_0+i)}, u^{(\xi,\eta)}_{(T_0+i)}) = \pm i\delta_{\tau,\eta}\delta_{\sigma,1-\xi} \tag{3}$$

where $\sigma,\xi = 0,1;\tau;\eta = 1,\ldots,\dim \Pi_0$ *(see Lemma 1.2(*2^0*)) and one of two signs is realized.*

3.2 Formalism of asymptotics construction.

Asymptotics of a solution of the problem in a domain with an edge contains terms of the form $U^{(\sigma,\tau)}_{(\nu)}\mathcal{C}c^{(\sigma,\tau)}_\nu$ where $c^{(\sigma,\tau)}_\nu$ is a function on the edge M, depending on data of the problem on the whole, Γ is the extension operator (from M to G), obtained from the operators 2.(9) by partition of unity. $U^{(\sigma,\tau)}_{(\nu)}(y,z,D_z)$ are some differntial operators on M, depending on $y \in \mathbb{K}(z)$. The main difficulty in the process of asymptotic construction consists of clearing up differential properties of terms in respect to $z \in M$. The situation, where this question is resolved entirely, is described by the following assumption:

1) the numbers T, J_ν and $\kappa_{\tau\nu}$ don't depend on a point $z^0 \in M$;

2)there are such Jordan chains subject to Theorem 3.11 that the functions $z^0 \to \varphi^{(\sigma,\tau)}_{(\nu)}(\cdot;z^0)$ $\in C^\infty(\overline{\Omega(z^0)})$ are smooth on M;

3) if q is an integer and $\lambda_\nu(z^0) = \lambda_\zeta(z^0) + iq$ for certain z^0 the this equality holds for every $z^0 \in M$;

4) there is a number β such that the line $\mathbb{R} + i(\beta - l + q + (n-d)/2)$ doesn't contain eigenvalues $\lambda_1(z^0),\ldots,\lambda_T(z^0)$ for each $z^0 \in M$ and for every integer q.

We shall explain the construction algorithm of the operators $U^{(\sigma,\tau)}_{(\nu)}(y,z,D_z)$ for the problem $\{\mathcal{L},\mathfrak{N}\}$ in the wedge $\mathbb{D}$ with coefficients depending on $z \in \mathbb{R}^d$ and $y \in \mathbb{K}$. The operator $\{\mathcal{L},\mathfrak{N}\}$ admits the expansion

$$\{\mathcal{L}(y,z,D_{(y,z)}),\mathfrak{N}(y,z,D_{(y,z)})\} \sim \sum_{p\geq 0} r^P\{\mathcal{L}^{(p)}(y,z,D_{(y,z)}),\mathfrak{N}^{(p)}(y,z,D_{(y,z)})\}$$

while matrix elements $\mathcal{L}^{(p)}_{hj}$ and $\mathfrak{N}^{(p)}_{qj}$ are differential operators with respect to D_z with order not larger than p and have the following homogeneity property of degree $-t_h - t_j$ and $-\sigma_q - t_j$:

$$\mathcal{L}^{(p)}_{hj}(ay,z,a^{-1},D_y,D_z) = a^{-t_h-t_j}\mathcal{L}^{(p)}_{hj}(y,z,D_y,D_z) \;\; \forall a > 0.$$

The operators $U^{(\sigma,\tau)}_{(\nu)}$ are partial sums of formal series

$$\sum_{p\geq 0} u^{(\sigma,\tau)}_{(\nu,p)}(y,z,D_z) \tag{4}$$

while $u^{(\sigma,\tau)}_{(\nu,0)}(y,z)$ is a solution of the equation $\{\mathcal{L}^{(0)},\mathfrak{N}^{(0)}\}u = 0$ in $\mathbb{K}$ and remaining members of the series are determined by recurrent sequence of problems in $\mathbb{K}$:

$$\{\mathcal{L}^{(0)},\mathfrak{N}^{(0)}\}u^{(\sigma,\tau)}_{(\nu,p)} = -\sum_{h=1}^{p}\{\mathcal{L}^{(h)},\mathfrak{N}^{(h)}\}u^{(\sigma,\tau)}_{(\nu,p-h)} \tag{5}$$

It isn't difficult to check the representation

$$(u^{(\sigma,\tau)}_{(\nu,p)}(y,z,D_z) = r^{i\lambda_\nu(z)+t_j+p} \sum_{|\alpha|\le p} \mathcal{U}^{(\sigma,\tau,\alpha)}_{(\nu,p,j)}(z,\omega,\log r) D^\alpha_z \tag{6}$$

where $\mathcal{U}^{(\dots)}_{(\dots)}$ is a polynomial in $\log r$ with coefficients from $C^\infty(\mathbb{R}^d \times \overline{\Omega})$.

Remark 3.12 *If $\lambda_\nu(z) = \lambda_\zeta(z) + iq$ and $q = 1, 2, \dots$ then there is some arbitrariness in determination of $u^{(\sigma,\tau)}_{(\nu,q)}$ at (4) because of the solution of (5) is determined up to the term*

$$\sum_{\eta=1}^{J_\zeta} \sum_{\xi=0}^{\kappa_{\eta\zeta}-1} u^{(\xi,\eta)}_{(\zeta)}(y,z) K^{(\xi,\eta)}_{(\zeta)}(z,D_z) \tag{7}$$

with arbitrary operator coefficients $K^{(\xi,\eta)}_{(\zeta)}$, ord $K^{(\xi,\eta)}_{(\zeta)} = q$.

3.3 Theorem about asymptotics of solutions.

Let $u \in \mathcal{D}^l_\gamma V_l(G,M)$ be a solution of the problem 1.(3) with the right-hand side $\{f,g\} \in \mathcal{R}^l_\beta V(G,M)$, while $\gamma > \beta$ and there are no eigenvalues of $\mathfrak{a}(\cdot, z^0)$ on the line $\mathbb{R} + i(\gamma - l + (n-d)/2)$ (for every $z^0 \in M$).

The following Theorem is valid when the operator $\overset{0}{A}{}^l_\beta\,(\Theta; z)$ from 1.(13) is an monomorphism everywhere on M. In view of Theorems 1.3 and 1.6 this requirement is satisfied for each $\beta < l$.

Theorem 3.13 [10]. *Under the conditions formulated before the solution u admits the asymptotic expansion*

$$u = \sum_\nu \sum_{\tau=1}^{J_\nu} \sum_{\sigma=0}^{\kappa_{\tau\nu}-1} U^{(\sigma,\tau)}_{(\nu)} C c^{(\sigma,\tau)}_\nu + w. \tag{8}$$

Here summation extends to subscripts ν whose eigenvalues $\lambda_\nu(z^0)$ belong to the strip $\{\lambda : Im\lambda + l - (n-d)/2 \in (\beta,\gamma)\}$ are partial sums (with $p \le s_\nu(z^0)$) of the series (4) and the estimate

$$\|c^{(\tau)}_\nu; H^{s_\nu(\cdot)-\varepsilon}(M)\| + \|w; \mathcal{D}^l_{\beta+\varepsilon} V(G,M)\| \le c(\|\{f,g\}; \mathcal{R}^l_\beta V(G,M)\| + \|u; \mathcal{D}^l_\gamma V(G,M)\|) \tag{9}$$

holds for arbitrary positive ε, while $c^{(\tau)}_\nu = (c^{(0,\tau)}_\nu, \dots, c^{(\kappa_{\tau\nu}-1,\tau)}_\nu)$, $H^{s(\cdot)}(M)$ is the space of functions with variable smoothness $s \in C^\infty(M)$,

$$s_\nu(z^0) = Im\lambda_\nu(z^0) + l - \beta - (n-d)/2. \tag{10}$$

The estimate (9) becomes, generally speaking, noncorrect for $\varepsilon = 0$. It turns out that at $\varepsilon = 0$ the precise estimates are valid for other vectors $k^{(\tau)}_\nu = (k^{(0,\tau)}_\nu, \dots, k^{(\kappa_{\tau\nu}-1,\tau)}_\nu)$ determined by the formula

$$k^{(\tau)}_\nu = M(-i\Lambda; \kappa - \tau\nu) c^{(\tau)}_\nu \tag{11}$$

where

$$(\Lambda v)(z) = 2^{n-1}\Pi^{n/2}\Gamma(n/2) \int_M (v(z) - v(\zeta))|z-\zeta|^{-d} ds_\zeta + v(z) \tag{12}$$

$$M(t,m) = \begin{Vmatrix} 1 & t & (2!)^{-1}t^2 & \dots & ((m-1)!)^{-1}t^{m-1} \\ 0 & 1 & t & \dots & ((m-2)!)^{-1}t^{m-2} \\ \dots & \dots & \dots & \dots & \dots \\ 0 & 0 & 0 & \dots & 1 \end{Vmatrix} \tag{13}$$

Theorem 3.14 *The arbitrariness mentioned in Remark 3.12 may be used in order to make the following assertions valid:*

1. *the sum* $||k_\nu^{(\tau)}; H^{s_\nu(\cdot)}(M)|| + ||w; \mathcal{D}^l_\beta V(G, M)||$ *doesn't exceed the right-hand side of the inequality (8);*

2. *for each function* $k_\nu^{(\sigma,\tau)} \in H^s_\nu(\cdot)(M)$ *the inequality*

$$||U^{(\sigma,\tau)}_{(\nu)} C k_\nu^{(\sigma,\tau)}; \mathcal{D}^l_\gamma V(G, M)|| + ||\{\mathcal{L}, \mathfrak{N}\} U^{(\sigma,\tau)}_{(\nu)} C k_\nu^{(\sigma,\tau)}; \mathcal{R}^l_\beta V(G, M)|| \leq c||k_\nu^{(\sigma,\tau)}; H^{s_\nu(\cdot)}(M)||$$

is satisfied

3.4 Contribution of singularities of solutions to the Green formula.

The equation 1.(1) and symmetry of the form 1.(2) imply the Green formula

$$(\mathcal{L}u, v)_G + (\mathfrak{N}u, \mathcal{D}v)_{\partial G} = (u, \mathcal{L}v)_G + (\mathcal{D}u, \mathfrak{N}v)_{\partial G} \tag{14}$$

which spreads for $u \in \mathcal{D}^l_\beta V(G, M)$and $v \in \mathcal{D}_{2l-\beta} V(G, M)$ by continuity. Let β be the number, chosen in Section @, and let $\gamma = 2l - \beta$. If both functions belong to $\mathcal{D}^l_\gamma V(G, M)$, then (14) loses its meaning because of divergence of the integrals.

Suppose the function $u \in \mathcal{D}^l_\gamma V(G, M)$ satisfies the relation

$$u - \sum_{\nu=1}^{T} \sum_{\tau=1}^{J_\nu} \sum_{\sigma=0}^{\kappa_{\tau\nu}-1} U^{(\sigma,\tau)}_{(\nu)} C c_\nu^{(\sigma,\tau)} \in \mathcal{D}^l_\beta V(G, M) \tag{15}$$

where $c_\nu^{(\sigma,\tau)} \in C^\infty(M)$. Introduce the vectors $c_\nu^{(\tau)}$ and $k_\nu^{(\tau)}$ by equalities (11). Supposing the same properties of v, we obtain the vectors $b_\nu^{(\tau)}$ and $l_\nu^{(\tau)}$ analoguos to $c_\nu^{(\tau)}$ and $k_\nu^{(\tau)}$. To formulate this result it is convenient to enumerate the coefficients $k_\nu^{(\sigma,\tau)}$ by one index $h = 1, \dots, 2\rho$ (we recall that 2ρ is a total multiplicity of numbers $\lambda_i, \dots, \lambda_T$). We write $k = (k_1, \dots, k_{2\rho})$ so that $k_\nu^{(\sigma,\tau)}$ precedes $k_\zeta^{(\xi,\eta)}$ if $\nu < \zeta$ or $\nu = \zeta$, $\tau < \eta$ or $\nu = \zeta$, $\tau = \eta$, $\sigma < \xi$. The analogous vector of coefficients $c_\nu^{(\sigma,\tau)}$ is denoted by c. The relations $c_\nu^{(\tau)} = M(i\Lambda; \kappa_{\tau\nu}) k_\nu^{(\tau)}$ arising from (11)- (13) will be written by $c = \mathcal{M}k$. The operator $U^{(\sigma,\tau)}_{(\nu)}$ are enumerated by an index $h = 1, \dots, 2\rho$, respectively.

Theorem 3.15 *Suppose the functions* u, v *are subject to inclusions*

$$u - \sum_{h=1}^{2\rho} U^{(h)} C(\mathcal{M}k)_h, v - \sum_{h-1}^{2\rho} U^{(h)} C(\mathcal{M}l)_h \in \mathcal{D}^l_\beta \mathfrak{N}(G, M) \tag{16}$$

where $k, l \in C^\infty(M)$. *Then the formula*

$$(\mathcal{L}u, v)_G + (\mathfrak{N}u, \mathcal{D}v)_{\partial G} - (u, \mathcal{L}v)_G - (\mathcal{D}u, \mathfrak{N}v)_{\partial G} = i(kJl) \tag{17}$$

holds while J *is the involution in* $\mathbb{C}^{2\rho}$ *with the matrix*

$$\left\| \begin{array}{ccc} 0 & 0 & \mathcal{F}_s \\ 0 & \mathcal{T} & 0 \\ \mathcal{F}_s & 0 & 0 \end{array} \right\| \text{, where } \left\{ \begin{array}{lll} s = \rho, & \mathcal{T} = \emptyset & \text{at } T = 2T_0, \\ s = \rho - J_{T_0+1}, & \mathcal{T} = \pm diag\{\mathcal{F}_2, \dots, \mathcal{F}_2\} & \text{at } T = 2T_0 + 1, \end{array} \right.$$

$\mathcal{F}_s = ||\delta_{j,s-1-i}||_{j,i=0}^{s-1}$ *and the sign is the same as one in (3).*

Corollary 3.16 *Let $n-d$ be odd and let the operator A^l_l from 1.(16) be an isomorphism (for simplicity). We write a vector $a \in \mathbb{C}^{2p}$ in the form (a', a''), where $a', a'' \in \mathbb{C}^p$. By $s' \in C^\infty(M)^p$ we denote the vector, which consists of the functions (10), corresponding toe the eigenvalues $\lambda_1, \dots, \lambda_{T_0}$, and put $s'' = (-s'_p, -s'_{p-1}, \dots, s'_1)$. The infinite-dimensional kernel of A^l_β is isomorphic to the space $H^{s''(\cdot)}(M)$; more precisely, for each $l \in H^{s''(\cdot)}(M)$ there exists the solution $\zeta(l''; \cdot)$ of the problem $\{\mathcal{L}(x, D_x), \mathfrak{N}(x, D_x)\}u = 0$ such that $||\zeta(l''; \cdot); \mathcal{D}^l_{2l-\beta}V(G, M)|| \leq c||l''; H^{s''(\cdot)}(M)||$ and*

$$(f, \zeta(l''; \cdot))_G + (g, \mathcal{D}_\zeta(l''; \cdot))_{\partial G} = i(k', \mathcal{F}_p l'')_M \tag{18}$$

here $(k', 0)$ is a vector of coefficients in asymptotics (16) of the solution $u \in \mathcal{D}^l_l V(G, M)$ of the problem 1.(3) with the right-hand side $\{f, g\} \in \mathcal{R}^l_\beta V(G, M)$.

Remark 3.17 *If $s'_h(z_0) > (n-d)/2$ the the coefficient k_j is a continuous function near $z^0 \in M$. Let l'' be a vector with components $l = 0$ for $j \neq p - h + 1$ and $l_{p-h+1} = \delta(z - z^0)$. The right-hand side of (??) is equal to $ik_j(z^0)$, i.e. the formula (??) provides a value of the coefficient at the point z^0.*

4 Radiation condition on edge

In account of Theorem 3.15 and formulas of type 3.(8), 3.(16) it is possible to give new self-adjoint realizations of the problem in domain with an edge M, prescribing an asymptotic expansion to a solution near M, i.e. defining radiation conditions and connect the original operator with the operator 2.(3).

Let $\vec{s} = (s^{(1)}, \dots, s^{(2p)})$ be such vector that $s^{(j)}$ is the function 3.(10) while $k_j - k_\nu^{(\sigma,\tau)}$ (with respect to mentioned enumeration). Put $H^{\vec{s}(\cdot)}(M) = H^{s^{(1)}(\cdot)}(M) \times \dots \times H^{s^{(2p)}(\cdot)}(M)$. The right-hand side of 3.(17) contains scalar products in $L_2(M)$ of coefficients k_j and $(Jl)_j$ from 3.(16). Introduce the subspace $\tilde{H} \subset H^{\vec{s}(\cdot)}(M)$ so that for an arbitrary $k \in \tilde{H}$ we have $k_j = 0$ or $(Jk)_j = 0$. Then for the solutions u, v with asymptotics 3.(16) where $k, l \in \tilde{H}$ the formula 3.(??) becomes the ordinary Green formula 3.(17)

Let $N = \max t_j + (d-n)/2 > 0$ and let h be one of the numbers $0, \dots, N$. Put $\beta_h = l + h - N - \varepsilon$ where ε is small positive. Let j be an element of the set $\{1, \dots, 2p\}$, $k_j = k_\nu^{(\sigma,\tau)}$ and $k^{(j)}$ be an vector with the components $k_h^{(j)} = \delta_{j,h} k_j$. We include j in the set Ξ_h ($h = 0, \dots, N$ see Section 2.1) if the eigenvalue λ_ν belongs to the closed strip between the lines $\mathbb{R} + i(n-d)/2$, $\mathbb{R} + i(\beta^h - l + (n-d)/2)$ and one of two following conditions is satisfied:

1) the sum $\sum U^{(h)} \mathcal{C}(\mathcal{M} k^{(j)})$ coincides with $p\mathcal{C}\sigma$ modulo $\mathcal{D}^l_l V(G, M)$ where $\sigma \in H^{s_\nu(\cdot)}(M)$, $y \to p(y, z)$ is a polynomial from the lineal P_μ, $\mu = 0, \dots, \max t_j - \kappa - 1 - h$ (compare 2.(8) and 3.(18));

2) the sum $\sum U^{(h)} \mathcal{C}(\mathcal{M} J k^j)$ doesn't coincide with $p\mathcal{C}\sigma$ modulo $\mathcal{D}^l_l V(G, M)$ for each $\sigma \in H^{s_\nu(\cdot)}(M)$ and $p \in P_\mu$

Put $\tilde{H}_h = \{k \in H^{\vec{s}(\cdot)}(M) : (Jk)_j = 0 \text{ under } j \in \Xi_h\}$; it is the subspace $\tilde{H}_h$ satisfies the demand mentioned before. Determine the operator Π on the set $\mathcal{D}^l_{\beta^h} V(G, M) \times H^{\vec{s}(\cdot)}(M)$ by the equality $\Pi\{u^0, k\} = u^0 + \sum U^{(h)} \mathcal{C}(\mathcal{M} k)_h$; the inclusions

$$\Pi\{u^0, k\} \in \mathcal{D}^l_{2l-\beta^h} V(G, M), \{\mathcal{L}, \mathfrak{N}\}\Pi\{u^0, k\} \in \mathcal{R}^l_{\beta^h} V(G, M) \tag{1}$$

are valid.

Now expending the duality $(\cdot, J\cdot)_M$ in 3.(??) to the conjugate spaces $H^{\vec{s}(\cdot)}(M)$ and $H^{-\vec{s}(\cdot)}(M) = JH^{\vec{s}(\cdot)-2l(l-\beta)}(M)$ we obtain

Lemma 4.18 *The mapping* $\Pi : \mathcal{D}^l_{\beta h}V(G,M) \to \mathcal{D}^l_{\beta h}V(G,M)$ *is continuous.*

Theorem 4.19 *The operator*

$$\mathcal{D}^l_{\beta h}V(G,M) \times \tilde{H}_h \ni \{u^0,k\} \longrightarrow \mathcal{A}^l_{\beta,h}\{u^0,k\} \equiv \{\mathcal{L},\mathfrak{N}\}\Pi\{u^0,k\} \in \mathcal{R}^l_{\beta h}V(G,M) \tag{2}$$

is Fredholm and $Ind\mathcal{A}^l_{\beta,h} = 0$.

It is the mapping 4.(2) wich is connected with the generalized statement of the problem 1.(3), considered in Section 3 as the Sobolev problem with a number $h = 0,\dots,[N]$ (in particulary at $h = [N]$) there exists an obvious connection with the operators 1.(16) and 1.(??). It is possible to check that under $\{u^0,k\} \subset \mathcal{D}^l_{\beta h}V(G,M) \times \tilde{H}_h$ the function $\Pi\{u^0,k\}$ belongs to the space $\mathcal{D}^{0,h}H(G,M)$ and satisfies the equality 2.(1) where $Y = Y(\mathcal{A}^l_{\beta,h}\{u^0,k\})$. So the assertion is valid, playing the role of Theorem 2.9 in the case $h < [N]$.

Theorem 4.20 *At* $h = 0,\dots,[N]$ *the operators* $A_h\Pi$ *and* $Y\mathcal{A}^l_{\beta h,h}$, *determined on* $\mathcal{D}^l_{\beta h}V(G,M)\times\tilde{H}_h$, *coincide.*

We shal give sketchy description of general radiation conditions on the edge. We have the generalized Green's formula

$$(\mathcal{L}u,v)_G + (\mathfrak{N}u,\mathcal{D}v)_{\partial G} + (\mathcal{B}k,\mathcal{T}l)_M = (u,\mathcal{L}v)_G + (\mathcal{D}u,\mathfrak{N}v)_{\partial G} + (\mathcal{T}k,\mathcal{B}l)_M, \tag{3}$$

wher $\mathcal{B} = ||\mathcal{B}_{hj}||$ and $\mathcal{T} = ||\mathcal{T}_{hj}||$ are $\rho \times 2\rho$-matrices of pseudodifferential operators. Let p be the vector with coordinates $p^{(1)},\dots,p^{(2\rho)} \in C^\infty(M)$, $p^{(j)} = s^{(j)} \geq 0$ or $p^{(j)} = s^{(j)} - 2(l-\beta)$ while $k \in H^{\vec{p}(\cdot)}(M) \Leftrightarrow Jk \in H^{\vec{p}(\cdot)}(M)$. Let $q = (q_1,\dots,q_\rho) \in C^\infty(M)$, *ord* $\mathcal{B}_{hj} = q_h + p^{(j)}$, *ord* $\mathcal{T}_{hj} = -q_h + p^{(j)}$. Introduce the $\rho\times\rho$-matrix $\mathcal{B}'$ consisting of such lines of $\mathcal{B}_j$ that $p^{(j)} < 0$ and put $X = \left\|\begin{matrix}\mathcal{B}\\ \mathcal{T}\end{matrix}\right\|$ (i.e. X is a $2\rho\times 2\rho$-matrix). If the formula 4.(3) holds the the operator X is elliptic. Suppose besides that $\mathcal{B}'$ is elliptic, too. Then the operator

$$\begin{aligned}&\mathcal{D}^l_\beta V(G,M) \times H^{\vec{p}(\cdot)}(M) \ni \{u^0,k\} \longrightarrow \mathcal{A}_\mathcal{B}\{u^0,k\}\\ \equiv\ &(\{\mathcal{L},\mathfrak{N}\}\Pi\{u^0,k\},\mathcal{B}k) \in \mathcal{R}^l_\beta V(G,M) \times H^{-\vec{q}(\cdot)}(M)\end{aligned} \tag{4}$$

is Fredholm and $Ind\mathcal{A}_\mathcal{B} = 0$. This operator corresponds to the problem 1.(3) with the radiation conditions $\mathcal{B}k = g$ on M.

We denote the complement of Ξ_h with respect to the set $\{1,\dots,2\rho\}$ by $\{n_1,\dots,n_\rho\}$. If the matrix $\mathcal{B}$ consists of the lines $(\mathcal{B}_{i,1},\dots,\mathcal{B}_{i,2\rho})$, $i = 1,\dots,\rho$, where $\mathcal{B}_{ij} = \delta_{j,n_i}$, then the restriction of $\mathcal{A}_\mathcal{B}$ to the set $\mathcal{D}^l_\beta V(G,M) \times \tilde{H}_h$ coincides with the operator $(\mathcal{A}_h, 0)$ from 4.(2).

5 Self-adjoint extension of the Neumann problem operator

Let $\mathcal{L}$ be an elliptic system and $t \equiv t_1 =,\dots = t_k$. Choose any function $R \in C^\infty(\overline{G}\setminus M)$ which is equivalent to the distance from M near the edge and denote by $\mathcal{H}_\sigma$ the Hilbert space with scalar product $< u,v >_\sigma = (R^\sigma u, R^\sigma v)_G$. We consider the unbounded operator $\mathcal{U}_\sigma$ in $\mathcal{H}_\sigma$, given by the expression $R^{-2\sigma}\mathcal{L}(x,D_x)$ with the domain $\mathcal{D}(\mathcal{U}_\sigma) = \{u \in \mathcal{D}^t_{-\sigma}V(G,M) : \mathfrak{N}u = 0 \text{ on } \partial G\setminus M\}$.

Theorem 5.21 *Let* β *be a number submitting to the conditions 1-4 of Section 3.2, where* $l = t$ *while* $\beta < l$. *Then the operator* $\mathcal{U}_{-\beta}$ *is closed and symmetric.*

In view of the following Theorem 5.22 the operator $\mathcal{U}_{-\beta}$ is not self-adjoint when there are eigenvalues of the pencil 1.(14) between the lines $\mathbb{R} + i(\beta - l + (n-d)/2)$, $\mathbb{R} + i(l - \beta + (n-d)/2)$. According to 3.(17) functions v submitting to to 3.(16) with $l \in C^\infty(M)^{2\rho}$ must be included to $\mathcal{D}(\mathcal{U}^*_{-\beta})$. However, generally speaking, the necessary condition $\mathfrak{N}\Pi\{0, l\} = 0$ on $\partial G \setminus M$ is not fulfilled. It is clear that this disagreement may be eliminated by adding the operator : $H^{\vec{s}(\cdot)}(M) \to \mathcal{D}^l_\beta G(G, M)$, compensating the residual $\mathfrak{N}\Pi\{0, l\}$ in the boundary condition, to the operator Π.

Theorem 5.22 *Under the condition of Theorem 5.21 the operator $\mathcal{U}^*_{-\beta}$ adjoint to $\mathcal{U}_{-\beta}$ is given by the same differential expression and has the domain*

$$\begin{aligned} \mathcal{D}(\mathcal{U}^*_{-\beta}) &= \big\{u = \Pi\{u^0, k\} : u^0 \in \mathcal{D}(\mathcal{U}_{-\beta}), k = (k_1, \dots, k_{2\rho}), \\ &\qquad k_j \in H^{2(t-\beta)-s^{(j)}(\cdot)}(M)\big\} \subset \mathcal{D}^t_{2t-\beta} V(G, M). \end{aligned}$$

The enumeration question of self-adjoint extensions for the operator $\mathcal{U}_{-\beta}$ is solved by Theorem 5.22, with the help of general theory results [11]. We restrict ourselves to the formulation of a particular case.

Theorem 5.23 *Suppose the condition of Theorem 5.21 and the formula 4.(3) are satisfied, where thew operators $\mathcal{B}$ and $\mathcal{T}$ have the properties mentioned at the end of Section 4. Then the restriction of $\mathcal{U}^*_{-\beta}$ to the set*

$$\Big\{u = \Pi\{u^0, k\} : u^0 \in \mathcal{D}(\mathcal{U}_{-\beta}), k \in H^{\vec{p}(\cdot)}(M), \mathcal{B}k = 0 \text{ on } M\Big\} \tag{1}$$

is a self-adjoint extension $\mathcal{U}_{-\beta}(\mathcal{B})$ of $\mathcal{U}_{-\beta}$.

Now we are ready to compare $\mathcal{U}_{-\beta}(\mathcal{B})$ with $\mathcal{A}_\mathcal{B}$. Let $l = t$ in 4.(4) and let δ_β be the bounded operator:

$$\mathcal{R}^t_\beta V(G, M) \times H^{-\vec{q}(\cdot)}(M) \ni \{f, g, \mathcal{G}\} \longrightarrow R^{2\beta} f \in V^0_{-\beta}(G, M)^*.$$

Then the restriction of $\delta_\beta \mathcal{A}_\mathcal{B}$ to the set 5.(1) coincides with $\mathcal{U}_{-\beta}(\mathcal{B})$. This fact admits the interpretation of the statements of the problem 1.(3) from Section 2 as special self-adjoint extension of $\mathcal{U}_{-\beta}$. At least if the line $\mathbb{R} - i(l - (n-d)/2)$ doesn't contain eigenvalues of the pencil 1.(14) then we can put $\beta = 0$. In this case the Theorem 5.23 gives self-adjoint extensions of $\mathcal{L}$ in $L_2(G)$.

References

[1] Arutyunyan N.Kh., Nasarov S.A., Shoĭkhet B.A. Estimates and asymptotics of weight-deformed states of the three-dimensional body with tension in the elasticicty thory and the theory of creep flow. Dokl. AN SSSR, 1892, t.266, No 6, pp 1365-1369.

[2] Zaĭonchkovskiĭ V., Solonnikov V.A. On the Neumann problem for elliptic equations of second order in domains with edges on the boundary. Zapiski nauchnykh seminarov Leningr. otd. Matem. in-ta AN SSSR, 1983, t.36, No 4,pp 12-25.

[3] Dauge M. Elliptic boundary value problems on corner domains. Lecture Notes in Mathematics. Vol. 1341, 1988.

[4] Grachev N.V., Maz'ja V.G. On the Fredholm radius of integral operators in the potential theory. Problemy matem. analiza, 1990, vyp. II, pp 109-133, nzd-vo Leningr. univesiteta.

[5] Nazarov S.A. Estimates near the edge of solutions of the Neumann problem for elliptic systems. Vestnik Leningr. universiteta, 1988, No 1,pp 37-32.

[6] Nazarov S.A., Plamenevskiĭ B.A. The radiation principle for elliptic self - adjoint problems. Problemy matem. fiziki, vyp. 13, izd-vo Leningr. universiteta (to appear)

[7] Nazarov S.A., Plamenevskiĭ B.A. The Neumann problem for self-adjoint elliptic systems in domains with piece-wise smooth boundary. Trudy Leningr. matem. obshchestva, 1990, t.1.

[8] Nazarov S.A., Plamenevskiĭ B.A. Elliptic problems in domains with piece-wise smooth boundary. "Nauka", 1991 (to appear)

[9] Roĭtberg Ya.A., Sheftel' Z.G. The theorem on homeomorphisms for elliptic systems and applications. Matem. sbornik 1969, t. 78, No 3, pp 446-472.

[10] Maz'ja V.G., Rossman J. Über die Asymptotik der Lösungen elliptischer Randwertaufgaben in der Umgebung von Kanten. - Math. Nachr., 1988, Bd 138, S.27-53.

[11] Akhiezier N.I., Glazman I.M. Theory of linear operators in the Hilbert space. "Nauka" 1966.

[12] Gorbachuk V.I., Gorbachuk M.L. Boundary problems for differential-operator equations. "Naukova Dumka" 1984.

S.A. Nazarov, B.A. Plamenevskiĭ
Electrotechnical Institute
Nabereshnaya moĭki 61
St. Petersburg 191065, USSR

Spectral and Scattering Theory for Acoustic Operators in Non-Homogeneous Fluids. Continuous and Discrete Models

V.S.Rabinovich (Rostov-on-Don Univ.)

1 Introduction

In this paper we formulate the results on the spectral property and scattering theory for the linearized Navier-Stokes system, discribing the process of sound propagation in the non-homogeneous halfspace $\mathbb{R}^3_+ = \{x = (x_1, x_2, x_3) : x_3 > 0\}$

$$\begin{aligned} \frac{\partial \vec{v}}{\partial t} &= -\rho^{-1}\nabla p + \rho^{-1}\vec{f}, \\ \frac{\partial p}{\partial t} &= -\rho C^2 \nabla \cdot \vec{v} + c^2 \rho a. \end{aligned} \tag{1}$$

Here ∇ denotes the gradient operator and $\nabla\cdot$ denotes the divergence operator, ρ denotes the fluid density, c is the sound speed, $\vec{v} = (v_1, v_2, v_3)$ is the vector of vibrating volocity, p is the acoustic pressure, $\vec{f} = (f_1, f_2, f_3)$ and a are solid densities corresponding the source of force and solid velocity respectively.

We asume $\rho(x)$ and $c(x)$ to be measurable and such that

$$0 < \alpha_1 \leq \rho(x) \leq \alpha_2 < \infty, \ \alpha_1 \leq c(x) \leq \alpha_2 \tag{2}$$

for some constants $\alpha_1, \alpha_2 > 0$. The derivations in (1) are taken in the distributional sense.

Let $\rho_0 = \rho_0(x_3)$ and $c_0 = c_0(x_3)$ be measurable functions on $\mathbb{R}_+$, which satisfy the same conditions stabilization by $x_3 \to \infty$ and such that

$$|\rho(x) - \rho_0(x_3)| \leq c(1 + |x|)^{-2\delta_0}, \ |c(x) - c_0(x_3)| \leq c(1 + |x|)^{-2\delta_0}, \ \delta_0 > 1. \tag{3}$$

The condition (3) means that the non-homogeneous fluid stabilize to stratified fluid as $x \to \infty$. Such conditions are intrinsic for mathematical models of underwater sound propagation [1].

The spectral property of the acoustic operator

$$Ap = -c^2 \rho \nabla \cdot \rho^{-1} \nabla p$$

in the halfspace $\mathbb{R}^3_+$ under the condition (3) and in additional smoothnaess conditions on $\rho(x)$ were considered by the author in [2], [3] and independently by Ben-Artizy M. ,Dermjan J. ,Guliot J. in [4] (see also [5], [6] where the case $\rho(x) = const$ was treated). The spectral and scattering theory in deformed optical wave guides was studied by R. Weder in [7].

In spite of the fact that the system (1) is known to be reduced to a single acoustic equation $\frac{\partial^2 p}{\partial t^2} = Ap + F$ for pressure the development of spectral and scattering theory for the system (1) is preferable because it permits to avoid the additional smoothness condition on $\rho(x)$. We remark, that the investigation of the system (1) permits to exploit the standard scheme of scattering theory for short range perturbation [8]-[10].

We formulate the results on the spectral property for the difference scheme of the sound propagation problem in perturbed stratified fluids as well. It should be noted that the spectral properties of the continuous model of the acoustic operator are different from those of the discrete one.

2 The limiting absorbtion principle for (1) in the stratified fluids

. Let $\rho_0(x_3)$ and $c_0(x_3)$ be measurable on $\mathbb{R}_+$ and let the following limits

$$\rho_\infty = \lim_{y\to+\infty} \rho_0(y), \quad c_\infty = \lim_{y\to+\infty} c_0(y)$$

exist. Let also

$$\int_0^\infty |\rho_0(y) - \rho_\infty| dy < \infty, \quad \int_0^\infty y^2 |c_0(y) - c_\infty| dy < \infty$$

and

$$c_m = \inf_{\mathbb{R}_+} c_0(y) < c_\infty, \quad \lim_{y\to+\infty} y^2(c^{-2}(y) - c_\infty^{-2}) \leq 0.$$

Let H_0 be the Hilbert space of vectors $u = (u_1, u_2, u_3, u_4)$ with the following scalar product

$$< u, v >_0 = \int_{\mathbb{R}^3_+} (G_0^{-1} u, v)_{\mathbb{C}^4} dx,$$

where

$$G_0(x_3) = \begin{pmatrix} \rho_0^{-1} & 0 & 0 & 0 \\ 0 & \rho_0^{-1} & 0 & 0 \\ 0 & 0 & \rho_0^{-1} & 0 \\ 0 & 0 & 0 & \rho_0 c_0^2 \end{pmatrix}$$

and let

$$\begin{aligned} \overset{0}{\mathcal{H}^1} &= \{u \in \mathcal{D}'(\mathbb{R}^3_+, \mathbb{C}^4) : ||u||_{\mathcal{H}^1} \\ &= (||u||^2_{L_2^{(4)}(\mathbb{R}^3_+)} + ||\nabla \cdot u||^2_{L_2(\mathbb{R}^3_+)} + ||\nabla u_4||^2_{L_2^{(3)}(\mathbb{R}^3_+)})^{1/2} < \infty, \\ &\quad u_4((x_1, x_2, 0) = 0\} \end{aligned}$$

where $u' = (u_1, u_2, u_3)$, $L_2^{(m)}(\mathbb{R}^3_+) = L_2(\mathbb{R}^3_+) \otimes \mathbb{C}^m$.

We define the unbounded operator A_0 in H_0 with domain $\overset{0}{\mathcal{H}^1}$ in the following way

$$A_0 u = -i G_0(x_3) \Lambda u,$$

where

$$\Lambda = \begin{pmatrix} 0 & 0 & 0 & \partial_{x_1} \\ 0 & 0 & 0 & \partial_{x_2} \\ 0 & 0 & 0 & \partial_{x_3} \\ \partial_{x_1} & \partial_{x_2} & \partial_{x_3} & 0 \end{pmatrix}.$$

The operator A_0 is selfadjoint in H_0 and the system (1) is equivalent to

$$i\frac{\partial u}{\partial t} = A_0 u + F, \quad F = (\rho_0^{-1} f_1, \rho_0^{-1} f_2, \rho_0^{-1} f_3, c_0^2 \rho_0 a).$$

It follows easily that A_0 has zero as an infinite dimensional eigenvalue and

$$ker\ A_0 = ker\ \Lambda = \{u \in L_2^{(4)}(\mathbb{R}^3_+) : \nabla \cdot u' = 0, u_4 = 0\}.$$

The operator A_0 has no other eigenvalues.

We shall study the generalized eigenfunctions of the operator A_0. Let $L_0(\mu)$ be the second order differential operator on the half-line $\mathbb{R}_+$ depending on the parameter $\mu \in \mathbb{R}$

$$L_0(\mu)v(y) = -c_0^2(y)\rho_0(y)(\rho_0^{-1}(y)v'(y))' + \mu^2 c_0^2(y)v(y), y > 0.$$

The operator $L_0(\mu)$ can be shown to have a self-adjoint realization $L(\mu)$ in the Hilbert space $L_2(\mathbb{R}^1_+, c_0^{-2}\rho_0^{-1}dy)$ with domain

$$\{u \in L_2(\mathbb{R}^1_+) : u' \in L_2(\mathbb{R}^1_+),\ (\rho_0^{-1}u')' \in L_2(\mathbb{R}^1_+),\ u(0) = 0\}.$$

$L(\mu)$ see ([11]) has the single continuous spectrum $[c_\infty^2\mu^2, +\infty)$ and the single discrete spectrum on $(c_m^2\mu^2, c_\infty^2\mu^2)$ $(c_m = inf\ c_0(y) < c_\infty)$.

Let $\kappa_1^2(\mu) < \kappa_2^2(\mu) < \ldots \kappa_N^2(\mu)$, $N = N(\mu)$, $(\kappa_j(\mu) > 0)$ be eigenvalues of $L(\mu)$. They are defined on $(\mu_j, +\infty)$, where μ_j is a critical wave number, i.e. that value of the parameter under which the eigenvalue $\kappa_j^2(\mu)$ is absorbed by the continuous spectrum. The μ_j form a monotonically increasing sequence and

$$\lim_{j\to\infty} \mu_j = \infty.$$

If $\kappa_j^2(\mu)$ are the eigenvalues of the operator $L(\mu)$, then $\varphi_j(y, \mu)$ are the normed eigenfunctions. Let $\varphi_0(y, \alpha, \mu)$ be the normed generalized eigenfunction of $L(\mu)$ depending on the parameter

$$\alpha = (\lambda^2 - c_\infty^2\mu^2)^{1/2},$$

where $\lambda^2 (\geq c_\infty^2\mu^2)$ is the spectral parameter of $L(\mu)$.

For any $\delta \in \mathbb{R}$ we define

$$L_{2,\delta}^{(m)} = \{u : ||u||_{2,\delta} = (\int_{\mathbb{R}^3_+} (1 + |x|)^{2\delta} ||u(x)||^2_{\mathbb{C}^m} dx)^{1/2} < \infty\},$$

$$\overset{0}{\mathcal{H}}{}^1_\delta = \{u : ||u||_{\mathcal{H}^1_\delta} = (||u||^2_{L^{(4)}_{1,\delta}} + ||\Lambda u||^2_{L^{(4)}_{2,\delta}})^{1/2}, u_4(x_1, x_2, 0) = 0\}.$$

For $\zeta \in \mathbb{C} \setminus \mathbb{R}$ we define the resolvent of A_0 as $R_0(\zeta) = (A_0 - \zeta I)^{-1}$.

Given the pair of Banach spaces, X, Y, we denote by $B(X, Y)$ the Banach space of all bounded linear operators from X into Y.

The main result of this section is

Theorem 1 *i) Let $\delta > 1$, $\lambda \in \mathbb{R} \setminus \{0\}$, then the limit*

$$R_0^\pm(\lambda) = \lim_{\varepsilon\to+0} R_0(\lambda \pm i\varepsilon)$$

exists in $B(L_{2,\delta}^{(4)}, \overset{0}{\mathcal{H}}{}^1_{-\delta})$. The operator-valued functions $R_0^\pm(\lambda)$ are locally Hölder continuous on $\mathbb{R} \setminus \{0\}$ with exponent $\gamma = \min\{1, \delta - 1\}$.

ii) The spectrum of the operator A_0 is absolutely continuous on $\mathbb{R}\setminus\{0\}$ and the generalized eigenfunctions of A_0 are defined by

$$\begin{aligned}
\psi_{j,\pm}(x,\xi') &= \frac{1}{2\pi}\left(\frac{i}{\rho_0\kappa_j(|\xi'|)}\nabla_x(e^{ix'\cdot\xi'}\varphi_j(x_3,|\xi'|)), e^{ix'\cdot\xi'}\varphi_j(x_3,|\xi'|)\right),\\
& x'=(x_1,x_2),\ \xi'=(\xi_1,\xi_2),\ x'\cdot\xi'=x_1\xi_1+x_2\xi_2;\ |\xi'|>\mu_j,\\
& \kappa_j(|\xi'|)=|\lambda|,\ j=1,2,\dots,N(|\lambda|).\\
\psi_{0,\pm}(x,\xi) &= \frac{1}{2\pi}\left(\pm\frac{i}{\rho_0\kappa_0(\xi)}\nabla_x(e^{ix'\cdot\xi'}\varphi_0(x_3,\xi_3,|\xi'|)), e^{ix'\cdot\xi'}\varphi_0(x_3,\xi_3,|\xi'|)\right),\\
\kappa_0(\xi) &= (c_\infty^2|\xi'|^2+\xi_3^2)^{1/2}=|\lambda|,
\end{aligned}$$

where λ is a spectral parameter.

iii) If $dE_0(\lambda)$ is a spectral measure of A_0, $\lambda\in\mathbb{R}\setminus\{0\}$, $\delta>1$, $f\in L_{2,\delta}^{(4)}$, then

$$\begin{aligned}
\frac{d}{d\lambda}< E_0(\lambda)f,f> &= \sum_{j=1}^{N(|\lambda|)}\int_{S_j(\lambda)}\frac{\|f_j^{\sim}(\xi')\|_{\mathbb{C}^2}^2 ds_j}{\kappa_j'(|\xi'|)}\\
&+ \int_{S_0^+(\lambda)}\frac{\|f_0^{\sim}(\xi)\|_{\mathbb{C}^2}^2 ds_0}{|\nabla\kappa_0(\xi)|},
\end{aligned}$$

where

$$\begin{aligned}
f_j^{\sim}(\xi') &= (f_{j,+}^{\sim}(\xi'),f_{j,-}^{\sim}(\xi')),\ j=1,\dots,\dots,\ f_0^{\sim}(\xi)=(f_{0,+}^{\sim}(\xi),f_{0,-}^{\sim}(\xi)),\\
& f_{j,\pm}^{\sim}(\xi')=< f(\cdot),\psi_{j,\pm}(\cdot,\xi')>_0,\ \xi\in\Omega_j=\{\xi'\in\mathbb{R}^2:|\xi'|>\mu_j\},\ j=1,\dots,\dots\\
& f_{0,\pm}^{\sim}(\xi)=< f(\cdot),\psi_{0,\pm}(\cdot,\xi)>_0,\ \xi\in\mathbb{R}^3_+,\\
S_j(\lambda) &= \{\xi'\in\mathbb{R}^2:\kappa_j(|\xi'|)=|\lambda|\},\ j=1,\dots,N(|\lambda|),\\
S_0^+ &= \{\xi\in\mathbb{R}^3_+:\kappa_0(\xi)=|\lambda|\},
\end{aligned}$$

and ds_j and ds_0 are the standard measures on $S_j(\lambda)$, $S_0^+(\lambda)$.

We define the operator $F_0(\lambda)$, $\lambda\in\mathbb{R}\setminus\{0\}$ on functions $f\in L_{2,\delta}^{(4)}$ $(\delta>1)$ by the formulas

$$F_0(\lambda)f=(f_1^{\sim}|_{S_1(\lambda)},\dots,f_N^{\sim}|_{S_N(\lambda)},f_0^{\sim}|_{S_0^+(\lambda)}).$$

The operator $F_0(\lambda)$ is bounded from $L_{2,\delta}^{(4)}$ in

$$\hat{H}_0(\lambda)=\bigoplus_{j=1}^{N(|\lambda|)}L_2^{(2)}(S_j(\lambda),\frac{ds_j}{\kappa_j(|\xi'|)})\bigoplus L_2^{(2)}(S_0^+(\lambda),\frac{ds_0}{|\nabla\kappa_0(\xi)|}).$$

The operator $F_0(\lambda)$ plays an important role in the construction of the scattering matrix.

3 Acoustic operator in perturbed stratified fluid

We assume that $\rho(x)$ and $c(x)$ are measurable and satisfy (3), where $\rho_0(x_3)$ and $c_0(x_3)$ were defined in Section 1. Let

$$G(x)=\begin{pmatrix}\rho_0^{-1}&0&0&0\\0&\rho_0^{-1}&0&0\\0&0&\rho_0^{-1}&0\\0&0&0&\rho_0c_0^2\end{pmatrix}$$

and let H be the Hilbert space with the scalar product

$$<u,v>=\int_{\mathbb{R}^3_+}(G^{-1}u,v)_{\mathbb{C}^4}dx.$$

We define the operator A in H by

$$Au=-iG(x)\Lambda u$$

with the domain $\overset{0}{\mathcal{H}}{}^1$.

Let $J: H_0 \to H$ be the unitary operator given by

$$H_0 \ni u \to Ju = u \in H.$$

The resolvents of the operators A_0 and A are connected in the following way:

$$\begin{aligned} R_0(\zeta) &= J^*R(\zeta)JQ_0(\zeta), \\ R(\zeta) &= JR_0(\zeta)J^*Q(\zeta), \zeta \in \mathbb{C}\setminus\mathbb{R}, \end{aligned}$$

where

$$\begin{aligned} Q_0(\zeta) &= \Phi^{-1}+\zeta(\Phi^{-1}-I)R_0(\zeta), \\ Q(\zeta) &= \Phi+\zeta(I-\Phi)R(\zeta),\ \Phi=G_0G^{-1}. \end{aligned}$$

It is clear, that $JQ_0(\zeta)J^*Q(\zeta)=I_H$, $J^*Q(\zeta)JQ_0(\zeta)=I_{H_0}$. Since the operators

$$T_0^{\pm}(\lambda)=\lambda(\Phi^{-1}-I)R_0^{\pm}(\lambda)$$

are compact in $L^{(4)}_{2,\delta}$, $1<\delta<\delta_0$, we may apply the standard scheme of stationary scattering theory for short range perturbation (see [8]-[10]).

Theorem 2 *i) The point spectrum $\sigma_p(A)$ of A consists of zero, which is an eigenvalue of infinite multiplicity and the set $\sigma_p(A)\setminus\{0\}$ of eigenvalues of finite multiplicity is discrete in $\mathbb{R}\setminus\{0\}$.*

ii) If $1<\delta<\delta_0$, $\lambda\notin\sigma_p(A)$, then the limit

$$R^{\pm}(\lambda)=\lim_{\varepsilon\to+0}R(\lambda\pm i\varepsilon)$$

exists in $B(L^{(4)}_{2,\delta},\overset{0}{\mathcal{H}}{}^1_{-\delta})$.

It follows from Theorem 1 that the limit

$$Q^{\pm}(\lambda)=\lim_{\varepsilon\to+0}Q(\lambda\pm i\varepsilon)$$

exists in $B(L^{(4)}_{2,\delta},L^{(4)}_{2,\delta})$.

We define the operators $F_{\pm}(\lambda)$ on $L^{(4)}_{2,\delta}$, $\lambda\mathbb{R}\setminus\sigma_p(A)$ by $F_{\pm}(\lambda)f=F_0(\lambda)J^*Q^{\pm}(\lambda)f$.

The generalized Fourier transform connected with A is defined by the following formulas:

$$F_{\pm}f(\xi)=((F_{\pm}f)_1(\xi'),\ldots,(F_{\pm}f)_N(\xi'),\ldots,(F_{\pm}f)_0(\xi)),$$

where

$$\begin{aligned} (F_{\pm}f)_j(\xi') &= (J^*Q^{\pm}(\lambda)f)_j^{\sim}(\xi'),\ |\lambda|=\kappa_j(|\xi'|),\ j=1,\ldots \\ (F_{\pm}f)_0(\xi) &= (J^*Q^{\pm}(\lambda)f)_0^{\sim}(\xi),\ |\lambda|=\kappa_0(\xi). \end{aligned}$$

Let H^d be the subspace of H generated by the eigenfunctions of A, $H^c=H\ominus H^d$ and let P^d be the orthogonal projector from H to H^d, $P^c=I_H-P^d$.

Theorem 3 *Let $f \in L^{(4)}_{2,\delta}$, $1 < \delta < \delta_0$ then $F_\pm f(\xi)$ are defined almost everywhere on $\mathbb{R}^3_+$ and are measurable. Moreover*

$$||P^c f||^2_H = \sum_{j=1}^{\infty} ||(F_\pm f)^\sim_j||^2_{L_2^{(2)}(\Omega_j)} + ||(F_\pm f)^\sim_0||^2_{L_2^{(2)}(\mathbb{R}^3_+)},$$

such that the operators $F_\pm$ extend to unitary operators from H^c to $\hat{H}_0 = \bigoplus_{j=1}^{\infty} L_2^{(2)}(\Omega_j) \oplus L_2^{(2)}(\mathbb{R}^3_+)$ and

$$F_\pm (AP^c) F^*_\pm g = \hat{A}_0 g,$$

where A_0 is defined on $\hat{H}_0$ by the formulas

$$\begin{aligned}\hat{A}_0 g &= (\kappa_1(|\xi'|) g_{1,+}(\xi'), -\kappa_1(|\xi'|) g_{1,-}(\xi'), \dots, \\ &\quad \kappa_0(\xi) g_{0,+}(\xi), -\kappa_0(\xi) g_{0,-}(\xi))\end{aligned}$$

The unitary groups e^{itA_0} and e^{itA} are described by the dynamics of the wave propagation in stratified wave guide and in its perturbation.

Theorem 4 *There exist the strong limits*

$$W_\pm = s - \lim_{t \to \pm\infty} e^{-itA} J e^{itA_0} P^0$$

where $P^0 : H_0 \to H_0 \ominus ker\ \Lambda$ is orthogonal projector. The wave operators $W_\pm$ are complete i.e.

$$\text{range } W_+ \;=\; \text{range } W_- \;=\; H^c.$$

Moreover

$$W_\pm^{-1} A P^c W_\pm = A_0 P^0.$$

Since the spectrum of $A_0 P^0$ is absolutely continuous, AP^c has the absolutely continuous spectrum, too.

The scattering operator S is defined by the formulas

$$S = W_+^* W_-.$$

The operator S is a unitary operator in $\tilde{H}_0 = H_0 \ominus ker\ \Lambda$ and it is unitary equivalent to the operator $\tilde{S} = F_+ F_-^*$.

The operator $\tilde{S}$ commutes with the operator $\hat{A}_0$, therefore the spaces $\hat{H}_0(\lambda)$ are invariant relative to $\tilde{S}$. We set

$$\hat{S}(\lambda) = \tilde{S}|_{\hat{H}_0(\lambda)}, \ \lambda \in \mathbb{R} \setminus \sigma_p(A).$$

The operator-function $\hat{S}(\lambda)$ is called a scattering matrix. The structure of the scattering matrix is described by following theorem.

Theorem 5 *The operators $\hat{S}(\lambda)$ with $\lambda \in \mathbb{R} \setminus \sigma_p(A)$ are unitary in $\hat{H}_0(\lambda)$ and*

$$\begin{aligned}\hat{S}(\lambda) &= I_{\hat{H}_0(\lambda)} - 2\pi i \lambda F_0(\lambda) J^* Q^+(\lambda) J(\Phi^{-1} - I) F_0^*(\lambda), \\ \hat{S}^{-1}(\lambda) &= I_{\hat{H}_0(\lambda)} + 2\pi i \lambda F_0(\lambda) J^* Q^-(\lambda) J(\Phi^{-1} - I) F_0^*(\lambda),\end{aligned}$$

where the operators

$$T_\pm(\lambda) = F_0(\lambda) Q^\pm(\lambda)(\Phi^{-1} - I) F_0^*(\lambda)$$

are compact in $\hat{H}_0(\lambda)$.

The operators $\hat{S}(\lambda)$ have the intrinsic matrix structure

$$\hat{S}(\lambda) = (\hat{S}_{jk}(\lambda))_{j,k=0}^{N(|\lambda|)},$$

where

$$\begin{aligned}
\hat{S}_{jk}(\lambda) &= F_{0j}(\lambda)(I_{\hat{H}_0(\lambda)} - 2\pi i\lambda J^* Q^+(\lambda) J(\Phi^{-1} - I)) F_{0k}^*(\lambda) \\
F_{0j}(\lambda) f &= f_j^{\sim}(\xi')|_{S_j(\lambda)}, \; j = 1, 2, \ldots, N(|\lambda|), \\
F_{00}(\lambda) f &= f_0^{\sim}(\xi)|_{S_0^+(\lambda)}.
\end{aligned}$$

4 Spectral property of discrete acoustic operator for perturbed stratified fluids

Let $\mathbb{Z}^n$ be integer lattice in $\mathbb{R}^n$ and $\mathbb{Z}_+^n = \{x \in \mathbb{Z}^n : x_n \geq 0\}$ be the discrete half-space; $l_{2,\delta}(\mathbb{Z}^n)$ denote the Banach space of complex-valued functions on $\mathbb{Z}_+^n$ with the norm

$$||f||_{2,\delta} = (\sum_{\mathbb{Z}_+^n} |f(x)|^2 (1 + |x|)^{2\delta})^{1/2}, \; \delta \in \mathbb{R}.$$

Let

$$\begin{aligned}
\tau_j f(x) &= f(x_1, \ldots, x_j - 1, x_n), \; \tau_j^{-1} f(x) = f(x_1, \ldots, x_j + 1, \ldots, x_n), \\
\nabla_{x_j} f(x) &= \tau_j f(x) - f(x), \; \nabla_{x_j}^* f = \tau_j^{-1} f - f, \; \nabla = (\nabla_{x_1}, \ldots, \nabla_{x_n}), \; \nabla^* = (\nabla_{x_1}^*, \ldots, \nabla_{x_n}^*).
\end{aligned}$$

We consider the difference scheme of acoustic operator for the stratified fluid

$$\begin{aligned}
A_0 u(x) &= h^{-2} c_0^2(x_n) \rho_0(x_n) \nabla^* \cdot \rho_0(x_n) \nabla u(x), \; x \in \mathbb{Z}_+^n, \\
u(x) &= 0, \; x_n < 0, \; (h > 0)
\end{aligned}$$

The operator A_0 is bounded and self-adjoint in the Hilbert space H_0 with the scalar product

$$< u, v >_0 = \sum_{\mathbb{Z}_+^n} c_0^{-2}(x_n) \rho_0^{-1}(x_n) u(x) \overline{v}(x).$$

We assume that

$$\begin{aligned}
\rho_0(x_n) &= \rho_+, \; c_0(x_n) = c_+, \; x_n > H (> 0), \\
0 < c_m &= \inf_{0 \leq x_n \leq H} c_0(x_n) < c_+, \; 0 < \rho_m = \inf_{0 \leq x_n \leq H} \rho_0(x_n) \leq \rho_+.
\end{aligned}$$

Let $L(\mu)$ be the second order difference operator on $\mathbb{Z}_+^1$ depending on the parameter $\mu \in \overline{\mathbb{R}_+}$

$$L(\mu) v(y) = h^{-2} c_0^2(y) \rho_0(y) \nabla_y^* \rho_0^{-1}(y) \nabla_y + \mu c_0^2(y) v(y), \; y > 0, \; v(y) = 0, \; y < 0.$$

The operator $L(\mu)$ (see [12]) has the single continuous spectrum $[c_+^2\mu, c_+^2\mu(1 + 4h^{-2})]$ and the single discrete spectrum on the interval $(c_m^2\mu, c_+^2\mu)$. Let $\kappa_1(\mu) < \kappa_2(\mu) < \ldots < \kappa_N(\mu)$, $N = N(\mu)$ be eigenvalues of $L(\mu)$. They are defined on $(\mu_j, +\infty)$, where μ_j is that μ under which the eigenvalue $\kappa_j(\mu)$ is absorbed by the continuous spectrum. The μ_j are monotonically increasing and $\lim \mu_j = \infty$. We denote by $\varphi_j(y, \mu)$, $\mu > \mu_j$, $\varphi_0(y, \alpha, \mu)$ the normed eigenfunction and the normed generalized eigenfunction of the continuous spectrum, respectively. Let $\mu_1(\xi') = 4h^{-2}\sum_{k=1}^{n-1}\sin^2\xi_k/2$, $\xi_k \in [-\pi,\pi]$,

$$\begin{aligned}\mu(\xi) &= 4h^{-2}\sum_{k=1}^{n}\sin^2\xi_k/2,\\ K_j &= \{4h^{-2},\ldots,4h^{-2}(n-1)\}\cap\{\mu:\mu>\mu_j\}\\ \Lambda_j &= \kappa_j(K_j),\ j=1,\ldots,N,\\ \Lambda_0 &= \{0,4h^{-2}c_+^2,\ldots,4h^{-2}nc_+^2\},\ \Lambda=\cup_{j=0}^N\Lambda_j.\end{aligned}$$

For $\zeta \in \mathbb{C}\setminus\mathbb{R}$ we define the resolvent of A_0 as $(A_0-\zeta I)^{-1}$.

Theorem 6 *i) The operator A_0 has the continuous spectrum $[0,4h^{-2}nc_+^2]$ only.*

ii) Let $\delta>1$, $\lambda\in[0,4h^{-2}nc_+^2]\setminus\Lambda$, then the limit

$$R_0^\pm(\lambda)=\lim_{\epsilon\to+0}R_0(\lambda\pm i\epsilon)$$

exists in $B(l_{2,\delta},l_{2,-\delta})$. The operator-valued functions $R_0^\pm(\lambda):[0,4h^{-2}nc_+^2]\setminus\Lambda\to B(l_{2,\delta},l_{2,-\delta})$ are locally Hölder continuous with exponent $\gamma=\min\{1,\delta-1\}$.

iii) The normed generalized eigenfunctions of A_0 are defined by

$$\begin{aligned}\psi_j(x,\xi') &= (2\pi)^{-\frac{n-1}{2}}e^{ix'\cdot\xi'}\varphi_j(x_n,\mu_1(\xi')),\ \xi'=(\xi_1,\ldots\xi_{n-1}),\\ \tilde\kappa_j(\xi') &= \kappa_j(\mu_1(\xi'))=\lambda,\ j=1,\ldots,N,\\ \psi_0(x,\xi) &= (2\pi)^{-\frac{n-1}{2}}e^{ix'\cdot\xi'}\varphi_0(x_n,\xi_n,\mu_1(\xi')),\\ \kappa_0(\xi) &= c_+^2\mu(\xi)=\lambda,\end{aligned}$$

where λ is a spectral parameter.

iv) If $E_0(\lambda)$ is the canonical spectral family of the operator A_0 and $f\in l_{2,\delta}(\mathbb{Z}_+^n)$, $\delta>1$, $\lambda\in[0,4h^{-2}nc_+^2]\setminus\Lambda$, then

$$\frac{d}{d\lambda}<E_0(\lambda)f,f>_0=\sum_{j=1}^N\int_{s_j(\lambda)}\frac{|f_j^\sim(\xi')|^2ds_j}{|\nabla\tilde\kappa_j(\xi')|}+\int_{S_0^+(\lambda)}\frac{|f_0^\sim(\xi)|^2ds_0}{|\nabla\kappa_0(\xi)|},$$

where

$$\begin{aligned}f_j^\sim(\xi') &= <f(\cdot),\varphi_j(\cdot,\xi')>_{H_0},\ j=1,\ldots,N,\\ f_0^\sim(\xi) &= <f(\cdot),\varphi_0(\cdot,\xi)>_{H_0},\end{aligned}$$

and ds_j, ds_0 are the standard measures on the surfaces

$$S_j(\lambda)=\{\xi'\in[-\pi,\pi]^{n-1}:\kappa_j(\xi')=\lambda\},$$
$$S_0^+(\lambda)=\{\xi\in[-\pi,\pi]^{n-1}\times[0,\pi]:\kappa_0(\xi)=\lambda\},$$

respectively.

We assume now $c(x)$, $\rho(x)\in l_\infty(\mathbb{Z}_+^n)$ and

$$\inf_{\mathbb{Z}_+^n}\rho(x)>0\ ,\quad \inf_{\mathbb{Z}_+^n}c(x)>0\ .$$

Let

$$|\rho(x)-\rho_0(x_n)|\le c(1+|x|)^{-2\delta_0},\ |c(x)-c_0(x_n)|\le c(1+|x|)^{-2\delta_0},\ \delta_0>1,$$

where $\rho_0(x_n)$ and $c_0(x_n)$ were defined earlier.

We define the discrete acoustic operator in the perturbed stratified fluid by

$$Au(x) = c^2(x)\rho(x)\nabla^* \cdot \rho^{-1}(x)\nabla u(x),$$

where $x \in \mathbb{Z}^n_+$ and $u(x) = 0$ for $x_n < 0$.

The operator A is bounded and self-adjoint in the Hilbert space H with the scalar product

$$< u, v >= \sum_{\mathbb{Z}^n_+} c^{-2}(x)\rho(x)u(x)\overline{v}(x).$$

Theorem 7 *i) The operator A is non-negative in H and its spectrum is discrete outside of* $[0, 4h^{-2}nc_+^2 H]$.

ii) The point spectrum $\sigma_p(A)$ of operator A is discrete in $[0, 4h^{-2}nc_+^2]\backslash\Lambda$ and has finite multiplicity.

iii)Let $1 < \delta < \delta_0$, $\lambda \in [0, 4h^{-2}nc_+^2] \setminus (\Lambda \cup \sigma_p(A))$;then the the limit

$$R^{\pm}(\lambda) \lim_{\varepsilon \to +0} R(\lambda \pm i\varepsilon)$$

exists in $B(l_{2}, \delta, l_{2,-\delta})$.

iv) The operator A has no singular continuous spectrum and the wave oprators

$$W_{\pm} = s - \lim e^{-itA} J e^{itA_0}$$

exist and are complete.

(J is the identification operator : $Jf(x) = f(x)$ from H_0 to H).

Let $Q_m = \{x \in \mathbb{Z}^n : |x_1| \leq m, \ldots, |x_{n-1}| \leq m, 0 \leq x_n \leq m\}$

$$P_m f(x) = \begin{cases} f(x), & x \in Q_m \\ 0, x \notin Q_m \end{cases}$$

Theorem 8 *Let $1 < \delta < \delta_0$, $\lambda \in [0, 4nc_+^2 h^{-2}]\backslash(\Lambda \cup \sigma_p(A))$ then $\forall \varepsilon > 0 \exists m_0 \in \mathbb{N}$ such that $\forall m \geq m_0$ the equation*

$$P_m(A - (\lambda + i\varepsilon))P_m u = P_m f,\ f \in l_{2,\delta}(\mathbb{Z}^n_+)$$

has the unique solution $u_{m,\varepsilon}$. If $u_+ = R^+(\lambda)f$ is the solution of the equation

$$(A - \lambda)u = f,$$

then

$$\lim_{\varepsilon \to +0} \lim_{m \to +\infty} u_{\varepsilon,m} = u_+ \text{ in } l_{2,-\delta}(\mathbb{Z}^n_+).$$

This Theorem follows from Theorem 7 and of the results of [13].

References

[1] Brekhovskkh L.M. Waves in Lauerd Media. Academic Press, 1960.

[2] Rabinovich V.S. On the Solvability of the Acoustic Problems in a Open Waveguide. (In Russian). Rostov Univ.— Rostov-on-Don, VINITI, 15.03.89, N1687-89, 43 p.

[3] Rabinovich V.S. On the Solvability of the Acoustic Problems in a open Waveguide.(In Russian). 14 School on Theory of Operators in a Functional Space. Thesis. v.3,Novgorod, 1989.

[4] Ben-Artizi M.,Dermenjan J.,Guliot J.C. Acoustic Waves in Perturbed Stratified Fluids, a Spectral Theory. Comm. Part. Diff. Equ. 1989, 14, N14, p.479-517.

[5] Dermenjan Y.,Guliot J,C. Theorie Spectrale de la Propagation des Ondes Acoustic dans un Milieu Stratific Perturbé. J. Diff. Equ. 1986, v, 62, 357-409.

[6] Weder R. Spectral and Scattering Theory in Perturbed Stratified Fluids. J. Math. Pures et Appl. 1985,v.64,149-173.

[7] Weder R. Spectral and Scattering Theory in Derformed Optical Wave Guide. J. Reine Angew. Math., 1984, v.390, p.130-169.

[8] Reed M. and Simon B. Methods of Modern Mathematical Physics. v.3-4. Academic Press.

[9] Sigal I.M. Scattering Theory for Many-Body Quantum Mechanical Systems. Lect. Notes in Math, 1983, v..1011 p.130.

[10] Hörmander L. The Analysis of Liner Partial Differential Operators. II. Springer-Verlag, Berlin, Heidelberg, New-York, Tokio 1983.

[11] Vilcox C.H. Sound Propagation in Stratified Fluids. Appl. Math. Sci, v.50,Springer-Verlag, 1984.

[12] Nikishin E.M. Discrete Sturm-Liouville Operator and some Problem of the Function Theory. (In Russian). Trudy Seminara I.G. Petrovskogo, v. 10, 1984. p.3-77.

[13] Kozak A.V. On the Reduction Method for Many-Dimensional Discrete Convolution (In Russian). Math. Zam. Kishenev, 1973, v. 8,N3, p.157-160.

V.S. Rabinovich
Rostov University
Department of Mathematics
Rostov 344022, USSR

The Cauchy Problem, Boundary Value and Mixed Problems for Hyperbolic systems in the Complete Scale of Sobolev Type Spaces

Ja.A. Roitberg
Chernigov, USSR

1 Introduction

This article continues the author's papers [1]-[3]. Here the Cauchy problem, boundary value and mixed problems are studied for the strictly hyperbolic in the sense of Leray-Volevich systems in the complete scale of Sobolev type spaces depending on real parameters s and τ; s characterizes the smoothness of solutions in all variables and τ characterizes the additionall smoothness of the solutions in the tangential directions. When s and τ decrease the solution becomes "more generalized", for big enough s, τ the solution is an usual classical solution of the problem in consideration. In [1]-[3] such problems were studied in the case of a single equation. About former investigations see references in [1]-[3].

In earlier papers of Lions, Magenes, Berezansky, S. Krein, Roitberg and others (see [4]-[9] and the references contained there) the elliptic problems have been studied in the scales of Sobolev type spaces depending on parameter $s \in \mathbb{R}$. In these papers so-called theorems on the complete collection of isomorphisms; these theorems have many different applications. For parabolic problems similar results were obtained by Žitarashu [10].

2 General strictly hyperbolic systems. Formulation of problems

Let $(t,x) = (t,x_1,\ldots,x_n) \in \mathbb{R}^{n+1}$, $(\sigma,\xi) = (\sigma,\xi_1,\ldots,\xi_n)$ be the dual variable,

$$l = l(D_t,D_x) = (l_{kj}(D_t,D_x) : k,j = 1,\ldots,N) \tag{1}$$

is a matrix differential expression, $ord\, l_{kj} = s_k+t_j$ (may be $l_{kj} = 0$ for some j,k), $s_1,\ldots,s_N,t_1,\ldots,t_N$ are integers, $t_1 \geq \ldots \geq t_N \geq 0 = s_1 \geq \ldots \geq s_N$, $l_{kj} = 0$ if $s_k + t_j < 0$; $l_{kj}(D_t,D_x)$ is a homogeneous linear differential expression with constant complex coefficients. Here $D_t = i\partial/\partial t$, $D_x = (D_1,\ldots,D_n)$, $D_j = i\partial/\partial x_j$. Let $s_1 + \ldots + s_N + t_1 + \ldots = t_N = r$ and

$$L(\sigma,\xi) = det(l(\sigma,\xi)) = \sum_{j+|\alpha|=r} a_{j\alpha}\sigma^j\xi^\alpha. \tag{2}$$

The expression (1) is called strictly hyperbolic (in the sense of Leray - Volevich) if the polynomial (2) is strictly hyperbolic. It means that a) the coefficient $a_{(r,0,\ldots,0)}$ at σ^r in (2) is not equal to zero; b) for any $\xi \in \mathbb{R}^n \setminus \{0\}$ all roots of the equation $L(\sigma,\xi) = 0$ with respect to σ are real and different.

In this paper the expression (1) is assumed to be strictly hyperbolic. Studying the boundary value and mixed problems we shall assume in addition that the hyperplane $x_n = 0$ is not characteristic for the expression (1), i.e. the coefficient $a_{(0,\ldots,0,r)}$ of (2) at ξ_n^r is not zero.

It follows from the strict hyperbolicity of $l(D_t, D_x)$ that for any $\gamma > 0$ the equation

$$L(\sigma + i\gamma, \xi', \xi_n) = 0 \tag{3}$$

has no real ξ_n-roots. Let

$$\zeta_1(\sigma + i\gamma, \xi'), \ldots, \zeta_r(\sigma + i\gamma, \xi') \tag{4}$$

$((\sigma + i\gamma, \xi') \neq (0,0), \gamma \geq 0)$ be the ξ_n-roots of the equation (3). We shall assume that the first m roots (4) have negative imaginary parts. We put

$$\begin{aligned} L(\sigma + i\gamma, \xi', \xi_n) &= L_-(\sigma + i\gamma, \xi', \xi_n) L_+(\sigma + i\gamma, \xi', \xi_n), \\ L_-(\sigma + i\gamma, \xi', \xi_n) &= \prod_{1 \leq j \leq m} (\xi_n - \xi_j(\sigma + i\gamma, \xi')). \end{aligned} \tag{5}$$

In this paper we investigate the solvability in $\mathbb{R}^{n+1}$ of the problem

$$l(D_t, D_x)u = f \ (u = (u_1, \ldots, u_N), f = (f_1, \ldots, f_N)) \tag{6}$$

and the solvability of the Cauchy problem in the half-space

$$\begin{aligned} \Omega &= \left\{(x,t) \in \mathbb{R}^{n+1} : t > 0\right\} : l(D_t, D_x)u = f \text{ in } \Omega; \\ & D_t^{k-1} u_j|_{t=0} = u_{jk} \ \ (\forall j : t_j \geq 1, k = 1, \ldots, t_j), \end{aligned} \tag{7}$$

we study also the solvability of both these problems for the system

$$(l(D_t, D_x) + l'(t, x, D_t, D_x))u = f \tag{8}$$

obtained by the perturbation of the system (6) by lower terms with infinitely smooth coefficients having all bounded derivatives.

We consider also the boundary value problem

$$(l(D_t, D_x)u = f \text{ in } G, \ (bu)_h = \sum_{1 \leq j \leq N} b_{hj}(D_t, D_x)u|_{x_n=0} = \varphi_h \ (h = 1, \ldots, m) \tag{9}$$

in a half-space $G = \{(t,x) = (t, x', x_n) \in \mathbb{R}^{n+1} : x_n > 0\}$. Here $b = (b_{hj}(D_t, D_x) : h = 1, \ldots, m; j = 1, \ldots N)$, *ord* $b_{hj} = \sigma_h + t_j$ (some of b_{hj} may be zero); $b_{hj} = 0$ if $\sigma_h + t_j < 0$; $b_{hj}(D_t, D_x)$ is a linear homogeneous differential expression with constant complex coefficients.

The problem (9) is called hyperbolic if the expression (1) is strictly hyperbolic, the hyperplane $x_n = 0$ is not characteristic, the number of boundary conditions equals the number m of the roots of (4) having negative imaginary parts and the Lopatinsky condition holds. The latter means that for any $(\sigma + i\gamma, \xi') \neq (0,0), \gamma \geq 0$ the rows of the matrix

$$L(\sigma + i\gamma, \xi', \xi_n)(b(\sigma + i\gamma, \xi', \xi_n))(l^{-1}(\sigma + i\gamma, \xi', \xi_n)) \tag{10}$$

with elements regarded as ξ_n-polynomials, are linearly independent modulo $L_-(\xi_n) = L_-(\sigma + i\gamma, \xi', \xi_n)$.

In this paper we study the hyperbolic problem (9) and the problem

$$\begin{aligned} (l(D_t, D_x) + l'(t, x, D_t, D_x))u &= f \text{ in } G, \\ (b(D_t, D_x) + b'(t, x, D_t, D_x))u|_{x_n=0} &= \varphi \end{aligned} \tag{11}$$

obtained by perturbation of (9) by the lower terms having infinitely smooth coefficients with all bounded derivates.

It will be shown that if the right-hand sides of (9), (11) vanish for $t \leq 0$ then the solutions of these problems also vanish for $t \leq 0$. Therefore from the theorems on solvability of the problems (9), (11) there follow the assertions on the solvability of corresponding mixed problems in $G_+ = \{(t,x) \in G : t > 0\}$ with homogeneous (zero) initial data for $t = 0$.

The solvability of all problems in consideration will be stated in the complete scale of Sobolev type spaces.

3 Functional spaces

3.1

Let $s, t, \gamma \in \mathbb{R}$, $\Omega = \{(t,x) \in \mathbb{R}^{n+1} : t > 0\}$, $G = \{(t, x', x_n) \in \mathbb{R}^{n+1} : x_n > 0\}$. We denote $H^{s,\tau}(\mathbb{R}^{n+1}, \gamma)$ both the space distributions f with the norm

$$||f, \mathbb{R}^{n+1}, \gamma||'_{s,\tau} = (\int (1 + \gamma^2 + \sigma^2 + |\xi|^2)^s (1 + \gamma^2 + |\xi|^2)^\tau |\tilde{f}(\sigma, \xi)|^2 d\sigma d\xi)^{1/2} \qquad (12)$$

and the space of distributions f with the norm

$$||f, \mathbb{R}^{n+1}, \gamma||''_{s,\tau} = (\int (1 + \gamma^2 + \sigma^2 + |\xi|^2)^s (1 + \gamma^2 + \sigma^2 + |\xi'|^2)^\tau |\tilde{f}(\sigma, \xi)|^2 d\sigma d\xi)^{1/2}. \qquad (13)$$

Here $\tilde{f}(\sigma, \xi)$ is the Fourier transform of f, $\tilde{f}(\sigma, \xi) = \int f(t, x) \exp i(t\sigma + x\xi) dt dx$ for sufficiently regular f. Studying the Cauchy problem in Ω we shall use the space with the norm (12) and studying the boundary value and mixed problems in G we shall use the space with the norm (13). The primes in the notation of norms will be mostly omitted.

3.2

We denote $H^{s,\tau}(\Omega, \gamma)(s, \tau, \gamma \in \mathbb{R}, s \geq 0)$ the set of restrictions on Ω of the functions from $H^{s,\tau}(\mathbb{R}^{n+1}, \gamma)$ with the norm of the factor space, $H^{-s,-\tau}(\Omega, \gamma)$ is the dual space of $H^{s,\tau}(\Omega, \gamma)$ with respect to the extension of the inner product $(.,.) = (.,.)_\Omega$ in $L_2(\Omega)$; $||., \Omega, \gamma||_{s,\tau}$ is the norm in $H^{s,\tau}(\Omega, \gamma)$ $(s, \tau \in \mathbb{R})$.

Replacing here Ω by G, space with the norm (12) by the space with the norm (13) we define the space $H^{s,\tau}(G, \gamma)$ $(s, \tau \in \mathbb{R})$ with the norm $||., G, \gamma||_{s,\tau}$.

3.3

The space $H^s(\partial\Omega, \gamma)$ $(s, \gamma \in \mathbb{R})$ is the space of such distributions g that

$$||g, \partial\Omega, \gamma||_s = (\int_{\partial\Omega} |\hat{g}(\xi)|^2 (1 + \gamma^2 + |\xi|^2)^s d\xi)^{1/2} < \infty.$$

Here $\hat{g}(\xi) = (F'_{x\to\xi} g)(\xi)$ is the partial Fourier transform of g.

In a similar way we can define the space $H^s(\partial G, \gamma)$ $(s, \gamma \in \mathbb{R})$. It is the space with the norm

$$||g, \partial G, \gamma||_s = (\int_{\partial G} |\hat{g}(\sigma, \xi')|^2 (1 + \sigma^2 + \gamma^2 + |\xi'|^2)^s d\sigma d\xi')^{1/2}.$$

3.4

Let $r > 0$ be a fixed integer, $s, \tau, \gamma \in \mathbb{R}$, $s \neq k + 1/2$ $(k = 0, \ldots, r-1)$. We denote $\tilde{H}^{s,\tau,(r)}(\Omega, \gamma)$ the completion of $C_0^\infty(\overline{\Omega})$ with respect to the norm

$$|||u, \Omega, \gamma|||_{s,\tau,(r)} = (||u, \Omega, \gamma||^2_{s,\tau} + \sum_{1 \leq j \leq r} ||D_t^{j-1} u, \partial\Omega, \gamma||^2_{s-j+1/2+\tau})^{1/2}. \qquad (14)$$

Similar spaces were introduced by the author in [6] and studied in [8] (see also [5], Ch III, Section 6.8). For $s = k + 1/2$ $(k = 0, \ldots, r-1)$ the spaces $\tilde{H}^{s,\tau,(r)}(\Omega, \gamma)$ and norms $|||., \Omega, \gamma|||_{s,\tau,(r)}$ are defined by means of interpolation. It follows from (14) that the closure of S of the mapping

$$u \longrightarrow (u|_\Omega, u|_{\partial\Omega}, \ldots, D_t^{r-1} u|_{\partial\Omega}) \ \ (u \in C_0^\infty(\overline{\Omega}))$$

realizes the isometric isomorphism between $\tilde{H}^{s,\tau,(r)}(\Omega,\gamma)$ and the subspace of the direct product

$$H^{s,\tau}(\Omega,\gamma) \times \prod_{1\le j\le r} H^{s+t-j+1/2}(\partial\Omega,\gamma).$$

Therefore we shall identify the element $u \in \tilde{H}^{s,\tau,(r)}(\Omega,\gamma)$ with the element $Su = (u_0,\dots,u_r)$; we shall write $u = (u_0,\dots,u_r)$ $(\forall u \in \tilde{H}^{s,\tau,(r)}(\Omega,\gamma))$.

Replacing in the above definitions ω by G and D_t by D_n in (14) we shall obtain the definition of the space $\tilde{H}^{s,\tau,(r)}(\Omega,\gamma)$ and the corresponding norm.

3.5

We define now the spaces $\mathcal{H}^{s,\tau}(\mathbb{R}^{n+1},\gamma)$, $\mathcal{H}^{s,\tau}(\Omega,\gamma)$, $\tilde{\mathcal{H}}^{s,\tau,(r)}(\Omega,\gamma)$ with the the norms respectively $|u,\mathbb{R}^{n+1},\gamma|_{s,\tau}$, $|u,\Omega,\gamma|_{s,\tau}$, $|u,\Omega,\gamma|_{s,\tau,(r)}$; similar spaces and norms are introduced by by replacing Ω by G. We put

$$\begin{aligned} \mathcal{H}^{s,\tau}(\mathbb{R}^{n+1},\gamma) &= \{u : e^{-\gamma t}u \in H^{s,\tau}(\mathbb{R}^{n+1},\gamma)\}; \\ |u,\mathbb{R}^{n+1},\gamma|_{s,\tau} &= ||e^{-\gamma t}u,\mathbb{R}^{n+1},\gamma||_{s,\tau}. \end{aligned} \tag{15}$$

Replacing in (15) $\mathbb{R}^{n+1}$ by Ω we obtian the definition of the space $\mathcal{H}^{s,\tau}(\Omega,\gamma)$ and corresponding norm. Similary we put

$$\begin{aligned} \tilde{\mathcal{H}}^{s,\tau,(r)}(\Omega,\gamma) &= \{u : e^{-\gamma t}u \in \tilde{H}^{s,\tau,(r)}(\Omega,\gamma)\}; \\ |u,\Omega,\gamma|_{s,\tau,(r)} &= |||e^{-\gamma t}u,\Omega,\gamma|||_{s,\tau,(r)}. \end{aligned} \tag{16}$$

In similar way can define the space $\mathcal{H}^s(\partial G,\gamma)$ with the norm $|\cdot,\partial G,\gamma|_s$.

3.6

Lemma 1 *Let $M(t,x,D_t,D_x)$ be a linear differential expression of order m with infinitely smooth Coefficients having all bounded derivatives. Then for any $s,\tau \in \mathbb{R}$ there exists such constant $c > 0$ independent of u and γ that*

$$|Mu,\mathbb{R}^{n+1},\gamma|_{s-m,\tau} \le c|u,\mathbb{R}^{n+1},\gamma|_{s,\tau} \quad (u \in C_0^\infty(\mathbb{R}^{n+1})); \tag{17}$$

if $m \le r$ then

$$|Mu,\Omega,\gamma|_{s-m,\tau} \le c|u,\Omega,\gamma|_{s,\tau,(r)} \quad (u \in C_0^\infty(\overline{\Omega})); \tag{18}$$

$$|Mu,\Omega,\gamma|_{s-m,\tau,(r-m)} \le c|u,\Omega,\gamma|_{s,\tau,(r)} \quad (u \in C_0^\infty(\overline{\Omega})); \tag{19}$$

if $m \le r-1$ then

$$|Mu,\Omega,\gamma|_{s-m+\tau-1/2} \le c|u,\Omega,\gamma|_{s,\tau,(r)} \quad (u \in C_0^\infty(\overline{\Omega})). \tag{20}$$

All these assertions are valid also after replacing Ω by G.

Proof. Since

$$M(t,x,D_t,D_x)u = e^{\gamma t}M(t,x,D_t+i\gamma,D_x)(e^{-\gamma t}u)$$

we have

$$\begin{aligned} |Mu,\mathbb{R}^{n+1},\gamma|_{s-m,\tau} &= ||e^{-\gamma t}Mu,\mathbb{R}^{n+1},\gamma||_{s-m,\tau} \\ &= ||M(t,x,D_t+i\gamma,D_x)(e^{-\gamma t}u),\mathbb{R}^{n+1},\gamma||_{s-m,\tau} \\ &\le c||(e^{-\gamma t}u,\mathbb{R}^{n+1},\gamma||_{s,\tau} = c|u,\mathbb{R}^{n+1},\gamma|_{s,\tau} \end{aligned}$$

and we have proved the estimate (17). The others can be proved similary (cf. [8], [9]; [5] ch. III, Section 6.8). □

4 Hyperbolic systems in $\mathbb{R}^{n+1}$

It follows from the Lemma 1 that for any $s, \tau \in \mathbb{R}$ the closure l of the mapping $u \to lu$, $u \in (C_0^\infty(\mathbb{R}^{n+1}))^N$ acts continuously in the pair of spaces

$$\mathcal{H}^{T+s,\tau}(\mathbb{R}^{n+1},\gamma) \equiv \prod_{1\le j\le N} \mathcal{H}^{t_j+s,\tau}(\mathbb{R}^{n+1},\gamma) \longrightarrow$$
$$\prod_{1\le k\le N} \mathcal{H}^{s-s_k,\tau}(\mathbb{R}^{n+1},\gamma) \equiv \mathcal{H}^{s-S,\tau}(\mathbb{R}^{n+1},\gamma).$$

The norms in the spaces of the preimages and images in (21) will be denoted by $|u,\mathbb{R}^{n+1},\gamma|_{T+s,\tau}$ and $|f,\mathbb{R}^{n+1},\gamma|_{s-S,\tau}$, respectively. Naturally the question arises if the operator l is invertible.

Theorem 1 *For any $f \in \mathcal{H}^{s-S,\tau}(\mathbb{R}^{n+1},\gamma)$ $(s,\tau,\gamma \in \mathbb{R}, |\gamma| \ge \gamma_0 > 0)$ there exists one and only one element $u \in \mathcal{H}^{T+s,\tau-1}(\mathbb{R}^{n+1},\gamma)$ such that $lu = f$. There exists such a constant $c > 0$ independent of f, u, γ $(|\gamma| > \gamma_0 > 0)$ that*

$$|u,\mathbb{R}^{n+1},\gamma|_{T+s,\tau-1} \le c|\gamma|^{-1}|f,\mathbb{R}^{n+1},\gamma|_{s-S,\tau}. \tag{21}$$

If in addition supp $f \subset \overline{\Omega}$ then also supp $u \subset \overline{\Omega}$.

To prove this theorem we replace the equation $lu = f$ by the equivalent equation

$$l(D_t + i\gamma, D_x)(e^{-\gamma t}u) = e^{-\gamma t} f$$

and to the last equation we apply the Fourier transform. Then the proof follows easily from the next lemma.

Lemma 2 *There exists such a constant $c > 0$ independent of $(\sigma,\gamma,\xi) \in \mathbb{R}^{n+2}$ that*

$$|L(\sigma + i\gamma,\xi)| \ge c|\gamma|(\sigma^2+\gamma^2+|\xi|^2)^{(\tau-1)/2} \quad ((\sigma,\gamma,\xi) \in \mathbb{R}^{n+2});$$

there exists a such constant $M > 0$ that for $(\gamma,\xi) \in \mathbb{R}^{n+1}\setminus\{0\}$ and $|\sigma| \ge M(\gamma^2+|\xi|^2)^{1/2}$ the estimate

$$|L(\sigma + i\gamma,\xi)| \ge c|\gamma|(\sigma^2+\gamma^2+|\xi|^2)^{\tau/2} \tag{22}$$

holds. If the hyperplane $x_n = 0$ is not characteristic for the expression l then the constant $M > 0$ can be chosen in such a way that the estimate (22) is valid also for $|\xi_n| \ge M(\sigma^2+\gamma^2+|\xi'|^2)^{1/2}$.

The last assertion of the theorem follows from certain Paley-Wiener type theorem.

Comparison of the formulas (21) and (21) shows that going from f to u "we loose one unit of smoothness in the tangential directions". But the norm of the operator $f \to u$ is estimated by $c|\gamma|^{-1}$ and thus is small for great $|\gamma|$. Therefore the next Theorem 2 easily follows from Theorem 1.

Theorem 2 *There exists such a number $\gamma_0 > 0$ that for $|\gamma| \ge \gamma_0$ the problem (8) with any $f \in \mathcal{H}^{s-S,\tau}(\mathbb{R}^{n+1},\gamma)$ has one and only one solution $u \in \mathcal{H}^{T+s,\tau-1}(\mathbb{R}^{n+1},\gamma)$. There exists such a constant $c > 0$ independent of f, u, γ $(|\gamma| \ge \gamma_0)$ that the estimate (21) holds. If supp $f \subset \overline{\Omega}$ then also supp $u \subset \overline{\Omega}$.*

5 The Cauchy problem

With the Cauchy problem (7) we connect the closure $A = A_{s,\tau}$ $(s, \tau \in \mathbb{R})$ of the mapping

$$u \longrightarrow (lu, \{D_t^{k-1}u_j|_{t=0}(j : t_j \geq 1; k = 1, \ldots, t_j)\}) \quad (u \in (C_0^\infty(\Omega))^N)$$

acting continuously (according to Lemma 1) from the whole

$$\tilde{\mathcal{H}}^{T+s,\tau,(T)} \equiv \prod_{1 \leq j \leq N} \tilde{\mathcal{H}}^{t_j+s,\tau,(t_j)}(\Omega, \gamma)$$

into

$$\begin{aligned} K^{s,\tau} &\equiv \prod_{1 \leq j \leq N} \tilde{\mathcal{H}}^{s-s_j,\tau,(-s_j)}(\Omega, \gamma) \times \prod_{j:t_j \geq 1} \prod_{1 \leq k \leq t_j} H^{t_j+s+\tau-k+1/2}(\partial\Omega, \gamma) \\ &\equiv \tilde{\mathcal{H}}^{s-S,\tau,(-S)}(\Omega, \gamma) \times B^{s,\tau}(\partial\Omega, \gamma). \end{aligned}$$

Let in (7) $f = (f_1, \ldots, f_N) \in \tilde{\mathcal{H}}^{s-S,\tau,(-S)}(\Omega, \gamma)$, $f_k = (f_{k0}, \ldots, f_{k,-s_k}) \in \tilde{\mathcal{H}}^{s+s_k,\tau,(-s_k)}(\Omega, \gamma)$,

$$U = \{u_{jk} | j : t_j \geq 1, k = 1, \ldots, t_j\} \in B^{s,\tau}(\partial\Omega, \gamma) \quad (s, \tau \in \mathbb{R}).$$

The element $u = (u_1, \ldots u_N) \in \tilde{\mathcal{H}}^{T+s,\tau,(T)}(\Omega, \gamma)$ satisfying $Au = (f, U)$ is called (generalized) solution of the Cauchy problem (7). Similarly there can be defined the solution of the Cauchy problem for the system (8).

The Cauchy problem (7) to be solvable it is necessarry that f and U satisfy certain compatibility condition. In fact if l_k $(k = 1, \ldots, N)$ is a row of the matrix l and $u \in \tilde{\mathcal{H}}^{T+s,\tau,(T)}(\Omega, \gamma)$ is a solution of the problem (7) then

$$D_t^{m-1} l_k u|_{t=0} = D_t^{m-1} f_k|_{t=0} = f_{km} \in H^{s+\tau-s_k+1/2-m}(\partial\Omega, \gamma) \ (k : -s_k \geq 1; m = 1, \ldots, -s_k). \quad (23)$$

The left-hand side of (23) is completely defined by the vector U of initial data and the right-hand side is defined by vector f. Therefore the conditions (23) are necessary for the solvability of the Cauchy problem.

Theorem 3 *Let $s, \tau, \gamma \in \mathbb{R}$, $|\gamma| > \gamma_0 > 0$, $F = (f, U) \in K^{s,\tau}$ and the compatibility condition holds. Then the Cauchy problem (7) has one and only one solution $u \in \tilde{\mathcal{H}}^{T+s,\tau-1,(T)}(\Omega, \gamma)$. There exists such constant $c > 0$ independet on F, u, γ $(|\gamma| \geq \gamma_0 > 0)$ that*

$$||u||_{\tilde{\mathcal{H}}^{T+s,\tau-1,(T)}(\Omega,\gamma)} \leq c|\gamma|^{-1}||F||_{K^{s,\tau}}. \quad (24)$$

To prove this theorem we extend the solution u by zero on $\mathbb{R}^{n+1} \setminus \Omega$ and the Cauchy problem is reduced to the problem in $\mathbb{R}^{n+1}$ (cf. [1]-[2]). After such reduction we use the Theorem 1.

In the Theorem 3 going from F to u we loose one unit of smoothness in tangential directions. But the norm of the operator $F \rightarrow u$ is estimated by $c|\gamma|^{-1}$ and thus is small for great $|\gamma|$. Therefore the next Theorem 4 easily follows from the Theorem 3.

Theorem 4 *Let $s, \tau \in \mathbb{R}$. There exists such number γ_0 that for $|\gamma| \geq \gamma_0$ for any $F = (f, u) \in K^{s,\tau}$ satisfying the compatibility condition the Cauchy problem for (8) has one and only one solution $u \in \tilde{\mathcal{H}}^{T+s,\tau-1,(T)}(\Omega, \gamma)$ and the estimate (24) holds.*

Remark 1 *Obviously the compatibility condition does not arise if $s_1 = \ldots = s_N = 0$. In general cases these conditions can not be omitted. In some cases these conditions can be omitted by changing the formulation of the problem. Let in (7) $f = f_0 = (f_{10}, \ldots, f_{N0}) \in \prod_{1 \leq K \leq N} \mathcal{H}^{s-s_k,\tau}(\Omega, \gamma)$*

$\equiv \mathcal{H}^{s-S,\tau}(\Omega,\gamma)$ *and the solution is being thought as before in* $\tilde{\mathcal{H}}^{T+s,\tau,(T)}(\Omega,\gamma)$. *The possibility of such consideration follows from the estimates (18), (20). Now we define the elements* $f_{km} \in H^{s+\tau-s_k+1/2-m}(\partial\Omega,\gamma)$ *by the formulas (23). Then the conditions*

$$(f_{k0},\ldots,f_{k,-s_k}) \in \tilde{\mathcal{H}}^{s-s_k,\tau,(-s_k)}(\Omega,\gamma) \quad (\forall k : -s_k \geq 1) \tag{25}$$

are necessary for the Cauchy problem (7) with $f = f_0 \in \mathcal{H}^{s-s,\tau}(\Omega,\gamma)$ *to be solvable in* $\tilde{\mathcal{H}}^{T+s,\tau,(T)}(\Omega,\gamma)$. *If* $|\gamma| \geq \gamma_0 > 0$ *then it follows from the Theorem 3 that the conditions (25) are also sufficient for the solvability of this problem in* $\tilde{\mathcal{H}}^{T+s,\tau-1,(T)}(\Omega,\gamma)$. *Thus in this case the compatibility condition is replaced by the condition (25). But if*

$$s < s_j + 1/2 \quad (j = 1,\ldots,N) \tag{26}$$

then (25) always holds. Therefore in the classes of "not very smooth" functions the Cauchy problem is always solvable. But the smoothness of such a solution does not always increase with the increasing of the smoothness of the right-hand sides.

A similar assertion holds also for the Cauchy problem for the system (8).

6 Boundary value and mixed problems

Now we study the hyperbolic problem (9). Let $\kappa = \max\{0,\sigma_1,\ldots,\sigma_m\}$ Then for any $s,\tau \in \mathbb{R}$ the closure $\mathfrak{a} = \mathfrak{a}_{s,\tau}$ of the mapping $u \to (lu, bu|_{\partial G})$ $(u \in (C_0^\infty(\overline{G}))^N)$ acts continuously in the pair of spaces

$$\begin{aligned}
\tilde{\mathcal{H}}^{T+s,\tau,(T+\kappa)}(G,\gamma) &\equiv \prod_{1\leq j\leq N} \tilde{\mathcal{H}}^{t_j+s,\tau,(t_j+\kappa)}(G,\gamma) \longrightarrow \\
&\equiv \prod_{1\leq j\leq N} \tilde{\mathcal{H}}^{s-s_j,\tau,(\kappa-s_j)}(G,\gamma) \times \prod_{1\leq h\leq m} \mathcal{H}^{s-\sigma_h+\tau-1/2}(\partial G,\gamma) \\
&\equiv \tilde{\mathcal{H}}^{s-S,\tau,(\kappa-S)}(G,\gamma) \times B^{s-\sigma+\tau-1/2}(\partial G,\gamma) \\
&\equiv \tilde{K}^{s,\tau}.
\end{aligned} \tag{27}$$

The norms in the spaces of the preimages and images will be denoted by

$$|u,G,\gamma|_{T+S,\tau,(T+\kappa)} \text{ and } |f,G,\gamma|_{s-S,\tau,(\kappa-S)} + |\varphi,\partial G,\gamma|_{s-\sigma+\tau-1/2},$$

repectively. The question arises if the operator $\mathfrak{a}$ is invertible.

Theorem 5 *Let* $\gamma \geq \gamma_0 > 0$, $s,\tau \in \mathbb{R}$. *Then for any*

$$f \in \tilde{\mathcal{H}}^{s-S,\tau+1,(\kappa-S)}(G,\gamma),\ \varphi \in B^{s-\sigma+\tau}(\partial G,\gamma)$$

there exists one and only one element $u \in \tilde{\mathcal{H}}^{T+s,\tau,(T+\kappa)}(G,\gamma)$ *such that* $\mathfrak{a}u = (f,\varphi)$. *There exists such a constant* $c > 0$ *independent of* u, f, γ *and* $(\gamma \geq \gamma_0 > 0)$ *that*

$$|u,G,\gamma|_{T+s,\tau,(T+\kappa)} \leq c(\frac{1}{\gamma}|f,G,\gamma|_{s-S,\tau,(\kappa-S)} + \frac{1}{\sqrt{\gamma}}|\varphi,\partial G,\gamma|_{s-\sigma+\tau}). \tag{28}$$

If in addition supp $(f,\varphi) \subset \overline{G}_+ = \{(t,x) \in \overline{G}, t \geq 0\}$ *then also supp* $u \subset \overline{G}_+$.

Since the norm of the operator $(f,\varphi) \to u$ is small for great enough γ the next theorm also holds.

Theorem 6 *Let* $s,\tau \in \mathbb{R}$. *There exists such number* $\gamma_0 > 0$ *that for* $\gamma \geq \gamma_0$ *for any* $f \in$

$\tilde{\mathcal{H}}^{s-S,\tau+1,(\kappa-S)}(G,\gamma)$, $\varphi \in B^{s-\sigma+\tau}(\partial G,\gamma)$ *there exists one and only one solution* $u \in \tilde{\mathcal{H}}^{T+s,\tau,(T+\kappa)}(G,\gamma)$ *of the problem (11) and the estimate (24) holds. If supp* $(f,\varphi) \subset \overline{G}_+$ *then also supp* $u \subset \overline{G}_+$.

In the particular case of a single equation ($N = 1$) and $\kappa = 0$, $f = 0$ the Theorems 5 and 6 were proved in [3]. The transfer to the case $f \neq 0$ without additional loss of smoothness still in the case $N = 1$ is not trivial: the transposition and interpolation are used.

7 Some applications

The obtained results are applied to the studying of the considered problems in the case of right-hand sides having arbitrary power singularities along manifolds of various dimensions. After regularization the right-hand sides are embedded in certain negative Sobolev spaces. The above theorems make it possible finding the spaces containing the solutions.

The similar procedure allowes to study the Green functions of the problems in consideration and to study a certain class of degenerate hyperbolic problems.

References

[1] Roitberg. Ja.A. The Cauchy problem for hyperbolic equations in the full scale of spaces of Sobolev type. Dokl. AN SSSR 1986, t.290, No 2,pp. 296-300

[2] Roitberg. Ja.A. Solvability of the Cauchy problem for hyperbolic equations in the full scale of spaces of Sobolev type. In the book: "Spectral theory for differential - operator equations", Kiev, In-t matematiki AN USSR, 1986, pp. 33-52

[3] Roitberg. Ja.A. Boundary and mixed problems for homogeneous equations in the full scale of spaces of Sobolev type. Ukr. mat. žurn. t.41, No 5,pp. 685-690

[4] Lions J.-L. Magenes E. Problèmes aux limites non homogènes et applications. V.I.- Paris: Dunod. 1968

[5] Berezanskiĭ Ju.M. Decomposition into eigenfunctions of self - adjoint operators. Kiev, Naukova dumka 1965

[6] Roitberg. Ja.A. Elliptic problems with non-homogeneous boundary conditions and local gain of smoothness of generalized solutions along the boundary. Dokl. AN USSR 1964, t.157, No 4, pp.798-801 (Soviet Math. Dokl. 5 (1964) 1034-1037)

[7] Roitberg. Ja.A. The theorem on homoeomorphisms and Green's formula for general elliptic boundary conditions. Mat. sb. 1970, t.83, No 2, pp.181-213 (Math. USSR Sbornik 12 (1970), 177-212)

[8] Roitberg. Ja.A. On the values on the boundary of a domain of generalized solutions of elliptic equations. Mat. sb. 1971, t.86, No 2, pp.248-267 (Math. USSR Sbornik 15,2 (1971) 241-260)

[9] Roitberg. Ja.A., Serdyuk V.A. Elliptic problems with parameters in L_2-spaces of generalized functions for general systems of equations. Kiev, In-t matem. AN USSR 1982; preprint 82-30, p.62

[10] Žitarashu N.V. The theorem on complete choice of isomorphism in the L_2 theory of generalized solutions of boundary problems for a petrovskiĭ-parabolic equation. Mat. sb. 1985, t.189, No 4, pp.451-473

Ja.A. Roĭtberg, Chernigov State Ped. Inst., ul. Sverdlova 53
Chernigov 250012, USSR

On the Density of the Solutions of Elliptic Boundary Value Problems in the Sense of Petrovsky Systems in Functional Spaces on Manifolds

Ja.A. Roitberg and Z.G. Sheftel Chernigov, USSR

1 Introduction

In a bounded domain $G \subset \mathbb{R}^n$ with infinitely smooth boundary ∂G, let us consider the boundary problem

$$lu = f \text{ in } G, \quad bu = \varphi \text{ on } \partial G. \tag{1}$$

Here $l = l(x, D) = (l_{jk}(x, D) : j, k = 1, \dots, N)$ is properly elliptic in the sense of Petrovsky matrix differential expression of order $T = (t_1, \dots, t_n)$, $|T| = t_1 + \dots + t_N = 2m$, $t_1 \geq \dots \geq t_N$; u, f are functional columns of length N; $l_{jk}(x, D)$ is a linear differential expression with complex valued infinitely smooth coefficients in $\overline{G}$, $ord\ b_{hk} \leq t_k + \sigma_h$, σ_h are given negative integers; the expression $b_{hk}(x, D)$ are differential in the normal direction to ∂G and, in general, pseudo-differential along ∂G. Extracting the main parts of the expressions l_{jk} and b_{hk} we leave only the members of orders, t_k and $t_k + \sigma_h$, respectively. The ellipticity of the problem (1) means that l is properly elliptic and that on ∂G the Lopatinsky condition holds.

Let $\Gamma_1 \subset G$ be a smooth $(n-1)$-dimensional orientable manifold without boundary, Γ is an open subset of Γ_1. For any smooth solution $u = (u_1, \dots, u_N)$ of the problem (1) we define

$$\nu_T u = (u_1|_\Gamma, \dots, D_\nu^{t_1 - 1} u_1|_\Gamma, \dots, u_N|_\Gamma, \dots, D_\nu^{t_N - 1} u_N|_\Gamma); \tag{2}$$

here $D_\nu = i\partial/\partial\nu$, ν is the normal to Γ. We shall arbitrarily change f in an arbitrary small fixed subdomain $G_0 \subset G$ (or change φ on an arbitrary small open patch $\gamma \subset \partial G$). We investigate the possibility approximating any given vector-function on Γ by vectors $\nu_T u$ obtained by this change. Similar problems for a single equation have been studied since 1960 by Beckert, Göpfert, Beyer, Wildenhain, Schulze, Hamann, Wanka and others (see [1], [2] and the refereces there). In all these papers the boundary expressions were assumed to be normal in the Aronszain-Milgram-Schechter sense. Such results were obtained by authors without normality assumption in [3], [4]. In this paper similar results are obtained for the problem (1) (also without normality assumption for boundary conditions). Before this investigation we have studied the formal adjoint to (1) in a complete scale of Sobolev type spaces.

2 Functional spaces

For $p, t \in \mathbb{R}$, $1 < p < \infty$ and $t \geq 0$, let $H^{t,p}(G)$ be the space of Bessel potentials and $H^{-t,p'}(G)$, where $1/p + 1/p' = 1$, relative to the extension $(\cdot, \cdot)$ of the inner product in $L_2(G)$; $\|\cdot\|_{t,p}$ is the norm in $H^{t,p}(G)$, $t \in \mathbb{R}$; $(.,.)$ will denote the inner product (or its extension) both in $L_2(G)$ and the direct product $(L_2(G))^k$.$B^{t,p}(\Gamma_1)$, $t \in \mathbb{R}$, denotes the Besov space; $B^{t,p}(\Gamma_1)$ and $B^{-t,p'}(\Gamma_1)$ are mutually adjoint with respect to the extension $< .,. >$ of the inner product in $L_2(\Gamma_1)$; $<< . >>_{t,p}$ is the norm in $B^{t,p}(\Gamma_1)$.

Lemma 1 *Let be $v \in H^{-s,p'}(G)$, $s > 0$, and let supp $v \subset \Gamma_1$. Then*

a) if $s \leq 1/p$ then $v = 0$;

b) if $s > 1/p$ then $v = \sum_{0\leq j\leq\kappa} D_\nu^j(\tau_j \times \delta_{\Gamma_1})$, where $\tau_j \in B^{-(s-j-1/p),p'}(\Gamma_1)$, δ_{Γ_1} is the Dirac measure concentrated on Γ_1 and κ is the biggest integer less than $s - 1/p$.

Proof. It follows from the assumptions that $(v, w) = 0$ $(\forall w \in C_0^\infty(\overline{G} \setminus \Gamma_1))$. If $s \leq 1/p$ then $C_0^\infty(\overline{G} \setminus \Gamma_1)$ is dense in $H^{s,p}(G)$; therefore $(v, w) = 0$ $(\forall w \in H^{s,p}(G))$ and hence $v = 0$.

If $s > 1/p$ then the closure of $C_0^\infty(\overline{G} \setminus \Gamma_1)$ consists of such and only such $w \in H^{s,p}(G)$ for which $D_\nu^j w|_{\Gamma_1} = 0$, $j = 0, \ldots, \kappa$. Therefore the antilinear functional $L(w) = (v, w)$, $w \in H^{s,p}(G)$, depends only on the collection of the normal derivates $(w|_{\Gamma_1}, \ldots, D_\nu^\kappa w|_{\Gamma_1}) \in \prod_{0\leq j\leq\kappa} B^{s-j-1/p,p}(\Gamma_1)$. Hence there exist $\tau_j \in B^{-(s-j-1/p),p'}(\Gamma_1)$, $j = 0, \ldots, \kappa$, such that

$$(v, w) = L(w) = \sum_{0\leq j\leq\kappa} < \tau_j, D_\nu^j w >= (\sum_{0\leq j\leq\kappa} D_\nu^j(\tau_j \times \delta_{\Gamma_1}), w) \ (\forall w \in H^{s,p}(G)).$$

3 The case of connected $G \setminus \Gamma$

3.1

In the following we assume that for the formal adjoint l^+ of l the property of uniqueness of the Cauchy problem holds: if $l^+ v = 0$ in a domain $G_1 \subset G$ and $v = 0$ in a subdomain $G_0 \subset G_1$ the $v = 0$ in G_1.

Let G_0 be an open subset of G, $\overline{G}_0 \subset G \setminus \Gamma$; the diameter of G_0 can be arbitrary small. We put

$$M(G_0) = \{u \in (C^\infty(\overline{G}))^N : supp\ lu \subset G_0, bu|_{\partial G} = 0\}$$

and

$$\nu_T M(G_0) = \{\nu_T u : u \in M(G_0)\},$$

where $\nu_T u$ is defined by formula (2).

Theorem 1 *If the set $G_0 \setminus \overline{\Gamma}$ is connected then $\nu_T M(G_0)$ is dense in the direct product $\prod_{1\leq j\leq|T|} B^{s_j,p}(\Gamma)$ for any $s_j \geq 0$ $(j = 1, \ldots, |T|)$ and $1 < p < \infty$.*

Proof. It is enough to show that for any $s \geq 0$ and $1 < p < \infty$ the set $\nu_T M(G_0)$ is dense in $\prod_{j:t_j\geq1} \prod_{1\leq r\leq t_j} B^{t_j+s-r+1-1/p,p}(\Gamma)$. To prove this density we have to verify that if

$$\mu_{jr} \in B^{-(t_j+s-r+1-1/p),p'}(\Gamma)$$

and

$$\sum_{j:t_j\geq1} \sum_{1\leq r\leq t_j} < \mu_{jr}, D_\nu^{r-1} u_j >= 0 \quad (\forall u \in M(G_0)) \tag{3}$$

then all $\mu_{jr} = 0$. The equality (3) can be rewritten in the form

$$\sum_{j:t_j\geq1} (\sum_{1\leq r\leq t_j} D_\nu^{r-1}(\mu_{jr} \times \delta_\Gamma), u_j) = 0 \quad (\forall u \in M(G_0)) \tag{4}$$

where δ_Γ is the Dirac measure concentrated on Γ. Let us prove now that

$$w_j = \sum_{1\leq r\leq t_j} D_\nu^{r-1}(\mu_{jr} \times \delta_\Gamma) \in H^{-t_j-s,p'}(G) \ (\forall j : t_j \geq 1).$$

We have

$$\begin{aligned}|(D_\nu^{r-1}(\mu_{jr}\times\delta_\Gamma),u)| &= |<\mu_{jr},D_\nu^{r-1}u>|\leq<<\mu_{jr}>>_{-(t_j+s-r+1-1/p),p'}\\ &\times <<D_\nu^{r-1}u>>_{t_j+s-r+1-1/p,p}\leq<<\mu_{jr}>>_{-(t_j+s-r+1-1/p),p'}||u||_{t_j+s,p},\end{aligned}$$

hence $D_\nu^{r-1}(\mu_{jr}\times\delta_\Gamma)\in H^{-t_j-s,p'}(G)$ and the required inclusion is proved.

From (4) we see that the N-dimensional vector $w=(w_1,\dots,w_{N_1},0,\dots,0)$ (N_1 is the number of $t_j\neq 0$) is orthogonal to the kernel $\mathfrak{N}\subset(C^\infty(\overline{G}))^N$ of the problem (1).

We denote $C^\infty(bd)^+$ the set of such $v\in(C^\infty(\overline{G}))^N$ that

$$(lu,v)=(u,l^+v)\ \ (\forall u\in(C^\infty(\overline{G}))^N: bu=0 \text{ on } \partial G). \tag{5}$$

A constructive description of the set $C^\infty(bd)^+$ can be obtained by using the Green's formula (cf. [5], where this formula is proved under additional assumptions; these assumptions always hold if the problem (1) is elliptic). The problem

$$l^+v=g,\ v\in C^\infty(bd)^+ \tag{6}$$

is formally adjoint to the problem (1) with $\varphi=0$. It follows from (5) that the condition $(g,\mathfrak{N})=0$ is necessary for the solvability of the problem (6). This condition is also sufficient (cf. for the case $N=1$ [6], theorem 4.7). Considering the probem (6) after closing the corresponding operator in the spaces of distributions we shall write (6) in the form $l^+v=g,\ v\in(bd)^+$. The condition $(g,\mathfrak{N})=0$ as before is necessary and sufficient for the solvability of this problem.

Since $(w,\mathfrak{N})=0$ the problem

$$l^+v=w\in\prod_{1\leq j\leq N}H^{-t_j-s,p'}(G),\ \ v\in(bd)^+ \tag{7}$$

has a solution $v\in(H^{-s,p'}(G))^N\cap(C^\infty(\overline{G}\setminus\overline{\Gamma}))^N$. It follows from (4)-(7) that this v satisfies

$$(lu,v)=(u,l^+v)=(u,w)=0\ \ (\forall u\in M(G_0)). \tag{8}$$

We denote by $\mathfrak{N}^+$ the space of solutions of the problem (6) with $g=0$; $\mathfrak{N}^+$ is finite-dimensional and the problem (1) with $\varphi=0$ is solvable if and only if $(f,\mathfrak{N}^+)=0$ (cf. [5], [6]). If $\mathfrak{N}^+=0$ it follows immediately from (8) that $v=0$ in G_0.

Let be $\mathfrak{N}^+\neq 0$. If v_1 is a solution of the problem (7) then $v=v_1+v_0$ ($v\in\mathfrak{N}^+$) is also solution of the problem. We shall show that v_0 can be selected in such way that $v=0$ in G_0. Let $e_1,\dots,e_k$ be the basis in $\mathfrak{N}^+$; we can find such $h_1,\dots,h_k\in(C_0^\infty(G_0))^N$ that $(e_i,h_j)=\delta_{ij}$ ($i,j=1,\dots,k$). For any $g\in(C_0^\infty(G_0))^N$ we put

$$g_1=g-\sum_{1\leq j\leq k}(g,e_j)h_j. \tag{9}$$

It is clear that *supp* $g\subset G_0$ and $(g,e_j)=0$, $i=1,\dots,k$, i.e. $(g,\mathfrak{N}^+)=0$. Then there exists such $u\in(C^\infty(\overline{G}))^N$ that

$$lu=g_1,\ bu=0 \text{ on } \partial G.$$

It means that $u\in M(G_0)$. Then it follows from(8) and (9) that

$$0=(v_1,g-\sum_{1\leq j\leq k}(g,e_j)h_j)=(v_1-\sum_{1\leq j\leq k}(v,h_j)e_j,g)\ \ (\forall g\in(C_0^\infty(G))^N).$$

Therefore $v=v_1-\sum_{1\leq j\leq k}(v,h_j)e_j=0$ in G_0. Now it follwos from the uniqueness of the Cauchy problem for l^+ and the connectedness of $G\setminus\overline{\Gamma}$ that $v=0$ in $\overline{G}\setminus\overline{\Gamma}$, i.e. *supp* $v\subset\overline{\Gamma}\subset\Gamma_1$. Since $v=(v_1,\dots,v_N)$, $v_i\in H^{-s,p'}(G)$, we have from the lemma 1 that

$$v_i=\sum_{0\leq j\leq\kappa}D_\nu^j(\tau_{ij}\times\delta_\Gamma),\ \ i=1,\dots,N,$$

where

$$\tau_{ij} \in B^{-(s-j-1/p),p'}(\Gamma), \quad j = 1, \ldots, \kappa.$$

Substituting these expressions for v in (7) and taking into account the form of w, the linear independentce of the derivatives of the Dirac measures and the ellipticity of l^+ we obtain that all $\tau_{ij} = 0$ and hence all $\mu_{ir} = 0$.

3.2

Let γ be an open subset of ∂G;we put

$$\begin{aligned} M(\gamma) &= \{u \in (C^\infty(\overline{G}))^N : lu = 0, supp\ bu|_{\partial G} \subset \gamma\}; \\ \nu_T M(\gamma) &= \{\nu_T u : u \in M(\gamma)\}. \end{aligned}$$

Theorem 2 *If the set $G \backslash \Gamma$ is connected then $\nu_T M(\gamma)$ is dense in $\prod_{1 \leq j \leq |T|} B^{s_j,p}(\Gamma)$ for any $s_j \geq 0$, $j = 1, \ldots, N$ and $1 < p < \infty$.*

Proof. Let us consider a domain $G' \subset G$ with the smooth boundary $\partial G' \supset \partial G \setminus \gamma$. The domain G' is constructed in such way that the coefficients of l and b could be extended into G' and on $\partial G'$, respectively with preservation of smoothness, ellipticity and Lopatinski's conditions. Let now G_0 be an open subset of $G' \setminus \overline{G}$. Then it follows from the Theorem 1 applied to the domain G' that

$$\overline{\nu_T M(G_0)} = \prod_{1 \leq j \leq |T|} B^{s_j,p}(\Gamma). \tag{10}$$

Since $bu = 0$ on $\partial G'$ for any $u \in M(G_0)$ we have that $supp\ bu|_{\partial G} \subset \gamma$ for such u. Therefore $M(G_0) \subset M(\gamma)$ and the assertion follows from (10).

3.3

The propositions similar to the Theorems 1, 2 are valid also in the case when Γ_1 is a k-dimensional manifold, $1 \leq k < n - 1$.

4 The case of non-connected $G \setminus \overline{\Gamma}$

4.1

Let $\Gamma = \Gamma_1$ be the boundary of a subdomain $G_1 \subset G$ and $\overline{G}_0 \cap \overline{G}_1 = \emptyset$.Let $T' = (t'_1, \ldots, t'_N)$ be a multi-index, $0 \leq t'_j \leq t_j$; $|T'| = t'_1 + \ldots t'_N = m$, i.e. $2|T'| = |T|$; $\nu_{T'} u$ is defined by the formula (2) where T is replaced by T'. We put also

$$\nu_{T'} M(G_0) = \{\nu_{T'} u : u \in M(G_0\}.$$

Theorem 3 *Let the problem*

$$l^+ v = 0 \text{ in } G_1,\ D_\nu^{r_j - 1} v_j = 0 \text{ on } \Gamma\ (\forall j : t_j - t'_j \geq 1, r = 1, \ldots, t_j - t'_j) \tag{11}$$

have no more than one solution. Then $\nu_{T'} M(G_0)$ is dense in the set

$$\prod_{j : t'_j \geq 1} \prod_{1 \leq r \leq t'_j} B^{t'_j - r + 1 - 1/p, p}(\Gamma).$$

Proof. It is enough to prove that if $\mu_{jr} \in B^{-(t'_j-r+1-1/p),p'}(\Gamma)$ and

$$\sum_{j:t'_j\geq 1}\sum_{1\leq r\leq t'_j} < \mu_{jr}, D_\nu^{r-1}u_j >= 0 \quad (\forall u \in M(G_0)) \tag{12}$$

then all $\mu_{jr} = 0$. We rewrite (12) in the form

$$\sum_{j:t'_j\geq 1}\left(\sum_{1\leq r\leq t'_j} D_\nu^{r-1}u_j(\mu_{jr}\times\delta_\Gamma), u_j\right) = 0.$$

It is easily seen that $w_j = \sum_{1\leq r\leq t'_j} D_\nu^{r-1}(\mu_{jr}\times\delta_\Gamma) \in H^{-t'_j,p'}(G)$. After changing (if it is necessary the order of the functions we put $w = (w_1,\ldots,w_{N'_1},0,\ldots,0)$, where N'_1 is the number of $t'_j \neq 0$. Then as in the proof of Theorem 1 we see that there exists a solution $v \in \prod_{1\leq j\leq N} H^{t_j-t'_j,p}(G)$ of the problem $l^+v = w$, $v \in (bd)^+$. This solution can be selected in such way that $v = 0$ in G_0. Then it follows from the uniqueness of the Cauchy problem for l^+ that $v = 0$ in $G \setminus \overline{G}_1$. Hence it follows in particular that $D_\nu^{r_j-1}v_j = 0$ on Γ $(r_j = 1,\ldots,t_j - t'_j; j : t_j - t'_j \geq 1)$. It follows now from the uniqueness of solution of the problem (11) that $v = 0$ in G_1 and thus $v = 0$ in $\overline{G}$.

4.2

In the considered case the analogue of the Theorm 2 also holds. Let, like in Section 3.2, γ be an open subset of ∂G, $M(\gamma)$ is defined like in Section 3.2.

Theorem 4 *If the assumption of the Theorem 3 are satisfied then the set $\nu_{T'}M(\gamma)$ is dense in the set*

$$\prod_{j:t'_j\geq 1}\prod_{1\leq r\leq t'_j} B^{t'_j-r+1-1/p,p}(\Gamma).$$

References

[1] Hamann U. Wildenhain G. Approximation by solutions of general boundary value problems for elliptic equations of arbitray order // Partial differential equations, Banach center publications.- 1987.-Vol. 19.-P. 113-119.

[2] Haman U. Approximation durch Normalableitungen von Lösungen elliptischer Randwertprobleme in beliebigen Sobolev-Räumen // Math. Nachr.-1986.-128.-S. 199-214.

[3] Roitberg Ja.A. Sheftel Z.G. On the density of solutions of elliptic problems with localized right hand sides in function spaces on manifolds. Dokl. AN SSSR 1989, t 305, No 6, pp 1317-1320 (Soviet math Dokl. 39,2 (1989) 411-414)

[4] Roitberg Ja.A. Sheftel Z.G. On the approximation of functions on manifolds by solutions of elliptic boundary problems. Nelineige granichnye zadachi 1989, byp.1, pp. 86-90

[5] Roitberg Ja.A. Sheftel Z.G. Greens formula and the theorem of homeomorphisms for elliptic systems with non-normal boundary conditions. Mat. issled. 1972, 7, byp.2(24) pp. 143-157

[6] Roitberg Ja.A. The theorem on homeomorphisms and Greens formula for general elliptic boundary problems with non-normal boundary conditions. (Math. USSR Sbornik 12 (1970) 177-212)

Ja.A. Roĭtberg, Z.G. Sheftel , Chernigov State Ped. Inst., ul. Sverdlova 53
Chernigov 250012, USSR

Singularities of Non-Rotationally Symmetric Solutions of Boundary Value Problems for the Lamé Equations in a 3 dimensional Domain with Conical Points

A.-M.Sändig*and R.Sändig**

1 Introduction

It is well known that singularities are present in solutions of boundary value problems for the Lamé equations in conical domains. It follows from the general theory [5, 9] that the solutions consist of singular terms of the form $r^{\alpha}(\ln r)^{q}\,\mathbf{F}(\alpha,\phi,\theta)$ (r is the distance to the vertex of the cone, ϕ and θ are the spherical angles) and a more regular term.

Rotationally symmetric solutions of the Lamé equations under zero boundary displacements or stress free boundary conditions are investigated in [1, 2], where the values of α and q have been computed.Here we are concerned with the more general case, namely that the volume and surface forces of our problems are non rotationally symmetric.That means that the solutions depend not only on r and θ, but on the polar angle ϕ too. Using a monotonicity principle of V.A.KOZLOV , V.G.MAZ'JA and C.SCHWAB [6] one can get regularity results for polyhedral domains too.

2 Formulation of the problem

Let Ω be a three-dimensional bounded domain with an only circular conical point on its boundary (see Fig.1). Assume that the displacement field $\mathbf{u}(\mathbf{x})$ of this isotropic elastic body satisfies the linear system of equations

$$\begin{aligned} \mathbf{L}\,\mathbf{u} &= \mathbf{L}(D_{\mathbf{x}})\,\mathbf{u}(\mathbf{x}) \\ &= \mu\Delta\mathbf{u}(\mathbf{x}) + (\lambda+\mu)\nabla(\nabla\cdot\mathbf{u}(\mathbf{x})) = -\mathbf{f}(\mathbf{x}) \quad \text{for} \quad \mathbf{x}\in\Omega \end{aligned} \tag{1}$$

and the boundary conditions

$$\mathbf{L}_1\,\mathbf{u} = \mathbf{L}_1(D_{\mathbf{x}})\,\mathbf{u}(\mathbf{x}) = \mathbf{g}_1(\mathbf{x}) \quad \text{for} \quad \mathbf{x}\in\partial\Omega \tag{2}$$

or

$$\mathbf{L}_2\,\mathbf{u} = \mathbf{L}_2(D_{\mathbf{x}})\,\mathbf{u}(\mathbf{x}) = \mathbf{S}(\mathbf{u}(\mathbf{x}))\cdot\mathbf{n}(\mathbf{x}) = \mathbf{g}_2(\mathbf{x}) \quad \text{for} \quad \mathbf{x}\in\partial\Omega \tag{3}$$

where λ and μ are the Lamé constants, $\mathbf{f}(\mathbf{x})$ is the vector of the volume forces, $\mathbf{g}_1(\mathbf{x})$ is a prescribed displacement (Dirichlet conditions), $\mathbf{g}_2(\mathbf{x})$ is a traction (Neumann conditions). $\mathbf{S}(\mathbf{u}(\mathbf{x}))$ denotes the stress tensor, with Cartesian components

$$S_{ij}(\mathbf{u}(\mathbf{x})) = \mu\left(\frac{\partial u_i}{\partial x_j} + \frac{\partial u_j}{\partial x_i}\right) + \delta_{ij}\lambda\nabla\cdot\mathbf{u}(\mathbf{x}) \tag{4}$$

*Department of Mathematics
**Department of Physics, University of Rostock, Rostock, Germany

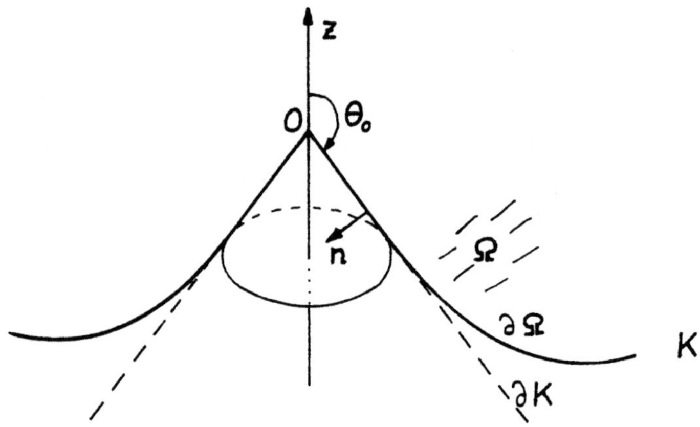

Figure 1: A body with a rotationally symmetric conical vertex

where u_i is the ith component of $\mathbf{u}(\mathbf{x})$ and δ_{ij} is the Kronecker symbol. $\mathbf{n}(\mathbf{x})$ is the unit vector of the outward normal to $\partial\Omega$ in the point $\mathbf{x}$.

In general, one can consider right hand sides $\mathbf{f} \in [W_p^k(\Omega)]^3$ and $\mathbf{g}_i \in [W_p^{k+3-i-1/p}(\partial\Omega)]^3; i = 1,2; k = 0,1,\ldots; 1 \leq p < \infty$, but we can restrict to boundary forces $\mathbf{g}_i = \mathbf{0}$ subtracting from $\mathbf{u}$ a displacement field $\mathbf{v}$ with $\mathbf{L}_i\mathbf{v} = \mathbf{g}_i$ on $\partial\Omega$.

In order to investigate the behavior of $\mathbf{u}$ near the conical point $\mathbf{0}$ we can use the regularity results of V.A.KONDRATJEV, V.G.MAZ'JA and B.A.PLAMENEVSKIJ in the weighted Sobolev spaces

$$V_p^k(\Omega,\beta) = \left\{ u \in W_p^k(\Omega) : \| u \| = \left(\sum_{|\alpha|\leq k} \int_\Omega r^{p(|\alpha|+\beta-k)} |D^\alpha u| d\mathbf{x} \right)^{\frac{1}{p}} < \infty \right\} \tag{5}$$

Thus it is proved in a paper of V.G.MAZ'JA and B.A.PLAMENEVSKIJ [8] (theorem 2.1) that every element f from $[W_p^k(\Omega)]^3$ can be written as a sum

$$\mathbf{f} = \mathbf{f}^o + \mathbf{p} \tag{6}$$

where $\mathbf{f}^o$ is from $[V_p^k(\Omega,0)]^3$ and $\mathbf{p}$ is a polynomial vector of degree $\leq k-1$ provided $p > 3$, of degree $\leq k-2$ provided $3/2 < p < 3$ and of degree $\leq k-3$ provided $1 < p < 3/2$.

This decomposition yields an expansion of $\mathbf{u}$ near the conical point $\mathbf{0}$ in two kinds of singular terms; one is coming from $\mathbf{f}^o$ and one from the polynomial vector $\mathbf{p}$ (see [4, 5]). Thus we get for a weak solution $\mathbf{u}$ from $[W_2^1(\Omega)]^3$ of the equations

$$\begin{aligned} \mathbf{Lu} &= -\mathbf{f} \quad \text{in} \quad \Omega \\ \mathbf{L}_i\mathbf{u} &= \mathbf{0} \quad \text{on} \quad \partial\Omega \qquad i = 1 \text{ or } 2 \end{aligned} \tag{7}$$

that

$$\mathbf{u} = \sum_{-1/2<\Re\alpha_i<-3/p+2+k} \eta\, c_i\, r^{\alpha_i} \left(\mathbf{F}_i(\phi,\theta,\alpha_i) + \ln r\, \mathbf{H}_i(\phi,\theta,\alpha_i) + \cdots \right) +$$

$$\sum_{-1/2<\beta_i<-3/p+2+k} \eta\, r^{\beta_i} \left(\mathbf{G}_i(\phi,\theta,\beta_i) + \ln r\, \mathbf{J}_i(\phi,\theta,\beta_i) + \cdots\right) + \mathbf{w} \tag{8}$$

where β_i is an integer, $\mathbf{w}$ is from $[W_p^{k+2}(\Omega)]^3$, $\eta = \eta(r)$ is a cut-off function near $\mathbf{0}$ and α_i are eigenvalues of a parameter depending boundary value problem. Here $\Re$ denotes the real and $\Im$ the imaginary part. In the following we consider only the first type of singularities, that means we assume that $\mathbf{f}$ in eq.(7) is from $[V_p^k(\Omega,0)]^3$. These singular terms are determined by the nontrivial solutions of the following "model" problem in the infinite cone K (see Fig.1).

$$\begin{aligned} \mathbf{Lu} &= \mathbf{0} \quad \text{in } K \\ \mathbf{L}_i\mathbf{u} &= \mathbf{0} \quad \text{on } \partial K \qquad i = 1 \text{ or } 2 \end{aligned} \tag{9}$$

3 Solutions of the Lamé equations in an infinite cone

When studying problems over bodies with circular conical points (such as that shown in Fig.1) it is natural to use spherical coordinates (r,θ,ϕ) with origin at the apex $\mathbf{0}$. In these coordinates the local orthonormal basis vectors are

$$\begin{aligned} \mathbf{e}_r &= (\sin\theta\cos\phi, \sin\theta\sin\phi, \cos\theta)^T \\ \mathbf{e}_\theta &= (\cos\theta\cos\phi, \cos\theta\sin\phi, -\sin\theta)^T \\ \mathbf{e}_\phi &= (-\sin\phi, \cos\phi, 0)^T \end{aligned} \tag{10}$$

Then any vector $\mathbf{u}$ can be written as $\mathbf{u} = u_r\mathbf{e}_r + u_\theta\mathbf{e}_\theta + u_\phi\mathbf{e}_\phi$ or shortly

$$\mathbf{u} = \mathbf{u}(r,\theta,\phi) = \begin{pmatrix} u_r(r,\theta,\phi) \\ u_\theta(r,\theta,\phi) \\ u_\phi(r,\theta,\phi) \end{pmatrix} \tag{11}$$

Now we have to look for solutions of $\mathbf{Lu} = \mathbf{0}$ of the shape $r^\alpha\, \mathbf{F}(\phi,\theta,\alpha)$ (see section 2). Since the domain K is rotationally symmetric we are only interested in solutions with $\mathbf{F}(\phi,\theta,\alpha) = \mathbf{F}(\phi+2\pi,\theta,\alpha)$. Therefore we consider the expansion of $\mathbf{u}$ in a Fourier series

$$\begin{aligned} \mathbf{u} &= r^\alpha \mathbf{F} \\ &= r^\alpha\{\mathbf{A}_0(\theta,\alpha) + \mathbf{A}_1(\theta,\alpha)\cos\phi + \mathbf{A}_2(\theta,\alpha)\cos 2\phi \\ &\quad + \mathbf{B}_1(\theta,\alpha)\sin\phi + \mathbf{B}_2(\theta,\alpha)\sin 2\phi + \cdots\} \end{aligned} \tag{12}$$

In order to construct the unknown coefficients $\mathbf{A}_i(\theta,\alpha)$ and $\mathbf{B}_i(\theta,\alpha)$ we use the Papkovich - Neuber representation of the displacement fields through harmonic functions. It holds that

$$\mathbf{u} = (3-4\nu)\begin{pmatrix} \mathbf{B}\cdot\mathbf{e}_r \\ \mathbf{B}\cdot\mathbf{e}_\theta \\ \mathbf{B}\cdot\mathbf{e}_\phi \end{pmatrix} - \begin{pmatrix} r\mathbf{e}_r\cdot\partial\mathbf{B}/\partial r + \partial B_4/\partial r \\ \mathbf{e}_r\cdot\partial\mathbf{B}/\partial\theta + r^{-1}\partial B_4/\partial\theta \\ \sin^{-1}\theta\, \mathbf{e}_r\cdot\partial\mathbf{B}/\partial\phi + (r\sin\theta)^{-1}\partial B_4/\partial\phi \end{pmatrix} \tag{13}$$

where $\mathbf{B} = (B_1, B_2, B_3)$, $\triangle B_i = 0$ in K, $i = 1...4$ and $\nu = \lambda/2(\lambda+\mu)$ is the Poisson ratio. The general form of the harmonic functions B_i, which yield displacement fields of the type eq.(12), is known [3], namely

$$\begin{aligned} B_1 &= r^\alpha \sum_{k=0}^{\infty} P_\alpha^{-k}(\cos\theta)\,(C_{k1}\cos k\phi + C_{k2}\sin k\phi) \\ B_2 &= r^\alpha \sum_{k=0}^{\infty} P_\alpha^{-k}(\cos\theta)\,(C_{k3}\cos k\phi + C_{k4}\sin k\phi) \end{aligned}$$

$$
\begin{aligned}
B_3 &= r^\alpha \sum_{k=0}^{\infty} P_\alpha^{-k}(\cos\theta)\,(C_{k5}\cos k\phi + C_{k6}\sin k\phi)\\
B_4 &= r^{\alpha+1} \sum_{k=0}^{\infty} P_{\alpha+1}^{-k}(\cos\theta)\,(C_{k7}\cos k\phi + C_{k8}\sin k\phi)
\end{aligned}
\tag{14}
$$

Here $P_\alpha^{-k} = P_\alpha^{-k}(\cos\theta)$ are the associated Legendre functions of first kind [7] with the special notation $P_\alpha = P_\alpha^0(\cos\theta)$.

Inserting eq.(14) into eq.(13) and using recurrence formulae for the Legendre functions [7] we get for $i = 0$

$$
\begin{aligned}
\mathbf{A}_0(\theta,\alpha) &= c_{10}(\alpha)\begin{pmatrix} (3-4\nu-\alpha)\,P_\alpha\cos\theta \\ -(3-4\nu)\,P_\alpha\sin\theta + (P_\alpha)'\sin\theta\cos\theta \\ 0 \end{pmatrix} + \\
&\quad c_{20}(\alpha)\begin{pmatrix} (\alpha+1)P_{\alpha+1} \\ -(P_{\alpha+1})'\sin\theta \\ 0 \end{pmatrix} + \\
&\quad c_{30}(\alpha)\begin{pmatrix} 0 \\ 0 \\ (1-\nu)P_\alpha^{-1} \end{pmatrix}
\end{aligned}
\tag{15}
$$

and for $i > 0$

$$
\begin{aligned}
\mathbf{A}_i(\theta,\alpha) &= c_{1i}(\alpha)\begin{pmatrix} (3-4\nu-\alpha)P_\alpha^{-i}\cos\theta \\ -(3-4\nu)P_\alpha^{-i}\sin\theta + (P_\alpha^{-i})'\sin\theta\cos\theta \\ -iP_\alpha^{-i}\cos\theta/\sin\theta \end{pmatrix} + \\
&\quad c_{2i}(\alpha)\begin{pmatrix} (\alpha+1)P_{\alpha+1}^{-i} \\ -(P_{\alpha+1}^{-i})'\sin\theta \\ iP_{\alpha+1}^{-i}/\sin\theta \end{pmatrix} + \\
&\quad c_{3i}(\alpha)\begin{pmatrix} (3-4\nu-\alpha)P_\alpha^{-i+1}\sin\theta \\ (3-4\nu)P_\alpha^{-i+1}\cos\theta + (P_\alpha^{-i+1})'\sin^2\theta \\ (4-4\nu-i)P_\alpha^{-i+1} \end{pmatrix}
\end{aligned}
\tag{16}
$$

$$
\begin{aligned}
\mathbf{B}_i(\theta,\alpha) &= d_{1i}(\alpha)\begin{pmatrix} (3-4\nu-\alpha)P_\alpha^{-i}\cos\theta \\ -(3-4\nu)P_\alpha^{-i}\sin\theta + (P_\alpha^{-i})'\sin\theta\cos\theta \\ iP_\alpha^{-i}\cos\theta/\sin\theta \end{pmatrix} + \\
&\quad d_{2i}(\alpha)\begin{pmatrix} (\alpha+1)P_{\alpha+1}^{-i} \\ -(P_{\alpha+1}^{-i})'\sin\theta \\ -iP_{\alpha+1}^{-i}/\sin\theta \end{pmatrix} + \\
&\quad d_{3i}(\alpha)\begin{pmatrix} (3-4\nu-\alpha)P_\alpha^{-i+1}\sin\theta \\ (3-4\nu)P_\alpha^{-i+1}\cos\theta + (P_\alpha^{-i+1})'\sin^2\theta \\ -(4-4\nu-i)P_\alpha^{-i+1} \end{pmatrix}
\end{aligned}
\tag{17}
$$

Here we have denoted $(P_\alpha^{-i})' = (P_\alpha^{-i}(\cos\theta))' = dP_\alpha^{-i}(\cos\theta)/d\cos\theta$. We remark that in the rotationally symmetric case [1, 2], where $u_\phi = 0$, $u_r = u_r(r,\theta)$ and $u_\theta = u_\theta(r,\theta)$ only $\mathbf{A}_0(\theta,\alpha)$ with $c_{30}(\alpha) = 0$ occurs.

4 Dirichlet conditions

In section 3 we have constructed the solution eq.(12) of the Lamé equation system in an infinite cone K. Now we have to determine the complex eigenvalues α in such a way, that the Dirichlet conditions $\mathbf{u} = \mathbf{0}$ on ∂K are satisfied. That means, we demand $\mathbf{u}(r, \theta_0, \phi) = \mathbf{0}$ and therefore we have to calculate those eigenvalues $\alpha_i \, (i = 0, 1, \ldots)$, for which $\mathbf{A}_0(\theta_0, \alpha_0) = \mathbf{0}$ or $\mathbf{A}_i(\theta_0, \alpha_i) = \mathbf{B}_i(\theta_0, \alpha_i) = \mathbf{0} \, (i = 1, \ldots)$.

Let us begin with the case $i = 0$:
$\mathbf{A}_0(\theta, \alpha)$ consists of three linearly independent solutions (see eq.(15)). A nontrivial solution $\mathbf{u}_0 = r^{\alpha_0} \mathbf{A}_0(\theta, \alpha_0)$ with $\mathbf{u}_0(\theta_0, \alpha_0) = \mathbf{0}$ exists, if the determinate $\tilde{D}_0(\alpha, \theta_0)$ of the corresponding linear system of equations for the unknowns $c_{i0}(\alpha)$ $(i = 1, 2, 3)$ vanishes for $\alpha = \alpha_0$.

Using the recurrence formulae for the Legendre functions we get from $\tilde{D}_0$ the following transcendental equation

$$\begin{vmatrix} (3 - 4\nu - \alpha) P_\alpha \cos\theta_0 & P_{\alpha+1} & 0 \\ \begin{array}{l} P_\alpha (-3 + 4\nu \\ +(4 - 4\nu + \alpha) \cos^2\theta_0) \\ -(\alpha+1) P_{\alpha+1} \cos\theta_0 \end{array} & \begin{array}{l} P_{\alpha+1} \cos\theta_0 \\ -P_\alpha \end{array} & 0 \\ 0 & 0 & \begin{array}{l} P_\alpha \cos\theta_0 \\ -P_{\alpha+1} \end{array} \end{vmatrix} = 0 \tag{18}$$

The real parts of the zeros $\alpha_0 = \alpha_0(\theta_0, \lambda, \mu)$, where $\alpha_0 \neq 0$ and $\neq 1$, are shown in fig.2 for $\nu = 0.3$ and $0 < \theta_0 < \pi$. Real α_0's are drawn as solid lines, while the real part of complex α_0's is given by dashed ones. The "eigenfunctions" $\mathbf{F}_0(\theta, \alpha_0) = \mathbf{A}_0(\theta, \alpha_0)$ (see eq.(8) and eq.(12)) can be calculated inserting the nontrivial solutions $c_{i0}(\alpha_0) \, (i = 1, 2, 3)$ in eq.(15) . Comparing with the rotationally symmetric case ([2] fig.2) we find in fig.2 new solutions coming from the factor $P_\alpha \cos\theta_0 - P_{\alpha+1}$.
Now we consider the case $i > 0$:
Since $\mathbf{A}_i(\theta, \alpha)$ and $\mathbf{B}_i(\theta, \alpha)$ consist of three linearly independent solutions as in the case $i = 0$, we get a linear system of equations for the unknowns $c_{ij}(\alpha)$ and $d_{ij}(\alpha)$ in eqs.(16, 17). The zeros α_i of the corresponding determinates $\tilde{D}_i(\alpha, \theta_0)$ for $\mathbf{A}_i(\theta_0, \alpha) = \mathbf{0}$ or $\mathbf{B}_i(\theta_0, \alpha) = \mathbf{0}$ coincide. Now the following transcendental equation for α may be derived from $\tilde{D}_i$

$$\begin{vmatrix} (3 - 4\nu - \alpha) P_\alpha^{-i} \cos\theta_0 & -(\alpha+1) P_{\alpha+1}^{-i} & i P_{\alpha+1}^{-i} \\ \begin{array}{l} P_\alpha^{-i} (-3 + 4\nu + \\ (4 - 4\nu + \alpha) \cos^2\theta_0) - \\ (\alpha + 1 + i) P_{\alpha+1}^{-i} \cos\theta_0 \end{array} & \begin{array}{l} (\alpha + 1 - i) P_\alpha^{-i} - \\ (\alpha+1) P_{\alpha+1}^{-i} \cos\theta_0 \end{array} & i P_{\alpha+1}^{-i} \cos\theta_0 \\ -i P_\alpha^{-i} \cos\theta_0 & -i P_{\alpha+1}^{-i} & \begin{array}{l} (\alpha+1) P_{\alpha+1}^{-i} - \\ (\alpha + 1 - i) P_\alpha^{-i} \cos\theta_0 \end{array} \end{vmatrix}$$

$$= 0 \tag{19}$$

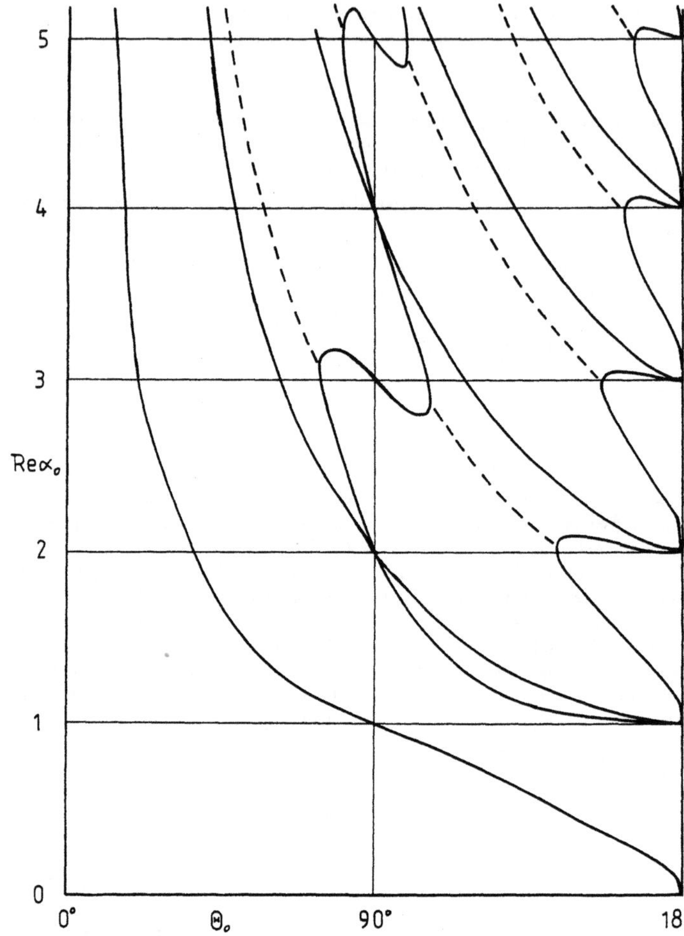

Figure 2: Dependence of the eigenvalues α_0 on θ_0 for Dirichlet conditions at $\nu = 0.3$ (——— $\Im\alpha = 0$, – – – – $\Im\alpha \neq 0$)

For the numerical calculation of the Legendre functions we used the Mehler - Dirichlet representation and the Hermite quadrature approximation

$$\begin{aligned} P_\alpha(\cos\theta) &= \frac{\sqrt{2}}{\pi}\int_0^\theta \frac{\cos(\alpha+\frac{1}{2})t}{\sqrt{\cos t - \cos\theta}}\, dt \\ &\approx \frac{2\sqrt{2}}{n}\sum_{j=1}^{[\frac{n+1}{2}]} G(x_j) \\ G(x) &= \frac{\sqrt{t}\cos(\alpha+\frac{1}{2})t}{\sqrt{\sin\frac{t+\theta}{2}\,\frac{2}{t-\theta}\,\sin\frac{t-\theta}{2}}} \end{aligned} \tag{20}$$

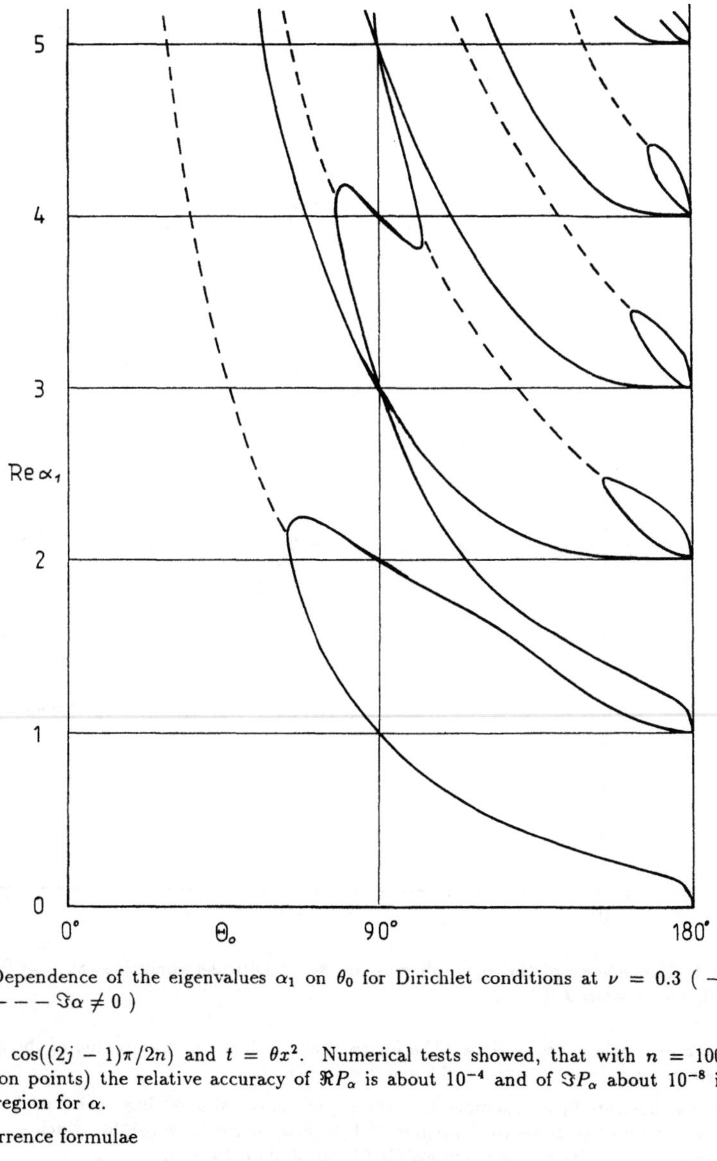

Figure 3: Dependence of the eigenvalues α_1 on θ_0 for Dirichlet conditions at $\nu = 0.3$ (——— $\Im\alpha = 0$, – – – – $\Im\alpha \neq 0$)

where $x_j = \cos((2j-1)\pi/2n)$ and $t = \theta x^2$. Numerical tests showed, that with $n = 100$ (i.e. 50 integration points) the relative accuracy of $\Re P_\alpha$ is about 10^{-4} and of $\Im P_\alpha$ about 10^{-8} in the interesting region for α.

The recurrence formulae

$$P_\alpha^{-i}(\cos\theta) = \left[\cos\theta P_\alpha^{-i+1} - P_{\alpha+1}^{-i+1}\right] / (\alpha - i + 1)\sin\theta \tag{21}$$
$$P_{\alpha+1}^{-i}(\cos\theta) = \left[P_\alpha^{-i+1} - \cos\theta P_{\alpha+1}^{-i+1}\right] / (\alpha + i + 1)\sin\theta \tag{22}$$

for $\alpha - i + 1 \neq 0$ or $\alpha + i + 1 \neq 0$ are sufficient stable for $i = 1, 2$. To calculate the zeros $\alpha_i(\theta_0)$ of the above given nonlinear equations we used a downhill simplex method [10], which searches for all

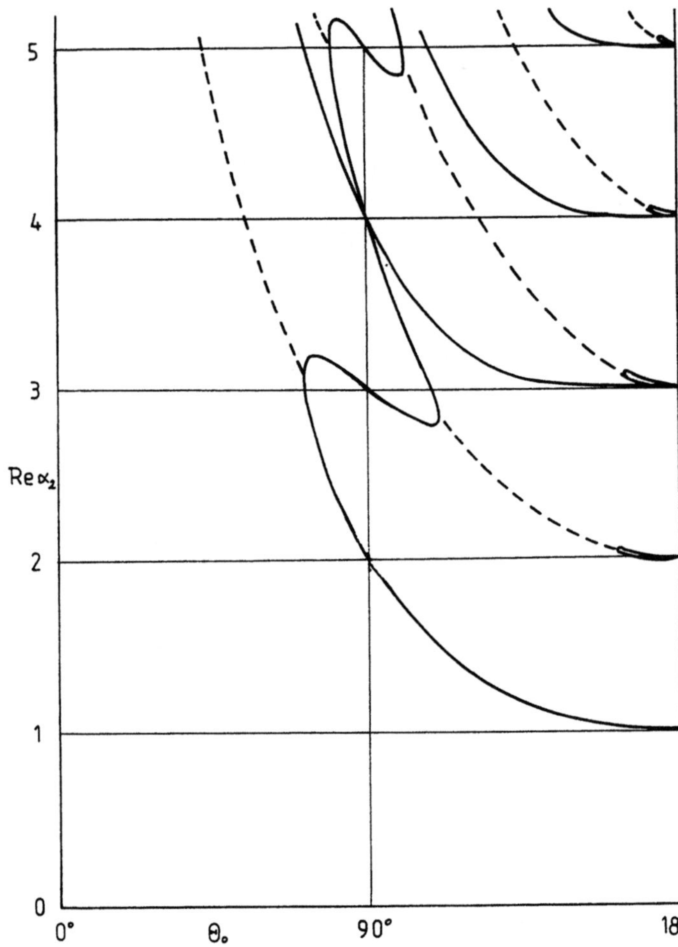

Figure 4: Dependence of the eigenvalues α_2 on θ_0 for Dirichlet conditions at $\nu = 0.3$ (——— $\Im\alpha = 0$, – – – – $\Im\alpha \neq 0$)

zeros in a rectangle in the α-plane. The figures 3 and 4 show the distribution of the real parts of the eigenvalues α_i $(i = 1, 2)$ for $\nu = 0.3$ and $0 < \theta_0 < \pi$.

Now we illustrate by an example, how the expansion eq.(8) looks like. Let $\mathbf{f}$ be from $[L_2(\Omega)]^3$. Then only those singular vector functions $r^{\alpha_i}\,\mathbf{F}_i(\phi, \theta, \alpha_i)$ occur in eq.(8), for which $-1/2 < \Re\alpha_i < 1/2$, whilst the singular vector functions $\mathbf{H}_i, \mathbf{G}_i$ and $\mathbf{J}_i$ do not occur.

Let us denote by θ_0^i those angles with $\alpha_i(\theta_0^i) = 1/2$, $i = 0, 1$. The numerical calculation yields $\theta_0^1 \approx 124.22°$ and $\theta_0^0 \approx 143.66°$ for $\nu = 0.3$.

Lemma 1 : Let $\mathbf{u}$ be a weak solution from $[\overset{\circ}{W}{}^1_2(\Omega)]^3$ of the Dirichlet problem

$$\mathbf{L}\,\mathbf{u} = -\mathbf{f} \quad \text{in } \Omega \tag{23}$$
$$\mathbf{u} = \mathbf{0} \quad \text{on } \partial\Omega \tag{24}$$

where $\Omega \subset R^3$ is a bounded domain with only one circular conical point $\mathbf{0}$ on the boundary, $\mathbf{f} \in [L_2(\Omega)]^3$ and $\nu = 0.3$. Then the following expansions near the conical point $\mathbf{0}$ hold:
For $0 < \theta_0 < \theta_0^1$ we have $\mathbf{u} = \mathbf{w} \in [W_2^2(\Omega)]^3$.
For $\theta_0^1 < \theta_0 < \theta_0^0$ we have $\mathbf{u} = r^{\alpha_1}(c_1\mathbf{A}_1(\theta,\alpha_1)\cos\phi + c_2\mathbf{B}_1(\theta,\alpha_1)\sin\phi) + \mathbf{w}$.
For $\theta_0^0 < \theta_0 < \pi$ we have $\mathbf{u} = r^{\alpha_1}(d_1\mathbf{A}_1(\theta,\alpha_1)\cos\phi + d_2\mathbf{B}_1(\theta,\alpha_1)\sin\phi) + r^{\alpha_0}c_0\mathbf{A}_0(\theta,\alpha_0) + \mathbf{w}$.

The eigenvalues $\alpha_0 = \alpha_0(\theta_0,\nu)$ and $\alpha_1 = \alpha_1(\theta_0,\nu)$ are given by the lowest lines of fig.(2) and fig.(3), respectively.

5 Stress boundary conditions

We start again from the general solutions eq.(12) of the Lamé equation system $\mathbf{L}\,\mathbf{u} = \mathbf{0}$ in the infinite cone and determine the complex numbers α in such a way, that $\mathbf{S}(\mathbf{u}) = \mathbf{0}$ on ∂K. Since $\mathbf{n} = \mathbf{e}_\theta$, the normal stresses are $\mathbf{S}(\mathbf{u})\cdot\mathbf{n} = (S_{r\theta}, S_{\theta\theta}, S_{\phi\theta})^T$. After some calculations we get a linear system of equations for the unknowns $c_{ij}(\alpha)$ and $d_{ij}(\alpha)$ $(i = 0,1,\ldots; j = 1,2,\ldots)$ in eqs.(15, 16, 17). Nontrivial solutions exist, if the corresponding determinates vanish. We get for the calculation of the zeros α_i $(i = 0,1,\ldots)$ of the determinates the following transcendental equations

$$\begin{aligned}
N_i(\alpha,\theta_0) &\equiv det(\mathbf{N}) = 0 \qquad (25)\\
N_{11} &= P_\alpha^{-i}(-\alpha(1-2\nu) + (\alpha^2 - 2 + 2\nu)\cos^2\theta_0) + \\
&\quad P_{\alpha+1}^{-i}(\alpha + i + 1)(2 - 2\nu - \alpha)\cos\theta_0\\
N_{12} &= (\alpha - 1)iP_\alpha^{-i}/2\\
N_{13} &= P_{\alpha+1}^{-i}\,\alpha\cos\theta_0 + P_\alpha^{-i}(-\alpha + i/2)\\
N_{21} &= P_\alpha^{-i}\cos\theta_0\,(2\alpha + 2 - 2\nu + \alpha^2 - i^2 + \\
&\quad \cos^2\theta_0\,(-3\alpha - 3 + 2\nu - \alpha^2)) + \\
&\quad P_{\alpha+1}^{-i}((\alpha + i + 1)((-3 + 2\nu)\sin^2\theta_0 + 1))\\
N_{22} &= i(P_\alpha^{-i}\cos\theta_0\,(-\alpha - 2) + (\alpha + i + 1)P_{\alpha+1}^{-i})\\
N_{23} &= P_\alpha^{-i}\cos\theta_0\,(i + 1) + P_{\alpha+1}^{-i}(-\alpha\sin^2\theta_0 - i - 1)\\
N_{31} &= i(P_\alpha^{-i}((-1 + 2\nu)(\alpha + 3 - 2\nu)\cos^2\theta_0) + \\
&\quad P_{\alpha+1}^{-i}\cos\theta_0(-\alpha - 1 - i))\\
N_{32} &= P_\alpha^{-i}(\alpha + 1 + i^2 - \sin^2\theta_0\,(\alpha + 1)(\alpha + 2)/2) - \\
&\quad P_{\alpha+1}^{-i}(\alpha + 1 + i)\cos\theta_0\\
N_{33} &= P_{\alpha+1}^{-i}\cos\theta_0\,(i + 1) + P_\alpha^{-i}((-i - 1) + \sin^2\theta_0\,(1 + \alpha/2))
\end{aligned}$$

Figures 5, 6 and 7 show the distribution of the eigenvalues α_i for $i = 0,1,2$, $\nu = 0.3$ and $0 < \theta_0 < \pi$.

Comparing with the results for rotationally symmetric solutions [1] and [2] (fig.3) we have new lines for $i = 0$ coming from the new factor $P_{\alpha+1}\cos\theta_0 + P_\alpha(-1 + \sin^2\theta_0(\alpha + 2)/2)$ in the nonrotationally symmetric case.

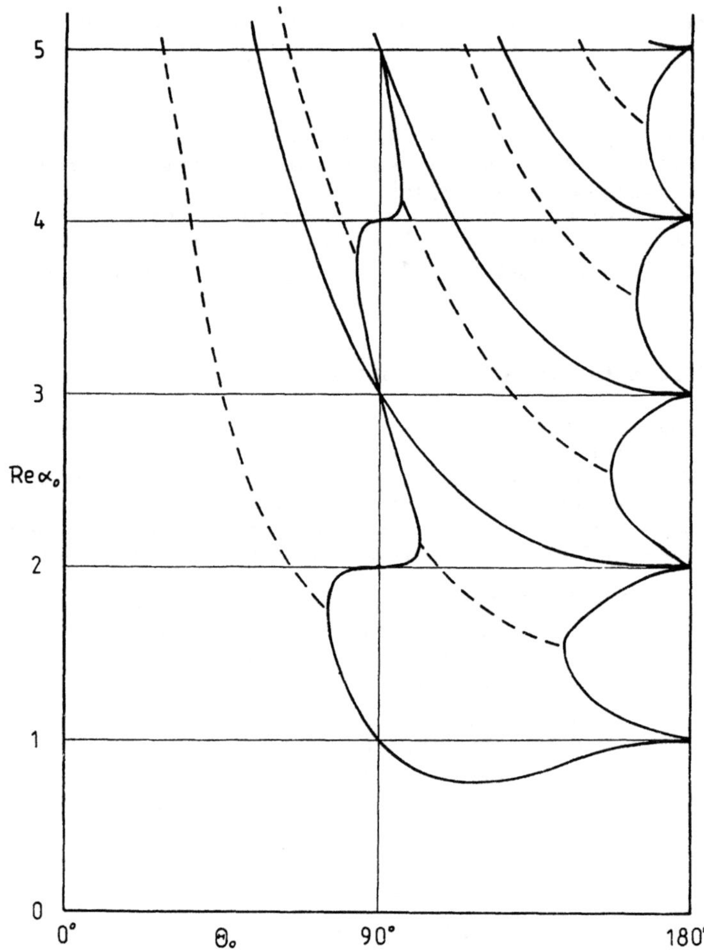

Figure 5: Dependence of the eigenvalues α_0 on θ_0 for Neumann conditions at $\nu = 0.3$ (——— $\Im\alpha = 0$, - - - - $\Im\alpha \neq 0$)

Analogously to the Dirichlet problem in section 4 we consider now a solution $\mathbf{u} \in [W_2^1(\Omega)]^3$ of $\mathbf{L}\,\mathbf{u} = -\mathbf{f}$ in Ω and $\mathbf{S}(\mathbf{u}) \cdot \mathbf{n} = \mathbf{0}$ on $\partial\Omega$ for $\nu = 0.3$ and $\mathbf{f} \in [L_2(\Omega)]^3$. Since only the eigenvalues in the strip $-1/2 < \Re\alpha_\iota < 1/2$ are of interest in the asymptotic expansion eq.(8), we get the following

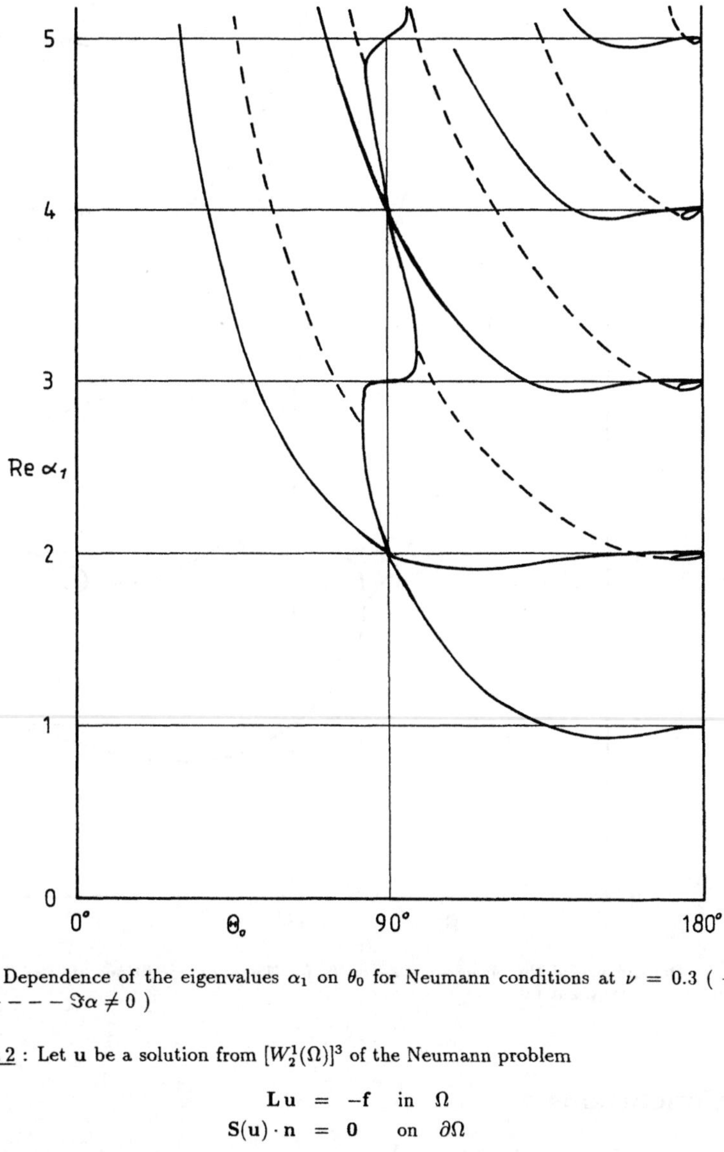

Figure 6: Dependence of the eigenvalues α_1 on θ_0 for Neumann conditions at $\nu = 0.3$ (——— $\Im\alpha = 0$, – – – – $\Im\alpha \neq 0$)

Lemma 2 : Let $\mathbf{u}$ be a solution from $[W_2^1(\Omega)]^3$ of the Neumann problem

$$\begin{aligned} \mathbf{L}\,\mathbf{u} &= -\mathbf{f} \quad \text{in } \Omega \\ \mathbf{S}(\mathbf{u})\cdot\mathbf{n} &= \mathbf{0} \quad \text{on } \partial\Omega \end{aligned} \tag{26}$$

where Ω is a bounded domain in R^3 with circular conical points on the boundary, $\mathbf{f} \in [L_2(\Omega)]^3$ and $\nu = 0.3$. Then $\mathbf{u}$ is from $[W_2^2(\Omega)]^3$.

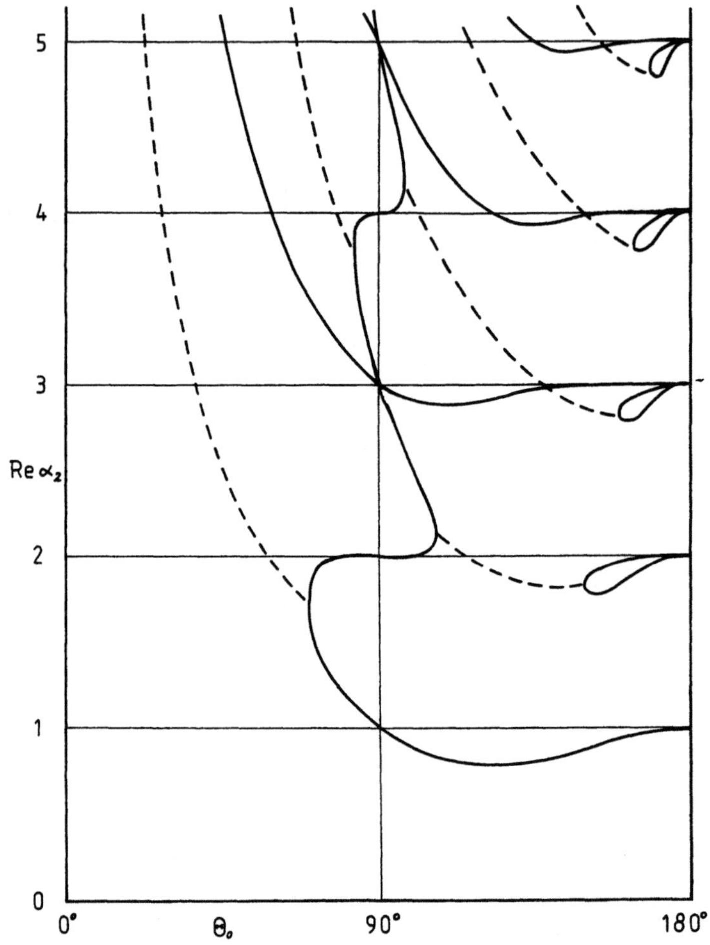

Figure 7: Dependence of the eigenvalues α_2 on θ_0 for Neumann conditions at $\nu = 0.3$ (——— $\Im\alpha = 0$, – – – – $\Im\alpha \neq 0$)

6 Conclusions

In this paper we have considered bounded domains with a circular conical point and calculated the corresponding singular terms. For polyhedral domains the situation is much more complicated. However we can estimate the "eigenvalues" α_k for the polyhedral corner singularities by the "eigenvalues" of axial symmetric conical corners in some cases, using a result of V.A.Kozlov, V.G.Maz'ja and C.Schwab. The corollary 1 of theorem 3.2 in [6] states for Dirichlet conditions:

Let be $K_i = (0,\infty) \times \Omega_i \, (i = 1,2)$ cones in R^3, where $\Omega_i \subset S^2$ and K_i can have edges. Then the "eigenvalues" α_k depend monotonically on Ω in the interval $(-1/2, \Lambda_\gamma(\Omega))$, i.e. if $\alpha_k \in (-1/2, \Lambda_\gamma(\Omega))$ and $\Omega_1 < \Omega_2$ then $\alpha_k(\Omega_2) \geq \alpha_k(\Omega_1)$. Here $\gamma = \mu/(\lambda + \mu)$ and $\Lambda_\gamma(\Omega) \geq 1$ is a real number.

References

[1] Z.P.BAŽANT and L.M.KEER: "Singularities of Elastic Stresses and of Harmonic Functions at Conical Notches or Inclusions", Int.J.Solids Structures **10**, 957 (1974)

[2] A.E.BEAGLES and A.-M.SÄNDIG: "Singularities of Rotationally Symmetric Solutions of Boundary Value Problems for the Lamé Equations", ZAMM (in press)

[3] A.E.BEAGLES and J.R.WHITEMAN: "General Conical Singularities in Three-dimensional Poisson Problems", Math.Methods in the Applied Sciences **11**, 215 (1989)

[4] P.GRISVARD: "Elliptic Problems in Non-smooth Domains", Pitman, (1985)

[5] V.A.KONDRAT'EV: "Boundary Value Problems for Elliptic Equations in Domains with Conical or Angular Points", Trudy Moskovskogo Math. Obschestva **16**, 209 (1967) and Trans. Moscow Math. Soc. **16**, 227 (1967)

[6] V.A.KOZLOV, V.G.MAZ'JA and C.SCHWAB: "On the First Boundary Value Problem of 3-D Elasticity on Conical Domains", (to appear)

[7] A.KRATZER and W.FRANZ: "Transzendente Funktionen", Akademische Verlagsgesellschaft, Leipzig (1960)

[8] V.G.MAZ'JA and B.A.PLAMENEVSKIJ: "Weighted Spaces with inhomogeneous Norms and Boundary Value Problems in Domains with Conical Points", Proceedings of the Conference "Elliptische Differentialgleichungen", Rostock (1977)

[9] V.G.MAZ'JA and B.A.PLAMENEVSKIJ: "Estimates in L_p and in Hölder Classes and the Miranda - Agmon Maximum Principle for Solutions of Elliptic Boundary Value Problems in Domains with Singular Points on the Boundary", Math. Nachr. **81**, 25 (1978) and Amer. Math. Soc. Transl. (2), 123, 1 (1984)

[10] W.H.PRESS, B.P.FLANNERY, S.A.TEUKOLSKY and W.T.VETTERLING: "Numerical Recipes", Cambridge Univ. Press (1986)

A.-M. Sändig, R. Sändig
FB Mathematik
Universität Rostock
Universitätsplatz 1
O-2500 Rostock, Deutschland

A collocation method for potential problems with a mixed Dirichlet-Signorini boundary condition

Hermann Schmitz, Universität Stuttgart, Math. Institut A, Pfaffenwaldring 57, 7000 Stuttgart 80, FRG

1. Introduction

The application of boundary integral equation methods to potential problems with unilateral boundary condition was developped in [5] , [6] and [14]. Spann considered in his thesis the problem

$$\begin{aligned} -\Delta u &= 0 \quad \text{in } \Omega \\ u|_{\Gamma_D} &= g|_{\Gamma_D} \\ u|_{\Gamma_S} &\geq g|_{\Gamma_S} \\ \left.\frac{\partial u}{\partial n}\right|_{\Gamma_S} &\geq f|_{\Gamma_S} \\ \left(u - g, \frac{\partial u}{\partial n} - f\right)_{\Gamma_S} &= 0 \end{aligned} \tag{1}$$

with a smoothly bounded domain $\Omega \subset \mathbb{R}^2$, $\partial\Omega =: \Gamma = \Gamma_S \cup \Gamma_D$. $\Gamma_S \cap \Gamma_D$ is supposed to be a set of isolated points. g and f are given, u is unknown. In the situation of $\Gamma_S \cap \Gamma_D = \emptyset$ and Γ smooth the $H^2(\Omega)$-regularity of the solution is known (for $g \in H^{\frac{3}{2}}(\Gamma)$, $f \in H^{\frac{1}{2}}(\Gamma)$) (see [2],[7]). In the case of $\Gamma_S \cap \Gamma_D \neq \emptyset$ or Γ_S nonsmooth very interesting questions concerning the singular behaviour of solutions arise which are still incompletely solved. In [7] one can find regularity estimates for the solution in weighted spaces but they don't seem to be very sharp. So it is not possible at the moment to do numerical analysis in this situation. Thus, we restrict our development to the case $\Gamma_S \cap \Gamma_D = \emptyset$.

In [13] the results on the convergence of penalty solutions were improved to asymptotic error estimates in the energy norm. Then the results of [10] and [11] were modified to the nonlinear integral equation for the penalty solutions given here in order to analize the convergence of a Galerkin scheme with piecewise linear splines as test and trial functions.

In this paper we investigate a collocation method for the solution of Dirichlet-Signorini problems. After recalling the necessary background from [13] we show the unique solvability of the corresponding discrete equations by using an inverse stability estimate. Coupling of meshsize and penalty parameter leads to a convergent scheme of order $\frac{1}{2}$.

2. The Dirichlet-Signorini problem for the Laplacian

We consider a domain $\Omega \subset \mathbb{R}^2$. The boundary Γ which is assumed to be smooth splits into two disjoint parts Γ_S, Γ_D. We are looking for a solution of

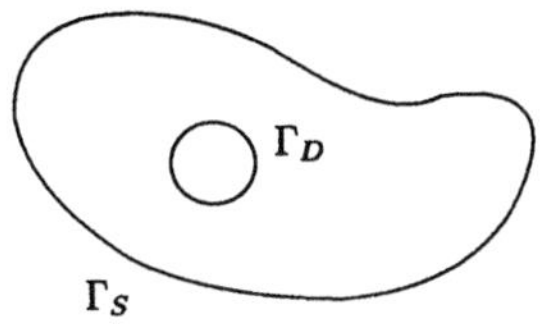

$$
\begin{aligned}
-\Delta u &= f && \text{in } \Omega, \\
u &= 0 && \text{on } \Gamma_D, \\
\frac{\partial u}{\partial n} \cdot u = 0\,, \ \frac{\partial u}{\partial n} &\geq 0\,, \ u \geq 0 && \text{on } \Gamma_S.
\end{aligned} \tag{2}
$$

The variational formulation of this problem is (see [4])

Problem P: Find $u \in K := \left\{ v \in H^1(\Omega) \mid v|_{\Gamma_D} = 0\,,\ v|_{\Gamma_S} \geq 0 \right\}$ with

$$\int_\Omega \nabla u \, \nabla (v - u)\, dx \geq \int_\Omega f\,(v-u)\, dx \quad \forall v \in K.$$

This variational inequality is uniquely solvable due to the coerciveness of

$$a(u,v) := \int_\Omega \nabla u \, \nabla v \, dx \tag{3}$$

on

$$H_0^1(\Omega) := \left\{ v \in H^1(\Omega) \mid v|_{\Gamma_D} = 0 \right\}$$

and can be discretized directly by finite element schemes (see [3], [4]).
In order to introduce boundary element methods we have to remove the right hand side f. For that let u_f be the solution of the Poisson problem

$$
\begin{aligned}
-\Delta u_f &= f && \text{in } \Omega, \\
u_f &= 0 && \text{on } \Gamma.
\end{aligned} \tag{4}
$$

We introduce the boundary-integral-operators

$$K\varphi(x) := \frac{1}{\pi} \int_\Gamma \varphi(y) \frac{\partial}{\partial n_y} \log|x-y| \, ds_y\,, \tag{5}$$

$$V\psi(x) := -\frac{1}{\pi} \int_\Gamma \psi(y) \log|x-y| \, ds_y\,. \tag{6}$$

Let $T : H^{\frac{1}{2}}(\Gamma) \to H^{-\frac{1}{2}}(\Gamma)$ denote the continuous linear map which assigns to τ the normal derivative of the harmonic function v with $v|_\Gamma = \tau$. The operator $V : H^{-\frac{1}{2}}(\Gamma) \to H^{\frac{1}{2}}(\Gamma)$ is invertible under the assumption that the diameter of Ω is less than 1 which we still presuppose in the sequel. Hence $T = V^{-1}(I - K)$. By Greens representation formula we get

Theorem 1. *If $u \in H^1(\Omega)$ solves problem P then the boundary data $u|_\Gamma$ solve*

Problem Q: *Find $t \in C := \left\{\tau \in H^{\frac{1}{2}}(\Gamma) \mid \tau|_{\Gamma_D} = 0,\ \tau|_{\Gamma_S} \geq 0\right\}$ such that*

$$(T\,t, \tau - t)_\Gamma \geq -\left(\frac{\partial u_f}{\partial n}, \tau - t\right)_\Gamma \qquad \forall \tau \in C.$$

If $t \in H^{\frac{1}{2}}(\Gamma)$ is a solution of problem Q then the formula

$$u(x) = u_f(x) + \frac{1}{2\pi}\int_\Gamma \frac{\partial}{\partial n_y} \log|x-y|\, t(y)\, ds_y - \frac{1}{2\pi}\int_\Gamma \log|x-y|\, T\,t(y)\, ds_y \quad , x \in \Omega \tag{7}$$

represents the solution of problem P.

Proof. See [13].

A discretisation of the variational inequality of problem Q can be found in [14]. In order to get an alternative approach to solve problem P by boundary elements we introduce $P\,\tau$ to be the L^2–Projection of τ onto the L^2–closure of C. P satisfies the identity

$$P\,\tau(x) = \begin{cases} \tau^+(x) := \sup\{0, \tau(x)\} & \text{for } x \in \Gamma_S\,, \\ 0 & \text{for } x \in \Gamma_D\,. \end{cases} \tag{8}$$

Define the nonlinear operator B by $B\,\tau := \tau - P\,\tau$. The penalty solutions t_ε , $\varepsilon > 0$ are defined to solve

Problem Q$_\varepsilon$: Find $t_\varepsilon \in H^{\frac{1}{2}}(\Gamma)$ with

$$(T\,t_\varepsilon, \tau)_\Gamma + \frac{1}{\varepsilon}\,(B\,t_\varepsilon\,, \tau)_\Gamma = -\left(\frac{\partial u_f}{\partial n}, \tau\right)_\Gamma \qquad \forall \tau \in H^{\frac{1}{2}}(\Gamma).$$

By the identity

$$\frac{1}{2}\,(I-K)\,u_f\,(x) - \frac{1}{2}\,V\,\frac{\partial u_f}{\partial n}\,(x) = \frac{1}{2\pi}\int_\Omega f(y)\,\log|x-y|\,dy \quad , x \in \Gamma \tag{9}$$

Problem Q$_\varepsilon$ corresponds to the nonlinear boundary integral equation

$$(I-K)\,t_\varepsilon\,(x) + \frac{1}{\varepsilon}\,V\,B\,t_\varepsilon\,(x) = \frac{1}{\pi}\int_\Omega f(y)\,\log|x-y|\,dy\ . \tag{10}$$

Proposition 2. *Problem Q_ε is uniquely solvable and t_ε converges to the solution t_0 of problem Q weakly in $H^{\frac{1}{2}}(\Gamma)$ for $\varepsilon \to 0$. Furthermore the functions t_ε are uniformly bounded in $H^{\frac{3}{2}}(\Gamma)$ and converge weakly to t_0 in this space.*

Proof. The proof of this proposition can be found in [7].

Theorem 3. *The penalty solutions t_ε of problem Q_ε satisfy the asymptotic error estimate*

$$\|t_\varepsilon - t_0\|_{\frac{1}{2}} \leq C\,\varepsilon \tag{11}$$

where t_0 denotes the solution of problem Q.

Proof. The proof is based on proposition 1.2, the fact that P maps $H^{\frac{1}{2}}(\Gamma)$ into $H^{\frac{1}{2}}(\Gamma)$ (see [8]) and a Garding inequality for T

$$(T\tau, \tau)_\Gamma \geq \gamma\,\|\tau\|^2_{\Gamma_D} - c(\tau,\tau)_{\Gamma_D} \quad \forall \tau \in H^{\frac{1}{2}}(\Gamma).$$

The arguments are written down precisely in [13].

3. Convergence of the collocation solutions

Problem Q_ε involves the solution of the nonlinear integral equation

$$A\,t_\varepsilon(x) := (I-K)\,t_\varepsilon(x) + \frac{1}{\varepsilon}\,V\,B\,t_\varepsilon(x) = \frac{1}{\pi}\int_\Omega f(y)\,\log|x-y|\,dy \;=:\, g(x). \tag{12}$$

So we have to solve a nonlinear equation in the Banach space $H^{\frac{1}{2}}(\Gamma)$. We consider the collocation method in order to find an approximate solution of (12). For this purpose let Δ_n be a sequence of grids with meshsize h_n, $S(\Delta_n)$ the corresponding space of piecewise linear continuous splines with respect to a given parameter-representation of Γ. We suppose Δ_n to be smoothly graded in the sense of [1]. Hence, the interpolation operator $I_n : C(\Gamma) \to S(\Delta_n)$ (interpolating in the nodal points) has the following approximation property [1], [12]

$$\|I_n v - v\|_t \leq c\,h^{s-t}\,\|v\|_s\,, \tag{13}$$

$0 \leq t < \frac{3}{2}$, $\frac{1}{2} < s \leq 2$, $t \leq s$. For $g \in H^{\frac{3}{2}}(\Gamma)$ the collocation equations are

$$A_n\,t^n_\varepsilon := I_n\,A\,P_n\,t^n_\varepsilon = I_n\,g \tag{14}$$

where $P_n : S(\Delta_n) \to H^{\frac{1}{2}}(\Gamma)$ denotes the canonical embedding. In order to show the unique solvability of (14) we follow the approach in [9], [10]. First, we introduce some auxiliary results.

Lemma 4. *$A_n : S(\Delta_n) \to S(\Delta_n)$ where $S(\Delta_n)$ is equipped with the $H^{\frac{1}{2}}(\Gamma)$-norm is continuous for all $n \in \mathbb{N}$.*

Proof.

$$A_n = I_n\,(I - K + \varepsilon^{-1}\,V\,B)\,P_n = I - I_n\,K\,P_n + \varepsilon^{-1}\,I_n\,V\,B\,P_n$$

is continuous by the mapping properties of K, V (order -1) and B (in $L^2(\Gamma)$).

Lemma 5. *For $\varepsilon \leq 1$ the sequence A_n is asymptotically inversely stable, i.e. there are constants $c, d > 0$ such that*

$$\|A_n u - A_n v\|_{\frac{1}{2}} \geq c\, \|u - v\|_{\frac{1}{2}} \tag{15}$$

for all n with $h_n^{\frac{1}{2}} \leq d\varepsilon$ and $u, v \in S(\Delta_n)$.

Proof. The proof rests upon the strongly V^{-1}-monotonicity of A

$$\left(Au - Av\,,\, V^{-1}(u - v)\right)_0 \geq c\, \|u - v\|_{\frac{1}{2}}^2$$

which was shown in [13] to hold for $\varepsilon \leq 1$. We conclude

$$\left(A_n u - A_n v\,,\, V^{-1}(u - v)\right)_0 \geq c\, \|u - v\|_{\frac{1}{2}} - \|(I - I_n)(Au - Av)\|_{\frac{1}{2}}\, \|u - v\|_{\frac{1}{2}}\,.$$

Using the approximation property of the interpolation operator I_n we get

$$\begin{aligned}
\|(I - I_n)(Au - Av)\|_{\frac{1}{2}} &= \|(I - I_n)\,K(u - v)\|_{\frac{1}{2}} + \varepsilon^{-1}\, \|(I - I_n)\,V\,(Bu - Bv)\|_{\frac{1}{2}} \\
&\leq c\, h_n\, \|u - v\|_{\frac{1}{2}} + c\,\varepsilon^{-1}\, h_n^{\frac{1}{2}}\, \|u - v\|_0 \\
&\leq c\left(h_n + \frac{h_n^{\frac{1}{2}}}{\varepsilon}\right) \|u - v\|_{\frac{1}{2}}
\end{aligned}$$

by using the mapping properties of K, V and B. Hence,

$$\|A_n u - A_n v\|_{\frac{1}{2}}\, \|u - v\|_{\frac{1}{2}} \geq C\, \|u - v\|_{\frac{1}{2}}^2$$

for $h_n + \dfrac{h_n^{\frac{1}{2}}}{\varepsilon}$ small enough.

Proposition 6. *There exists a constant $d > 0$ such that for $\varepsilon \leq 1$ the collocation equation (14) is uniquely solvable as long as $h_n^{\frac{1}{2}} \leq d\varepsilon$.*

Proof. A_n is continuous and satisfies the a-priori estimate (15). Let $\{y_m\} \subset S(\Delta_n)$ be a sequence of points $y_m = A_n x_m$ converging to $y \in S(\Delta_n)$. By (15) x_m is a Cauchy sequence converging to $x \in S(\Delta_n)$. By the continuity of A_n we get $A_n x = y$ and hence range A_n closed. Range A_n is also open. Indeed, let $y_0 \in$ range A_n. Then the map $x \to A_n x - y_0$ has a critical value x_0. For $\|y - y_0\|$ sufficiently small the estimate (15) yields the surface $\|x - x_0\|_{\frac{1}{2}} = 1$ to be free of critical values for all maps $x \to A_n x - y$ which are clearly homotopic. Hence, $y \in$ range A_n too by mapping degree theory. Since range $A_n \neq \emptyset$ we have range $A_n = S(\Delta_n)$ and the solvability of (14) follows. By (15) again we obtain uniqueness.

Proposition 7. *The sequence t_ε^n converges strongly to t_ε and satisfies the asymptotic error estimate*

$$\|t^\varepsilon - t_n^\varepsilon\|_{\frac{1}{2}} \leq c\left(h_n + \frac{h_n^{\frac{3}{2}}}{\varepsilon}\right) \|t^\varepsilon\|_{\frac{3}{2}}\,.$$

Proof. By (15) we get

$$\|I_n t^\varepsilon - t_n^\varepsilon\|_{\frac{1}{2}} \leq c\,\|A_n I_n t^\varepsilon - A_n t_n^\varepsilon\|_{\frac{1}{2}} = c\,\|I_n A I_n t^\varepsilon - I_n A t^\varepsilon\|_{\frac{1}{2}}$$

$$\leq c\,\|I_n K (I_n - I) t^\varepsilon\|_{\frac{1}{2}} + c\varepsilon^{-1}\,\|I_n (V B I_n t^\varepsilon - V B t^\varepsilon\|_{\frac{1}{2}}$$

$$\leq c\left(h_n + \frac{h_n^{\frac{3}{2}}}{\varepsilon}\right)\,\|t_\varepsilon\|_{\frac{3}{2}}\ .$$

Coupling the meshsize h_n and the penalty parameter ε we end up with

Theorem 8. *There is a constant $d > 0$ such that for $h_n^{\frac{1}{2}} = d\,\varepsilon$ and $\varepsilon \leq 1$ the collocation equations (15) are uniquely solvable. The solution satisfies the asymptotic error estimate*

$$\left\|t_n^\varepsilon - t^0\right\|_{\frac{1}{2}} \leq \|t_n^\varepsilon - t^\varepsilon\|_{\frac{1}{2}} + \left\|t^\varepsilon - t^0\right\|_{\frac{1}{2}} \leq c\,h_n^{\frac{1}{2}}.$$

References

[1] D. N. Arnold and W. L. Wendland. The convergence of spline collocation for strongly elliptic equations on curves. *Numer. Math.*, 47:317–343, 1985.

[2] H. Brezis. Problemes unilateraux. *J. Math. Pures Appl.*, 51:1–168, 1972.

[3] R. S. Falk. Error estimates for the approximation of a class of variational inequalities. *Math. Comp.*, 28:963–971, 1974.

[4] R. Glowinski. *Numerical methods for nonlinear variational problems.* Springer, New York, 1980.

[5] H. Han. The boundary finite element method for Signorini problem. In *Lecture notes in mathematics 1297*, pages 38–49. Springer, 1987.

[6] H. Han and G. Hsiao. The boundary element method for a contact problem. In *Proceedings of the 2nd China-Japan symposium on boundary element methods 1988*, 1988.

[7] B. Kawohl. On nonlinear mixed boundary value problems for second order elliptic differential equations on domains with corners. *Proceedings of the Royal Society of Edinburgh*, 87 A:35–51, 1980.

[8] D. Kinderlehrer and G. Stampacchia. *An Introduction to Variational Inequalities and their Applications.* Academic Press, New York, 1980.

[9] W. V. Petryshyn. Projection methods in nonlinear numerical functional analysis. *J. Math. Mech.*, 17:353–373, 1967.

[10] K. Ruotsalainen and J. Saranen. On the collocation method for a nonlinear boundary integral equation. *Jour. Comp. Appl. Math.*, 29, 1989.

[11] K. Ruotsalainen and W. Wendland. On the boundary element method for some nonlinear boundary value problems. *Num. Math 53*, pages 299–314, 1988.

[12] G. Schmidt. On spline collocation methods for boundary integral equations in the plane. *Math. Meth. in the Appl. Sci.*, 7:74–89, 1985.

[13] H. Schmitz and G. Schneider. Boundary element solution of the Dirichlet-Signorini problem by a penalty method. In preparation.

[14] W. Spann. *Fehlerabschätzungen zur Randelementmethode beim Signorini-Problem für die Laplace-Gleichung.* PhD thesis, Universität München, 1989.

The structure of branching asymptotics for Elliptic boundary value problems in domains with edges

Bernd Schmutzler
Technical University Chemnitz

1 Introduction

Elliptic boundary value problems in domains with edges were considered by many authors. Beside the general theory of Rempel/Schulze [16] and Melrose [14] there are papers of Kondrat'ev [10], Maz'ja/Roßmann [13], Grisvard [7], Dauge [4], Costabel/Dauge [3] and other authors. There the edge problem is attributed to the cone theory, established by Kondrat'ev in the fundamental paper [9] and further developed by Maz'ja/Plamenevskij [11]. The asymptotics in a cone $K = \mathbb{R}_+ \times \Omega$ with $\Omega \subset S^N$ of the form

$$t^{p_i} \ln^j t \, \varphi_i^{(j)}(x) \tag{1}$$

($t \in \mathbb{R}_+, x \in \Omega$) generate edge asymptotics of the form

$$t^{p_i(y)} \ln^j t \, \varphi_i^{(j)}(x,y) \tag{2}$$

($y \in \mathbb{R}$). In this connection it was used, that the generalized eigenvalues and eigenvectors of a generalized eigenvalue problem $p_i(y)$ and $\varphi_i^{(j)}(x,y)$ smoothly depend on y, which is a consequence of the smooth coefficients in the boundary value problem. But this is true only if the functions $p_i(y)$ do not cross each other and the generalized Jordan-structure does not change along these curves. In most of the papers mentioned above having the aim to find a representation of the asymptotics as explicit as possible the case of such branching was excluded.

A great progress was achieved by Rempel and Schulze. In their papers [16]-[19] asymptotics were represented by analytic functionals acting on the function t^{-z}, which can be described by complex curve integrals as well. This is practicable even in the most general cases. [18] contains the branching edge asymptotic terms in Sobolev spaces. Based on this idea the exact structure of asymptotics in a neighbourhood of branching points will be investigated in the present paper. We shall establish the connection between the presentation by analytic functionals and the solutions of the generalized eigenvalue problem mentioned above. Although we do not study the solutions themselves but parameter-dependent objects, our results may be used for a corresponding refinement of the branching behaviour of asymptotics of the solutions from [17].

Other cases of branching arise from so-called shadows. These are asymptotics of the form

$$t^{p_i(y)+k} \ln^j t \, \varphi_{i,k}^{(j)}(x,y)$$

with $k \in \mathbb{N}$, which have to be considered in addition to those of expression (2), and Taylor asymptotics

$$t^k \psi_k(x,y),$$

which arise from polynomial parts of the right-hand side. In this connection smooth curves $q_{i,k}(y) = p_i(y) + k$ and $q_k(y) = k$ intersect each other.

A problem, where both cases do occur at the same point, was already considered by Costabel and Dauge [3]. In their paper the branching asymptotics were represented also by complex curve integrals.

2 Spectral theory of holomorphic operator bundles

Let B be a Banach space, $D \subset \mathbb{C}$ and $I \subset \mathbb{R}$ open sets. A function $e(z) : D \to B$ is called a holomorphic vector bundle (denoted by $e(z) \in \mathcal{A}(D, B)$), if for each $z_0 \in D$ there exists a sequence $e^{(j)}, j = 0, 1, \dots$ in B and real positive numbers c, q such that

$$||e^{(j)}||_B \leq c \cdot q^{-j} \text{ and } e(z) = \sum_{j=o}^{\infty} e^{(j)} \cdot (z - z_o)^j$$

for z in a neighbourhood $U_q(z_0)$ holds.

A function $a(y) : I \to B$ is said to be strongly continuous (denoted by $a(y) \in C(I, B)$), if for all $y_0 \in I$

$$\lim_{y \to y_0} ||a(y) - a(y_0)||_B = 0$$

holds. A function $a'(y)$ is called strong derivative of $a(y) \in C(I, B)$, if for all $y_0 \in I$

$$\lim_{y \to y_0} ||\frac{a(y) - a(y_0)}{y - y_0} - a'(y_0)||_B = 0$$

if a' belongs to $C(I, B)$, $a(y)$ is said to be stronglydifferentiable and the derivative $a'(y)$ is denoted by $\frac{d}{dy}a(y)$. The k-th derivative $\frac{d^k}{dy^k}a(y)$ is defined as the derivative of the $(k-1)$-th derivative inductively analogously to real-valued functions.

The space of all functions $a(y) : I \to B$ with $\frac{d^k}{dy^k}a(y) \in C(I, B)$ for all $k \in \mathbb{N}$ is denoted by $C^\infty(I, B)$.

The function $a(y, z) : I \times D \to B$ is called a holomorphic vector bundle depending smoothly on a parameter y, if for each $z_0 \in D$ there exist functions $a_j(y) \in C^\infty(I, B), j = 1, 0, \dots$ and for each compact $K \subset I$ there exist real positive numbers q and c_k, $k = 0, 1, \dots$ such that

$$||\frac{d^k}{dy^k} a_j(y)||_B \leq c_k \cdot q^{-j}$$

for all $y \in K, j, k = 0, 1, \dots$ and

$$a(y, z) = \sum_{j=0}^{\infty} a_j(y) \cdot (z - z_0)^j$$

for $y \in K$ and $z \in U_q(z_0)$.

In the case that B is the space $\mathcal{L}(B_1, B_2)$ of all linear continuous operators from a Banach space B_1 to a Banach space B_2 , vector bundles are also called operator bundles.

Let $L(z) \in \mathcal{A}(D, \mathcal{L}(B_1, B_2))$ be a holomorphic operator bundle, $p_0 \in D$ and

$$L(z) = \sum_{j=0}^{\infty} L^{(j)} \cdot (z - p_0)^j$$

with $L^{(j)} \in \mathcal{L}(B_1, B_2)$ in a neighbourhood of p_0. The number p_0 is called eigenvalue or characteristic number of $L(z)$ if there exists a $\varphi \in B_1$ with $\varphi \neq 0$ and $L^{(0)}\varphi = 0$.

A sequence $\varphi^{(0)}, \varphi^{(1)}, \dots, \varphi^{(m-1)}$ is called Keldyš chain with length m to $L(z)$ in p_0, if $\varphi^{(0)} \neq 0$ and

$$\sum_{j=0}^{\mu} L^{(j)} \varphi^{(\mu-j)} = 0 \tag{3}$$

for $\mu = 0, 1, \dots, m-1$. The vector $\varphi^{(0)}$ is called eigenvector and the vectors $\varphi^{(j)}, j = 1, \dots, m-1$ associated vectors. The dimension $k(p_0)$ of the space $Ker\ L(p_0)$ of all eigenvectors is called geometric multiplicity and the maximal length $\nu(p_0)$ of all Keldyš chains to $L(z)$ in p_0 with φ as the eigenvector is called the rank of φ. If $k(p_0)$ and $\nu(p_0)$ are finite, then there exists a so-called canonical system of Keldyš chains. A system of Keldyš chains $\varphi_i^{(0)}, \varphi_i^{(0)}, \dots, \varphi_i^{(m_i-1)}$, $i = 1, 2, \dots, k(p_0)$, to $L(z)$ in p_0 is said to be canonical, if $m_1 = \nu(p_0)$ and m_i coincides with the maximal rank of the vectors of all complementary subspaces in $Ker\ L(p_0)$ of the subspace generated by $\varphi_1^{(0)}, \varphi_2^{(0)}, \dots, \varphi_{i-1}^{(0)}$ for $i = 2, \dots, k(p_0)$. The number

$$n(p_0) = m_1 + m_2 + \dots + m_{k(p_0)}$$

is called algebraic multiplicity of p_0 with respect to $L(z)$. Now, let $L(y,z) \in C^\infty(I, \mathcal{A}(D_0, \mathcal{L}(B_1, B_2)))$ satisfy the following conditions:

(I) For all $(y,z) \in I \times D_0$ the operator $L(y,z)$ is a Fredholm operator with index 0.

(II) For each fixed $y \in I$ the holomorphic operator bundle $L(y,z)$ has isolated eigenvalues with finite multiplicity in D_0 only.

The set of all eigenvalues of $L(y,z)$ in D_0 is denoted by $S(y)$. For each fixed y and each eigenvalue p the numbers $n(p), k(p), \nu(p)$ defined above are denoted by $n(y,p), k(y,p), \nu(y,p)$. For $z \notin S(y)$ we set $n(y,z) = k(y,z) = \nu(y,z) = 0$. Then the follwing Lemma holds:

Lemma. *Let $L(y,z) \in C^\infty(I, \mathcal{A}(D_0, \mathcal{L}(B_1, B_2)))$ satisfy the conditions (I) and (II), let D be a simply connected domain with $\overline{D} \subset\subset D_0$, $\partial D \in C^\infty$, $D \cap S(y_0) = \{p_0\}$ for a number $y_0 \in I$ and $\partial D \cap S(y) = \emptyset$ for $y \in I$. Then*

$$p_D(y,z) := \prod_{p \in S(y) \cap D} (z-p)^{n(y,p)} \in C^\infty(U(y_0), \mathcal{A}(\mathbb{C}))$$

for a neighbourhood $U(y_0)$ of y_0.

For the proof there were used ideas of Gohberg and Sigal [5].

3 Branching asymptotics

Let us consider the boundary value problem

$$\left.\begin{array}{rcll} A(\hat{x}, y, D_{\hat{x}}, D_y)u(\hat{x}, y) &=& f(\hat{x}, y), & (\hat{x}, y) \in G \\ B_i(\hat{x}, y, D_{\hat{x}}, D_y)u(\hat{x}, y) &=& g_i(\hat{x}, y), & (\hat{x}, y) \in G \\ && \hat{x} \neq 0, & i = 1, \dots, M \end{array}\right\} \quad (4)$$

in a wedge $G = K \times \mathbb{R} \subset \mathbb{R}^{N+2}$. Here A and B_i are differential operators of order μ_0 and μ_i, respectively, with coefficients in $C^\infty(\overline{G})$ and $K = \{\hat{x} = t \cdot x, t \in \mathbb{R}_+, x \in \Omega \subset S^N\}$ is a cone with $\partial\Omega \in C^\infty$. We assume that the operator A is uniformly elliptic in G and at each point $(\hat{x}, y)$ of the boundary ∂G with $\hat{x} \neq 0$ the Shapiro-Lopatinskij condition is satisfied. The main contribution to the asymptotics along the edge $\{(\hat{x}, y) \in \overline{G} : \hat{x} = 0\}$ comes from solutions of

$$\left.\begin{array}{rcll} \mathcal{A}(x, y, t\frac{\partial}{\partial t}, D_x)v(t, x, y) &=& 0, & (t, x, y) \in \mathbb{R}_+ \times \Omega \times \mathbb{R} \\ \mathcal{B}_i(x, y, t\frac{\partial}{\partial t}, D_x)v(t, x, y) &=& 0, & (t, x, y) \in \mathbb{R}_+ \times \partial\Omega \times \mathbb{R} \\ && i = 1, \dots, M & \end{array}\right\} \quad (5)$$

where $t^{-\mu_0}\mathcal{A}(x, y, t\frac{\partial}{\partial t}, D_x)$ and $t^{-\mu_i}\mathcal{B}_i(x, y, t\frac{\partial}{\partial t}, D_x)$ are the principal parts of $A(0, y, D_{\hat{x}}, 0)$ and $B_i(0, y, D_{\hat{x}}, 0)$, respectively, after the transformation of $\hat{x}$ into polar coordinates (t, x).

For the systems (5), written in the form

$$\mathcal{L}(y,-t\frac{\partial}{\partial t},x,D_x)v(t,x,y)=0 \tag{6}$$

we have $L(y,z):=\mathcal{L}(y,z,x,D_x)\in C^\infty(\mathbb{R},\mathcal{A}(\mathbb{C},\mathcal{L}(B_1,B_2)))$ with

$$\begin{aligned} B_1 &= H^\mu(\Omega) \\ B_2 &= H^{\mu-\mu_0}(\Omega)\times \mathsf{X}_{i=1}^M H^{\mu-\mu_i-1/2}(\partial\Omega) \end{aligned}$$

for ordinary Sobolev spaces $H^s(\Omega)$, $H^s(\partial\Omega)$ and $\mu-\mu_0>0$, $\mu-\mu_i-1/2>0, i=1,\dots,M$. We seek such solutions of (6), which belong to $C^\infty(I,\mathcal{H}^\mu_{\gamma_1,\gamma_2}(K))$ for a given interval $I\subset\mathbb{R}$. Here $\mathcal{H}^\mu_{\gamma_1,\gamma_2}(K))$ is the space of all functions $v(x,t):K\to\mathbb{C}$, for which the norm

$$||u(x,t)||_{\mathcal{H}^\mu_{\gamma_1,\gamma_2}(K)}:=\left[\int_0^\infty\sum_{m=0}^\mu||(t\frac{\partial}{\partial t})^m v(x,t)||^2_{H^{\mu-m}(\Omega)}g^2_\gamma(t)t^N dt\right]^{1/2}$$

with $g_\gamma(t)=\omega(t)t^{-\gamma_1}+(1-\omega(t))t^{-\gamma_2}$ is finite, where $\omega(t)\in C^\infty(\mathbb{R}_+)$, $\omega(t)=1$ for $t\le 1$, $\omega(t)=0$ for $t\ge 2$ and $\gamma_2<\gamma_1$.

It follows from the ellipticity of the problem (4), that $\mathcal{L}(y,z,x,D_x)$ is elliptic with parameter z in the sense of Agranovich-Vishik for each fixed y, e.g. $L(y,z)$ satisfies the conditions (I) and (II).

For each fixed y the space of solutions of problem (5) is spanned by all functions

$$v(t,x)=\frac{1}{m!}\frac{d^m}{dz^m}[t^z\cdot\varphi(z,x)]\bigg|_{z=p} \tag{7}$$

with

$$\varphi(z,x)=\sum_{j=0}^\infty\varphi^{(j)}(x)\cdot(z-p)^j\in\mathcal{A}(U(p),B_1)$$

where $\varphi^{(0)}(x),\dots,\varphi^{(m-1)}(x)$ is a Keldyš chain to $L(y,z)$ at the eigenvalue p and p belongs to

$$S^N_{\gamma_1,\gamma_2}:=\{z\in\mathbb{C}:\ \gamma_1-\frac{N+1}{2}<Re\ z<\gamma_2-\frac{N+1}{2}\}.$$

Now it is possible to state the following

Theorem. *Let $L(y,z)\in C^\infty(\mathbb{R},\mathcal{A}(\mathbb{C},\mathcal{L}(B_1,B_2)))$ be the holomorphic operator bundle depending smoothly on the parameter y defined by (6) for the boundary value problem (4). Let $S(y_0)\cap S^N_{\gamma_1,\gamma_2}=\{p_0\}$, $S(y)\cap\partial S^N_{\gamma_1,\gamma_2}=\emptyset$ for y in a neighbourhood of y_0, $n=n(y_0,p_0)$, $k=k(y_0,p_0)$, $\nu=\nu(y_0,p_0)$ and let $\varphi_i^{(0)},\dots,\varphi_i^{(m_i-1)}$, $i=1,\dots,k$ be a canonical system of Keldyš chains to $L(y,z)$ at $z=p_0$. Then there exists a neighbourhood $U(y_0)$ of y_0, a smoothly bounded simply connected domain D with $p_0\in D$, $\overline{D}\subset\subset S^N_{\gamma_1,\gamma_2}$, $\partial D\cap S(y)=\emptyset$ for $y\in U(y_0)$ and functions $\varphi_i(y,z,x)\in C^\infty(U(y_0),\mathcal{A}(D,B_1))$, such that the following statement holds :*

If the function

$$p_D(y,z)=\prod_{p\in S(y)\cap D}(z-p)^{n(y,p)}$$

is written in the form

$$p_D(y,z)=\prod_{r=1}^{\kappa}(z-p_r(y))^{n_r}$$

for $y\in U(y_0)\setminus\{y_0\}$ with $\kappa=\kappa(y)$, $n_r=n(y,p_r(y))$ then the partial fraction expansion of the functions

$$q^{(j)}(y,z):=\binom{n}{j+1}^{-1}\frac{1}{(j+1)!}\left[\frac{d^{j+1}}{dz^{j+1}}p_D(y,z)\right]p_D^{-1}(y,z),$$

$j = 0, 1, \ldots, n-1$ gives

$$q^{(j)}(y,z) = \sum_{r=1}^{\kappa} \sum_{l=0}^{\min((n_r-1,j)} \frac{c_{j,r,l}(y)}{(z-p_r(y))^{l+1}}$$

and the functions

$$v_i^{(j)}(t,x,y) := \frac{1}{2\pi i} \oint_{\partial D} q^{(j)}(y,z) t^z \varphi_i(y,z,x) dz$$

$$= \begin{cases} \sum_{m=0}^{j+m_i-n} \frac{1}{m!} t^{p_0} \ln^m t \varphi_i^{(j-m+m_i-n)}(x) & y = y_0 \\ \sum_{r=1}^{\kappa} \sum_{l=0}^{\min(n_r-1,j)} c_{j,r,l}(y) \sum_{m=0}^{l} \frac{1}{m!} t^{p_r(y)} \ln^m t \frac{1}{(l-m)!} \frac{d^{l-m}}{dz^{l-m}} \varphi_i(y,z,x)|_{z=p_r(y)} & y \in U(y_0) \setminus \{y_0\} \end{cases}$$

$i = 1, \ldots, k$; $j = n - m_i, \ldots, n-1$ belong to $C^\infty(U(y_0), \mathcal{H}^{\mu}_{\gamma_1,\gamma_2}(K))$ and generate a basis of the space of all solutions of (5) in $\mathcal{H}^{\mu}_{\gamma_1,\gamma_2}(K)$ at each $y \in U(y_0)$.

4 Example

We consider the boundary value problem

$$\left.\begin{array}{rcll} \Delta\Delta\ u(\hat{x}',y) & = & f'(\hat{x}',y) & (\hat{x}',y) \in G' \\ u(\hat{x}',y) & = & g_1'(\hat{x}',y) & \\ \frac{\partial}{\partial n} u(\hat{x}',y) & = & g_2'(\hat{x}',y) & (\hat{x}',y) \in \partial G', \ \hat{x}' \neq 0 \end{array}\right\} \tag{8}$$

for the biharmonic operator

$$\Delta\Delta = \left[\frac{\partial^2}{\partial x_1'^2} + \frac{\partial^2}{\partial x_2'^2} + \frac{\partial^2}{\partial y^2}\right]^2$$

in a wedge $G' = \{(\hat{x}',y) = (x_1',x_2',y) \in \mathbb{R}^3 : x_2' \geq |x_1'| \cot\omega(y)\}$ with a varying opening angle $2\omega(y) \in C^\infty(\mathbb{R})$ with $3/4\pi + \varepsilon \leq \omega(y) \leq \pi - \varepsilon$ for some $\varepsilon > 0$.

We define the mapping $\hat{x}' \to \hat{x} = (x_1, x_2)$ by $x_1 = x_1'$, $x_2 = \alpha(y) x_2'$ where

$$\alpha(y) = \frac{\cot\overline{\omega}}{\cot\omega(y)} \in C^\infty(\mathbb{R})$$

for some $\overline{\omega} \in (0, \pi/2)$. Then (8) is transformed into a problem (8') in a wedge G with constant angle $2\overline{\omega}$, which has the form (4). If we introduce polar coordinates (t', x') with respect to $\hat{x}'$ such that $x_1' = -t' \sin x'$, $x_2' = t' \cos x'$ and (t, x) with respect to $\hat{x}$ in the same way, then the connection between these coordinates may be described by $t' = t \cdot s(x,y)$, $x' = x'(x,y)$, where s and x' depend smoothly on x and y.

The solution of the problem (5) corresponding to (8') can be determined after the transformation $(t,x) \to (t',x')$. We get

$$\left.\begin{array}{ll} a(-t'\frac{\partial}{\partial t'}, \frac{\partial}{\partial x'}) v'(t',x',y) = 0 & x' \in \Omega(y) \\ v'(t',x',y) = \frac{\partial}{\partial x'} v'(t',x',y) = 0 & x' \in \partial\Omega(y) \\ & t \in \mathbb{R}_+, y \in \mathbb{R} \end{array}\right\} \tag{9}$$

where $a(z,\xi) = (z^2, \xi^2)[(z-2)^2 + \xi^2]$ and $\Omega(y) = (-\omega(y), \omega(y))$. Returning to Cartesian coordinates $\hat{x}' = (x_1', x_2')$ the problem (9) may be transformed into a problem similar to (8), but in a two-dimensional cone with an opening angle of $2\omega(y)$ for each fixed y. Thus the solutions of (9) are well-known ([2], [6]). If we define

$$\begin{aligned} \varphi'(y,z,x') := {} & [\sin((z-2)\omega(y)) \sin(zx') - \sin(z\omega(y)) \sin((z-2)x')] \\ & [\sin(2(z-1)\omega(y)) - (z-1)\sin(2\omega(y))] + \\ & [\cos((z-2)\omega(y)) \cos(zx') - \cos(z\omega(y)) \cos((z-2)x')] \\ & [\sin(2(z-1)\omega(y)) - (z-1)\sin(2\omega(y))] \end{aligned}$$

and $p(y,z) = \sin^2(2(z-1)\omega(y)) - (z-1)^2\sin^2(2\omega(y))$, then the solutions of (9) for a fixed y are

$$t'^z\varphi'(y,z,x')|_{z=p_0}$$

if p_0 is a zero of $p(y,z)$ and futhermore

$$\frac{d}{dz}[t'^z\varphi'(y,z,x')]|_{z=p_0}$$

is an additional solution if p_0 is a double zero of $p(y,z)$. Hence for $\varphi(y,z,x) = s^z(x,y)\varphi'(y,z,x'(x,y))$ the functions

$$t^z\varphi(y,z,x)|_{z=p_0} \text{ and } \frac{d}{dz}[t^z\varphi(y,z,x)]|_{z=p_0}$$

are the corresponding solutions of (5) and $\varphi(y,z,x)$ has all properties of the functions $\varphi_i(y,z,x)$ used in the theorem.

Now there exists an angle $\omega_0 \approx 0,4\pi$ so that for $\omega(y) = \omega_0$ the function $p(y,z)$ has a double zero $p_0 \approx 2,76$. We assume $\omega(y_0) = \omega_0$ and $\omega(y) \neq \omega_0$ in a neighbourhood of y_0 for $y \neq y_0$. We choose γ_1 and γ_2 so, that $p_0 \in S^N_{\gamma_1,\gamma_2}$ and no other zeros of $p(y_0,z)$ belong to $S^N_{\gamma_1,\gamma_2}$. Moreover let D satisfy the conditions of the theorem. Then we get

$$p_D(y,z) = (z - p_1(y))(z - p_2(y))$$

with $p_1(y) \neq p_2(y)$ for $y \in U(y_0) \setminus \{y_0\}$ and

$$p_1(y_0) = p_2(y_0) = p_0,$$

$$\begin{aligned}
q^{(0)}(y,z) &= \frac{1}{2}\left[\frac{1}{z-p_1(y)} + \frac{1}{z-p_2(y)}\right] \\
q^{(1)}(y,z) &= \frac{1}{p_1(y)-p_2(y)}\left[\frac{1}{z-p_1(y)} - \frac{1}{z-p_2(y)}\right] \\
v_1^{(0)}(t,x,y) &= \frac{1}{2}[t^{p_1(y)}\varphi(y,p_1(y),x) + t^{p_2(y)}\varphi(y,p_2(y),x)] \\
v_1^{(1)}(t,x,y) &= \frac{1}{p_1(y)-p_2(y)}[t^{p_1(y)}\varphi(y,p_1(y),x) - t^{p_2(y)}\varphi(y,p_2(y),x)]
\end{aligned}$$

for $y \neq y_0$ and

$$\begin{aligned}
v_1^{(0)}(t,x,y) &= t^{p_0}\varphi(y,p_0,x) \\
v_1^{(1)}(t,x,y) &= \frac{d}{dz}[t^z\varphi(y,z,x)]|_{z=p_0}
\end{aligned}$$

for $y = y_0$. The functions $v_1^{(0)}(t,x,y)$ and $v_1^{(1)}(t,x,y)$ belong to $C^\infty(U(y_0), \mathcal{H}^\mu_{\gamma_1,\gamma_2}(K))$ and generate a basis of the space of asymptotics for the problem (8'), which may be represented as solutions of (5) in $\mathcal{H}^\mu_{\gamma_1,\gamma_2}(K)$. By transforming φ, x, t into φ', x', t' we get the asymptotics for the original problem (8).

References

[1] H. Baumgärtel. Analytic Perturbation Theory for Matrices and Operators. Akademie-Verlag, Berlin 1984.

[2] H. Blum, R. Rannacher. Zum Randwertproblem des biharmonischen Operators auf Gebieten mit Ecken. Preprint, Universität Bonn 1979.

[3] M. Costabel, M. Dauge. Edge asymptotics on a skew cylinder. Proc. of the Intern. Workshop Analysis in Domains and on Manifolds with Singularities, Breitenbrunn 1990, Teubner,

[4] M. Dauge. Elliptic Boundary Value Problems in Corner Domains - Smoothness and Asymptotics of Solutions. Lecture Notes in Mathematics, Vol.1341. Springer-Verlag, Berlin 1988.

[5] I. Gohberg, E. I. Sigal. An operator generalization of the logarithmic residue theorem and Rouche's theorem. Mat. Sb. (N.S.) 84(126) (1971)607-629.

[6] P. Grisvard. Problèmes aux limites dans les polygones, Mode d'emploi. EDF Bull. Direction Etuder Rech. Sér. C. Math. Inform. No.1, 1986, 21-59.

[7] P. Grisvard. Boundary Value Problems in Non-Smooth Domains. Pitmann, London 1985.

[8] T. Kato. Perturbation Theory for Linear Operators. Springer-Verlag, Berlin-Heidelberg-New York 1976.

[9] V. A. Kondrat'ev. Boundary value problems for elliptic equations in domains with conical or angular points. Trans. Moscow Mat. Soc. 16(1967) 227-313.

[10] V. A. Kondrat'ev. Singularities of a solution of Dirichlet's problem for a second order elliptic equation in a neighbourhood of an edge. Differential Equations 13 (1970) 1411-1415

[11] V. G. Maz'ja, B. A. Plamenevskij. On coefficients in the asymptotics of solutions of elliptic boundary value problems in domains with conical points. Math. Nachr. 76 (1977), 29-60.

[12] V. G. Maz'ja, B. A. Plamenevskij. On boundary value problems for a second order elliptic equation in a domain with edges. Vestnik Leningrad Univ. Mathematics 8 (1980) 99-106.

[13] V. G. Maz'ja, J. Roßmann. Über die Asymptotik der Lösungen elliptischer Randwertaufgaben in der Umgebung von Kanten. Math. Nachr. 138 (1988) 27-53.

[14] R. Melrose. Pseudo-differential operators on manifolds with corners. Manuscript MIT, Boston 1987.

[15] A. Nikishin. Singularities of the solution of the Dirichlet problem for a second order equation in a neighbourhood of an edge. Moscow Univ. Math. Bull. 34(2) (1979) 53-64.

[16] S. Rempel, B.-W. Schulze. Asymptotics for Elliptic Mixed Boundary Problems. Akademie-Verlag, Berlin 1989.

[17] B. W. Schulze. Regularity with continuous and branching asymptotics for elliptic operators on manifolds with edges. Int.Equ. and Operator Theory 11 (1988) 557-602.

[18] B.-W. Schulze. Elliptic regularity with continuous and branching edge asymptotics. Int. Series of Numer. Math. Vol. 94, Birkhäuser Verlag Basel 1990, pp. 275-284.

[19] B.-W. Schulze. Pseudo -differential operators on manifolds with edges. Symp. "Part. Diff. Equations" Holzhau 1988, Teubner Texte zur Mathematik 112, Leipzig 1989, pp. 259-288.

B. Schmutzler
FB Mathematik
Technische Universität Chemnitz
Reichenhainer Str. 39-41
O-9022 Chemnitz, Deutschland

The Mellin Pseudo-Differential Calculus on Manifolds with Corners

Bert-Wolfgang Schulze
Karl-Weierstraß-Institut für Mathematik, Berlin

Contents

0 Introduction

It is well-known that the parametrices of elliptic partial differential equations on C^∞ manifolds (say closed compact or compact with C^∞ boundary) can be expressed by pseudo-differential operators. This implies the elliptic regularity in terms of (for instance) the standard Sobolev spaces.

The same sort of questions is interesting on manifolds (or more generally on 'branched spaces') with a piece-wise smooth geometry e.g., with conical points, edges, corners, ..., higher singularities or non-compact 'exits'. Analytically this corresponds to operators with 'degenerate' symbols.

In recent years this area of problems found growing interest in the literature, though the applications (e.g., of applied sciences, differential geometry, topology) are often classical. The vast variety of special investigations suggests a general approach for sufficiently wide classes of such problems.

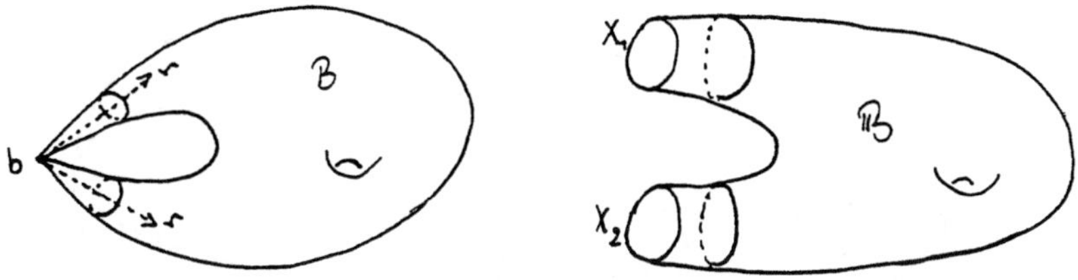

Figure 1:

The present paper is devoted to a corresponding pseudo-differential calculus. It is a survey on results of the author from the theory for corners, locally being regarded as cones, where the bases also have conical singularities. More details may be found in [S3], [S4] or will be published in [S8].

Corners of higher orders (as they occur, for instance, in problems of 'quarter plane type') are also of interest. Our theory will be organized in such a way that it can be iterated for those cases.

Remember that a 'manifold' B with conical singularities is a compact topological space with a finite subset $B_0 \subset B$ of 'conical points' such that $B \setminus B_0$ is a C^∞ manifold, and every $b \in B_0$ has an open neighbourhood U in B that allows a homeomorphism

$$U \cong (X \times [0,1))/(X \times \{0\}) \tag{1}$$

(which is kept fixed), with X being a closed compact C^∞ manifold (the 'base of the cone' close to b), and (1) restricts to a diffeomorphism

$$U \setminus \{b\} = X \times (0,1) \ni (x,r). \tag{2}$$

r is called the cone axis variable. Figure 1 shows a case where X has two components X_1, X_2. The picture on the right is the stretched manifold to B (cf. the explanation below). It is clear that the cone is a special case of a corner. Some elements of the cone theory are to be applied for the general corners, again (cf. [R1], [M1];[S4], II).

A 'manifold' K with corners in the present set-up will be a compact topological space K containing compact subspaces $K_1 \supset K_0$, where $K_0 = \{v_1, \dots, v_N\}$is a finite set of 'corners', $K_1 \setminus K_0$ a C^∞ manifold of dimension 1 (consisting of the edges, emanated from the corners), and $K \setminus K_1$ is a C^∞ manifold of dimension $n+2$. Every $v \in K_0$ has an open neighbourhood V in K for which there is a homeomorphism

$$V \cong (B \times [0,1))/(B \times \{0\}) \tag{3}$$

with B being a manifold with conical singularities (the 'base of the corner' close to v) and (3) restricts to a homeomorphism

$$V \setminus \{v\} \cong B \times (0,1) \ni (\widetilde{x}, t) \tag{4}$$

and to diffeomorphisms

$$V \setminus K_1 \cong (B \setminus B_0) \times (0,1)\ , \ V \cap (K_1 \setminus K_0) \cong B_0 \times (0,1).$$

(3) is kept fixed. t is the variable on the edge emanated from the corner. It is called the corner axis variable.

A further assumption on K is that every $y \in K_1 \setminus K_0$ has an open neighbourhood W in K for which there is a homeomorphism

$$W \cong \{(X \times [0,1))/(X \times \{0\})\} \times (\alpha, \beta) \tag{5}$$

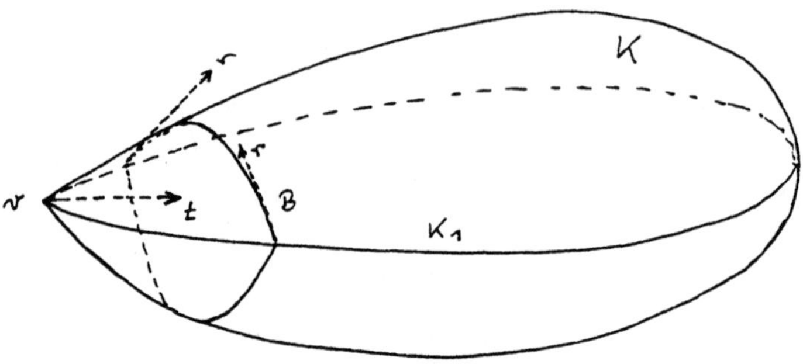

Figure 2:

for certain reals α, β, with a closed compact C^∞ manifold X, and (5) restricts to a diffeomorphism

$$W \setminus K_1 \cong X \times (0,1) \times (\alpha, \beta) \ni (x, r, t). \tag{6}$$

(6) is also kept fixed. The set

$$X^\wedge = X \times \mathbb{R}_+ \ni (x, r) \tag{7}$$

will also be called (open stretched) model cone for the wedge with edge through y. Moreover the various maps are assumed to be compatible in an obvious manner over intersections of neighbourhoods. For brevity we want to avoid here a lengthy discussion of axioms. We could easily introduce the categories of spaces with conical points, edges, corners and evident notions of morphisms, in particular isomorphisms.

The space K in Figure 2 has one corner v, i.e. $K_0 = \{v\}$, the base B of the corner has two conical points.

Note that K can be viewed again as the base of a corner of 'third order'

$$(K \times \overline{\mathbb{R}_+})/(K \times \{0\}). \tag{8}$$

The theory for (8) covers the problems of quarter plane type as well as of boundary value problems for domains where the boundary has 'second order' corners as they were described for K itself.

Let Ω be a C^∞ manifold and denote by $Diff^\mu(\Omega)$ the space of all differential operators over Ω of order μ with C^∞ coefficients in local coordinates. $Diff^\mu(\Omega)$ is equipped with a natural locally convex topology (which is Fréchet in our concrete cases). In particular, for

$$\Omega = K \setminus K_1 \tag{9}$$

we look at operators $A \in Diff^\mu(\Omega)$ of the form

$$A = t^{-\mu} \sum_{k=0}^{\mu} a_k(t)(-t\frac{\partial}{\partial t})^k \tag{10}$$

on $V \setminus K_1$ in the 'coordinates' from (4) with certain $a_k(t) \in C^\infty([0,1), Diff^{\mu-k}(B \setminus B_0))$, and close to some $b \in B_0$ in the coordinates $(x, r, t) \in X \times (0,1) \times (0,1) \cong (U \setminus \{b\}) \times (0,1)$ it is assumed that

$$A = t^{-\mu} r^{-\mu} \sum_{j+k \leq \mu} a_{jk}(r,t)(-r\frac{\partial}{\partial r})^j (-rt\frac{\partial}{\partial t})^k \tag{11}$$

with certain $a_{jk} \in C^\infty([0,1) \times [0,1), Diff^{\mu-(j+k)}(X))$, further on $W \setminus K_1$ in the coordinates from (6)

$$A = r^{-\mu} \sum_{j+k \leq \mu} c_{jk}(r,t)(-r\frac{\partial}{\partial r})^j (r\frac{\partial}{\partial t})^k \tag{12}$$

with certain $c_{jk} \in C^\infty([0,1) \times (\alpha,\beta), Diff^{\mu-(j+k)}(X))$. The property (10) will be referred to as Fuchs type, (11) as corner degeneracy, (12) as edge degeneracy.

Differential operators of this form follow, for instance, by repeatedly substituting polar coordinates in operators

$$\overset{\approx}{A} = \sum_{|\alpha| \leq \mu} a_\alpha(\overset{\approx}{x}) D^\alpha_{\overset{\approx}{x}}$$

in $\mathbb{R}^{n+2} \ni \overset{\approx}{x}, a_\alpha \in C^\infty(\mathbb{R}^{n+2})$. Then $\overset{\approx}{x} \longrightarrow (\tilde{x}, t), \tilde{x} = \overset{\approx}{x} / | \overset{\approx}{x} |, t = | \overset{\approx}{x} |$, leads to operators like (10). Moreover, if $\tilde{x}$ is shrinked from S^{n+1} to a coordinate neighbourhood U with local coordinates in $\mathbb{R}^{n+1}$, around the origin, for brevity now called $\tilde{x}$, again, $\overset{\approx}{x} \longrightarrow (\tilde{x}, t) \longrightarrow (x, r, t), x = \tilde{x} / | \tilde{x} |, r = | \tilde{x} |$, gives rise to (11). Note that (11) is in particular of the form (12).

Now the programme of the calculus is to introduce

(i) symbolic structures,

(ii) a ψDO calculus,

(iii) adequate Sobolev spaces,

such that for the operators A under an ellipticity condition in terms of (i) it follows the Fredholm property in (iii) and the form of a parametrix by means of (ii). In addition it is desirable to study

(iv) subspaces with corner and edge asymptotics

such that the elliptic regularity holds not only for (iii) but also for (iv).

The solution will necessarily employ the analogous theory on manifolds with conical singularities and edges. It is known (cf. [S1]) that the calculus for edges covers the case of boundary value problems (where the edge is the boundary and the model cone the inner normal $\mathbb{R}_+$). Here, for getting the Fredholm property, there will be posed additional edge conditions (in general of trace and potential type, similarly to Boutet de Monvel's theory, cf. [B1], satisfying an analogue of the Shapiro-Lopatinski condition). This has to be combined with procedures from the cone theory where an extra operator-valued Mellin symbolic level is required to be bijective along a weight line, cf. [K1]. In other words we will need preparations on those aspects, and there will arise several leading symbolic levels.

The goal of this paper is a ψDO calculus on manifolds with corners including ellipticity, Fredholm property and the asymptotics of solutions. The exposition is not completely self-contained. Details and complete proofs may be found elsewhere in the papers mentioned in the beginning. But the technicalities are arranged in such a way that manything appears straightforward after the knowledge of [S1], [S4].

Chapter 1 will repeat some material on conical singularities and edges. The results of Section 1.4. are motivated by the application in the corner theory. The cone operators from Section 1.2. are the model for the shape of the corner $\psi DO's$ of Chapter 3.

The new ingredients such as parameter-dependent cone operators, corner asymptotics, Mellin symbols will be given in Chapter 2.

Chapter 3 will establish the full corner operator class including the symbolic structures and the concept of ellipticity.

The reader will realize that many points suggest further investigations. The style of exposition was dictated by the desire to formulate the theory as a whole and to confirm the expectation that (in spite of the complexity of tools) there does actually exist an operator algebra containing the ideas of Boutet de Monvel's theory as well as of 'higher operator' structures for conical and edge singularities.

Operators on spaces with singularities have been studied by many authors, in particular of Konrat'ev [K1], Maz'ja, Plamenevskij [M3], Plamenevskij [P1], Melrose, Mendoza [M1], Melrose [M2], Roßman [R2], Dauge [D1], Rempel, Schulze [R1], Schulze [S1]–[S5]. A more complete bibliography may be found in [S1].

As far as we know the problem of representing parametrices of differential equations near corners (in the present sense) in terms of symbols of operators in an algebra and of obtaining the corner asymptotics by means of the parametrices was not yet treated before. The experience with simpler problems with singularities or degeneracies for differential equations (e.g. mixed elliptic ones such as the Zaremba problem or with jumping oblique derivatives for the Laplacian) shows that the analogous questions lead at once to corresponding algebras of rather generality. For proving an analogue of the Atiyah-Singer index theorem such an approach would be necessary, anyway.

In carrying out the programme for corners then it turns out that even the functional analytic background of corner Sobolev spaces and subspaces with asymptotics (which is closely related to the character of smoothing operators) is to be prepared first. Also other in a sense non-standard elements of the ψDO calculus, e.g. the parameter- dependent version of a given operator algebra (here for conical singularities) and various sorts of operator conventions are to be established in advance. Our exposition is voluminous, since all these elements are needed here at the same time, though they deserve separate papers.

1 The cone and wedge pseudo-differential calculus

1.1 Motivation of the approach

This section will prepare the auxiliary material on cone ψDO's which is at the same time a motivation for the methods for corners.

First recall that every manifold B with conical singularities can be viewed as

$$B = (\ldots((\mathbb{B}/X_1)/X_2)\ldots/X_N),$$

where $\mathbb{B}$ is a compact C^∞ manifold with C^∞ boundary $\partial\mathbb{B} \cong X_1 \cup \ldots \cup X_N$ (disjoint union) for closed compact C^∞ manifolds X_i which are the cone base close to $b_i \in B_0$, $i = 1, \ldots, N$. Then $\partial\mathbb{B}$ has a collar neighbourhood V in $\mathbb{B}$ for which there is given a diffeomorphism

$$\overline{\varphi} \;:\longrightarrow (\bigcup_{i=1}^{N} X_i) \times [0,1)$$

which restricts to diffeomorphisms

$$\varphi_i : int\ V_i \longrightarrow X_i \times (0,1),\ i = 1, \ldots, N,$$

(with V_i being a corresponding collar neighbourhood of the component $\partial\mathbb{B}_i \cong X_i$, cf. Figure 1). $X_i \times (0,1)$ will be identified with the corresponding cartesian product in (2). We call $\mathbb{B}$ the stretched manifold associated with B. The analysis on B will be performed on $int\ \mathbb{B} \cong B \setminus B_0$.

Without loss of generality (up to trivial arguments) we may assume that B_0 consists of a single point, i.e., we have one cone base X, possibly with several connected components. It will be convenient to develop the analysis also on

$$X^{\wedge} = X \times \mathbb{R}_+.$$

$H^s_{loc}(\Omega)$ for a C^∞ manifold Ω ,$s \in \mathbb{R}$, is the space of all $f \in \mathcal{D}'(\Omega)$ with $\chi_*(f|_U) \in H^s_{loc}(\mathbb{R}^n)$ for every chart $\chi : U \longrightarrow \mathbb{R}^n$. Here $H^s_{loc}(\mathbb{R}^n)$ is the standard distribution class of all $u \in \mathcal{D}'(\mathbb{R}^n)$ with $\psi u \in H^s(\mathbb{R}^n)$ for every $\psi \in C_0^\infty(\mathbb{R}^n)$. Moreover $H^s_{comp}(\Omega)$ consists of all $f \in H^s_{loc}(\Omega)$ with compact support. Here and in the sequel Ω is assumed to be paracompact. For closed compact Ω we write $H^s(\Omega)$ instead of $H^s_{comp}(\Omega)$ $(= H^s_{loc}(\Omega))$.

$H^s_{loc}(\Omega)$ is a Fréchet space in a canonical way, $H^s_{comp}(\Omega)$ the inductive limit of Banach spaces.

If E is a locally convex vector space which is a module over an algebra A we write

$$[a]E = \textit{completion of} \{ae : e \in E\} \tag{1}$$

for any fixed $a \in A$.

If E_0, E_1 are Fréchet spaces which are subspaces of a topological Hausdorff space F then

$$E := E_0 + E_1 = \{e_0 + e_1 : e_0 \in E_0, e_1 \in E_1\} \tag{2}$$

is also Fréchet space in the topology induced by the bijection $E \cong E_0 \oplus E_1/D$, with $D = \{(e, -e) : e \in E_0 \cap E_1\}$ being equipped with the topology induced by $E_0 \oplus E_1$.

Incidentally we shall employ analogous sums of more general topological vector spaces.

Let $U \subseteq \mathbb{R}^p$ be open. Then $S^\mu(U \times \mathbb{R}^n)$ for $\mu \in \mathbb{R}$ is the space of all $a(x,\xi) \in C^\infty(U \times \mathbb{R}^n)$ with $|D_x^\alpha D_\xi^\beta a(x,\xi)| \le c(1 + |\xi|)^{\mu - |\beta|}$ for every $\alpha \in \mathbb{N}^p$, $\beta \in \mathbb{N}^n$, $x \in K \subset\subset U$, $\xi \in \mathbb{R}^n$, with constants $c = c(\alpha, \beta, K) > 0$. By $S^\mu_{cl}(U \times \mathbb{R}^n)$ we denote the subspace of classical elements. Both $S^\mu(U \times \mathbb{R}^n)$ and $S^\mu_{cl}(U \times \mathbb{R}^n)$ are Fréchet spaces in a natural way. The convergence in $S^\mu_{cl}(U \times \mathbb{R}^n)$ implies the convergence in $C^\infty(U \times (\mathbb{R}^n \setminus \{0\}))$ of all homogeneous components of the elements of a given sequence. By $L^\mu(U)(L^\mu_{cl}(U))$ for open $U \subseteq \mathbb{R}^n$ we denote the spaces of ψDO's (classical ψDO's) in U, i.e. the space of all $A = F^{-1}a(x,\xi)F + C$ with $a(x,\xi) \in S^\mu(U \times \mathbb{R}^n)$ $(\in S^\mu_{cl}(U \times \mathbb{R}^n))$, and C being an operator with kernel in $C^\infty(U \times U)$, and F the Fourier transform in $\mathbb{R}^n$.

Let $\Lambda = \mathbb{R}^l$ for some $l \in \mathbb{N}$. Then $L^\mu(U;\Lambda)$ $(L^\mu_{cl}(U;\Lambda))$ will denote the space of all operator families $A(\lambda) = F^{-1}a(x,\xi,\lambda)F^{-1} + C(\lambda), \lambda \in \Lambda$, with $a(x,\xi,\lambda) \in S^\mu(U \times (\mathbb{R}^n \times \Lambda))$ $(\in S^\mu_{cl}(U \times (\mathbb{R}^n \times \Lambda)))$ and $C(\lambda) \in \mathcal{S}(\Lambda, C^\infty(U \times U))$, where $\mathcal{S}(\Lambda, E)$ is the Schwartz space on Λ of E–valued functions.

By using invariance of all these classes under diffeomorphisms we can define

$$L^\mu(\Omega), L^\mu_{cl}(\Omega), L^\mu(\Omega;\Lambda), L^\mu_{cl}(\Omega;\Lambda) \tag{3}$$

for any C^∞ manifold Ω. In particular $L^{-\infty}(\Omega;\Lambda)$ will consist of $\mathcal{S}(\Lambda, L^{-\infty}(\Omega))$, based on the canonical Fréchet topology on $L^{-\infty}(\Omega) = \bigcap_\mu L^\mu(\Omega)$. Also the spaces (3) are all Fréchet in a natural way (cf. e.g. [S4],I). The Fréchet topologies in the operator spaces with *cl* are stronger than those induced by the spaces without *cl*. In order to avoid too much comment on topologies (and for other reasons) we shall mainly deal with classical operators.

Let X be a closed compact C^∞ manifold and consider the class $L^\mu_{cl}(X;\Lambda)$. Denote by $\sigma^\mu_\psi(A)(x,\xi,\lambda)$ the parameter dependent homogeneous principal symbol of $A(\lambda) \in L^\mu_{cl}(X;\Lambda)$ of order μ. This is a (unique) function in

$$S^{(\mu)}(T^*X \times \Lambda \setminus 0). \tag{4}$$

(4) is defined as the space of all $p(x,\xi,\lambda) \in C^\infty(T^*X \times \Lambda \setminus 0)$ (0 indicates $(\xi,\lambda) = 0$) which are (positively) homogeneous in (ξ,λ) of order μ. The map

$$\sigma^\mu_\psi : L^\mu_{cl}(X;\Lambda) \longrightarrow S^{(\mu)}(T^*X \times \Lambda \setminus 0) \tag{5}$$

is surjective.

Then $A(\lambda)$ is called parameter dependent elliptic if $\sigma_\psi^\mu(A)(x,\xi,\lambda) \neq 0$ for all x and $(\xi,\lambda) \neq 0$.

Let $A(\lambda) \in L_{cl}^\mu(X;\Lambda)$ be parameter dependent elliptic. Then there exists a $c_1 > 0$ such that

$$A(\lambda) : H^s(X) \longrightarrow H^{s-\mu}(X)$$

is an isomorphism for all $|\lambda| > c_1$ and every $s \in \mathbb{R}$.

On X we fix once and for all a Riemannian metric. Then we have the space $L^2(X)$ with respect to a corresponding fixed measure dx, where $L^2(X) \cong H^0(X)$.

An example of an element in (4) is $(|\xi|^2 + |\lambda|^2)^{\mu/2}$. In view of the surjectivity of (5) we find an $R^\mu(\lambda) \in L_{cl}^\mu(X;\Lambda)$ with $\sigma_\psi^\mu(R^\mu)(x,\xi,\lambda) = (|\xi|^2 + |\lambda|^2)^{\mu/2}$. By applying the construction for $|\lambda|^2 + c^2$ instead of $|\lambda|^2, c > 0$ sufficiently large, we obtain an $R^\mu(\lambda)$ which induces isomorphisms

$$R^\mu(\lambda) : H^s(X) \longrightarrow H^{s-\mu}(X) \tag{6}$$

for all $\lambda \in \Lambda, s \in \mathbb{R}$.

Let us now turn to some elementary things on the Mellin transform, first for scalar functions on $\mathbb{R}_+$. The Mellin transform

$$Mu(w) = \int_0^\infty r^{w-1} u(r)dr,$$

$w \in \mathbb{C}, u \in C_0^\infty(\mathbb{R}_+)$, gives rise to a continuous operator $M : C_0^\infty(\mathbb{R}_+) \to \mathcal{A}(\mathbb{C})$. $\mathcal{A}(U)$ for any open $U \subseteq \mathbb{C}$ is the space of all holomorphic functions in U in the (Fréchet) topology of uniform convergence on compact subsets. Set

$$\Gamma_\sigma = \{w \in \mathbb{C} : Re\, w = \sigma\} \tag{7}$$

for any $\sigma \in \mathbb{R}$. Denote by $\mathcal{S}(\Gamma_\sigma)$ the Schwartz space on Γ_σ (i.e. the pull-back of $\mathcal{S}(\mathbb{R})$ under $\Gamma_\sigma \to \mathbb{R}$, $\sigma + i\rho \to \rho$). Then M induces also continuous operators

$$M_\beta : C_0^\infty(\mathbb{R}_+) \longrightarrow \mathcal{S}(\Gamma_{1/2-\beta}) \tag{8}$$

for every $\beta \in \mathbb{R}$, where $M_\beta u = Mu|_{\Gamma_{1/2-\beta}}$ for $u \in C_0^\infty(\mathbb{R}_+)$ and the inverse follows by

$$M_\beta^{-1} g(r) = \frac{1}{2\pi i} \int_{\Gamma_{1/2-\beta}} r^{-w} g(w) dw.$$

It is well-known that M_β extends by continuity to an isomorphism

$$M_\beta : r^\beta L^2(\mathbb{R}_+) \longrightarrow L^2(\Gamma_{1/2-\beta})$$

with obvious notation (the L^2 spaces are equipped with the scalar products associated with the corresponding Lebesgue measures). We will often write $M = M_0$. Then $(M_\beta u)(w) = (T^\beta M u)(w)$ on $C_0^\infty(\mathbb{R}_+)$ with

$$(T^\beta h)(w) = h(w + \beta). \tag{9}$$

It is clear that

$$-r\partial/\partial r = M^{-1} w M,$$

defined, for instance, on the subspace of all $u \in L^2(\mathbb{R}_+)$ with $r\partial/\partial r\, u \in L^2(\mathbb{R}_+)$. This gives rise to Mellin ψDO's

$$op_M(a)u(r) = M_{w\to r}^{-1} a(r,r',w) M_{r'\to w} u(r') \tag{10}$$

with Mellin amplitude functions $a(r,r',w)$ to be defined more precisely below. We shall also use the notation

$$op_M^\beta(a) = r^\beta op_M(T^{-\beta}a) r^{-\beta} \tag{11}$$

$(T^{-\beta}a)(r, r', w) = a(r, r', w - \beta)$.

The notions around the Mellin transform have a straightforward extension to E–valued functions for any (say Fréchet) space E. This will tacitly be used in the sequel. We set $\mathcal{A}(U, E) = \mathcal{A}(U) \otimes_\pi E$ ($\otimes_\pi$ is the completed projective tensor product), $\mathcal{S}(\mathbb{R}^q, E) = \mathcal{S}(\mathbb{R}^q) \otimes_\pi E$ and so on.

Distributions u on $X^\wedge$ will often be regarded as vector-valued ones and then written as $u(r)$ with the Mellin transform $(Mu)(w)$.

We define $L^\mu_{cl}(X; \Gamma_\sigma)$ as the space of all $A(w), w$ varying on Γ_σ, such that $A(\sigma + i\rho) \in L^\mu_{cl}(X; \mathbb{R}_\rho)$. For every $s \in \mathbb{R}$ we choose a parameter dependent elliptic $R^s(w) \in L^\mu_{cl}(X; \Gamma_\rho)$ with $R^s(w) : H^s(X) \overrightarrow{\cong} L^2(X)$ for all w.

Definition 1.1.1 *$\mathcal{H}^{s,\beta}(X^\wedge)$ for $s, \gamma \in \mathbb{R}$ is the completion of $C_0^\infty(X^\wedge)$ with respect to the norm*

$$||u||_{\mathcal{H}^{s,\beta}(X^\wedge)} = \left\{ \frac{1}{2\pi i} \int_{\Gamma_{\sigma-\beta}} ||R^s(w) Mu(w)||^2_{L^2(X)} dw \right\}^{1/2} \tag{12}$$

with $\sigma = (n+1)/2$, $n = dimX$, and $R^s(w)$ being an order reducing family.

Clearly (12) is independent of the concrete choice of $R^s(w)$, modulo equivalence of norms. We have

$$\mathcal{H}^{s,\beta}(X^\wedge) = r^\beta \mathcal{H}^s(X^\wedge)$$

with $\mathcal{H}^s(X^\wedge) := \mathcal{H}^{s,0}(X^\wedge)$. Note that $\mathcal{H}^{s,\beta}(X^\wedge) \subset H^s_{loc}(X^\wedge)$ for all $s, \beta \in \mathbb{R}$.

On $X^\wedge$ we can also define the standard Sobolev spaces $H^s(X^\wedge)$ which are the restriction to $X^\wedge$ of the space $H^s(X \times \mathbb{R})$, where the behaviour on $U^\wedge = U \times \mathbb{R}_+ \ni (x, r)$ for $r \to \infty$ for any coordinate neighbourhood U on X with local coordinates $x = \tilde{x}/|\tilde{x}|$ from a corresponding open set $U_1 \subset S^n \subset \mathbb{R}^{n+1} \ni \tilde{x}$, $U_1 \cong U$, with $r = |\tilde{x}|$, is the same as $H^s(\mathbb{R}^{n+1})|_{U_1^\wedge}$, $U_1^\wedge = \{\tilde{x} : \tilde{x}/|\tilde{x}| \in U_1\}$ (for more explanations, cf. [S4], I, Section 1.1.1.). Both $\mathcal{H}^{s,\beta}(X^\wedge)$ and $H^s(X^\wedge)$ are Banach spaces, and modules over the algebra of all functions in $C^\infty(\mathbb{R}_+)$ which are constant for $r < c_0$, $r > c_1$ for certain $0 < c_0 < c_1 < \infty$.

A cut-off function $\omega(r)$ in this paper will allways mean some $\omega \in C^\infty(\overline{\mathbb{R}_+})$ with $\omega \equiv 1$ for $r < c_0$, $\omega \equiv 0$ for $r > c_1$ for certain $0 < c_0 < c_1 < \infty$.

$$\mathcal{K}^{s,\beta}(X^\wedge) = [\omega]\mathcal{H}^{s,\beta}(X^\wedge) + [1 - \omega]H^s(X^\wedge) \tag{13}$$

cf. (2), (1) equipped with the topology of the sum. (13) is independent of the concrete choice of ω. Moreover define

$$\mathcal{H}^{s,\beta}(\mathbb{B}) = \varphi^*[\omega]\mathcal{H}^{s,\beta}(X^\wedge) + [1 - \varphi^*\omega]H^s_{loc}(int\ \mathbb{B}). \tag{14}$$

Here it is assumed that ω vanishes for $r > c_1$ with $c_1 > 0$ small enough such that the pull-back of ωu for $u \in \mathcal{H}^{s,\beta}(X^\wedge)$ to $int\ \mathbb{B}$ under the fixed diffeomorphism

$$\varphi : int\ V \longrightarrow X \times (0, 1) \tag{15}$$

exists (V is the collar neighbourhood of $\partial\mathbb{B}$ in $\mathbb{B}$). In the sequel we shall often drop pull-backs if the identifications of objects with their pull-backs are evident. Then, in simplified notation,

$$\mathcal{H}^{s,\beta}(\mathbb{B}) = [\omega]\mathcal{H}^{s,\beta}(X^\wedge) + [1 - \omega]H^s_{loc}(int\ \mathbb{B}). \tag{16}$$

More pedantically we should also insert *int* $\mathbb{B}$ on the left of (14) and (16). But we prefer to employ the shorter notation (we hope this will not cause confusion).

The spaces $\mathcal{H}^{s,\beta}(\mathbb{B})$, $\mathcal{H}^{s,\beta}(X^\wedge)$, $\mathcal{K}^{s,\beta}(X^\wedge)$ can be viewed as Hilbert spaces after having chosen corresponding scalar products. We will only need fixed scalar products in the cases $s = \beta = 0$. In

$\mathcal{H}^0(X^\wedge) = \mathcal{K}^0(X^\wedge)$ we take the scalar product from $\mathcal{K}^0(X^\wedge) \cong r^{-n/2}L^2(X \times \mathbb{R}_+)$ with the L^2 space with respect to $dxdt$, dx being associated with a Riemannian metric on X. In $\mathcal{H}^0(\mathbb{B})$ we get a scalar product from

$$\mathcal{H}^0(\mathbb{B}) = [\omega]\mathcal{H}^0(X^\wedge) + [1-\omega]L^2(int\ \mathbb{B}),$$

where $L^2(int\ \mathbb{B})$ refers to a Riemannian metric on $int\ \mathbb{B}$.

Now let $D \in Diff^\mu(int\ \mathbb{B})$ be an arbitrary differential operator which is close to $\partial\mathbb{B}$ in the coordinates $(x,r) \in V$ of the form

$$D = r^{-\mu}\sum_{j=0}^{\mu} d_j(r)(-r\frac{\partial}{\partial r})^j \tag{17}$$

with certain $d_j \in C^\infty([0,1), Diff^{\mu-j}(X))$. Then D induces continuous operators

$$D : \mathcal{H}^{s,\beta}(\mathbb{B}) \longrightarrow \mathcal{H}^{s-\mu,\beta-\mu}(\mathbb{B}) \tag{18}$$

for every $s, \beta \in \mathbb{R}$.

The cone operator theory says that we have to observe two leading symbolic levels for the Fredholm property of (18), namely the usual homogeneous principal symbol of D of order μ

$$\sigma_\psi^\mu(D) \in C^\infty(T^*(int\ \mathbb{B}) \setminus 0)$$

and the conormal symbol of D of 'conormal order' μ

$$\sigma_M^\mu(D)(w) = \sum_{j=0}^{\mu} d_j(0)w^j, \tag{19}$$

restricted to $w \in \Gamma_{(n+1)/2-\beta}$, and regarded as operator family

$$\sigma_M^\mu(D)(w) : H^s(X) \longrightarrow H^{s-\mu}(X) \tag{20}$$

for all $s \in \mathbb{R}$. In view of the particular form of $\sigma_\psi^\mu(D)$ we introduce the space

$$S_b^{(\mu)}(T^*(int\ \mathbb{B}) \setminus 0) \tag{21}$$

of all $p \in C^\infty(T^*(int\ \mathbb{B}) \setminus 0)$ which are homogeneous of order μ with respect to the canonical $\mathbb{R}_+$ action along the fibres of $T^*(int\ \mathbb{B}) \setminus 0$ and close to $\partial\mathbb{B}$ in the coordinates $(x,r) \in int\ V$ of the form $p(x,r,\xi,\rho) = q(x,r,\xi,r\rho)$ for some $q(x,r,\xi,\rho)$ which is C^∞ in r up to $r=0$. Then

$$\sigma_\psi^\mu(D) \in g_0^{-\mu}S_b^{(\mu)}(T^*(int\ \mathbb{B}) \setminus 0), \tag{22}$$

where $g_0^\eta \in C^\infty(int\ \mathbb{B})$ is an arbitrary strictly positive function with $g_0^\eta = r^\eta$ close to $\partial\mathbb{B}$.

Definition 1.1.2 *D is called elliptic (with respect to the weight $\beta \in \mathbb{R}$) if*

(i) *$\sigma_\psi^\mu(D) \neq 0$ on $T^*(int\ \mathbb{B})\setminus 0$ in the usual sense, and $g_0^\mu\sigma_\psi^\mu(D)(x,r,\xi,r^{-1}\rho) \neq 0$ in the coordinates $(x,r) \in int\ V$, up to $r=0$,*

(ii) *(20) is bijective for all $w \in \Gamma_{(n+1)/2-\beta}$ and any fixed $s \in \mathbb{R}$.*

Theorem 1.1.3 *The following conditions are equivalent*

(i) *D is elliptic (with respect to the weight β)*

(ii) *(18) is a Fredholm operator for any $s \in \mathbb{R}$.*

We now return to the differential operators A on $\Omega = K \setminus K_1$ mentioned in the beginning (cf. 0.(10), 0.(11)), where K is a compact manifold with corners K_0, and K_1 the skeleton of one-dimensional edges. It will be convenient to introduce a 'stretched manifold' $\mathbb{K}$, defined as a compact space with a projection

$$p : \mathbb{K} \longrightarrow K \tag{23}$$

where an open dense subset $\tilde{\Omega} \subset \mathbb{K}$ is mapped diffeomorphically to Ω, further $p^{-1}(\{v\}) \cong \mathbb{B}$ for any $v \in K_0$, with $\mathbb{B}$ being the stretched manifold of the base B of the corner near v, $p^{-1}(\{y\}) \cong X$ for any $y \in K_1 \setminus K_0$, where X is the base of the model cone of the wedge with edge through y. Moreover $p^{-1}(v) \cong \mathbb{B} \times [0,1)$ for a neighbourhood V of $y \in K_1 \setminus K_0$, with X, α, β as in 0.(6).

$\mathbb{K}$ can be viewed as another compactification of Ω. Therefore, we shall identify Ω and $\tilde{\Omega}$ and write also

$$\Omega = int\ \mathbb{K}. \tag{24}$$

We will also set

$$\mathbb{W} = p^{-1}(K \setminus K_0) \tag{25}$$

which can be regarded as the stretched manifold of the manifold $K \setminus K_0$ with edges.

Let us assume for a moment that we have already defined the corner Sobolev spaces

$$\mathcal{H}^{s,\gamma}(\mathbb{K})\ for\ s \in \mathbb{R}, \gamma = (\gamma_0,\ \gamma_1) \in \mathbb{R}^2 \tag{26}$$

analogously to the cone Sobolev spaces $\mathcal{H}^{s,\beta}(\mathbb{B})$ from Definition 1.1.1 (the definition of $\mathcal{H}^{s,\gamma}(\mathbb{K})$ will be given in Section 2.2). Here we have a couple of weights γ, where γ_0 refers to the cone axis variable r. Then, our differential operator A will induce continuous operators

$$A : \mathcal{H}^{s,\gamma}(\mathbb{K}) \longrightarrow \mathcal{H}^{s-\mu,\gamma-\mu}(\mathbb{K}) \tag{27}$$

for all $s \in \mathbb{R}$, $\gamma = (\gamma_0, \gamma_1) \in \mathbb{R}^2$, $\gamma - \mu := (\gamma_0 - \mu, \gamma_1 - \mu)$. In order to get an analogue of Theorem 3 we shall have to identify the leading symbolic levels of A, such that their bijectivity is responsible for the Fredholm property. This will be done next, i.e. we want to read off from A the adequate symbolic structure.

First we have, of course, the usual homogeneous principal symbol order μ

$$\sigma_\psi^\mu(A) \in C^\infty(T^*(int\ \mathbb{K}) \setminus 0).$$

We want to express it in an analogous manner as (22). To this end we introduce the space

$$S_{2b}^{(\mu)}(T^*(int\ \mathbb{K}) \setminus 0) \tag{28}$$

of all $p \in C^\infty(T^*(int\ \mathbb{K}) \setminus 0)$ which are homogeneous of order μ with respect to the canonical $\mathbb{R}_+$ action along the fibres of $T^*(int\ \mathbb{K}) \setminus 0$, and we assume

(i) close to $p^{-1}(\{v\})$ (cf. (23)) for any $v \in K_0$ in the coordinates $(x, r, t) \in X \times (0,1) \times (0,1)$

$$p(x, r, t, \xi, \rho, \tau) = q_1(x, r, t, \xi, r\rho, rt\tau) \tag{29}$$

for some $q_1(x, r, t, \xi, \rho, \tau)$ which is C^∞ up to $t = 0$, $r = 0$

(ii) close to $p^{-1}(\{v\})$ for any $v \in K_0$ in the coordinates $(\tilde{x}, t) \in int\ \mathbb{B} \times (0,1)$

$$p(\tilde{x}, t, \tilde{\xi}, \tau) = q_2(\tilde{x}, t, \tilde{\xi}, t\tau) \tag{30}$$

for some $q_2(\tilde{x}, t, \tilde{\xi}, \tau)$ which is C^∞ up to $t = 0$

(iii) close to $p^{-1}(\{y\})$ for any $y \in K_1 \setminus K_0$ in the coordinates $(x, r, t) \in X \times (0,1) \times (\alpha, \beta)$

$$p(x, r, t, \xi, \rho, \tau) = q_3(x, r, t, \xi, r\rho, r\tau) \tag{31}$$

for some $q_3(x, r, t, \xi, \rho, \tau)$ which is C^∞ up to $r = 0$.

Let g_2^η for $\eta \in \mathbb{R}$ be a strictly positive function in $C^\infty(int\ \mathbb{K})$ which is of the form

$$g_2^\eta = r^\eta t^\eta \text{ in the coordinates from (i)} \tag{32}$$
$$g_2^\eta = f(\tilde{x}) t^\eta \text{ in the coordinates from (ii)} \tag{33}$$
$$g_2^\eta = r^\eta h(x,t) \text{ in the coordinates from (iii)} \tag{34}$$

where $f(\tilde{x}) \in C^\infty(int\ \mathbb{B})$, $h(x,t) \in C^\infty(X \times (\alpha, \beta))$ are strictly positive. Any such g_2^η will be called a weight function on $int\ \mathbb{K}$, for the weight η. It can easily be proved that weight functions like this actually do exist. We have then obviously

$$\sigma_\psi^\mu(A) \in g_2^{-\mu} S_{2b}^{(\mu)}(T^*(int\ \mathbb{K}) \setminus 0). \tag{35}$$

For introducing an analogue of (19) for the operator A we choose a cut-off function $\tilde{\omega}(r)$ and write A close to $t = 0$ (which corresponds to $v \in K_0$) as

$$A = \tilde{\omega}(r) t^{-\mu} r^{-\mu} \sum_{j+k \le \mu} a_{jk}(r,t)(-r\frac{\partial}{\partial r})^j (-rt\frac{\partial}{\partial t})^k$$
$$+(1-\tilde{\omega}(r)) t^{-\mu} \sum_{k=0}^{\mu} a_k(t)(-t\frac{\partial}{\partial t})^k \tag{36}$$

where

$$a_{jk}(r,t) \in C^\infty([0,1) \times [0,1),\ Diff^{\mu-(j+k)}(X)), \tag{37}$$
$$a_k(t) \in C^\infty([0,1),\ Diff^{\mu-k}(B \setminus B_0)). \tag{38}$$

Then we set

$$\sigma_M^\mu(A)(z) = \tilde{\omega}(r) r^{-\mu} \sum_{j+k \le \mu} a_{jk}(r,0)(-r\frac{\partial}{\partial r})^j (rz)^k$$
$$+(1-\tilde{\omega}(r)) \sum_{k=0}^{\mu} a_k(0) z^k \tag{39}$$

regarded as a z dependent operator function with values in the class of operators of the type (17), z varying in $\mathbb{C}$ or along a weight line $\Gamma_{(n+2)/2-\gamma_1}$, $\gamma_1 \in \mathbb{R}$. For every z we get continuous operators

$$\sigma_M^\mu(A)(z) : \mathcal{H}^{s,\gamma_0}(\mathbb{B}) \longrightarrow \mathcal{H}^{s-\mu,\gamma_0-\mu}(\mathbb{B}) \tag{40}$$

for all $s, \gamma_0 \in \mathbb{R}$.

Now, as already announced in the beginning, a novelty for corners is the edge symbolic level along $K_1 \setminus K_0$, also being operator-valued. We shall adopt here the notations from the general theory of ψDO's on manifolds with edges (cf. [S1], [S5]). By assumption the operator A is in the coordinates (x, r, t) close to any $y \in K_1 \setminus K_0$ of the form

$$A = r^{-\mu} \sum_{j+k \le \mu} c_{jk}(r,t)(-r\frac{\partial}{\partial r})^j (r\frac{\partial}{\partial t})^k \tag{41}$$

with certain

$$c_{jk}(r,t) \in C^\infty([0,1) \times (\alpha,\beta),\ Diff^{\mu-(j+k)}(X)). \tag{42}$$

For y near $v \in K_0$ this is, of course, compatible with (36), such that for small r

$$c_{jk}(r,t) = t^{-\mu} a_{jk}(r,t)(-t)^k, \tag{43}$$

with the above $a_{jk}(r,t)$. We set

$$\sigma^\mu_\wedge(A)(t,\tau) = r^{-\mu} \sum_{j+k\leq\mu} c_{jk}(0,t)(-r\frac{\partial}{\partial r})^j (ir\tau)^k. \tag{44}$$

τ is treated here as a covariable to $t \in \mathbb{R}$ with respect to the Fourier transform along $\mathbb{R}$. Therefore $\partial/\partial t$ was replaced by $i\tau$. The role of inserting $r = 0$ in the coefficients is the same as of freezing of coefficients on the boundary for boundary value problems. (44) is an operator family

$$\sigma^\mu_\wedge(A)(t,\tau) : \mathcal{K}^{s,\gamma_0}(X^\wedge) \longrightarrow \mathcal{K}^{s-\mu,\gamma-\mu}(X^\wedge) \tag{45}$$

for every $s, \gamma_0 \in \mathbb{R}$, parametrized by (t,τ), regarded as point in $T^*(K_1 \setminus K_0) \setminus 0$. (45) is called the homogeneous principal edge symbol of A of order μ. The homogeneity refers to the action $(\kappa_\lambda u)(x,r) := \lambda^{(n+1)/2} u(x, \lambda r)$, $\lambda \in \mathbb{R}_+$, on the spaces $\mathcal{K}^{s,\beta}(X^\wedge)$. Then $\{\kappa_\lambda\}_{\lambda\in\mathbb{R}_+} \in C(\mathbb{R}_+, \mathcal{L}_\sigma(\mathcal{K}^{s,\beta}(X^\wedge)))$ for every $s, \beta \in \mathbb{R}$ (σ indicates the strong operator topology). (45) satisfies

$$\sigma^\mu_\wedge(A)(t,\lambda\tau) = \lambda^\mu \kappa_\lambda \sigma^\mu_\wedge(A)(t,\tau)\kappa_\lambda^{-1} \tag{46}$$

for all $\lambda \in \mathbb{R}_+$. The form (36) of A close to $v \in K_0$ suggests also to introduce a principal Mellin edge symbol of order μ, namely

$$\sigma^\mu_{b\wedge}(A)(t,\tau) = r^{-\mu} \sum_{j+k\leq\mu} a_{jk}(0,t)(-r\frac{\partial}{\partial r})^j (ir\tau)^k. \tag{47}$$

This is also an operator family

$$\sigma^\mu_{b\wedge}(A)(t,\tau) : \mathcal{K}^{s,\gamma_0}(X^\wedge) \longrightarrow \mathcal{K}^{s-\mu,\gamma_0-\mu}(X^\wedge) \tag{48}$$

for every $s, \gamma_0 \in \mathbb{R}$, parametrized by t close to $t = 0$ and $\tau = Im\ z$. It satisfies an analogous homogeneity as (46).

Clearly we have

$$\sigma^\mu_\wedge(A)(t,\tau) = t^{-\mu} \sigma^\mu_{b\wedge}(A)(t,\tau_1)|_{\tau_1=-t\tau}. \tag{49}$$

The components of

$$(\sigma^\mu_\psi(A),\ \sigma^\mu_M(A),\ \sigma^\mu_\wedge(A))$$

satisfy obvious compatibility conditions.

Definition 1.1.4 *A is called elliptic (with respect to the weights $\gamma = (\gamma_0, \gamma_1) \in \mathbb{R}^2$) if*

(i) *$\sigma^\mu_\psi(A) \neq 0$ on $T^*(int\ \mathbb{K}) \setminus 0$ in the usual sense and if $p = g_2^\mu \sigma^\mu_\psi(A)$ with q_1, q_2, q_3 being defined by (29), (30) and (31), respectively, satisfies $q_1(x,r,t,\xi,\rho,\tau) \neq 0$ up to $r = 0$, $t = 0$, $q_2(\tilde{x},t,\tilde{\xi},\tau) \neq 0$ up to $t = 0$, $q_3(x,r,t,\xi,\rho,\tau) \neq 0$ up to $r = 0$,*

(ii) *(40) is bijective for all $z \in \Gamma_{(n+2)/2-\gamma_1}$ and any fixed $s \in \mathbb{R}$,*

(iii) *(45) is bijective for all $(t,\tau) \in T^*(K_1 \setminus K_0) \setminus 0$ and any fixed $s \in \mathbb{R}$.*

Remark 1.1.5 *Let (ii) of Definition 1.1.4 be satisfied for fixed $s_0 \in \mathbb{R}$. Then it follows for all $s \in \mathbb{R}$. Analogously (iii) of Definition 1.1.4 for $s_0 \in \mathbb{R}$ implies the same for all $s \in \mathbb{R}$.*

A result of the precise theory of corner operators will be the following

Theorem 1.1.6 *The following conditions are equivalent*

(i) *A is elliptic (with respect to* $\gamma = (\gamma_0, \gamma_1)$*)*

(ii) *(27) is a Fredholm operator for any* $s \in \mathbb{R}$.

Similarly to the theory of cone ψDO's it is interesting to perform a parametrix construction within a suitable class of corner ψDO's and to obtain the elliptic regularity with asymptotics close to $K_1 \setminus K_0$ in the sense of edge asymptotics and close to K_0 in the appropriate corner sense. In other words we shall have to introduce corner ψDO's and the corresponding distribution classes with asymptotics.

Since outside K_0 the theory coincides with the calculus of operators on manifolds with edges, we will have in general additional (trace and potential) conditions along $K_1 \setminus K_0$. They will also take part in the ellipticity and the edge symbolic structure. Even for our differential operators A it may happen that the operator has to be filled to a matrix

$$\mathcal{A} = \begin{pmatrix} A & K \\ T & Q \end{pmatrix} : \begin{array}{c} \mathcal{H}^{s,\gamma}(\mathbb{K}) \\ \oplus \\ \mathcal{H}^{s,\delta}(K_1 \setminus K_0, \mathbb{C}^N) \end{array} \longrightarrow \begin{array}{c} \mathcal{H}^{s-\mu,\gamma-\mu}(\mathbb{K}) \\ \oplus \\ \mathcal{H}^{s-\mu,\delta-\mu}(K_1 \setminus K_0, \mathbb{C}^M) \end{array}$$

for obtaining the Fredholm property. A convenient choice of the weight δ will be fixed below.

1.2 Operators on manifolds with conical singularities

The theory of operators on manifolds with corners will employ some material from the cone theory. Therefore, we shall briefly recall here the necessary things. More details and proofs may be found in [R1], [S1], [S4].

First we shall recall the asymptotics of distributions on *int* $\mathbb{B}$ close to $\partial\mathbb{B}$ (as above $\mathbb{B}$ is the stretched manifold belonging to a manifold B with conical singularities). This will be formulated in terms of $X^\wedge$. The discrete asymptotics of an $u(x,r) \in \mathcal{K}^{s,\beta}(X^\wedge)$ for $r \to 0$ are defined as

$$u(x,r) \sim \sum_{j=0}^{\infty} \sum_{k=0}^{m_j} \zeta_{jk}(x) r^{-p_j} \log^k r. \tag{1}$$

We talk about the discrete asymptotic type

$$P = \{(p_j, m_j, L_j)\}_{j \in \mathbb{N}} \tag{2}$$

of u, if the coefficients ζ_{jk} belong to finite-dimensional subspaces $L_j \subset C^\infty(X)$, $0 \le k \le m_j \in \mathbb{N}$, and the exponents $p_j \in \mathbb{C}$ are fixed with $Re\ p_j \to -\infty$. Clearly we assume $Re\ p_j < (n+1)/2 - \beta$ for all j. The precise meaning of (1) is that for any cut-off function $\omega(r)$ and every $\beta \in \mathbb{R}$ there is an $N = N(\beta)$ with

$$u(x,r) - \sum_{j=0}^{N} \sum_{k=0}^{m_j} \zeta_{jk}(x) r^{-p_j} \log^k r\ \omega(r) \in \mathcal{K}^{s,\beta}(X^\wedge) \tag{3}$$

We set

$$\pi_{\mathbb{C}} P = \{p_j\}_{j \in \mathbb{N}}.$$

Denote by $\mathcal{K}_P^{s,\beta}(X^\wedge)$ the subspace of all $u \in \mathcal{K}^{s,\beta}(X^\wedge)$ with asymptotics of type P. This is a Fréchet space in a natural way. The semi-norms are given by $|\zeta_{jk}(u)|$ for all j,k, by the $\mathcal{K}^{s,\beta_N}(X^\wedge)$ norms of the left hand sides of (3) for all N and the norm of u in $\mathcal{K}^{s,\beta}(X^\wedge)$.

The asymptotic types are associated with a weight β, the infinite weight interval $\Theta = (-\infty, 0]$ on the left of the weight line $\Gamma_{(n+1)/2-\beta}$, and with X, the base of the cone, including $n = dim\ X$.

Let

$$\mathfrak{a} = (\beta, \Theta) \tag{4}$$

be the weight data, and denote by

$$As(X, \mathfrak{a}^{\bullet}) \tag{5}$$

the set of all discrete asymptotic types P belonging to (4).

The parameter-dependent asymptotics as they occur along edges will lead also to non-constant P. Those variable discrete asymptotics can be formulated by means of the continuous ones. They are defined as follows. Instead of $\{p_j\}_{j\in\mathbb{N}}$ we choose an arbitrary sequence of compact sets $\{K_j\}_{j\in\mathbb{N}}$ in the complex w-plane with

$$\begin{aligned} &sup\ \{Re\ w : w \in K_j\} \to -\infty \text{ as } j \to \infty \\ &K_j \subset \{w : Re\ w < (n+1)/2 - \beta\} \text{ for all } j. \end{aligned} \tag{6}$$

Let $\{A_i\}_{i\in I}$ be a system of subsets of $\mathbb{C}$ and U be the union of all unbounded connected components of $\bigcup_{i\in I} \overline{A_i}$. We then define

$$\sum_{i\in I} A_i := \mathbb{C} \setminus U. \tag{7}$$

If the system consists only of a single set A then we write A^c instead of (7). Set

$$T = \sum K_j$$

for the compact sets K_j of (6).

Choose a sequence $\zeta_j \in \mathcal{A}'(K_j, C^\infty(X))$, $j \in \mathbb{N}$. Here $\mathcal{A}'(K, F) := \mathcal{A}(K) \otimes_\pi F$ for a Fréchet space F is the space of all F-valued analytic functionals, carried by K. Any $\zeta \in \mathcal{A}'(K, F)$ can be applied to r^{-w} with respect to $w \in \mathbb{C}$ for every $r \in \mathbb{R}_+$. The result belongs to $C^\infty(\mathbb{R}_+, F)$. Now the continuous asymptotics of $u(x,r) \in \mathcal{K}^{s,\beta}(X^\wedge)$ for $r \to 0$ are defined as

$$u(x,r) \sim \sum_{j=0}^{\infty} < \zeta_j, r^{-w} > .$$

This means that for any $\omega(r)$ and every $\beta \in \mathbb{R}$ there is an $N = N(\beta)$ with

$$u(x,r) - \sum_{j=0}^{N} < \zeta_j, r^{-w} > \omega(r) \in \mathcal{K}^{s,\beta}(X^\wedge).$$

We talk about the continuous asymptotic type D, defined as the set of pairs

$$\begin{aligned} D \ = \ &\{(T, [\{\zeta_j\}_{j\in\mathbb{N}}]) : \zeta_j \in \mathcal{A}'(K_j, C^\infty(X)) \text{ for arbitrary} \\ &\text{compact sets } K_j,\ j \in \mathbb{N}, \text{ satisfying (6), } T \supseteq K_j \text{ for all } j\}. \end{aligned}$$

T was defined above. It is called the carrier of D,

$$T = carrier\ (D).$$

$[\{\zeta_j\}_{j\in\mathbb{N}}]$ is the equivalence class of the sequence $\{\zeta_j\}_{j\in\mathbb{N}}$ under the relation

$$\begin{aligned} \{\zeta_j\}_{j\in\mathbb{N}} \ \sim \ &\{\tilde{\zeta}_j\}_{j\in\mathbb{N}} \longleftrightarrow \\ &\omega\{\sum_{j=0}^{N} < \zeta_j - \tilde{\zeta}_j,\ r^{-w} >\} \in \mathcal{K}^{\infty,\beta}(X^\wedge) \end{aligned}$$

for every $\beta \in \mathbb{R}$ with some $N = N(\beta)$. We allow $\tilde{K}_j \neq K_j$ for the carriers $\tilde{K}_j$ of $\tilde{\zeta}_j$.

Denote by $\mathcal{K}_D^{s,\beta}(X^\wedge)$ the subspace of all $u \in \mathcal{K}^{s,\beta}(X^\wedge)$ with the asymptotics of type D. This is a Fréchet space in a natural way.

Let

$$As(X, \mathfrak{a}) \tag{8}$$

denote the set of all continuous asymptotic types d, associated with the weight data (4).

It will be convenient to have the sets (5), (8) also for finite $\Theta = (\vartheta, 0]$, $\vartheta > -\infty$.

First (5) for finite Θ is the set of all finite sequences

$$P = \{(p_j, m_j, L_j)\}_{0 \leq j \leq N}$$

with $N = N(P)$,

$$(n+1)/2 - \beta + \vartheta < Re\ p_j < (n+1)/2 - \beta$$

for all j, and L_j as above. Moreover (8) for finite Θ is the set of all pairs

$$D = \{(K, [\zeta]) : \zeta \in \mathcal{A}'(K, C^\infty(X))\}$$

where K is a compact set,

$$K \subset \{w \in \mathbb{C} : (n+1)/2 - \beta + \vartheta \leq Re\ w < (n+1)/2 - \beta\}$$

with $\mathbb{C} \setminus K$ being connected, and $[\zeta]$ is the equivalence class of ζ under the relation

$$\zeta \sim \tilde{\zeta} \longleftrightarrow \omega < \zeta - \tilde{\zeta},\ r^{-w} > \in \mathcal{K}_\Theta^{\infty,\beta}(X^\wedge).$$

Here

$$\mathcal{K}_\Theta^{s,\beta}(X^\wedge) := \varprojlim_{\epsilon > 0} \mathcal{K}^{s,\beta-\vartheta-\epsilon}(X^\wedge)$$

K is called the carrier of of D,

$$K = carrier\ (D) \tag{9}$$

Let Θ be finite and introduce the spaces

$$\mathcal{E}_P(X^\wedge) = \left\{ \sum_{j=0}^{N} \sum_{k=0}^{m_j} \zeta_{jk} r^{-p_j} \log^k r\ \omega(r) : \zeta_{jk} \in L_j,\ 0 \leq k \leq m_j \right\} \tag{10}$$

for $P \in As(X, \mathfrak{a}^\bullet)$, $N = N(P)$,

$$\mathcal{E}_D(X^\wedge) = \{< \zeta, r^{-w} > \omega(r) : \zeta \in \mathcal{A}'(K, C^\infty(X))\} \tag{11}$$

for $D \in As(X, \mathfrak{a})$, with a fixed cut-off function ω.

(10) and (11) are Fréchet spaces in a natural way ((10) is finite-dimensional). They are continuously embedded in $\mathcal{K}^{\infty,\beta}(X^\wedge)$. Set for finite Θ

$$\mathcal{K}_P^{s,\beta}(X^\wedge) = \mathcal{K}_\Theta^{s,\beta}(X^\wedge) + \mathcal{E}_P(X^\wedge) \tag{12}$$

for $P \in As(X, \mathfrak{a}^\bullet)$, and

$$\mathcal{K}_D^{s,\beta}(X^\wedge) = \mathcal{K}_\Theta^{s,\beta}(X^\wedge) + \mathcal{E}_D(X^\wedge) \tag{13}$$

for $D \in As(X, \mathfrak{a})$. We endow (12) and (13) with corresponding Fréchet topologies of the sums.

Note that the occurring carrier sets in the continuous cases are linked uniquely to a corresponding asymptotic type. The reason for distinguishing the notations of the objects is that below for corners there are more complicated spaces of coefficients for the asymptotics. Then the present notations correspond to a simpler special case. Now we define (for any cut-off function ω with $\omega \equiv 0$ for $r > 1/2$)

$$\mathcal{H}_P^{s,\beta}(\mathbb{B}) = [\omega]\mathcal{K}_P^{s,\beta}(X^\wedge) + [1-\omega]H^s_{loc}(int\ \mathbb{B}) \tag{14}$$

for $P \in As(X, \mathfrak{a}^\bullet)$ or $P \in As(X, \mathfrak{a})$ (pull-backs to $int\ \mathbb{B}$ from $X \times (0,1)$ are again dropped for abbreviation, cf. also 1.1.(16)). Further we set

$$\mathcal{S}_P^\beta(X^\wedge) = [\omega]\mathcal{K}_P^{\infty,\beta}(X^\wedge) + [1-\omega]H^s_{loc}(int\ \mathbb{B}) \tag{15}$$

for $P \in As(X, \mathfrak{a}^\bullet)$ or $P \in As(X, \mathfrak{a})$. Here $\mathcal{S}(\overline{X}^\wedge) = \mathcal{S}(\overline{\mathbb{R}}_+) \otimes_\pi C^\infty(X)$, $\mathcal{S}(\overline{\mathbb{R}}_+) = \mathcal{S}(\mathbb{R})|_{\overline{\mathbb{R}}_+}$.

Incidentally we will also use the space

$$\mathcal{S}^\beta(X^\wedge) = [\omega]\mathcal{H}^{\infty,\beta}(X^\wedge) + [1-\omega]\mathcal{S}(\overline{X}^\wedge). \tag{16}$$

Definition 1.2.1 *Let $\gamma, \delta \in \mathbb{R}$, $\Theta = (\vartheta, 0]$, $-\infty \leq \vartheta < 0$, $\mathfrak{g} = (\gamma, \delta, \Theta)$. Then $C_G(X^\wedge, \mathfrak{g}^\bullet)$ is defined as the space of all*

$$G \in \bigcap_{s \in \mathbb{R}} \mathcal{L}(\mathcal{K}^{s,\gamma}(X^\wedge), \mathcal{K}^{\infty,\delta}(X^\wedge)) \tag{17}$$

which induce continuous operators

$$G : \mathcal{K}^{s,\gamma}(X^\wedge) \longrightarrow \mathcal{S}_P^\delta(X^\wedge), \tag{18}$$

$$G^* : \mathcal{K}^{s,-\delta}(X^\wedge) \longrightarrow \mathcal{S}_Q^{-\gamma}(X^\wedge) \tag{19}$$

for all $s \in \mathbb{R}$, with certain $P \in As(X, (\delta, \Theta)^\bullet)$, $Q \in As(X, (-\gamma, \Theta)^\bullet)$ which may depend on G.

Here G^* is the formal adjoint of G with respect to the fixed scalar product of $\mathcal{K}^0(X^\wedge)$. Moreover $C_G(X^\wedge, \mathfrak{g})$ is defined as the set of all (17) such that (18), (19) hold for all $s \in \mathbb{R}$ with certain $P \in As(X, (\delta, \Theta)^\bullet)$, $Q \in As(X, (-\gamma, \Theta)^\bullet)$.

The operators in $C_G(X^\wedge, \mathfrak{g}^\bullet)(C_G(X^\wedge, \mathfrak{g}))$ are called Green operators on $X^\wedge$ with discrete (continuous) asymptotics.

Furthermore $C_G(\mathbb{B}, \mathfrak{g}^\bullet)$ is the space of all

$$G \in \bigcap_{s \in \mathbb{R}} \mathcal{L}(\mathcal{H}^{s,\gamma}(\mathbb{B}), \mathcal{H}^{\infty,\delta}(\mathbb{B}))$$

which induce continuous operators

$$G : \mathcal{H}^{s,\gamma}(\mathbb{B}) \longrightarrow \mathcal{H}_P^{\infty,\delta}(\mathbb{B}),$$

$$G^* : \mathcal{H}^{s,-\delta}(\mathbb{B}) \longrightarrow \mathcal{H}_Q^{\infty,-\gamma}(\mathbb{B})$$

for all $s \in \mathbb{R}$, with certain $P \in As(X, (\delta, \Theta)^\bullet)$, $Q \in As(X, (-\gamma, \Theta)^\bullet)$, dependent on G, where G^* is the formal adjoint with respect to the fixed $\mathcal{H}^0(\mathbb{B})$ scalar product. Analogously for $P \in As(X, (\delta, \Theta))$, $Q \in As(X, (-\gamma, \Theta))$ we get by definition the class $C_G(\mathbb{B}, \mathfrak{g})$.

Now we speak about Green operators on $\mathbb{B}$ with discrete and continuous asymptotics, respectively.

The cone ψDO's are defined by means of operator-valued Mellin symbols which also reflect the asymptotics.

Let us introduce the discrete and continuous asymptotic types of Mellin symbols.

$\underline{As}(X)^\bullet$ denotes the set of all sequences

$$P = \{(p_j, m_j, N_j)\}_{j \in \mathbb{Z}} \tag{20}$$

with $p_j \in \mathbb{C}$, $|Re\ p_j| \to \infty$ as $|j| \to \infty$, $m_j \in \mathbb{N}$, $N_j \subset C^\infty(X \times X) \cong L^{-\infty}(X)$ being a finite-dimensional subspace of operators of finite dimension for every $j \in \mathbb{N}$. We set

$$\pi_{\mathbb{C}} P = \{p_j\}_{j \in \mathbb{Z}}.$$

Furthermore a continuous asymptotic type $R \in \underline{As}(X)$ with quasi-discrete

$$carrier\ (R) = V$$

is the set of all sequences

$$R = \{(K_j, \zeta_j)_{j \in \mathbb{Z}} : \zeta_j \in \mathcal{A}'(K_j, C^\infty(X \times X)), j \in \mathbb{Z}\}$$

where

$$V = \sum_{j \in \mathbb{Z}} K_j$$

$$\begin{aligned} sup\{Re\ z : z \in K_j\} &\longrightarrow -\infty \text{ as } j \longrightarrow \infty \\ inf\{Re\ z : z \in K_j\} &\longrightarrow \infty \text{ as } j \longrightarrow -\infty \\ sup\{Re\ z : z \in K_{j+1}\} &< inf\{Re\ z : z \in K_j\} \end{aligned}$$

for all $j \in \mathbb{Z}$. An asymptotic type $R \in \underline{As}(X)$ with arbitrary carrier $V \subset \mathbb{C}$ (which is a closed set with $V^c = V$ and $V \cap \{\alpha \le Re\ w \le \beta\}$ compact for arbitrary reals $\alpha < \beta$) is the set of equivalence class of pairs (R_1, R_2) for $R_i \in \underline{As}(X)$ with quasi-discrete carriers, and

$$(R_1, R_2) \sim (\tilde{R}_1, \tilde{R}_2) \longleftrightarrow V = V_1 + V_2 = \tilde{V}_1 + \tilde{V}_2,$$

$V_i = carrier\ (R_i)$, $\tilde{V}_i = carrier\ (\tilde{R}_i)$, $i = 1, 2$. In this case we also write

$$R = R_1 + R_2. \tag{21}$$

If $A \subseteq \mathbb{C}$ is a set, a $\chi \in C^\infty(\mathbb{C})$ is called an A-excision function if $0 \le \chi \le 1$, $\chi(w) = 0$ for $dist(w, \overline{A}) < \varepsilon$, $\chi(w) = 1$ for $dist(w, \overline{A}) > 2\varepsilon$ with some $\varepsilon > 0$.

Definition 1.2.2 *$M_P^\mu(X)$ for $P \in \underline{As}(X)^\bullet$ being written as (20), $\mu \in \mathbb{R}$, denotes the space of all*

$$a(w) \in \mathcal{A}(\mathbb{C} \setminus \pi_{\mathbb{C}} P, L_{cl}^\mu(X))$$

which are meromorphic with poles at all p_j of multiplicities $m_j + 1$ and Laurent expansions

$$a(w) = \sum_{k=0}^{m_j} \nu_{jk}(w - p_j)^{-(k+1)} + \sum_{k=0}^{\infty} \nu_{jk}^+(w - p_j)^k \tag{22}$$

close to p_j, with $\nu_{jk} \in N_j$, $0 \le k \le m_j$, for all $j \in \mathbb{Z}$, and further

$$(\chi a)(\sigma + i\rho) \in L_{cl}^\mu(X; \mathbb{R}_\rho)$$

uniformly in ρ for $c \le \rho \le c'$ for all $c, c' \in \mathbb{R}$, for every $\pi_{\mathbb{C}} P$-excision function χ. For $\pi_{\mathbb{C}} P = \phi$ we will also write $M_O^\mu(X)$. Now $M_R^\mu(X)$ for $R \in \underline{As}(X)$ with quasi-discrete carrier V is the space of all

$$a(w) \in \mathcal{A}(\mathbb{C} \setminus V, L_{cl}^\mu(X))$$

with $a(w) = a_0(w) + a_1(w)$ for $a_0(w) \in M_O^\mu(X)$, $a_1 \in \mathcal{A}(\mathbb{C} \setminus V, L^{-\infty}(X))$, with $(\chi a)(\sigma + i\rho) \in L_{cl}^\mu(X; \mathbb{R}_\rho)$ uniformly in ρ for $c \le \rho \le c'$, $c, c' \in \mathbb{R}$, and any V-excision function χ. Moreover we set

$$M_R^\mu(X) = M_{R_1}^\mu(X) + M_{R_2}^\mu(X) \tag{23}$$

for arbitrary $R \in \underline{As}(X)$.

The spaces $M^\mu_P(X)$, $P \in \underline{As}(X)^\bullet$, and $M^\mu_R(X)$, $\mathbb{R} \in \underline{As}(X)$ are Fréchet in a natural way. (23) is regarded as a sum of Fréchet spaces. It is independent of the choice of the decomposition (21).

If $a \in M^\mu_P$ we write

$$sg(a) = \pi_{\mathbb{C}} P \text{ for } P \in \underline{As}(X)^\bullet, \ = \ carrier(P) \text{ for } P \in \underline{As}(X).$$

It is convenient to set

$$M^\mu_{as}(X)^\bullet \ = \ \varinjlim_P M^\mu_P(X), \tag{24}$$

$$M^\mu_{as}(X) \ = \ \varinjlim_R M^\mu_R(X) \tag{25}$$

where (24) means the inductive limit over $P \in \underline{As}(X)^\bullet$, (25) the inductive limit over $R \in \underline{As}(X)$.

We shall employ below a parameter-dependent analogue of $M^\mu_O(X)$ with parameters $\eta \in \mathbb{R}^q$.

$M^\mu_O(X;\mathbb{R}^q)$ denotes the space of all

$$a(w,\eta) \in \mathcal{A}(\mathbb{C}, L^\mu_{cl}(X;\mathbb{R}^q))$$

such that

$$a(\sigma + i\rho, \eta) \in L^\mu_{cl}(X;\mathbb{R}_\rho \times \mathbb{R}^q_\eta)$$

uniformly in ρ for $c \le \rho \le c'$ for all $c, c' \in \mathbb{R}$. The space $M^\mu_O(X;\mathbb{R}^q)$ is also Fréchet in a natural way.

Definition 1.2.3 *Let $\mu, \nu, \beta \in \mathbb{R}$, $\mu - \nu \in \mathbb{N}$, $\Theta = (-k, 0]$ for $k \in \mathbb{N}\setminus\{0\}$, and $\mathfrak{g} = (\beta, \beta-\mu, \Theta)$. Then $C^\nu(\mathbb{B}, \mathfrak{g}^\bullet)$ is the set of all operators*

$$A = A_0 + A_1 + M + G \tag{26}$$

with

$$A_0 \ = \ \omega r^{-\nu} op_M^{\beta - n/2}(h)\omega_0, \ h(r,w) \in C^\infty(\overline{\mathbb{R}}_+, M^\nu_O(X)), \tag{27}$$

$$A_1 \ = \ (1-\omega)P(1-\omega_1), \ P \in L^\nu_{cl}(int\ \mathbb{B}), \tag{28}$$

$$M \ = \ \omega_2 \left\{ r^{-\nu} \sum_{0 \le j \le k-\mu+\nu-1} r^j \, op_M^{\beta_j - n/2}(h_j) \right\} \omega_3 \tag{29}$$

where $h_j(w) \in M^{-\infty}_{as}(X)^\bullet$, $\beta_j \in \mathbb{R}$,

$$\beta - (\mu - \nu) - j \le \beta_j \le \beta, \tag{30}$$

$sg(h_j) \cap \Gamma_{(n+1)/2 - \beta_j} = \phi$ for all j, and $G \in C_G(\mathbb{B}, \mathfrak{g}^\bullet)$.

Moreover $C^\nu(\mathbb{B}, \mathfrak{g})$ is the set of all operators (26) with A_0, A_1 being given by (27), (28), further

$$M = \omega_2 \left\{ r^{-\nu} \sum_{0 \le j \le k-\mu+\nu-1} r^j \sum_{i=0}^{m} op_M^{\beta_{j,i} - n/2}(h_{j,i}) \right\} \omega_3 \tag{31}$$

with some $m = m(A) \in \mathbb{N}$, $h_{j,i}(w) \in M^{-\infty}_{as}(X)$, $\beta_{j,i} \in \mathbb{R}$ with

$$\beta - (\mu - \nu) - j \le \beta_{j,i} \le \beta,$$

$sg(h_{j,i}) \cap \Gamma_{(n+1)/2 - \beta_{j,i}} = \phi$ for all i, j, and $G \in C_G(\mathbb{B}, \mathfrak{g})$.

The cut-off functions ω, ω_i are arbitrary. We shall assume without loss of generality that

$$\omega\omega_0 = \omega, \ \omega\omega_1 = \omega_1 \tag{32}$$

and

$$\varphi^*(P|_{int\ V}) \sim r^{-\nu} op_M^{\beta-n/2}(h)|_{X\times(0,1)}, \tag{33}$$

where V is the tubular neighbourhood of $\partial\mathbb{B}$ in $\mathbb{B}$, $\varphi : int\ V \to X \times (0,1)$ being the fixed diffeomorphism. $\sim$ in (33) means equality *mod* $L^{-\infty}(X \times (0,1))$.

Finally we define also the classes

$$C^\nu(X^\wedge, \mathfrak{g}^\bullet)\ ,\ C^\nu(X^\wedge, \mathfrak{g}) \tag{34}$$

by inserting in (26) $G \in C_G(X^\wedge, \mathfrak{g}^\bullet)$ and $G \in C_G(X^\wedge, \mathfrak{g})$, respectively, and $P \in L^\nu_{cl}(X^\wedge)$. Here $L^\nu_{cl}(X^\wedge)$ are the classical ψDO's over $X^\wedge$ satisfying for $r \to \infty$ the 'exit condition', cf. [S1], [S4], II., [C1], [S7], and Definition 1.3.12 below. Let us recall the definition of the latter class for $X^\wedge = S^n \times \mathbb{R}_+ \cong \mathbb{R}^{n+1} \setminus \{0\} \ni \tilde{x}$. The analogue for arbitrary X then follows easily by applying restrictions to coordinate neighbourhoods on U on S^n, push-forwards under diffeomorphisms to corresponding neighbourhoods on X and a globalization in terms of a partition of unity with respect to an open covering of X. The specific things concern the region for large modules of variables and covariables. Thus it sufficies to look at symbols in $\mathbb{R}^{n+1}_{\tilde{x}} \times \mathbb{R}^{n+1}_{\tilde{\xi}} \ni (\tilde{x}, \tilde{\xi})$. First $S^{\mu,l}(\mathbb{R}^{n+1} \times \mathbb{R}^{n+1})$ for arbitrary $\mu, l \in \mathbb{R}$ is the subspace of all $a(\tilde{x}, \tilde{\xi}) \in S^\nu(\mathbb{R}^{n+1} \times \mathbb{R}^{n+1})$ such that

$$|D^\alpha_{\tilde{x}} D^\beta_{\tilde{\xi}} a(\tilde{x}, \tilde{\xi})| \leq c(1 + |\tilde{\xi}|)^{\nu-|\beta|}(1 + |\tilde{x}|)^{l-|\alpha|}$$

for all $\alpha, \beta \in \mathbb{N}^{n+1}$ with constants $c = c(\alpha, \beta)$ for all $\tilde{\nu} \in \mathbb{R}^{n+1}$, $\tilde{\xi} \in \mathbb{R}^{n+1}$. Moreover $S^{\nu,l}_{cl}(\mathbb{R}^{n+1} \times \mathbb{R}^{n+1})$ is the subspace of all

$$a(\tilde{x}, \tilde{\xi}) \in S^{\nu,l}(\mathbb{R}^{n+1} \times \mathbb{R}^{n+1}) \cap S^\nu_{cl}(\mathbb{R}^{n+1} \times \mathbb{R}^{n+1})$$

(where cl refers to $\tilde{\xi}$) such that

(i) there are elements $a_j(\tilde{x}/|\tilde{x}|, \tilde{\xi}) \in C^\infty(S^n, S^\nu_{cl}(\mathbb{R}^{n+1}))$ such that for all $N \in \mathbb{N}$

$$a(\tilde{x}, \tilde{\xi}) - \sum_{j=0}^{N} \chi(\tilde{x})|\tilde{x}|^{-j} a_j(\tilde{x}/|\tilde{x}|, \tilde{\xi}) \in S^{\nu,l-(N+1)}(\mathbb{R}^{n+1} \times \mathbb{R}^{n+1})$$

with any excision function $\chi(\tilde{x})$,

(ii) if

$$r(\tilde{x}, \tilde{\xi}) = \sum_{j=0}^{\infty} \chi(\tilde{x}/c_j)|\tilde{x}|^{-j} a_j(\tilde{x}/|\tilde{x}|, \tilde{\xi})$$

converges in $S^{\nu,l}(\mathbb{R}^{n+1} \times \mathbb{R}^{n+1}) \cap S^\nu_{cl}(\mathbb{R}^{n+1} \times \mathbb{R}^{n+1})$ for some choice of constants c_j (those c_j do always exist) then the homogeneous components of $a(\tilde{x}, \tilde{\xi}) - r(\tilde{x}, \tilde{\xi})$ in $\tilde{\xi}$ of orders $\nu - j$ are of order $-\infty$ in $\tilde{x}$ for all $j \in \mathbb{N}$

(iii) analogous conditions are satisfied for $b(\tilde{x}, \tilde{\xi}) := a(\tilde{x}, \tilde{\xi})$.

Now, the class with exit condition is by definition $S^{\nu,0}_{cl}$ over $(\mathbb{R}^{n+1} \setminus \{0\}) \times \mathbb{R}^{n+1}$ where the conditions apart from being classical in $\tilde{\xi}$ are required for $|\tilde{x}| > \varepsilon$ for every $\varepsilon > 0$. The element $a_0(\tilde{x}/|\tilde{x}|, \tilde{\xi})$ is called the exit symbol of a.

Remark 1.2.4 *The operator spaces*

$$C^\nu(\mathbb{B}, \mathfrak{g}^\bullet),\ C^\nu(\mathbb{B}, \mathfrak{g}),\ C^\nu(X^\wedge, \mathfrak{g}^\bullet),\ C^\nu(X^\wedge, \mathfrak{g}) \tag{35}$$

have natural locally convex topologies (they are inductive limits of Fréchet spaces), cf. [S1], [S4]; II, III. The definitions extend to the corresponding classes for $\mathfrak{g} = (\beta, \beta - \mu, \Theta)$ *with* $\Theta = (-\infty, 0]$, *namely as projective limit over those for* $\Theta_k = (-k, 0]$, $k \in \mathbb{N} \setminus \{0\}$.

We shall not repeat here once again all elements of the theory of the operator spaces (35). For references below let us only make the following

Remark 1.2.5 *From*

$$A \in C^{\nu}(\mathbb{B}, \mathfrak{g}) \subset L^{\nu}_{cl}(int\ \mathbb{B})$$

it follows the homogeneous principal symbol of order ν

$$\sigma^{\nu}_{\psi}(A) \in g_0^{-\nu} S^{(\nu)}_b(T^*(int\ \mathbb{B}) \setminus 0) \tag{36}$$

(cf. 1.1.(22)) and the sequence of conormal symbols

$$\sigma^{\nu-j}_M(A)(w) = \frac{1}{j!}(\frac{\partial^j h}{\partial r^j})(0, w) + h_j(w) \tag{37}$$

$j = 0, \dots k - \mu + \nu - 1$ *(cf. 1.1.(19) for* $j = 0$*). These symbolic levels are compatible with formal adjoints and composition of operators. An analogous remark holds for* $X^\wedge$ *instead of* $\mathbb{B}$ *where we have in addition the exit symbolic level (that with respect* $r \to \infty$*), cf. [S1], [S4], II, IV.*

Remark 1.2.6 *Definition 1.1.2 has a natural generalization to* $C^{\mu}(\mathbb{B}, \mathfrak{g}^\bullet)$, $C^{\mu}(\mathbb{B}, \mathfrak{g})$*. Then, an analogue of Theorem 1.1.3 holds, and the parametrices of elliptic operators belong to* $C^{-\mu}(\mathbb{B}, \mathfrak{h})$ *and* $C^{-\mu}(\mathbb{B}, \mathfrak{h})$*, respectively, for* $\mathfrak{h} = (\beta - \mu, \beta, \Theta)$*, cf. [S1], [S4], II, IV. An analogous remark holds for the corresponding operator classes over* $X^\wedge$*, where here the additional exit symbol for* $r \to \infty$ *is involved, cf. [S1], [S4], I, IV.*

1.3 Operators on manifolds with edges

This section shall present the material on ψDO's along the edges that are emanated from the corners. Since this theory is of independent interest, we will allow here the edges to be of arbitrary dimension q.

A "manifold" W with edges $Y \subset W$ (where Y is a C^∞ manifold of dimension $n + 1 + q$) is locally close to any $y \in Y$ of the form

$$\{(X \times [0, 1))/(X \times \{0\})\} \times \Omega \tag{1}$$

with an open neighbourhood Ω of y in Y, and a closed compact C^∞ manifold X, $n = dimX$, the base of the (open stretched) model cone. It is custom to pass to the stretched manifold $\mathbb{W}$ of W which is in this case a C^∞ manifold with C^∞ boundary. $\mathbb{W}$ is close to $\partial\mathbb{W}$ of the form

$$X \times [0, 1) \times \Omega, \tag{2}$$

there is a canonical "projection" $\mathbb{W} \to W$ which restricts to a diffeomorphism $\mathbb{W} \setminus \partial\mathbb{W} \to W \setminus Y$, under which (2) is just mapped to (1).

A global operator calculus with Fredholm property under ellipticity would suppose Y to be closed, compact (cf. [S1], [S5]). Our application for $q = 1$ will concern $\mathbb{W} = p^{-1}(K \setminus K_0)$, cf. 1.1.(25) where $Y = K_1 \setminus K_0$ is not compact. The main point in this case shall be the local picture. Close to corners the wedge calculus needs a Mellin variant along $K_1 \setminus K_0$. This will be subject of the following section, whereas now the theory is based on the Fourier transform along the edges. The ψDO's on $int\ \mathbb{W}$ shall form a class

$$Y^{\nu}(\mathbb{W}, \mathfrak{g}) \subset L^{\nu}_{cl}(int\ \mathbb{W})$$

($\mathfrak{g} = (\beta, \beta - \mu, \Theta)$ being the given weight data). In addition we will define a class of matrix-valued operators

$$\mathcal{Y}^{\nu}(\mathbb{W}, \mathfrak{g}; \mathbb{C}^N, \mathbb{C}^M) \ni \mathcal{A} = \begin{pmatrix} A & K \\ T & Q \end{pmatrix} \tag{3}$$

with $Y^{\nu}(\mathbb{W}, \mathfrak{g})$ being the class of occurring left upper corners. T and K play the role of additional trace and potential operators, respectively, along the edge. Q is a ψDO of order ν on the edge.

Since the specific information is located near $\partial\mathbb{W}$ and in view of the invariance it is allowed to look at the theory first on $X^{\wedge} \times \Omega$. This can be performed to some extent as a ψDO calculus along Ω with operator-valued symbols.

The technicalities for proving the assertions in this section may be found in [S1].

Let us now remind of the abstract wedge Sobolev spaces. Let E be a Banach space and $\{\kappa_\lambda\}_{\lambda \in \mathbb{R}_+}$ be a fixed representation of $\mathbb{R}_+$ in $\mathcal{L}(E)$ (i.e. $\kappa_\lambda \kappa_\mu = \kappa_{\lambda\mu}$, $\kappa_\lambda^{-1} = \kappa_{\lambda^{-1}}$ for all $\lambda, \mu \in \mathbb{R}_+$), where $\{\kappa_\lambda\} \in C(\mathbb{R}_+, \mathcal{L}_\sigma(E))$ (σ indicates the strong operator topology in $\mathcal{L}(E)$). Fix a strictly positive function $\eta \to [\eta]$ in $\mathbb{R}^q$ which is C^∞ and $[\eta] = |\eta|$ for $|\eta| > const$ for a *constant* > 0. Set $\kappa(\eta) = \kappa_{[\eta]}$. Let $F = F_{y \to \eta}$ denote the Fourier transform in $\mathbb{R}^q$.

Definition 1.3.1 *$\mathcal{W}^s(\mathbb{R}^q, E)$ for $s \in \mathbb{R}$ is defined as the completion of $\mathcal{S}(\mathbb{R}^q, E)$ with respect to the norm*

$$||u||_{\mathcal{W}^s(\mathbb{R}^q,E)} = \{\int [\eta]^{2s} ||\kappa(\eta)^{-1}(F_{y\to\eta}u)(\eta)||_E^2 d\eta\}^{1/2}. \tag{4}$$

Here $\mathcal{S}(\mathbb{R}^q, E)$ denotes the Schwartz space of E-valued functions on $\mathbb{R}^q$. For $\kappa_\lambda = identity$ for all $\lambda \in \mathbb{R}_+$ we will also write $H^s(\mathbb{R}^q, E)$.

Example 1.3.2 *Let $E = \mathcal{K}^{s,\beta}(X^\wedge)$, $s, \beta \in \mathbb{R}$, and $(\kappa_\lambda u)(x, r) = \lambda^{(n+1)/2} u(x, \lambda r)$, $\lambda \in \mathbb{R}_+$. Then we get the wedge Sobolev space (of smoothness s and weight β)*

$$\mathcal{W}^{s,\beta}(X^\wedge \times \mathbb{R}^q) := \mathcal{W}^s(\mathbb{R}^q, \mathcal{K}^{s,\beta}(X^\wedge)). \tag{5}$$

Here

$$\mathcal{W}^{s,\beta}(X^\wedge \times \mathbb{R}^q) \subset H^s_{loc}(X^\wedge \times \mathbb{R}^q) \tag{6}$$

(where loc on the right of (5) is to be interpreted in the standard sense). Moreover

$$\mathcal{W}^s(X^\wedge \times \mathbb{R}^q) := \mathcal{W}^{s,0}(X^\wedge \times \mathbb{R}^q)$$

satisfies for $s = 0$

$$\mathcal{W}^0(X^\wedge \times \mathbb{R}^q) = \mathcal{W}^0(\mathbb{R}^q, r^{-n/2}L^2(X \times \mathbb{R}_+)) = L^2(\mathbb{R}^q, r^{-n/2}L^2(X \times \mathbb{R}_+)). \tag{7}$$

Remark 1.3.3 *The operator $T = F^{-1}_{\eta\to y}\kappa(\eta)^{-1}F_{y\to\eta}$ induces an isomorphism*

$$T : \mathcal{W}^s(\mathbb{R}^q, E) \longrightarrow H^s(\mathbb{R}^q, E). \tag{8}$$

This allows an extension of the definition of $\mathcal{W}^s(\mathbb{R}^q, .)$ to vector subspaces $E_1 \subseteq E$ that are not necessarily preserved under κ_λ, namely

$$\mathcal{W}^s(\mathbb{R}^q, E_1) := T^{-1}H^s(\mathbb{R}^q, E_1). \tag{9}$$

Here E_1 is supposed to be a Banach space, continuously embedded in E. Then (9) is again a Banach (Fréchet) space in the topology induced by the bijection T. Moreover it follows for two such subspaces $E_0, E_1 \subseteq E$

$$\mathcal{W}^s(\mathbb{R}^q, E_0 + E_1) = \mathcal{W}^s(\mathbb{R}^q, E_0) + \mathcal{W}^s(\mathbb{R}^q, E_1) \tag{10}$$

with the (non-direct) sums, defined by 1.1.(2).

Example 1.3.4 *Let $P \in As(X, \mathfrak{a}^\bullet)$ or $P \in As(X, \mathfrak{a})$, $\mathfrak{a} = (\beta, \Theta)$, Θ finite, and insert in (10)*

$$E_0 = \mathcal{K}_\Theta^{s,\beta}(X^\wedge), \; E_1 = \mathcal{E}_P(X^\wedge). \tag{11}$$

Then (cf. 1.2.(12),(13))

$$\mathcal{W}^s(\mathbb{R}^q, \mathcal{K}_P^{s,\beta}(X^\wedge)) = \mathcal{W}^s(\mathbb{R}^q, \mathcal{K}_\Theta^{s,\beta}(X^\wedge)) + \mathcal{W}^s(\mathbb{R}^q, \mathcal{E}_P(X^\wedge)).$$

We shall also use the notations

$$\mathcal{W}_P^{s,\beta}(X^\wedge \times \mathbb{R}^q) \;=\; \mathcal{W}^s(\mathbb{R}^q, \mathcal{K}_P^{s,\beta}(X^\wedge)), \tag{12}$$

$$\mathcal{W}_\Theta^{s,\beta}(X^\wedge \times \mathbb{R}^q) \;=\; \mathcal{W}^s(\mathbb{R}^q, \mathcal{K}_\Theta^{s,\beta}(X^\wedge)). \tag{13}$$

(12) will be regarded as wedge Sobolev space with (edge-) asymptotics of type P, (13) as wedge Sobolev space of (edge-) flatness order $-\vartheta$ (relative to the weight β, $\Theta = (\vartheta, 0]$, $\vartheta < 0$). The space

$$\mathcal{W}^s(\mathbb{R}^q, \mathcal{E}_P(X^\wedge))$$

consists just of the "singular functions" of the edge asymptotics, namely

$$F^{-1}_{\eta \to y}\left\{[\eta]^{(n+1)/2} < \hat{\zeta}(\eta), (r[\eta])^{-w} > \omega(r[\eta])\right\} \tag{14}$$

for $P \in As(X, \mathfrak{a})$ where $\zeta(y)$ runs over $H^s(\mathbb{R}^q, \mathcal{A}'(K, C^\infty(X)))$, $K = carrier(P)$ and

$$F^{-1}_{\eta \to y}\left\{[\eta]^{(n+1)/2}(r[\eta])^{-p_j}\sum_{k=0}^{m_j}\hat{\zeta}_{jk}(x,\eta)\log^k(r[\eta])\omega(r[\eta])\right\} \tag{15}$$

for $P \in As(X, \mathfrak{a}^\bullet)$, $P = \{(p_j, M_j, L_j)\}_{j=0,\ldots,N(P)}$, $\zeta_{jk}(x,y) \in H^s(\mathbb{R}^q, L_j)$, $0 \le k \le m_j$, $j = 0, \ldots, N = N(P)$, $\tilde{\zeta}(x,\eta) = F_{y \to \eta}\zeta(x,y)$. For finite Θ we can write

$$u \sim \sum_{j=0}^{\infty} f_j \tag{16}$$

where for $P \in As(X, \mathfrak{a})$ the f_j are the singular functions associated with compact sets $K_j \subset \mathbb{C}_w$, satisfying 1.2.(6). For $P \in As(X, \mathfrak{a}^\bullet)$ the f_j are given by (15). The interpretation of (16) is that

$$u - \sum_{j=0}^{N} f_j \in \mathcal{W}^{s,\beta}(X^\wedge \times \mathbb{R}^q)$$

for every $\beta \in \mathbb{R}$ with some $N = N(\beta)$.

Remark 1.3.5 *For every $P \in As(X, \mathfrak{a}^\bullet)$ or $P \in As(X, \mathfrak{a})$ there exists a sequence of Banach spaces $\{E^j\}_{j \in \mathbb{N}}$ with $E^{j+1} \subseteq E^j \subseteq \mathcal{K}^{s,\beta}(X^\wedge)$, $\{\kappa_\lambda\} \in C(\mathbb{R}^q, \mathcal{L}_\sigma(E^j))$ for all j, and*

$$\mathcal{K}_P^{s,\beta}(X^\wedge) = \varprojlim_j E^j.$$

Then

$$\mathcal{W}_P^{s,\beta}(X^\wedge \times \mathbb{R}^q) = \varprojlim_j \mathcal{W}^s(\mathbb{R}^q, E^j).$$

It is necessary also to have the *comp, loc* versions of our spaces. The definitions rely on the fact that it is allowed to multiply by functions in $C_0^\infty(\mathbb{R}^q)$.

Definition 1.3.6 *Let $\Omega \subseteq \mathbb{R}^q$ be open. Then $\mathcal{W}^s_{loc}(\Omega, E)$ is defined as the subspace of all $u \in \mathcal{D}'(\Omega, E)$ such that $\varphi u \in \mathcal{W}^s(\mathbb{R}^q, E)$ for every $\varphi \in C_0^\infty(\Omega)$ (here φu is identified with the corresponding extension by zero to $\mathbb{R}^q \setminus \Omega$). Furthermore $\mathcal{W}^s_{comp}(\Omega, E)$ is the subspace of all $\mathcal{W}^s_{loc}(\Omega, E)$ which have compact support in Ω. In an analogous manner we obtain the spaces*

$$\mathcal{W}^s_{comp}(\Omega, E_1),\ \mathcal{W}^s_{loc}(\Omega, E_1)$$

in the sense of (9). For E_1 given by (11) we will also set

$$\mathcal{W}^{s,\beta}_{P,loc}(X^\wedge \times \Omega) := \mathcal{W}^s_{loc}(\Omega, \mathcal{K}^{s,\beta}_P(X^\wedge)) \tag{17}$$

and analogously with comp.

Moreover the invariance of our definitions under diffeomorphisms $\Omega \overset{\rightarrow}{\cong} \tilde{\Omega}$ and the property (6) allows to introduce the spaces

$$\mathcal{W}^{s,\beta}_{comp}(\mathbb{W}),\ \mathcal{W}^{s,\beta}_{loc}(\mathbb{W}) \tag{18}$$

as well as

$$\mathcal{W}^{s,\beta}_{P,comp}(\mathbb{W}),\ \mathcal{W}^{s,\beta}_{P,loc}(\mathbb{W}) \tag{19}$$

in a straightforward way. Here $P \in As(X, (\beta, \Theta))$ or $P \in As(X, (\beta, \Theta)^\bullet)$ are assumed to satisfy a "shadow condition" which means, for discrete asymptotic types, that

$$(p, m, L) \in P \longrightarrow (p - j, m(j), L(j)) \in P$$

for all $j \in \mathbb{N}$ with $Re\, p - j > (n+1)/2 - \beta + \vartheta$, $\Theta = (\vartheta, 0]$, and certain $m(j) \geq m$, $L(j) \supseteq L$. In the continuous case it means that

$$T^{-j}(carrier(P)) \cap \{(n+1)/2 - \beta + \vartheta \leq Re\, w < (n+1)/2 - \beta\} \subset carrier(P) \tag{20}$$

for all $j \in \mathbb{N}$; $T^{-j}A := \{w - j : w \in A\}$, (cf. 1.2.(9)).

Remark 1.3.7 *The identity (7) gives rise to a natural scalar product in $\mathcal{W}^0(X^\wedge \times \mathbb{R}^q)$. By restricting the Lebesgue measure on $\mathbb{R}^q$ to Ω we also get a scalar product in*

$$\mathcal{W}^0(X^\wedge \times \Omega) = \mathcal{W}^0(\Omega, r^{-n/2} L^2(X \times \mathbb{R}_+)). \tag{21}$$

If we fix a Riemannian metric on Y we obtain furthermore the space

$$\mathcal{W}^0(\mathbb{W}) \tag{22}$$

with a fixed scalar product.

Let $E, \tilde{E}$ be Banach spaces and $\{\kappa_\lambda\} \in C(\mathbb{R}_+, \mathcal{L}_\sigma(E)), \{\tilde{\kappa}_\lambda\} \in C(\mathbb{R}_+, \mathcal{L}_\sigma(\tilde{E}))$ be corresponding fixed groups of operators.

Definition 1.3.8 *$S^\mu(\Omega \times \mathbb{R}^q; E, \tilde{E})$ for open $\Omega \subseteq \mathbb{R}^p, \mu \in \mathbb{R}$, denotes the set of all $a(y, \eta) \in C^\infty(\Omega, \mathcal{L}(E, \tilde{E}))$ with*

$$||\tilde{\kappa}(\eta)^{-1}(D^\alpha_y D^\beta_\eta a(y, \eta))\kappa(\eta)||_{\mathcal{L}(E,\tilde{E})} \leq c[\eta]^{\mu - |\beta|} \tag{23}$$

for all $\alpha \in \mathbb{N}^p$, $\beta \in \mathbb{N}^q$, $y \in K \subset\subset \Omega$, with constants $c = c(\alpha, \beta, K)$.

The only occurring cases here are $\Omega \subseteq \mathbb{R}^q$ or $\Omega^2 := \Omega \times \Omega \subseteq \mathbb{R}^{2q}$. In the latter case we will also write $(y, y') \in \Omega^2$ when $y \in \Omega$, $y' \in \Omega$.

We will not repeat here the whole material on spaces of amplitude functions and ψDO's along Ω which is analogous to the standard calculus of ψDO's with scalar symbols. More specific things connected with the nature of $\{\kappa_\lambda\}_{\lambda\in\mathbb{R}_+}$ and the operator-valued case may be found in [S1], [S2], [S4].

For a Fréchet space $\tilde{E}$, written as projective limit of Banach spaces $\tilde{E}^j$, $j \in \mathbb{N}$, $\tilde{E}^{j+1} \subseteq \tilde{E}^j \subseteq \tilde{E}$, $\{\kappa_\lambda\}_{\lambda\in\mathbb{R}_+} \in C(\mathbb{R}_+, \mathcal{L}_\sigma(\tilde{E}^j))$ for all j we write

$$S^\mu(\Omega \times \mathbb{R}^q; E, \tilde{E}) = \varprojlim_j S^\mu(\Omega \times \mathbb{R}^q; E, \tilde{E}^j). \tag{24}$$

Analogous notations will be used for classical subspaces, indicated by *cl*. Here

$$S^\mu_{cl}(\Omega \times \mathbb{R}^q; E, \tilde{E}) \tag{25}$$

is the subspace of all $a(y,\eta) \in S^\mu(\Omega \times \mathbb{R}^q; E, \tilde{E})$ with

$$a(y,\eta) \sim \sum_{j=0}^{\infty} \chi(\eta) a_{(\mu-j)}(y,\eta), \tag{26}$$

$\chi(\eta)$ being an excision function (i.e. $\chi \in C^\infty$, $\chi = 0$ close to $y = 0$, $\chi = 1$ for $|\eta| > const$) and $a_{(\mu-j)}(y,\eta) \in C^\infty(\Omega \times (\mathbb{R}^q \setminus \{0\}))$ with

$$a_{(\mu-j)}(y,\lambda\eta) = \lambda^{\mu-j} \tilde{\kappa}_\lambda a_{(\mu-j)}(y,\eta)\kappa_\lambda^{-1} \tag{27}$$

for all $y \in \Omega$, $\eta \in \mathbb{R}^q \setminus \{0\}$, $\lambda \in \mathbb{R}_+$.

Set

$$Op(a)u(y) = (2\pi)^{-q} \int\int e^{i(y-y')\eta} a(y,y',\eta) u(y')\, dy'\, d\eta, \tag{28}$$

first regarded as (continuous) operator $C_0^\infty(\Omega, E) \to C^\infty(\Omega, \tilde{E})$.

$L^\mu(\Omega; E, \tilde{E})$ ($L^\mu_{cl}(\Omega; E, \tilde{E})$) is defined as the class of all operators (28) with arbitrary $a(y,y',\eta) \in S^\mu(\Omega^2 \times \mathbb{R}^q; E, \tilde{E})$ ($\in S^\mu_{cl}(\Omega^2 \times \mathbb{R}^q; E, \tilde{E})$). Then

$$L^{-\infty}(\Omega; E, \tilde{E}) \cong C^\infty(\Omega \times \Omega; \mathcal{L}(E, \tilde{E})).$$

For $a(y,y',\eta) \in S^\mu_{cl}(\Omega^2 \times \mathbb{R}^q; E, \tilde{E})$ we define

$$\sigma^\mu_\wedge(a)(y,\eta) = a_{(\mu)}(y,y,\eta), \ (y,\eta) \in T^*\Omega \setminus 0,$$

and we set

$$\sigma^\mu_\wedge(A)(y,\eta) = \sigma^\mu_\wedge(a)(y,\eta) \text{ for } A = Op(a). \tag{29}$$

This is called the (abstract) homogeneous principal edge symbol of A of order μ.

The distributional kernels of (28) may be understood as $\mathcal{L}(E, \tilde{E})$-valued ones. Then we can talk about properly supported operators with respect to (y,y')-variables.

Every $A \in L^\mu(\Omega; E, \tilde{E})$ induces continuous operators

$$A : \mathcal{W}^s_{comp}(\Omega, E) \longrightarrow \mathcal{W}^{s-\mu}_{loc}(\Omega, \tilde{E}) \tag{30}$$

for all $s \in \mathbb{R}$. We may write *comp* or *loc* on both sides of (30) if A is properly supported.

Definition 1.3.9 *Let $\mu, \beta \in \mathbb{R}$, $\Theta = (\vartheta, 0]$, $-\infty \leq \vartheta < 0$, $\mathfrak{g} = (\beta, \beta - \mu, \Theta)$. Then*

$$\mathcal{Y}_G^{-\infty}(X^\wedge \times \Omega, \mathfrak{g}; \mathbb{C}^N, \mathbb{C}^M)$$

for given fixed $M, N \in \mathbb{N}$ denotes the space of all

$$\mathcal{G} \in \bigcap_{s\in\mathbb{R}} \mathcal{L}(\mathcal{W}^{s,\beta}_{comp}(X^\wedge \times \Omega) \oplus H^s_{comp}(\Omega, \mathbb{C}^N), \mathcal{W}^{\infty,\beta-\mu}_{loc}(X^\wedge \times \Omega) \oplus H^\infty_{loc}(\Omega, \mathbb{C}^M))$$

which induce continuous operators

$$\begin{aligned} \mathcal{G} &: \mathcal{W}^{s,\beta}_{comp}(X^\wedge \times \Omega) \oplus H^s_{comp}(\Omega, \mathbb{C}^N) \longrightarrow \mathcal{W}^{\infty,\beta-\mu}_{P,loc}(X^\wedge \times \Omega) \oplus H^\infty_{loc}(\Omega, \mathbb{C}^M), \\ \mathcal{G}^* &: \mathcal{W}^{s,-\beta+\mu}_{comp}(X^\wedge \times \Omega) \oplus H^s_{comp}(\Omega, \mathbb{C}^M) \longrightarrow \mathcal{W}^{\infty,-\beta}_{Q,loc}(X^\wedge \times \Omega) \oplus H^\infty_{loc}(\Omega, \mathbb{C}^N) \end{aligned}$$

for all $s \in \mathbb{R}$, for certain $P \in As(X, \mathfrak{a})$ with $\mathfrak{a} = (\beta - \mu, \Theta)$, $Q \in As(X, \mathfrak{b})$ with $\mathfrak{b} = (-\beta, \Theta)$, dependent on $\mathcal{G}$.

$\mathcal{G}^*$ is the formal adjoint of $\mathcal{G}$ in the sense

$$(\mathcal{G}u, v)_{\mathcal{W}^0(X^\wedge \times \Omega) \oplus H^0(\Omega, \mathbb{C}^M)} = (u, \mathcal{G}^* v)_{\mathcal{W}^0(X^\wedge \times \Omega) \oplus H^0(\Omega, \mathbb{C}^N)}$$

for all $u \in C_0^\infty(X^\wedge \times \Omega) \oplus C_0^\infty(\Omega, \mathbb{C}^N)$, $v \in C_0^\infty(X^\wedge \times \Omega) \oplus C_0^\infty(\Omega, \mathbb{C}^M)$.

It will be convenient from time to time to set for abbreviation

$$\mathcal{K}^{s,\beta} = \mathcal{K}^{s,\beta}(X^\wedge), \tag{31}$$

$$\mathcal{S}^\beta_P = \mathcal{S}^\beta_P(X^\wedge), \ \mathcal{S}^\beta = \mathcal{S}^\beta(X^\wedge) \tag{32}$$

(c.f. 1.2.(16)).

Definition 1.3.10 *Let $\mu, \nu, \beta \in \mathbb{R}$, $\mu - \nu \in \mathbb{N}$, $\mathfrak{g} = (\beta, \beta - \mu, \Theta)$, $\Theta = (\vartheta, 0]$, $-\infty \le \vartheta < 0$. Then*

$$\mathcal{R}^\nu_G(\Omega^2 \times \mathbb{R}^q, \mathfrak{g}; \mathbb{C}^N, \mathbb{C}^M)$$

denotes the space of all

$$g(y, y', \eta) \in \bigcap_{s \in \mathbb{R}} S^\nu_{cl}(\Omega^2 \times \mathbb{R}^q; \mathcal{K}^{s,\beta} \oplus \mathbb{C}^N, \mathcal{K}^{\infty,\beta-\mu} \oplus \mathbb{C}^M)$$

with

$$\begin{aligned} g(y, y', \eta) &\in \bigcap_{s \in \mathbb{R}} S^\nu_{cl}(\Omega^2 \times \mathbb{R}^q; \mathcal{K}^{s,\beta} \oplus \mathbb{C}^N, \mathcal{S}^{\beta-\mu}_P \oplus \mathbb{C}^M), \\ g^*(y, y', \eta) &\in \bigcap_{s \in \mathbb{R}} S^\nu_{cl}(\Omega^2 \times \mathbb{R}^q; \mathcal{K}^{s,-\beta+\mu} \oplus \mathbb{C}^M, \mathcal{S}^{-\beta}_Q \oplus \mathbb{C}^N) \end{aligned}$$

for certain P, Q as in Definition 1.3.9, dependent on g, with g^* being the point-wise formal adjoint in the sense

$$(g\varphi, \psi)_{\mathcal{K}^0(X^\wedge) \oplus \mathbb{C}^M} = (\varphi, g^*\psi)_{\mathcal{K}^0(X^\wedge) \oplus \mathbb{C}^N}$$

for all $\varphi \in C_0^\infty(X^\wedge) \oplus \mathbb{C}^N$, $\psi \in C_0^\infty(X^\wedge) \oplus \mathbb{C}^M$. The y'-independent $g(y, \eta)$ constitute the class $\mathcal{R}^\nu_G(\Omega \times \mathbb{R}^q, \mathfrak{g}; \mathbb{C}^N, \mathbb{C}^M)$.

We always assume that κ_λ and $\tilde{\kappa}_\lambda$, ... act as the identity on $\mathbb{C}^N$ and $\mathbb{C}^M$, respectively, for all $\lambda \in \mathbb{R}_+$.

Let us set

$$\mathcal{R}^\nu_G(\Omega^2 \times \mathbb{R}^q, \mathfrak{g}) = \ l.u.c.\ \mathcal{R}^\nu_G(\Omega^2 \times \mathbb{R}^q, \mathfrak{g}; \mathbb{C}^N, \mathbb{C}^M) \tag{33}$$

where *l.u.c.* indicates the set of occurring left upper corners of the matrices.

The asssociated operators form the classes

$$\begin{aligned} &\mathcal{Y}^\nu_G(X^\wedge \times \Omega, \mathfrak{g}; \mathbb{C}^N, \mathbb{C}^M) := \\ &\{Op(g) + \mathcal{G}_0 : g \in \mathcal{R}^\nu_G(\Omega^2 \times \mathbb{R}^q, \mathfrak{g}; \mathbb{C}^N, \mathbb{C}^M), \mathcal{G}_0 \in \mathcal{Y}^{-\infty}_G(X^\wedge \times \Omega, \mathfrak{g}; \mathbb{C}^N, \mathbb{C}^M)\} \end{aligned} \tag{34}$$

and

$$Y_G^\nu(X^\wedge \times \Omega, \mathfrak{g}) := \ l.u.c.\ \mathcal{Y}_G^\nu(X^\wedge \times \Omega, \mathfrak{g}; \mathbb{C}^N, \mathbb{C}^M), \tag{35}$$

respectively.

The operators in (35) are called Green wedge operators. Those in (35) form matrices

$$\mathcal{G} = \begin{pmatrix} G & K \\ T & Q \end{pmatrix} : \begin{matrix} \mathcal{W}^{s,\beta}_{comp}(X^\wedge \times \Omega) \\ \oplus \\ H^s_{comp}(\Omega, \mathbb{C}^N) \end{matrix} \longrightarrow \begin{matrix} \mathcal{W}^{s-\nu,\beta-\mu}_{P,loc}(X^\wedge \times \Omega) \\ \oplus \\ H^{s-\nu}_{loc}(\Omega, \mathbb{C}^M) \end{matrix}, \tag{36}$$

(continuous for all $s \in \mathbb{R}$, with certain $P \in As(X, \mathfrak{a})$, $\mathfrak{a} = (\beta - \mu, \Theta)$). T is called an edge trace, K an edge potential operator. Furthermore we have $Q \in L^\nu_{cl}(\Omega; \mathbb{C}^N, \mathbb{C}^M)$. This notation indicates corresponding $M \times N$ systems with entries in $L^\nu_{cl}(\Omega)$.

Let

$$\bigcap_{s \in \mathbb{R}} \mathcal{L}^{(\nu)}(T^*\Omega \setminus 0; \mathcal{K}^{s,\beta} \oplus \mathbb{C}^N, \mathcal{K}^{s-\nu,\beta-\mu} \oplus \mathbb{C}^M) \tag{37}$$

be the space of all operator families

$$T^*\Omega \setminus 0 \ni (y,\eta) \longrightarrow a_{(\nu)}(y,\eta) : \begin{matrix} \mathcal{K}^{s,\beta}(X^\wedge) \\ \oplus \\ \mathbb{C}^N \end{matrix} \longrightarrow \begin{matrix} \mathcal{K}^{s-\nu,\beta-\mu}(X^\wedge) \\ \oplus \\ \mathbb{C}^M \end{matrix}$$

continuous for all $s \in \mathbb{R}$, C^∞ in (y, η), with

$$a_{(\nu)}(y, \lambda\eta) = \lambda^\nu \begin{pmatrix} \kappa_\lambda & 0 \\ 0 & 1 \end{pmatrix} a_{(\nu)}(y,\eta) \begin{pmatrix} \kappa_\lambda^{-1} & 0 \\ 0 & 1 \end{pmatrix}$$

for all $\lambda \in \mathbb{R}_+$. Then according to the above general notation, every $\mathcal{G} \in \mathcal{Y}_G^\nu(X^\wedge \times \Omega, \mathfrak{g}; \mathbb{C}^N, \mathbb{C}^M)$ has an homogeneous principal edge symbol of order ν

$$\sigma_\wedge^\nu(\mathcal{G})(y,\eta) \in \bigcap_{s \in \mathbb{R})} \mathcal{L}^{(\nu)}(T^*\Omega \setminus 0; \mathcal{K}^{s,\beta} \oplus \mathbb{C}^N, \mathcal{K}^{s-\nu,\beta-\mu} \oplus \mathbb{C}^M). \tag{38}$$

Clearly we have more precisely information on the image, but it is not used in this context.

Now let us introduce another sort of classical operator-valued symbols, namely

$$m(y,\eta) = \omega(r[\eta]) \left\{ r^{-\nu} \sum_{0 \le j \le k-\mu+\nu-1} r^j \sum_{|\alpha| \le j} \sum_{i=0}^{N} op_M^{\beta_{j,i}-n/2}(h_{j,i,\alpha})(y)\eta^\alpha \right\} \tilde{\omega}(r[\eta]). \tag{39}$$

Here $\mu, \nu \in \mathbb{R}$, $\mu - \nu \in \mathbb{N}$, $\Theta = (-k, 0]$, $k \in \mathbb{N} \setminus \{0\}$, $k - \mu + \nu \ge 1$ (for $k - \mu + \nu < 1$ the operator (39) disappears), $N = N(m)$, $\omega, \tilde{\omega}$ are arbitrary cut-off functions,

$$h_{j,i,\alpha}(y, w) \in C^\infty(\Omega, M^{-\infty}_{Q_{j,i,\alpha}}(X))$$

for certain $Q_{j,i,\alpha} \in \underline{As}(X)$, $sg(h_{j,i,\alpha}) \cap \Gamma_{(n+1)/2-\beta_{j,i}} = \emptyset$ and

$$\beta - (\mu - \nu) - j \le \beta_{j,i} \le \beta$$

for all i, j, α (cf. also 1.2.(31)). We then have

$$m(y,\eta) \in S^\nu_{cl}(\Omega \times \mathbb{R}^q; \mathcal{K}^{s,\beta}, \mathcal{S}^{\beta-\mu}), \tag{40}$$

(cf. 1.2.(16)). Then

$$\sigma_\wedge^\nu(m)(y,\eta) = \omega(r|\eta|) \left\{ r^{-\nu} \sum_{0 \le j \le k-\mu+\nu-1} r^j \sum_{|\alpha| = j} \sum_{i=0}^{N} op_M^{\beta_{j,i}-n/2}(h_{i,j,\alpha})(y)\eta^\alpha \right\} \tilde{\omega}(r|\eta|), \tag{41}$$

$$\sigma_{\wedge}^{\nu}(m)(y,\eta):\mathcal{K}^{s,\beta}(X^\wedge)\longrightarrow\mathcal{K}^{s-\nu,\beta-\mu}(X^\wedge) \tag{42}$$

for all $s\in\mathbb{R}$, $(y,\eta)\in T^*\Omega\setminus 0$. The operator

$$M=Op(m) \tag{43}$$

is called a smoothing Mellin operator of the class of wedge ψDO's.

Definition 1.3.11 *Let $\mu,\nu,\beta\in\mathbb{R}$, $\mu-\nu\in\mathbb{N}$, $\mathfrak{g}=(\beta,\beta-\mu,\Theta)$, $\Theta=(-k,0]$, $k\in\mathbb{N}\setminus\{0\}$. Then*

$$Y_{M+G}^{\nu}(X^\wedge\times\Omega,\mathfrak{g})$$

is the space of all operators $M+G$ with M being of the sort (43), and $G\in Y_G^{\nu}(X^\wedge\times\Omega,\mathfrak{g})$.

As further and main ingredient we now add the ψDO's themselves. Let us start with the symbols, first on $U^\wedge\times\Omega$ for a coordinate neighbourhood U on X, with local coordinates $x\in\mathbb{R}^n$.

Definition 1.3.12 *Denote by*

$$S_{cl}^{\nu}(U^\wedge\times\Omega\times\mathbb{R}^{n+1+q})_1, \tag{44}$$

$\nu\in\mathbb{R}$, the space of all $p(x,r,y,\xi,\rho,\eta)\in S_{cl}^{\nu}(U^\wedge\times\Omega\times\mathbb{R}^{n+1+q})$ that are independent of r for $r>1$. Moreover we set

$$\begin{aligned}&\tilde S_{cl}^{\nu}(U^\wedge\times\Omega\times\mathbb{R}^{n+1+q})_1=\\&\{p(x,r,y,\xi,r\rho,r\eta):p(x,r,y,\xi,\rho,\eta)\in S_{cl}^{\nu}(U^\wedge\times\Omega\times\mathbb{R}^{n+1+q})_1\\&\text{with } p \text{ being } C^\infty \text{ up to } r=o\}.\end{aligned} \tag{45}$$

Let

$$\tilde S^{(\nu)}(U^\wedge\times\Omega\times(\mathbb{R}^{n+1+q}\setminus\{0\}))_1 \tag{46}$$

denote the space of all $p_{(\nu)}(x,r,y,\xi,r\rho,r\eta)\in C^\infty(U^\wedge\times\Omega\times(\mathbb{R}^{n+1+q}\setminus\{0\}))$ satisfying

$$p_{(\nu)}(x,r,y,\lambda\xi,\lambda\rho,\lambda\eta)=\lambda^\nu p_{(\nu)}(x,r,y,\xi,\rho,\eta)$$

for all $\lambda>0$ and all $(x,r,y)\in U^\wedge\times\Omega$, $(\xi,\rho,\eta)\neq 0$, and $\chi(\xi,\rho,\eta)p_{(\nu)}(x,r,y,\xi,\rho,\eta)\in$ (44) for any (ξ,ρ,η)-excision function χ.

If $\kappa:U\to\tilde U$ is a diffeomorphism and $(\kappa,({}^t d\kappa)^{-1}):T^*U\to T^*\tilde U$ the associated map of cotangent bundles then $p_{(\nu)}(x,r,y,\xi,r\rho,r\eta)\in$ (45) implies $p_{(\nu)}(\kappa^{-1}(\tilde x),r,y,{}^t d\kappa(x(\tilde x))\tilde\xi,r\rho,r\eta)\in\tilde S^{(\nu)}(\tilde U^\wedge\times\Omega\times(\mathbb{R}^{n+1+q}\setminus\{0\}))_1$. This allows to define

$$\tilde S^{(\nu)}(T^*(X^\wedge\times\Omega)\setminus 0)_1 \tag{47}$$

as the space of all $p\in C^\infty(T^*(X^\wedge\times\Omega)\setminus 0)$ which belong over any coordinate neighbourhood U on X (in the local coordinates x) to the space (46).

For every $p\in\tilde S_{cl}^{\nu}(U^\wedge\times\Omega\times\mathbb{R}^{n+1+q})_1$, we set

$$op_{\psi,(x,r)}(p)(y,\eta)u(x,r)=(2\pi)^{-(n+1)}\int\!\!\int e^{i[(x-x')\xi+(r-r')\rho]}p(x,r,y,\xi,r\rho,r\eta)u(x',r')dx'dr'd\xi d\rho\ , \tag{48}$$

first for $u\in C_0^\infty(X^\wedge)$. This is a (y,η) -dependent family of ψDO's on $U^\wedge$ (belonging to $C^\infty(\Omega,L_{cl}^{\nu}(U^\wedge;\mathbb{R}^q))$, cf. the notations in 1.1.(3).

If $\omega(r)$, $\omega_1(r)$ are arbitrary cut-off functions and $\varphi(x)$, $\tilde\varphi(x)\in C_0^\infty(U)$, then we can form the family of operators

$$(1-\omega(r[\eta]))\varphi r^{-\nu}op_{\psi,(x,r)}(p)(y,\eta)\tilde\varphi(1-\omega_1(r[\eta])):\mathcal{K}^{s,\beta'}(X^\wedge)\longrightarrow\mathcal{K}^{s-\nu,\beta''}(X^\wedge), \tag{49}$$

which is continuous for every $s, \beta', \beta'' \in \mathbb{R}$. If we denote by $a_{(\nu)}(y,\eta)$ the analogous operator family for $c_\wedge p(x,y,\xi,r\rho,r\eta) := p(x,0,y,\xi,r\rho,r\eta)$ instead of p and $|\eta|$ instead of $[\eta]$, then it follows the homogeneity

$$a_{(\nu)}(y,\lambda\eta) = \lambda^\nu \kappa_\lambda a_{(\nu)}(y,\eta)\kappa_\lambda^{-1} \tag{50}$$

for all $\lambda \in \mathbb{R}_+$. As above, $(\kappa_\lambda u)(x,r) = \lambda^{(n+1)/2}\, u(x,\lambda r)$.

We now define an "operator convention", for brevity denoted by $op_{\psi,(x,r)}$, again, which assigns to any homogeneous symbol an operator family globally,

$$op_{\psi,(x,r)} : \tilde{S}^{(\nu)}(T^*(X^\wedge \times \Omega) \setminus 0)_1 \longrightarrow C^\infty(\Omega, L^\nu_{cl}(X^\wedge; \mathbb{R}^q)). \tag{51}$$

Choose an open covering $\{U_1, \ldots, U_N\}$ of X by coordinate neighbourhoods, a subordinated partition of unity $\{\varphi_1, \ldots, \varphi_N\}$ and a further system $\{\tilde{\varphi}_1, \ldots, \tilde{\varphi}_N\}$, $\tilde{\varphi}_j \in C_0^\infty(U_j)$, with $\varphi_j\tilde{\varphi}_j = \varphi_j$ for all j.

Let $p \in$(47) and

$$p_j := p|_{T^*(U_j^\wedge \times \Omega)\setminus 0},\ j = 1, \ldots, N.$$

Denote for abbreviation by (x,ξ) the variables and covariables on U_j for all j. Set

$$p_j(x,r,y,\xi,r\rho,r\eta) = \chi(\xi,r\rho,r\eta)p_{(j)}(x,r,y,\xi,r\rho,r\eta),$$

with some (ξ,ρ,η)-excision function $\chi(\xi,\rho,\eta)$, and define

$$op_{\psi,(x,r)}(p)(y,\eta) = \sum_{j=1}^{N} \varphi_j op_{\psi,(x,r)}(p_j)(y,\eta)\tilde{\varphi}_j \tag{52}$$

where $op_{\psi,(x,r)}$ on the right hand side is taken in the sense of (48).

Remark 1.3.13 *For $p \in \tilde{S}^{(\nu)}(T^*(X^\wedge \times \Omega) \setminus 0)_1$ we have $op_{\psi,(x,r)}(p)(y,\eta) \in C^\infty(\Omega, L^\nu_{cl}(X^\wedge; \mathbb{R}^q))$, and*

$$p = \sigma^\nu_\psi(op_{\psi,(x,r)}(p)),$$

where $\sigma^\nu_\psi(op_{\psi,(x,r)}(p))$ is the parameter-dependent homogeneous principal symbol of order ν, calculated point-wise for every $y \in \Omega$ (cf. also 1.1.(5)). From (49) it follows the family of continuous operators

$$(1-\omega(r[\eta]))r^{-\nu}op_{\psi,(x,r)}(p)(y,\eta)(1-\omega_1(r[\eta])) : \mathcal{K}^{s,\beta'}(X^\wedge) \longrightarrow \mathcal{K}^{s-\nu,\beta''}(X^\wedge)$$

for all $s, \beta', \beta'' \in \mathbb{R}$. Moreover

$$a_{(\nu)}(y,\eta) = (1-\omega(r|\eta|))r^\nu op_{\psi,(x,r)}(c_\wedge p)(y,\eta)(1-\omega_1(r|\eta|))$$

is homogeneous in the same sense as (50). Here

$$c_\wedge : \tilde{S}^{(\nu)}(T^*(X^\wedge \times \Omega) \setminus 0)_1 \longrightarrow \tilde{S}^{(\nu)}(T^*(X^\wedge \times \Omega) \setminus 0)_1$$

is the map of "freezing of coefficients" on the edge, defined by

$$c_\wedge(p)(x,y,\xi,r\rho,r\eta) = p(x,0,y,\xi,r\rho,r\eta).$$

Furthermore consider the space

$$C^\infty(\overline{\mathbb{R}}_+ \times \Omega, M^\nu_{\mathcal{O}}(X; \mathbb{R}^q)) \ni h(r,y,w,\eta)$$

(cf. the remarks after Definition 1.2.2) and define analogously

$$c_\wedge : C^\infty(\overline{\mathbb{R}}_+ \times \Omega, M^\nu_{\mathcal{O}}(X; \mathbb{R}^q)) \longrightarrow C^\infty(\Omega, M^\nu_{\mathcal{O}}(X; \mathbb{R}^q))$$

by

$$c_\wedge(h)(y,w,\eta) = h(0,y,w,\eta).$$

In the following we use systematically the notations 1.1.(11) for operator-valued Mellin symbols (i.e. the actions along X are automatically carried out), where the dependence on parameters such as y, η is indicated in brackets.

Theorem 1.3.14 *There exists a map*

$$m : \tilde{S}^{(\nu)}(T^*(X^\wedge \times \Omega) \setminus 0)_1 \longrightarrow C^\infty(\overline{\mathbb{R}}_+ \times \Omega, M_O^\nu(X; \mathbb{R}^q))$$

with

$$mc_\wedge = c_\wedge m, \tag{53}$$

such that for every $p \in \tilde{S}^{(\nu)}(T^*(X^\wedge \times \Omega) \setminus 0)_1$, *with* $f(r, y, w, \eta) := m(p)(r, y, w, r\eta)$

$$op_{\psi,(x,r)}(p)(y,\eta) - op_{M,r}^{\beta-n/2}(f)(y,\eta) \in C^\infty(\Omega, L^{-\infty}(X^\wedge; \mathbb{R}^q)) \tag{54}$$

for every $\beta \in \mathbb{R}$.

A proof of Theorem 1.3.14 may be found in [S3] (cf. Theorem 1.3.2 here employed in the improved version that m may obviously be chosen independently of β.) The assertion can be viewed as a Mellin operator convention which expresses the ψDO (-family) $op_{\psi,(x,r)}(p)$ in terms of Mellin operators, modulo controlled remainders. Since there is such a remainder, the map m is not canonical, but the final operator algebra will be independent of the concrete choice, for there will be involved also Y_{M+G}^ν terms.

Definition 1.3.15 *Let* $\mu, \nu, \beta \in \mathbb{R}$, $\mu - \nu \in \mathbb{N}$, $\mathfrak{g} = (\beta, \beta - \mu, \Theta)$, $\Theta = (-k, 0]$, $k \in \mathbb{N} \setminus \{0\}$. *Then*

$$Y^\nu(X^\wedge \times \Omega, \mathfrak{g})$$

is the space of all operators

$$A = A_0 + A_1 + M + G \tag{55}$$

with $M + G \in Y_{M+G}^\nu(X^\wedge \times \Omega, \mathfrak{g})$ *(cf. Definition 1.3.11) and* $A_i = Op(a_i)$, $i = 0, 1$, *with*

$$\begin{aligned} a_0(y,\eta) &= \omega(r[\eta])r^{-\nu} op_{M,r}^{\beta-n/2}(f)(y,\eta)\omega_0(r[\eta]), & (56)\\ a_1(y,\eta) &= (1-\omega(r[\eta]))r^{-\nu} op_{\psi,(x,r)}(p)(y,\eta)(1-\omega_1(r[\eta])) & (57)\end{aligned}$$

for arbitrary $p(x, r, y, \xi, r\rho, r\eta) \in \tilde{S}^{(\nu)}(T^*(X^\wedge \times \Omega) \setminus 0)_1$ *and* $f(r, y, w, \eta) = m(p)(r, y, w, r\eta)$ *(cf. Theorem 1.3.14).* $\omega, \omega_0, \omega_1$ *are arbitrary cut-off functions with*

$$\omega\omega_0 = \omega \ , \ \omega\omega_1 = \omega_1. \tag{58}$$

Finally

$$\mathcal{Y}^\nu(X^\wedge \times \Omega, \mathfrak{g}; \mathbb{C}^N, \mathbb{C}^M) \tag{59}$$

is the space of all

$$\mathcal{A} = \begin{pmatrix} A & 0 \\ 0 & 0 \end{pmatrix} + \mathcal{G} \tag{60}$$

with $A \in Y^\nu(X^\wedge \times \Omega, \mathfrak{g})$, $\mathcal{G} \in \mathcal{Y}_G^\nu(X^\wedge \times \Omega, \mathfrak{g}; \mathbb{C}^N, \mathbb{C}^M)$.

The analogous operator classes for $\Theta = (-\infty, 0]$ will be introduced as intersections of those for $\Theta_k = (-k, 0]$, $k \in \mathbb{N} \setminus \{0\}$.

Remark 1.3.16 *Every* $\mathcal{A} \in \mathcal{Y}^\nu(X^\wedge \times \Omega, \mathfrak{g}; \mathbb{C}^N, \mathbb{C}^M)$ *can be viewed as being properly supported with respect to the variables on* Ω, *modulo an operator in* $\mathcal{Y}_G^{-\infty}(X^\wedge \times \Omega, \mathfrak{g}; \mathbb{C}^N, \mathbb{C}^M)$.

Theorem 1.3.17 *An $\mathcal{A} \in \mathcal{Y}^\mu(X^\wedge \times \Omega, \mathfrak{g}; \mathbb{C}^N, \mathbb{C}^M)$ induces continuous operators*

$$\mathcal{A} : \mathcal{W}^{s,\beta}_{comp}(X^\wedge \times \Omega) \oplus H^s_{comp}(\Omega, \mathbb{C}^N) \longrightarrow \mathcal{W}^{s-\nu,\beta-\mu}_{loc}(X^\wedge \times \Omega) \oplus H^{s-\nu}_{loc}(\Omega, \mathbb{C}^M) \tag{61}$$

for every $s \in \mathbb{R}$, further

$$\mathcal{A} : \mathcal{W}^{s,\beta}_{P,comp}(X^\wedge \times \Omega) \oplus H^s_{comp}(\Omega, \mathbb{C}^N) \longrightarrow \mathcal{W}^{s-\nu,\beta-\mu}_{Q,loc}(X^\wedge \times \Omega) \oplus H^{s-\nu}_{loc}(\Omega, \mathbb{C}^M) \tag{62}$$

for every $P \in As(X, \mathfrak{a})$, $\mathfrak{a} = (\beta, \Theta)$, with some resulting $Q \in As(X, \mathfrak{b})$, $\mathfrak{b} = (\beta - \mu, \Theta)$, dependent on $P, \mathcal{A}$, not on s. For $\mathcal{A}$ being properly supported in the sense of Remark 1.3.16 we may set comp or loc on both sides of (61), (62).

Now let us express the leading sumbolic levels of the wedge calculus. From

$$Y^\nu(X^\wedge \times \Omega, \mathfrak{g}) \subset L^\nu_{cl}(X^\wedge \times \Omega)$$

we get the homogeneous principal interior symbolic map

$$\begin{aligned} \sigma^\nu_\psi &: Y^\nu(X^\wedge \times \Omega, \mathfrak{g}) \longrightarrow r^{-\nu}\tilde{S}^{(\nu)}(T^*(X^\wedge \times \Omega) \setminus 0)_1, \\ \sigma^\nu_\psi(A) &:= r^{-\nu}p. \end{aligned}$$

(cf. Remark 1.3.13). Furthermore we have the homogeneous principal edge symbolic level

$$\sigma^\nu_\wedge : Y^\nu(X^\wedge \times \Omega, \mathfrak{g}) \longrightarrow \bigcap_{s \in \mathbb{R}} \mathcal{L}^{(\nu)}(T^*\Omega \setminus 0; \mathcal{K}^{s,\beta}, \mathcal{K}^{s-\nu,\beta-\mu}). \tag{63}$$

The right hand side of (63) is defined as the space of all operator families

$$T^*\Omega \setminus 0 \ni (y, \eta) \longrightarrow a_{(\nu)}(y, \eta) : \mathcal{K}^{s,\beta}(X^\wedge) \longrightarrow \mathcal{K}^{s-\nu,\beta-\mu}(X^\wedge)$$

which are continuous for all $s \in \mathbb{R}$, satisfying (50) for all $\lambda \in \mathbb{R}_+$. Then

$$\begin{aligned} \sigma^\nu_\wedge(A_0)(y,\eta) &:= \omega(r|\eta|) r^{-\nu} op^{\beta-n/2}_{M,r}(c_\wedge f)(y,\eta)\omega_0(r|\eta|), \\ \sigma^\nu_\wedge(A_1)(y,\eta) &:= (1 - \omega(r|\eta|)) r^{-\nu} op_{\psi,(x,r)}(c_\wedge p)(y,\eta)(1 - \omega_1(r|\eta|)), \end{aligned}$$

$\sigma^\nu_\wedge(M)(y,\eta)$ for $M = Op(m)$ is defined by (41), $\sigma^\nu_\wedge(G)(y,\eta)$ by the left upper corner of (38). We set

$$\sigma^\nu_\wedge(A) = \sigma^\nu_\wedge(A_0) + \sigma^\nu_\wedge(A_1) + \sigma^\nu_\wedge(M) + \sigma^\nu_\wedge(G).$$

For $\mathcal{A} \in \mathcal{Y}^\nu(X^\wedge \times \Omega, \mathfrak{g}; \mathbb{C}^N, \mathbb{C}^M)$ we set

$$\sigma^\nu_\psi(\mathcal{A}) = \sigma^\nu_\psi(A), \tag{64}$$

$$\sigma^\nu_\wedge(\mathcal{A}) = \begin{pmatrix} \sigma^\nu_\wedge(A) & 0 \\ 0 & 0 \end{pmatrix} + \sigma^\nu_\wedge(\mathcal{G}) \tag{65}$$

for A as in (60). Then $\sigma^\nu_\wedge(\mathcal{A}) \in$(37). The operator spaces of Definition 1.3.15 after being localized close to $r = 0$, are preserved under natural operations, such as compositions, adjoints, push-forwards under diffeomorphisms $\Omega \to \tilde{\Omega}$. These are compatible with the symbolic levels. We will not formulate this explicitly here. It is rather evident anyway. In order to introduce (3) globally over $\mathbb{W}$ we first need the global negligible operators.

Definition 1.3.18 *Let $\mu, \beta \in \mathbb{R}$, $\Theta = (\vartheta, 0]$, $-\infty \leq \vartheta < 0$, $\mathfrak{g} = (\beta, \beta - \mu, \Theta)$. Then*

$$\mathcal{Y}^{-\infty}_G(\mathbb{W}, \mathfrak{g}; \mathbb{C}^N, \mathbb{C}^M)$$

denotes the space of all

$$\mathcal{G} \in \bigcap_{s\in\mathbb{R}} \mathcal{L}(\mathcal{W}^{s,\beta}_{comp}(\mathbb{W}) \oplus H^{s}_{comp}(Y,\mathbb{C}^N), \mathcal{W}^{\infty,\beta-\mu}_{loc}(\mathbb{W}) \oplus H^{\infty}_{loc}(Y,\mathbb{C}^M))$$

which induce continuous operators

$$\begin{aligned} \mathcal{G} : \mathcal{W}^{s,\beta}_{comp}(\mathbb{W}) \oplus H^{s}_{comp}(Y,\mathbb{C}^N) &\longrightarrow \mathcal{W}^{\infty,\beta-\mu}_{P,loc}(\mathbb{W}) \oplus H^{\infty}_{loc}(Y,\mathbb{C}^M), \\ \mathcal{G}^* : \mathcal{W}^{s,-\beta+\mu}_{comp}(\mathbb{W}) \oplus H^{s}_{comp}(Y,\mathbb{C}^M) &\longrightarrow \mathcal{W}^{\infty,-\beta}_{Q,loc}(\mathbb{W}) \oplus H^{\infty}_{loc}(Y,\mathbb{C}^N) \end{aligned}$$

for all $s \in \mathbb{R}$, for certain $P \in As(X,\mathfrak{a})$ with $\mathfrak{a} = (\beta-\mu,\Theta)$, $Q \in As(X,\mathfrak{b})$ with $\mathfrak{b} = (-\beta,\Theta)$, dependent on $\mathcal{G}$.

$\mathcal{G}^*$ is the formal adjoint of $\mathcal{G}$ in the sense

$$(\mathcal{G}u, v)_{\mathcal{W}^0(\mathbb{W})\oplus H^0(Y,\mathbb{C}^M)} = (u, \mathcal{G}^*v)_{\mathcal{W}^0(\mathbb{W})\oplus H^0(Y,\mathbb{C}^N)} \tag{66}$$

for all $u \in C_0^\infty(int\ \mathbb{W}) \oplus C_0^\infty(Y,\mathbb{C}^N)$, $v \in C_0^\infty(int\ \mathbb{W}) \oplus C_0^\infty(Y,\mathbb{C}^M)$. The interpretation of the scalar products in (66) for Y refers to (22). We set

$$Y_G^{-\infty}(\mathbb{W},\mathfrak{g}) = l.u.c.\ \mathcal{Y}_G^{-\infty}(\mathbb{W},\mathfrak{g};\mathbb{C}^N,\mathbb{C}^M). \tag{67}$$

Definition 1.3.19 $\mathcal{Y}^\nu(\mathbb{W},\mathfrak{g};\mathbb{C}^N,\mathbb{C}^M)$ *(with μ,ν, and the weight data $\mathfrak{g}$ being as in Definition 1.3.15) is the space of all*

$$\mathcal{A} \in \bigcap_{s\in\mathbb{R}} \mathcal{L}(\mathcal{W}^{s,\beta}_{comp}(\mathbb{W}) \oplus H^{s}_{comp}(Y,\mathbb{C}^N), \mathcal{W}^{s-\nu,\beta-\mu}_{loc}(\mathbb{W}) \oplus H^{s-\nu}_{loc}(Y,\mathbb{C}^M)),$$

$$\mathcal{A} = \begin{pmatrix} A & 0 \\ 0 & 0 \end{pmatrix} + \mathcal{G}$$

with $A \in L^\nu_{cl}(int\ \mathbb{W})$ and $\mathcal{G}$ having kernel in

$$\begin{pmatrix} C^\infty(int\ \mathbb{W} \times int\ \mathbb{W}) & C^\infty(int\ \mathbb{W} \times Y) \otimes \mathbb{C}^N \\ \mathbb{C}^M \otimes C^\infty(Y \times int\ \mathbb{W}) & \mathbb{C}^M \otimes C^\infty(Y \times Y) \otimes \mathbb{C}^N \end{pmatrix} \tag{68}$$

(with obvious notations) such that for arbitrary cut-off functions $\omega_1(r), \omega_2(r)$, supported by $[0,1)$ and $\varphi_1(y), \varphi_2(y) \in C_0^\infty(Y)$ the operator

$$\begin{pmatrix} \varphi_1\omega_1 & 0 \\ 0 & \varphi_1 \end{pmatrix} \mathcal{A} \begin{pmatrix} \varphi_2\omega_2 & 0 \\ 0 & \varphi_2 \end{pmatrix} \tag{69}$$

belongs to $\mathcal{Y}^\nu(X^\wedge \times \Omega,\mathfrak{g};\mathbb{C}^N,\mathbb{C}^M)$ once φ_1,φ_2 are supported by a coordinate neighbourhood Ω on Y, and to $\mathcal{Y}_G^{-\infty}(\mathbb{W},\mathfrak{g};\mathbb{C}^N,\mathbb{C}^M)$ whenever $supp\ \varphi_1 \cap supp\ \varphi_2 = \emptyset$.

Remark 1.3.20 *Every $\mathcal{A} \in \mathcal{Y}^\nu(\mathbb{W},\mathfrak{g};\mathbb{C}^N,\mathbb{C}^M)$ can be written as $\mathcal{A} = \mathcal{A}_0+\mathcal{G}$ where $\mathcal{A}_0 \in \mathcal{Y}^\nu(\mathbb{W},\mathfrak{g};\mathbb{C}^N,\mathbb{C}^M)$ is properly supported with respect to the variables on Y, and $\mathcal{G} \in \mathcal{Y}_G^{-\infty}(\mathbb{W},\mathfrak{g};\mathbb{C}^N,\mathbb{C}^M)$.*

Remark 1.3.21 *Theorem 1.3.17 has the corresponding global analogue for $\mathcal{A} \in \mathcal{Y}^\nu(\mathbb{W},\mathfrak{g};\mathbb{C}^N,\mathbb{C}^M)$.*

The above local symbolic levels σ^ν_ψ, $\sigma^\nu_\wedge$ can easily be generalized to the corresponding global ones. We can introduce

$$\tilde{S}^{(\nu)}(T^*(int\ \mathbb{W} \times Y) \setminus 0)$$

as the set of of all $p \in C^\infty(T^*(int\ \mathbb{W} \times Y) \setminus 0)$ which are positively homogeneous of order ν with respect to the canonical $\mathbb{R}_+$ action along the fibres such that p restricts to an element in $\tilde{S}^{(\nu)}(T^*(X^\wedge \times \Omega) \setminus 0)_1|_{X \times (0,1) \times \Omega}$ for every coordinate patch Ω on Y, (c.f. (2)).

Let g_0^β for arbitrary $\beta \in \mathbb{R}$ denote a strictly positive C^∞ function on $int\ \mathbb{W}$ which is close to Y in the coordinates (x, r, y) of the form r^β. It follows the homogeneous principal interior symbolic map

$$\sigma_\psi^\nu : \mathcal{Y}^\nu(\mathbb{W}, \mathfrak{g}; \mathbb{C}^N, \mathbb{C}^M) \longrightarrow g_0^{-\nu} \tilde{S}^{(\nu)}(T^*(int\ \mathbb{W} \times Y) \setminus 0),$$

acting only on

$$Y^\nu(\mathbb{W}, \mathfrak{g}) := l.u.c.\mathcal{Y}^\nu(\mathbb{W}, \mathfrak{g}; \mathbb{C}^N, \mathbb{C}^M). \tag{70}$$

Moreover we have an obvious generalization of (63), the homogeneous principal edge symbolic map, namely

$$\sigma_\wedge^\nu : \mathcal{Y}^\nu(\mathbb{W}, \mathfrak{g}; \mathbb{C}^N, \mathbb{C}^M) \longrightarrow \bigcap_{s \in \mathbb{R}} \mathcal{L}^{(\nu)}(T^*Y \setminus 0; \mathcal{K}^{s,\beta} \oplus \mathbb{C}^N, \mathcal{K}^{s-\nu,\beta-\mu} \oplus \mathbb{C}^M). \tag{71}$$

We will not carry out now the "standard elements" of the calculus as they are analogous to the corresponding ones in the more particular case of [S1] or in general of [S4], Chapter 4. Let us only mention the following

Theorem 1.3.22 *Let $\mathcal{A} \in \mathcal{Y}^\nu(\mathbb{W}, \mathfrak{g}; \mathbb{C}^N, \mathbb{C}^M)$ for $\mathfrak{g} = (\beta, \beta - \mu, \Theta)$, $\mathcal{B} \in \mathcal{Y}^{\tilde{\nu}}(\mathbb{W}, \tilde{\mathfrak{g}}; \mathbb{C}^M, \mathbb{C}^L)$ for $\tilde{\mathfrak{g}} = (\beta - \mu, \beta - \mu - \tilde{\mu}, \Theta)$ (and $\mathcal{A}$ or $\mathcal{B}$ be properly supported in the sense of Remark 1.3.20). Then $\mathcal{B}\mathcal{A} \in \mathcal{Y}^{\nu+\tilde{\nu}}(\mathbb{W}, \mathfrak{g}_0; \mathbb{C}^N, \mathbb{C}^L)$ for $\mathfrak{g}_0 = (\beta, \beta - \mu - \tilde{\mu}, \Theta)$, and*

$$\begin{aligned} \sigma_\psi^{\tilde{\nu}+\nu}(\mathcal{B}\mathcal{A}) &= \sigma_\psi^{\tilde{\nu}}(\mathcal{B})\sigma_\psi^\nu(\mathcal{A}), \\ \sigma_\wedge^{\tilde{\nu}+\nu}(\mathcal{B}\mathcal{A}) &= \sigma_\wedge^{\tilde{\nu}}(\mathcal{B})\sigma_\wedge^\nu(\mathcal{A}). \end{aligned}$$

Remark 1.3.23 *The whole theory of the operator classes $\mathcal{Y}^\nu(\mathbb{W}, \mathfrak{g}; \ldots)$ has a straightforward generalization to systems of left upper corners, or more generally to operators acting between distributional sections of vector bundles. This also concerns the trivial bundles $\mathbb{C}^N, \mathbb{C}^M$ over Y which can be replaced by $J_-, J_+ \in Vect(Y)$. Here $Vect(.)$ is the set of complex C^∞ vector bundles over the space in the brackets.*

In particular the latter case corresponds to operator spaces that we denote by

$$\mathcal{Y}^\nu(\mathbb{W}, \mathfrak{g}; J_-, J_+). \tag{72}$$

The left upper corners are here still scalar operators, and for simplicity we restrict ourselves to this case. But it may happen that non-trivial J_-, J_+ are actually necessary, as it is known, for instance, in the more particular case of boundary value problems.

Let us finish this section with the ellipticity that shall be formulated for general J_-, J_+.

Definition 1.3.24 *An operator $\mathcal{A} \in \mathcal{Y}^\mu(\mathbb{W}, \mathfrak{g}; J_-, J_+)$ for $\mathfrak{g} = (\beta, \beta - \mu, \Theta)$, $\Theta = (-k, 0]$, $k \in (\mathbb{N} \setminus \{0\}) \cup \{\infty\}$, is called elliptic if*

(i) $\sigma_\psi^\mu(\mathcal{A}) \neq 0$ on $T^*(int\ \mathbb{W}) \setminus 0$ in the usual sense and if $p = g_0^\mu \sigma_\psi^\mu(\mathcal{A})$ in the coordinates (x, r, y) close to $\partial\mathbb{W}$, where $p = p(x, r, y, \xi, r\rho, r\eta)$, satisfies $p(x, r, y, \xi, \rho, \eta) \neq 0$ up to $r = 0$,

(ii)

$$\sigma_\wedge^\mu(\mathcal{A})(y, \eta) : \begin{array}{c} \mathcal{K}^{s,\beta}(X^\wedge) \\ \oplus \\ J_{-,y} \end{array} \longrightarrow \begin{array}{c} \mathcal{K}^{s-\mu,\beta-\mu}(X^\wedge) \\ \oplus \\ J_{+,y} \end{array} \tag{73}$$

is an isomorphism for all $(y, \eta) \in T^*Y \setminus 0$, for some $s \in \mathbb{R}$.

Here $J_{\pm,y}$ is the fibre of $J_\pm$ over $y \in Y$.

Clearly the second condition for one particular $s = s_0$ implies the same for all s.

Remember that in our application $\mathbb{W}$ is not necessarily compact, where $Y = K \setminus K_0$. Therefore one might better speak about interior "edge ellipticity". The ellipticity up to non-compact "exits" of Y is just the content of the corner ellipticity below in Section 3.3.. If

$$\mathcal{A} = \begin{pmatrix} A & K \\ T & Q \end{pmatrix}$$

is elliptic, then T, K, Q are called elliptic edge conditions for A. The fibre dimensions of J_-, J_+ depend in general on the weight β.

Definition 1.3.25 *Let $\mathcal{A} \in \mathcal{Y}^\mu(\mathbb{W}, \mathfrak{g}; J_-, J_+)$, $\mathfrak{g} = (\beta, \beta - \mu, \Theta)$ and $\mathcal{P} \in \mathcal{Y}^{-\mu}(\mathbb{W}, \mathfrak{h}; J_+, J_-)$, $\mathfrak{h} = (\beta - \mu, \beta, \Theta)$, $\mathcal{A}$ or $\mathcal{P}$ being properly supported. Then $\mathcal{P}$ is called a parametrix of $\mathcal{A}$ if*

$$\begin{aligned} \mathcal{P}\mathcal{A} - 1 &\in \mathcal{Y}_G^{-\infty}(\mathbb{W}, \mathfrak{g}_l; J_-, J_-), \\ \mathcal{A}\mathcal{P} - 1 &\in \mathcal{Y}_G^{-\infty}(\mathbb{W}, \mathfrak{g}_r; J_+, J_+), \end{aligned}$$

with $\mathfrak{g}_l = (\beta, \beta, \Theta)$, $\mathfrak{g}_r = (\beta - \mu, \beta - \mu, \Theta)$, and 1 being the identical operator in the corresponding $\mathcal{Y}^0$-classes.

Theorem 1.3.26 *Let $\mathcal{A} \in \mathcal{Y}^\mu(\mathbb{W}, \mathfrak{g}; J_-, J_+)$ be elliptic, $\mathfrak{g} = (\beta, \beta - \mu, \Theta)$. Then $\mathcal{A}$ has a properly supported (with respect to edge variables) parametrix $\mathcal{P} \in \mathcal{Y}^{-\mu}(\mathbb{W}, \mathfrak{h}; J_+, J_-)$, $\mathfrak{h} = (\beta - \mu, \beta, \Theta)$. Moreover*

$$\mathcal{A}u = f \in \mathcal{W}^{s,\beta-\mu}_{loc}(\mathbb{W}) \oplus H^s_{loc}(Y, J_+)$$

for some fixed $s \in \mathbb{R}$, and

$$u \in \mathcal{W}^{-\infty,\beta}_{comp}(\mathbb{W}) \oplus H^{-\infty}_{comp}(Y, J_-) \tag{74}$$

imply

$$u \in \mathcal{W}^{s+\mu,\beta}_{comp}(\mathbb{W}) \oplus H^{s+\mu}_{comp}(Y, J_-).$$

Furthermore

$$\mathcal{A}u = f \in \mathcal{W}^{s,\beta-\mu}_{Q,loc}(\mathbb{W}) \oplus H^s_{loc}(Y, J_+)$$

for some $Q \in As(X, (\beta - \mu, \Theta))$ and (74) imply

$$u \in \mathcal{W}^{s+\mu,\beta}_{P,comp}(\mathbb{W}) \oplus H^{s+\mu}_{comp}(Y, J_+)$$

for a resulting $P \in As(X, (\beta, \Theta))$, dependent on $\mathcal{A}$ and Q.

Remark 1.3.27 *If $\mathbb{W}$ and Y are compact then the ellipticity of $\mathcal{A} \in \mathcal{Y}^\mu(\mathbb{W}, \mathfrak{g}; J_-, J_+)$ is equivalent to the Fredholm property of*

$$\mathcal{A} : \mathcal{W}^{s,\beta}(\mathbb{W}) \oplus H^s(Y, J_-) \longrightarrow \mathcal{W}^{s-\mu,\beta-\mu}(\mathbb{W}) \oplus H^{s-\mu}(Y, J_+)$$

for every $s \in \mathbb{R}$.

1.4 The edge asymptotics close to corners

The previous section has studied ψDO's on manifolds with (non-compact) edges, based on the Fourier transform along the edges. The corner ψDO's close to $t = 0$ (with $t \in \mathbb{R}_+$ as variable on a one-dimensional edge, where $t = 0$ corresponds to the corner itself) require an analogue of this calculus in terms of the Mellin transform along $\mathbb{R}_+ \ni t$. This yields in particular the edge contribution on the corner asymptotics for $t \to 0$. It is indispensable to formulate explicitly this Mellin edge calculus, though it induces a lot of material. But we shall see that manything can be red off from the Fourier version under employing information from the cone theory. On the other hand we get in this way a new interesting example of an operator algebra with a rich internal symbolic structure.

The concrete operators and distribution classes will be given on

$$X^\wedge \times \mathbb{R}_+ = X \times \mathbb{R}_+ \times \mathbb{R}_+ \ni (x, r, t),$$

where now mainly a neighbourhood of $t = 0$ is of interest. As for the above wedge theory we can first look at an abstract formulation. Similarly as in Definition 1.3.1 we start with a Banach space E and a group $\{\kappa_\lambda\}_{\lambda \in \mathbb{R}_+} \in C(\mathbb{R}_+, \mathcal{L}_\sigma(E))$. For the covariable $\tau \in \mathbb{R}$ of $t \in \mathbb{R}_+$ we employ the function $\tau \to [\tau]$, and set $\kappa(\tau) = \kappa_{[\tau]}$ Since we are in the Mellin set-up, τ will be at the same time interpreted as $Im\ z$, $z = \sigma + i\tau$. If $u(t) \in C_0^\infty(\mathbb{R}_+, E)$ then $Mu(z) = (M_{t \to z} u)(z) \in \mathcal{A}(\mathbb{C}, E)$ (= the space of E-valued entire functions). Then $Mu(z)$ may also be regarded as a function on Γ_σ for every $\sigma \in \mathbb{R}$.

Definition 1.4.1 *$\mathcal{V}^{s,\alpha}(\mathbb{R}_+, E)$ for $s, \alpha \in \mathbb{R}$ is the completion of $C_0^\infty(\mathbb{R}_+, E)$ with respect to the norm*

$$||u||_{\mathcal{V}^{s,\alpha}(\mathbb{R}_+,E)} = \left\{ \frac{1}{2\pi i} \int_{\Gamma_{1/2-\alpha}} [\tau]^{2s} ||\kappa(\tau)^{-1}(Mu)(z)||_E^2 dz \right\}^{1/2}. \tag{1}$$

We also set

$$\mathcal{V}^s(\mathbb{R}_+, E) = \mathcal{V}^{s,0}(\mathbb{R}_+, E).$$

For κ_λ = identity for all $\lambda \in \mathbb{R}_+$ we will also write

$$\mathcal{H}^{s,\alpha}(\mathbb{R}_+, E) \text{ and } \mathcal{H}^s(\mathbb{R}_+, E), \tag{2}$$

respectively

Example 1.4.2 *Let $E = \mathcal{K}^{s,\beta}(X^\wedge)$, $s, \beta \in \mathbb{R}$, and $(\kappa_\lambda u)(x, r) = \lambda^{(n+1)/2}\, u(x, \lambda r)$, $\lambda \in \mathbb{R}_+$. Then we get the corner Sobolev space (of smoothness s and weight (β, α))*

$$\mathcal{V}^{s,(\beta,\alpha)}(X^\wedge \times \mathbb{R}_+) := \mathcal{V}^{s,\alpha}(\mathbb{R}_+, \mathcal{K}^{s,\beta}(X^\wedge)). \tag{3}$$

The final corner Sobolev spaces over $\mathbb{B} \times \mathbb{R}_+$ for a manifold $\mathbb{B}$ with conical singularities (being the base of the corner) will contain (as it ought to be) a further contribution from $int\ \mathbb{B} \times \mathbb{R}_+$. It will be adequate then to revise the weight convention by a shift of α, dependent on $dim(int\ \mathbb{B})$, as it was already done for the cone Sobolev spaces.

Note that

$$\mathcal{V}^0(\mathbb{R}_+, \mathcal{K}^0(X^\wedge)) = r^{-n/2} L^2(X^\wedge \times \mathbb{R}_+) \tag{4}$$

with L^2 being defined in terms of $dx\ dr\ dt$ (cf. also 1.3.(7)).

Let us now return to the abstract spaces E, again.

Remark 1.4.3 *Let $\{E^j\}_{j \in \mathbb{N}}$ be a sequence of Banach spaces, $E^{j+1} \hookrightarrow E^j$, $\{\kappa_\lambda\} \in C(\mathbb{R}_+, \mathcal{L}_\sigma(E^j))$ for all j, and $E = \varprojlim_j E^j$ be the projective limit (which is Fréchet). Then*

$$\mathcal{V}^{s,\alpha}(\mathbb{R}_+, E) = \varprojlim_j \mathcal{V}^{s,\alpha}(\mathbb{R}_+, E^j). \tag{5}$$

This is also a Fréchet space.

As usual, the Mellin transform M will often be regarded as a map from distributions on $\mathbb{R}_+$ to distributions on $\Gamma_{1/2}$ (unless we do not observe at the same time analytic extensions). Analogously the "weighted Mellin transform"

$$M_\alpha := Mt^\alpha \text{ with } M_o = M \tag{6}$$

maps to distributions on $\Gamma_{1/2-\alpha}$, $\alpha \in \mathbb{R}$.

Remark 1.4.4 *The operator $T = M^{-1}_{z\to t}\kappa(\tau)^{-1}M_{t\to z}$, with z varying on any given weight line extends to an isomorphism*

$$T_{(\alpha)} : \mathcal{V}^{s,\alpha}(\mathbb{R}_+, E) \longrightarrow \mathcal{H}^{s,\alpha}(\mathbb{R}_+, E). \tag{7}$$

In other words

$$M_\alpha \mathcal{V}^{s,\alpha}(\mathbb{R}_+, E) = \{\kappa(\tau)f(z) : f(z) \in M_\alpha \mathcal{H}^{s,\alpha}(\mathbb{R}_+, E)\}. \tag{8}$$

Let us set

$$(\Phi_\alpha u)(\tilde{t}) = e^{(1/2-\alpha)\tilde{t}}u(e^{\tilde{t}}), \ (\Psi g)(\tau) = g(-\tau). \tag{9}$$

Then, for $\tau = Im\ z$, $z \in \Gamma_{1/2-\alpha}$, $u \in C_0^\infty(\mathbb{R}_+, E)$, it follows

$$(\Psi M_\alpha u)(\tau) = (F\Phi_\alpha u)(\tau) \tag{10}$$

with the one-dimensional Fourier transform $Fv(\tau) = \int e^{-i\tilde{t}\tau}v(\tilde{t})\,d\tilde{t}$. The notation on the left of (10) means that the function on the corresponding weight line is regarded as a function of τ. We obtain that

$$u \longrightarrow \left\{\int_{\mathbb{R}} [\tau]^{2s}\|\kappa(\tau)^{-1}(F\Phi_\alpha u)(\tau)\|_E^2 d\tau\right\}^{1/2} \tag{11}$$

is an equivalent norm on $\mathcal{V}^{s,\alpha}(\mathbb{R}_+, E)$. It follows that Φ_α induces an isomorphism

$$\Phi_\alpha : \mathcal{V}^{s,\alpha}(\mathbb{R}_+, E) \longrightarrow \mathcal{W}^s(\mathbb{R}, E),$$

cf. Definition 1.3.1. In other words

$$\mathcal{V}^{s,\alpha}(\mathbb{R}_+, E) = \{t^{-1/2+\alpha}u(\log t) : u(\tilde{t}) \in \mathcal{W}^s(\mathbb{R}, E)\}. \tag{12}$$

Thus

$$\mathcal{V}^{s,\alpha}(\mathbb{R}_+, E) = t^\alpha \mathcal{V}^s(\mathbb{R}_+, E). \tag{13}$$

We then have

$$T_{(\alpha)} = t^\alpha T_{(0)} t^{-\alpha}. \tag{14}$$

Remark 1.4.4 allows an extension of the definition of $\mathcal{V}^{s,\alpha}(\mathbb{R}_+, E)$ to vector subspaces $E_1 \subseteq E$ that are not necessarily preserved under κ_λ, namely

$$\mathcal{V}^{s,\alpha}(\mathbb{R}_+, E_1) := T_{(\alpha)}^{(-1)}\mathcal{H}^{s,\alpha}(\mathbb{R}_+, E_1). \tag{15}$$

Here E_1 is supposed to be a Banach (Fréchet) space, continuously embedded in E, and $\mathcal{H}^{s,\alpha}(\mathbb{R}_+, E_1)$ is taken in the sense of Remark 1.3.3. Similary as for the wedge spaces it follows for two such $E_0, E_1 \subseteq E$

$$\mathcal{V}^{s,\alpha}(\mathbb{R}_+, E_0 + E_1) = \mathcal{V}^{s,\alpha}(\mathbb{R}_+, E_0) + \mathcal{V}^{s,\alpha}(\mathbb{R}_+, E_1) \tag{16}$$

in the sense of 1.1.(2).

Remark 1.4.5 *From the equality of (13) and (11) (up to the factor $(2\pi)^{-1/2}$) it follows immediately*

$$(2\pi)^{-1/2}\|T_{(\alpha)}u\|_{\mathcal{H}^{s,\alpha}(\mathbb{R}_+,E)} = \|T\Phi_\alpha u\|_{H^s(\mathbb{R},E)}$$

with the operator T from 1.3.(8). Thus

$$T_{(\alpha)}^{-1}\mathcal{H}^{s,\alpha}(\mathbb{R}_+,E_1) + \Phi_\alpha^{-1}T^{-1}H^s(\mathbb{R},E_1),$$

in other words Φ_α induces an isomorphism

$$\Phi_\alpha : \mathcal{V}^{s,\alpha}(\mathbb{R}_+,E_1) \longrightarrow \mathcal{W}^s(\mathbb{R},E_1). \tag{17}$$

We are now in the position to introduce the subspaces of (3) with asymptotics. For $P \in As(X,\mathfrak{a}^\bullet)$ or $\in As(X,\mathfrak{a})$ for $\mathfrak{a} = (\beta,\Theta)$, Θ finite, we can insert the spaces 1.3.(11) and obtain

$$\mathcal{V}^{s,\alpha}(\mathbb{R}_+,\mathcal{K}_P^{s,\beta}(X^\wedge)) = \mathcal{V}^{s,\alpha}(\mathbb{R}_+,\mathcal{K}_\Theta^{s,\beta}(X^\wedge)) + \mathcal{V}^{s,\alpha}(\mathbb{R}_+,\mathcal{E}_P(X^\wedge)). \tag{18}$$

We will also set

$$\mathcal{V}_P^{s,(\beta,\alpha)}(X^\wedge \times \mathbb{R}_+) = \mathcal{V}^{s,\alpha}(\mathbb{R}_+,\mathcal{K}_P^{s,\beta}(X^\wedge)), \tag{19}$$

$$\mathcal{V}_\Theta^{s,(\beta,\alpha)}(X^\wedge \times \mathbb{R}_+) = \mathcal{V}^{s,\alpha}(\mathbb{R}_+,\mathcal{K}_\Theta^{s,\beta}(X^\wedge)). \tag{20}$$

The space (19) is an expression of the edge asymptotics close to the corner $t = 0$. The "singular functions" are just constituted by the elements of

$$\mathcal{V}^{s,\alpha}(\mathbb{R}_+,\mathcal{E}_P(X^\wedge)). \tag{21}$$

For $P \in As(X,\mathfrak{a})$ they are of the form

$$M_{\alpha,z\to t}^{-1}\left\{[\tau]^{(n+1)/2} < (M_{\alpha,t\to z}\zeta)(z), (r[\tau])^{-w} > \omega(r[\tau])\right\} \tag{22}$$

where $\zeta(t)$ runs over $\mathcal{H}^{s,\alpha}(\mathbb{R}_+,\mathcal{A}'(K,C^\infty(X)))$ for $K = carrier(P)$. If

$$P = \{(p_j,m_j,L_j)\}_{j=0,\ldots,N(P)} \in As(X,\mathfrak{a}^\bullet)$$

we obtain

$$M_{\alpha,z\to t}^{-1}\left\{[\tau]^{(n+1)/2}(r[\tau])^{-p_j}\sum_{k=0}^{m_j}(M_{\alpha,t\to z}\zeta_{jk})(x,z)\log^k(r[\tau])\omega(r[\tau])\right\} \tag{23}$$

for all j, with $\zeta_{jk}(x,t) \in \mathcal{H}^{s,\alpha}(\mathbb{R}_+,L_j)$, $0 \le k \le m_j$. For infinite Θ we can write

$$u \sim \sum_{j=0}^{\infty} f_j \tag{24}$$

where for $P \in As(X,\mathfrak{a})$ the f_j are the singular functions associated with compact sets $K_j \subset \mathbb{C}_w$, satisfying 1.2.(6). For $P \in As(X,\mathfrak{a}^\bullet)$ the f_j are given by (23). The interpretation of (24) is that

$$u - \sum_{j=0}^{N} f_j \in \mathcal{V}^{s,(\beta,\alpha)}(X^\wedge \times \mathbb{R}_+)$$

for every $\beta \in \mathbb{R}$ with some $N = N(\beta)$.

Remark 1.4.6 *If we choose $\{E^j\}_{j\in\mathbb{N}}$ as in Remark 1.3.5 then*

$$\mathcal{V}_P^{s,(\beta,\alpha)}(X^\wedge \times \mathbb{R}_+) = \varprojlim_j \mathcal{V}^{s,\alpha}(\mathbb{R}_+,E^j),$$

$P \in As(X,\mathfrak{a}^\bullet)$ or $P \in As(X,\mathfrak{a})$.

Theorem 1.4.7 *For every $s, \alpha, \beta \in \mathbb{R}$ we have*

$$\mathcal{V}^{s,(\beta,\alpha)}(X^\wedge \times \mathbb{R}_+) \subset \mathcal{W}^{s,\beta}_{loc}(X^\wedge \times \mathbb{R}_+),$$
$$\mathcal{V}^{s,(\beta,\alpha)}_P(X^\wedge \times \mathbb{R}_+) \subset \mathcal{W}^{s,\beta}_{P,loc}(X^\wedge \times \mathbb{R}_+),$$

for every $P \in As(X, \mathfrak{a}^\bullet)$ or $P \in As(X, \mathfrak{a})$, $\mathfrak{a} = (\beta, \Theta)$, satisfying the shadow condition (cf. 1.3.(20)).

This is an immediate consequence of the invariance of the $\mathcal{W}^{s,\beta}_{P,loc}$–spaces under diffeomorphisms and of (17) which implies

$$\Phi_\alpha : \mathcal{V}^{s,(\beta,\alpha)}_P(X^\wedge \times \mathbb{R}_+) \xrightarrow{\cong} \mathcal{W}^{s,\beta}_P(X^\wedge \times \mathbb{R}), \tag{25}$$

(cf. [S1], 3.2.5. Theorem 19).

The next step is to perform the Mellin ΨDO calculus with operator-valued symbols (cf. also [S2]). Choose $E, \{\kappa_\lambda\}, \tilde{E}, \{\tilde{\kappa}_\lambda\}$ as in Definition 1.3.8.

Definition 1.4.8 *Let*

$$Q = \overline{\mathbb{R}}_+ \times \overline{\mathbb{R}}_+ \ni (t, t'). \tag{26}$$

Then $S^\mu(Q \times \mathbb{R}; E, \tilde{E})$ is the space of all $a(t, t', \tau) \in S^\mu(int\ Q \times \mathbb{R}; E, \tilde{E})$ for which there is an open neighbourhood U of Q in $\mathbb{R}^2$ and an $\tilde{a}(t, t', \tau) \in S^\mu(U \times \mathbb{R}; E, \tilde{E})$ with $a = \tilde{a}|_{U \times \mathbb{R}}$. Analogously we define $S^\mu(\overline{\mathbb{R}}_+ \times \mathbb{R}; E, \tilde{E})$ as well as $S^\mu_{cl}(Q \times \mathbb{R}; E, \tilde{E})$, $S^\mu_{cl}(\overline{\mathbb{R}}_+ \times \mathbb{R}; E, \tilde{E})$.

Furthermore it will be convenient to denote by

$$S^\mu(Q \times \Gamma_\sigma; E, \tilde{E}) \ni a(t, t', z)$$

for $z = \sigma + i\tau$ the obtained space from $S^\mu(Q \times \mathbb{R}; E, \tilde{E})$ under pull-back with respect to $\Gamma_\sigma \to \mathbb{R}$, $\sigma + i\tau \to \tau$. Analogously we get

$$S^\mu(\overline{\mathbb{R}}_+ \times \Gamma_\sigma; E, \tilde{E}),\ S^\mu_{cl}(Q \times \Gamma_\sigma; E, \tilde{E}),\ S^\mu_{cl}(\overline{\mathbb{R}}_+ \times \Gamma_\sigma; E, \tilde{E}).$$

Finally we can talk about corresponding classes of amplitude functions for Fréchet spaces $\tilde{E}$ as in 1.3.(24).

As usual, we set for $h(t, t', z) \in S^\mu(Q \times \Gamma_{1/2-\alpha}; E, \tilde{E})$

$$op^\alpha_M(h)u(t) = t^\alpha op_M(T^{-\alpha}h)t^{-\alpha}u(t) \tag{27}$$

with $(T^{-\alpha}h)(\ldots, z) = h(\ldots, z - \alpha)$ and

$$op_M(a)u(t) = M^{-1}_{z\to t}a(t, t', z)M_{t'\to z}u(t'). \tag{28}$$

This means $op_M := op^0_M$. (27) is first regarded as continuous operator $C_0^\infty(\mathbb{R}_+, E) \to C^\infty(\mathbb{R}_+, \tilde{E})$. The generalities are needed only as a background information, where we restrict ourselves for convenience to Hilbert spaces $E, \tilde{E}$ (cf. also [S2]).

$ML^\mu(\overline{\mathbb{R}}_+; E, \tilde{E})$ $(ML^\mu_{cl}(\overline{\mathbb{R}}_+; E, \tilde{E}))$ is defined as the class of all operators

$$A = op_M(a) + G \tag{29}$$

for arbitrary $a(t, t', z) \in S^\mu(Q \times \Gamma_{1/2}; E, \tilde{E})$ $(S^\mu_{cl}(Q \times \Gamma_{1/2}; E, \tilde{E})$ and

$$G \in \bigcap_{s \in \mathbb{R}} \mathcal{L}(\mathcal{V}^s_{[comp)}(\mathbb{R}_+, E), \mathcal{V}^\infty_{[loc)}(\mathbb{R}_+, \tilde{E}))$$

for which also

$$G^* \in \bigcap_{s \in \mathbb{R}} \mathcal{L}(\mathcal{V}^s_{[comp)}(\mathbb{R}_+, E), \mathcal{V}^\infty_{[loc)}(\mathbb{R}_+, \tilde{E})).$$

G^* is the formal adjoint with respect to the scalar products in $\mathcal{V}^0(\mathbb{R}_+, E)$ and $\mathcal{V}^0(\mathbb{R}_+, \tilde{E})$, respectively. For instance, we define

$$(u, v)_{\mathcal{V}^0(\mathbb{R}_+,E)} = \frac{1}{2\pi i} \int (Mu(z), Mv(z))_E dz.$$

For the notations on *comp*, *loc*-spaces, cf. Definition 1.4.10 below. Clearly we can also define ML^{μ}-$\rightarrow$-$(\mathbb{R}_+;E,\tilde{E})$ $ML^{\mu}_{cl}(\mathbb{R}_+;E,\tilde{E}))$ as the class of all operators (29) with arbitrary $a(t,t',z) \in S^{\mu}(\mathbb{R}_+\times\mathbb{R}_+\times \Gamma_{1/2}; E, \tilde{E})$ $(S^{\mu}_{cl}(\mathbb{R}_+ \times \mathbb{R}_+ \times \Gamma_{1/2}; E, \tilde{E}))$ and G having a kernel in $C^\infty(X^\wedge \times \mathbb{R}_+ \times X^\wedge \times \mathbb{R}_+)$. It can be proved that then

$$ML^{\mu}(\mathbb{R}_+; E, \tilde{E}) = L^{\mu}(\mathbb{R}_+; E, \tilde{E}) \tag{30}$$

and the same for classical operators. Nevertheless the left hand side is needed because of the Mellin operator convention.

For $a(t,t',z) \in S^{\mu}_{cl}(\mathbb{R}_+ \times \mathbb{R}_+ \times \Gamma_\sigma; E, \tilde{E})$ we have a homogeneous principal part

$$a_{(\mu)}(t,t',\tau) \in C^\infty(\mathbb{R}_+ \times \mathbb{R}_+ \times (\mathbb{R} \setminus \{0\}), \mathcal{L}(E, \tilde{E}))$$

satisfying

$$a_{(\mu)}(t,t',\lambda\tau) = \lambda^{\mu} \tilde{\kappa}_\lambda a_{(\mu)}(t,t',\tau)\kappa_\lambda^{-1}$$

for all $\lambda \in \mathbb{R}_+$. If τ is interpreted as $Im\ z$ for $z \in \Gamma_\sigma$ then

$$a(t,t',z) - \chi(\tau)a_{(\mu)}(t,t',\tau) \in S^{\mu-1}_{cl}(\mathbb{R}_+ \times \mathbb{R}_+ \times \Gamma_\sigma; E, \tilde{E}).$$

We set

$$\sigma^{\mu}_{b\wedge}(a)(t,\tau) = a_{(\mu)}(t,t',\tau)_{t'=t} \tag{31}$$

and

$$\sigma^{\mu}_{b\wedge}(A)(t,\tau) = \sigma^{\mu}_{b\wedge}(a)(t,\tau) \tag{32}$$

for any classical operator A being given by (29).

The distributional kernels are $\mathcal{L}(E, \tilde{E})$-valued ones over $\mathbb{R}_+\times\mathbb{R}_+$. The notation properly supported refers to the variables (t, t').

Every $A \in ML^{\mu}(\overline{\mathbb{R}}_+; E, \tilde{E})$ induces continous operators

$$A : \mathcal{V}^s_{[comp)}(\mathbb{R}_+, E) \longrightarrow \mathcal{V}^{s-\mu}_{[loc)}(\mathbb{R}_+, \tilde{E}) \tag{33}$$

for all $s \in \mathbb{R}$. We may write *comp* or *loc* on both sides of (33) if A is properly supported.

We shall not repeat here too much from the general theory of ML^{μ} classes. The details may be found in [S2] and [S1]. Let us only mention the following

Proposition 1.4.9 *We have*

$$ML^{\mu}_{cl}(\overline{\mathbb{R}}_+; E, \tilde{E}) \subset L^{\mu}_{cl}(\mathbb{R}_+; E, \tilde{E}) \tag{34}$$

where the right hand side refers to the notations after 1.3.(28), and

$$\sigma^{\mu}_{\wedge}(t,\tau) = \sigma^{\mu}_{b\wedge}(A)(t,\tau_1)|_{\tau_1=-t\tau} \tag{35}$$

(cf. (32)).

Let us now complete the list of notations around the $\mathcal{V}^s$-spaces. The above $\mathcal{V}^s_{comp,loc}$-spaces are to be understood analogously to Definition 1.3.6.

Definition 1.4.10 *$\mathcal{V}^s_{[loc)}(\mathbb{R}_+,E)$ is defined as the subspace of all $u \in \mathcal{D}'(\mathbb{R}_+,E)$ such that $\varphi u \in \mathcal{V}^s(\mathbb{R}_+,E)$ for every $\varphi \in C_0^\infty(\overline{\mathbb{R}_+})$. Furthermore $\mathcal{V}^s_{[comp)}(\mathbb{R}_+,E)$ is the subspace of all $u \in \mathcal{V}^s_{[loc)}(\mathbb{R}_+,E)$ with bounded support in $t \in \overline{\mathbb{R}}_+$. In an analogous manner we obtain*

$$\mathcal{V}^s_{[comp)}(\mathbb{R}_+,E_1), \mathcal{V}^s_{[loc)}(\mathbb{R}_+,E_1)$$

in the sense of (15). In particular we get

$$\mathcal{V}^s_{[comp)}(\mathbb{R}_+,\mathcal{K}^{s,\beta}_P(X^\wedge)), \mathcal{V}^s_{[loc)}(\mathbb{R}_+,\mathcal{K}^{s,\beta}_P(X^\wedge))$$

for any asymptotic type $P \in As(X,(\beta,\Theta))$.

Furthermore, if we insert $E \oplus \mathbb{C}^N$ instead of E in (1) then κ_λ acts on $(e,c) \in E \oplus \mathbb{C}^N$ as $\kappa_\lambda(e,c) = (\kappa_\lambda e, c)$. Thus

$$\mathcal{V}^s(\mathbb{R}_+,E \oplus \mathbb{C}^N) = \mathcal{V}^s(\mathbb{R}_+,E) \oplus \mathcal{H}^s(\mathbb{R}_+,\mathbb{C}^N)$$

and analogously

$$\mathcal{V}^{s,\alpha}(\mathbb{R}_+,E \oplus \mathbb{C}^N) = \mathcal{V}^{s,\alpha}(\mathbb{R}_+,E) \oplus \mathcal{H}^{s,\alpha}(\mathbb{R}_+,\mathbb{C}^N),$$

cf. (2).

Definition 1.4.11 *Let $\mu,\beta \in \mathbb{R}$, $\Theta = (\vartheta,0]$, $-\infty \leq \vartheta < 0$, $\mathfrak{g} = (\beta,\beta-\mu,\Theta)$. Then*

$$M\mathcal{Y}^{-\infty}_G(X^\wedge \times \mathbb{R}_+,\mathfrak{g};\mathbb{C}^N,\mathbb{C}^M)$$

for given fixed $M,N \in \mathbb{N}$ denotes the space of all

$$\mathcal{G} \in \bigcap_{s\in\mathbb{R}} \mathcal{L}(\mathcal{V}^s_{[comp)}(\mathbb{R}_+,\mathcal{K}^{s,\beta}) \oplus \mathcal{H}^s_{[comp)}(\mathbb{R}_+,\mathbb{C}^N), \mathcal{V}^\infty_{[loc)}(\mathbb{R}_+,\mathcal{K}^{\infty,\beta-\mu}) \oplus \mathcal{H}^\infty_{[loc)}(\mathbb{R}_+,\mathbb{C}^M))$$

which induce continuous operators

$$\begin{aligned}\mathcal{G} : \mathcal{V}^s_{[comp)}(\mathbb{R}_+,\mathcal{K}^{s,\beta}) \oplus \mathcal{H}^s_{[comp)}(\mathbb{R}_+,\mathbb{C}^N) &\longrightarrow \mathcal{V}^\infty_{[loc)}(\mathbb{R}_+,\mathcal{K}^{\infty,\beta-\mu}_P) \oplus \mathcal{H}^\infty_{[loc)}(\mathbb{R}_+,\mathbb{C}^M)\\ \mathcal{G}^* : \mathcal{V}^s_{[comp)}(\mathbb{R}_+,\mathcal{K}^{s,-\beta+\mu}) \oplus \mathcal{H}^s_{[comp)}(\mathbb{R}_+,\mathbb{C}^M) &\longrightarrow \mathcal{V}^\infty_{[loc)}(\mathbb{R}_+,\mathcal{K}^{\infty,-\beta}_Q) \oplus \mathcal{H}^\infty_{[loc)}(\mathbb{R}_+,\mathbb{C}^N)\end{aligned}$$

for all $s \in \mathbb{R}$, for certain $P \in As(X,\mathfrak{a})$ with $\mathfrak{a} = (\beta-\mu,\Theta)$, $Q \in As(X,\mathfrak{b})$ with $\mathfrak{b} = (-\beta,\Theta)$, dependent on $\mathcal{G}$.

$\mathcal{G}^*$ is the formal adjoint of $\mathcal{G}$ in the sense

$$(\mathcal{G}u,v)_{\mathcal{V}^0(\mathbb{R}_+,\mathcal{K}^0)\oplus\mathcal{H}^0(\mathbb{R}_+,\mathbb{C}^M)} = (u,\mathcal{G}^*v)_{\mathcal{V}^0(\mathbb{R}_+,\mathcal{K}^0)\oplus\mathcal{H}^0(\mathbb{R}_+,\mathbb{C}^N)}$$

for all $u \in C_0^\infty(X^\wedge \times \mathbb{R}_+) \oplus C_0^\infty(\mathbb{R}_+,\mathbb{C}^N)$, $v \in C_0^\infty(X^\wedge \times \mathbb{R}_+) \oplus C_0^\infty(\mathbb{R}_+,\mathbb{C}^M)$.

Furthermore we define

$$\mathcal{R}^\nu_G(Q \times \mathbb{R},\mathfrak{g};\mathbb{C}^N,\mathbb{C}^M) \tag{36}$$

(cf. (26)) analogously to Definition 1.3.10, where Ω^2 is everywhere to be replaced by Q and $q=1$ (cf. also the above Definition 1.4.8). We also introduce

$$\mathcal{R}^\nu_G(Q \times \Gamma_\sigma,\mathfrak{g};\mathbb{C}^N,\mathbb{C}^M) \ni g(t,t',z) \tag{37}$$

with obvious notations. The classes of t'-indpendent g will be indicated by $\overline{\mathbb{R}}_+$ instead of Q, and we set

$$R^\nu_G(\overline{\mathbb{R}}_+ \times \Gamma_\sigma,\mathfrak{g}) = l.u.c.\mathcal{R}^\nu_G(\overline{\mathbb{R}}_+ \times \Gamma_\sigma,\mathfrak{g};\mathbb{C}^N,\mathbb{C}^M). \tag{38}$$

We then obtain the operator classes

$$M\mathcal{Y}_G^{\nu}(X^\wedge \times \mathbb{R}_+, \mathfrak{g}; \mathbb{C}^N, \mathbb{C}^M) := \{op_M(g) + \mathcal{G}_0 : g \in \mathcal{R}_G^{\nu}(Q \times \Gamma_{1/2}, \mathfrak{g}; \mathbb{C}^N, \mathbb{C}^M), \mathcal{G}_0 \in M\mathcal{Y}_G^{-\infty}(X^\wedge \times \mathbb{R}_+, \mathfrak{g}; \mathbb{C}^N, \mathbb{C}^M)\} \tag{39}$$

and

$$MY_G^{\nu}(X^\wedge \times \mathbb{R}_+, \mathfrak{g}) = l.u.c. M\mathcal{Y}_G^{\nu}(X^\wedge \times \mathbb{R}_+, \mathfrak{g}; \mathbb{C}^N, \mathbb{C}^M), \tag{40}$$

respectively. The operators in (40) are called Mellin Green wedge operators. The meaning of the other matrix elements in $\mathcal{G} \in$(39) is analogous to 1.3.(36). According to the above notations every $\mathcal{G} \in$(39) has a homogeneous principal "Mellin edge symbol" of order ν

$$\sigma_{b\wedge}^{\nu}(\mathcal{G})(t,\tau) \in \bigcap_{s \in \mathbb{R}} \mathcal{L}^{(\nu)}(T^*\overline{\mathbb{R}}_+ \setminus 0; \mathcal{K}^{s,\beta} \oplus \mathbb{C}^N, \mathcal{K}^{s-\nu,\beta-\mu} \oplus \mathbb{C}^M) \tag{41}$$

where the right hand side of (41) is analogously defined as 1.3.(37). Here we have smoothness in t up to $t = 0$.

Next we pass to the Mellin operators in t with smoothing Mellin operator-valued symbols on $X^\wedge \ni (x, r)$. Let

$$m(t,z) = \omega(r[\tau]) \left\{ r^{-\nu} \sum_{0 \le j \le k-\mu+\nu-1} r^j \sum_{l=0}^{j} \sum_{i=0}^{N} op_{M,r}^{\beta_{j,i}-n/2}(h_{j,i,l})(t) z^l \right\} \tilde{\omega}(r[\tau]) \tag{42}$$

Here $\mu, \nu \in \mathbb{R}$, $\mu - \nu \in \mathbb{N}$, $k \in \mathbb{N} \setminus \{0\}$, $k - \mu + \nu \geq 1$ (otherwise set $m = 0$), $N = N(m)$, $\omega, \tilde{\omega}$ are arbitrary cut-off functions, further

$$h_{j,i,l}(t,w) \in C^\infty(\overline{\mathbb{R}}_+, M_{Q_{j,i,l}}^{-\infty}(X))$$

for certain $Q_{j,i,l} \in \underline{As}(X)$, $sg(h_{j,i,l}) \cap \Gamma_{(n+1)/2-\beta_{j,i}} = \emptyset$ for all i, j, l,

$$\beta - (\mu - \nu) - j \le \beta_{j,i} \le \beta$$

for all i, j. We then have

$$m(t,z) \in S_{cl}^{\nu}(\overline{\mathbb{R}}_+ \times \Gamma_\sigma; \mathcal{K}^{s,\beta}, \mathcal{S}^{\beta-\mu})$$

for all $\sigma \in \mathbb{R}$ (cf 1.2.(16),1.3.(40)). Then

$$\sigma_{b\wedge}^{\nu}(m)(t,\tau) = \omega(r|\tau|) \left\{ r^{-\nu} \sum_{0 \le j \le k-\mu+\nu-1} r^j \sum_{i=0}^{N} op_M^{\beta_{j,i}-n/2}(h_{j,i,j})(t) \tau^j \right\} \tilde{\omega}(r|\tau|), \tag{43}$$

$$\sigma_{b\wedge}^{\nu}(m)(t,\tau) : \mathcal{K}^{s,\beta}(X^\wedge) \longrightarrow \mathcal{K}^{s-\nu,\beta-\mu}(X^\wedge) \tag{44}$$

for all $s \in \mathbb{R}$, $(t, \tau) \in T^*\overline{\mathbb{R}}_+ \setminus 0$. The operator

$$M = op_{M,t}(m) \tag{45}$$

is called a smoothing Mellin operator of the class of Mellin wedge ψDO's.

Definition 1.4.12 *Let $\mu, \nu, \beta \in \mathbb{R}$, $\mu - \nu \in \mathbb{N}$, $\mathfrak{g} = (\beta, \beta - \mu, \Theta)$, $\Theta = (-k, 0]$, $k \in \mathbb{N} \setminus \{0\}$. Then*

$$MY_{M+G}^{\nu}(X^\wedge \times \mathbb{R}_+, \mathfrak{g})$$

is the space of all operators $M + G$ with M being of the sort (45), and $G \in MY_G^{\nu}(X^\wedge \times \mathbb{R}_+, \mathfrak{g})$.

It remains to look at the interior ψDO's. We shall introduce the symbols first on $U^\wedge \times \mathbb{R}_+$ for a coordinate neighbourhood U on X, with local coordinates $x \in \mathbb{R}^n$.

$$S^{\nu}_{2b,cl}(U^\wedge \times \mathbb{R}_+ \times \mathbb{R}^{n+2})_1 \tag{46}$$

is the space of all $p(x, r, t, \xi, r\rho, rt\tau)$ where

$$p(x, r, t, \xi, \rho, \tau) \in S^{\nu}_{cl}(\overline{U^\wedge \times \mathbb{R}_+} \times \mathbb{R}^{n+2})_1. \tag{47}$$

Here $S^{\nu}_{cl}(\overline{U^\wedge \times \mathbb{R}_+} \times \mathbb{R}^{n+2})_1 := S^{\nu}_{cl}(\overline{U^\wedge \times \mathbb{R}_+} \times \mathbb{R}^{n+2}) \cap S^{\nu}_{cl}(U^\wedge \times \mathbb{R}_+ \times \mathbb{R}^{n+2})_1$ (cf. Definition 1.3.12). Let

$$S^{(\nu)}_{2b}(U^\wedge \times \mathbb{R}_+ \times (\mathbb{R}^{n+2} \setminus \{0\}))_1 \tag{48}$$

be the space of all $p_{(\nu)}(x, r, t, \xi, r\rho, rt\tau) \in C^\infty(U^\wedge \times \mathbb{R}_+ \times (\mathbb{R}^{n+2} \setminus \{0\}))$ with $p_{(\nu)}(x, r, t, \xi, \rho, \tau) \in C^\infty(\overline{U^\wedge \times \mathbb{R}_+} \times (\mathbb{R}^{n+2} \setminus \{0\}))$ and

$$p_{(\nu)}(x, r, t, \lambda\xi, \lambda\rho, \lambda\tau) = \lambda^\nu p_{(\nu)}(x, r, t, \xi, \rho, \tau)$$

for all $\lambda > 0$ and all $(x, r, t) \in \overline{U^\wedge \times \mathbb{R}_+}$, $(\xi, \rho, \tau) \neq 0$, and $\chi(\xi, \rho, \tau) p_{(\nu)}(x, r, t, \xi, \rho, \tau) \in S^{\nu}_{cl}(U^\wedge \times \mathbb{R}_+ \times \mathbb{R}^{n+2})_1$ for any (ξ, ρ, τ)-excision function χ. In view of the invariance of (48) under the diffeomorphisms $(\kappa, ({}^t d\kappa)^{-1}) : T^*U \setminus 0 \to T^*U \setminus 0$ we can also define the corresponding global class

$$S^{(\nu)}_{2b}(T^*(X^\wedge \times \mathbb{R}_+) \setminus 0)_1 \tag{49}$$

as the space of all $p \in C^\infty(T^*(X^\wedge \times \mathbb{R}_+) \setminus 0)$ which belong over any coordinate neighbourhood U on X (in the local coordinates x) to the space (48).

We now turn to the operator conventions which assign to any element of (49) a (t, τ)-dependent operator family.

Choose any open covering $\{U_1, \dots, U_N\}$ of X by coordinate neighbourhoods, a subordinated partition of unity $\{\varphi_1, \dots, \varphi_N\}$ and a further system $\{\tilde\varphi_1, \dots, \tilde\varphi_N\}$, $\tilde\varphi_j \in C_0^\infty(U_j)$, with $\varphi_j \tilde\varphi_j = \varphi_j$ for all j. If $p \in$(49) and

$$p_{(j)} := p|_{T^*(U_j^\wedge \times \mathbb{R}_+)\setminus 0} \quad, j = 1, \dots, N,$$

then we can form

$$p_j(x, r, t, \xi, r\rho, rt\tau) = \chi(\xi, r\rho, rt\tau) p_{(j)}(x, r, t, \xi, r\rho, rt\tau)$$

with some (ξ, ρ, τ)-excision function $\chi(\xi, \rho, \tau)$. We now define

$$op_{\psi,(x,r)}(p)(t, \tau) = \sum_{j=1}^{N} \varphi_j op_{\psi,(x,r)}(p_j)(t, \tau) \tilde\varphi_j \tag{50}$$

where $op_{\psi,(x,r)}$ on the right hand side is defined analogously to 1.3.(48). It follows analogously to Remark 1.3.13 that

$$op_{\psi,(x,r)}(p)(t, \tau) \in C^\infty(\overline{\mathbb{R}_+}, L^{\nu}_{cl}(X^\wedge; \mathbb{R}_\tau)),$$

and p is the parameter-dependent homogeneous principal symbol of order ν, calculated point-wise for every $t \in \overline{\mathbb{R}}_+$. By carrying out the ψD action in t by means of the Fourier transform on $\mathbb{R} \ni t$, we get

$$op_{\psi,(x,r,t)}(p) := op_{\psi,t}(op_{\psi,(x,r)}(p)(t, \tau)) \in L^{\nu}_{cl}(X^\wedge \times \mathbb{R}_+).$$

The final operator conventions in the present set-up will be obtained in several steps. First we switch to the Mellin convention with respect to r and t. Both reformulations of the original action will be glued together then by τ dependent cut-off functions in the r-variable.

Theorem 1.4.13 *Let the mentioned data $\{U_j\}$, $\{\varphi_j\}$, $\{\tilde{\varphi}_j\}$ be fixed and $p \in$(49) be given. Then there exists a system of symbols*

$$\underline{h}_j(x,r,t,\xi,\rho,z) \in S^{\nu}_{cl}(U_j \times \overline{\mathbb{R}}_+ \times \overline{\mathbb{R}}_+ \times \mathbb{R}^{n+1}_{\xi,\rho} \times \Gamma_{1/2-\alpha})_1$$

for every $\alpha \in \mathbb{R}$ (cf. Definition 1.3.12 in the version for $U_j \times \overline{\mathbb{R}}_+ \times \overline{\mathbb{R}}_+$ instead of $U \times \mathbb{R}_+ \times \Omega$ and $\Gamma_{1/2-\alpha} \ni z$ instead of $\mathbb{R}_\tau$) such that for

$$h_j(x,r,t,\xi,\rho,z) := \underline{h}_j(x,r,t,\xi,r\rho,1/2-\alpha+ir\tau)$$

and

$$f_1(t,z) := \sum_{j=1}^{N} \varphi_j op_{\psi,(x,r)}(h_j)(t,z)\tilde{\varphi}_j \tag{51}$$

it follows

$$op_{\psi,(x,r,t)}(p) = op^{\alpha}_{M,t}(f_1) \ mod \ L^{-\infty}_{cl}(X^\wedge \times \mathbb{R}_+). \tag{52}$$

Clearly $op^{\alpha}_{M,t}$ indicates the weighted Mellin action in t.

Note that

$$f_1(t,z) \in C^\infty(\overline{\mathbb{R}}_+, L^{\nu}_{cl}(X^\wedge;\Gamma_{1/2-\alpha})).$$

Theorem 1.4.14 *Let $\{U_j\}$, $\{\varphi_j\}$, $\{\tilde{\varphi}_j\}$ be fixed and $p \in$(49) be given. Then there exists an*

$$\underline{h}(r,t,w,z) \in C^\infty(\overline{\mathbb{R}}_+ \times \overline{\mathbb{R}}_+, \mathcal{A}(\mathbb{C}_z, M^{\nu}_{O}(X))) \tag{53}$$

with

$$\underline{h}(r,t,w,\sigma+i\tau) \in C^\infty(\overline{\mathbb{R}}_+ \times \overline{\mathbb{R}}_+, M^{\nu}_{O}(\dot{X};\Gamma_\sigma))$$

for all $\sigma \in \mathbb{R}$, such that for

$$f_0(r,t,w,z) := \underline{h}(r,t,w,rz)$$

it follows

$$op_{\psi,(x,r,t)}(p) = op^{\alpha}_{M,t} op^{\beta-n/2}_{M,r}(f_0) \ mod \ L^{-\infty}(X^\wedge \times \mathbb{R}_+) \tag{54}$$

for all $\alpha, \beta \in \mathbb{R}$. Furthermore

$$f_1(t,z) = op^{\beta-n/2}_{M,r}(f_0)(t,z) \ mod \ C^\infty(\overline{\mathbb{R}}_+, L^{-\infty}(X^\wedge;\Gamma_{1/2-\alpha})). \tag{55}$$

The proofs of Theorem 1.4.13 and Theorem 1.4.14 may be found in [S3], up to minor modifications (and the simple observation that f_0 can be chosen independently of α, β). Clearly f_0, f_1 are unique elements of the corresponding sorts of orders $-\infty$.

Let us define

$$(c_\wedge h_j)(x,r,t,\xi,\rho,\tau) := \underline{h}_j(x,0,t,\xi,r\rho,1/2-\alpha+ir\tau)$$

and set

$$(c_\wedge f_1)(t,\tau) := \sum_{j=1}^{N} \varphi_j op_{\psi,(x,r)}(c_\wedge h_j)(t,\tau)\tilde{\varphi}_j.$$

Furthermore we form

$$(c_\wedge f_0)(r,t,w,\tau) := \underline{h}(0,t,w,ir\tau)$$

and set

$$(c_\wedge op^{\beta-n/2}_{M,r}(f_0))(t,\tau) := op^{\beta-n/2}_{M,r}(c_\wedge f_0)(t,\tau).$$

Let us fix cut-off functions $\omega(r), \omega_0(r), \omega_1(r)$ satisfying 1.3.(58). Then we have the operator families

$$a_0(t,z) = \omega(r[\tau])r^{-\nu}op_{M,r}^{\beta-n/2}(f_0)(t,z)\omega_0(r[\tau]), \tag{56}$$
$$a_1(t,z) = (1-\omega(r[\tau]))r^{-\nu}f_1(t,z)(1-\omega_1(r[\tau])), \tag{57}$$

where

$$a_0(t,z)+a_1(t,z) : \mathcal{K}^{s,\beta}(X^\wedge) \longrightarrow \mathcal{K}^{s-\nu,\beta-\mu}(X^\wedge) \tag{58}$$

is continuous for all $s \in \mathbb{R}$ and $\mu - \nu \in \mathbb{N}$. Set

$$a_{(\nu)}(t,\tau) = \omega(r|\tau|)r^{-\nu}c_\wedge op_M^{\beta-n/2}(f_0)(t,\tau)\omega_0(r|\tau|) + (1-\omega(r|\tau|))r^{-\nu}c_\wedge f_1(t,\tau)(1-\omega_1(r|\tau|)). \tag{59}$$

This is a family of continuous operators

$$a_{(\nu)}(t,\tau) : \mathcal{K}^{s,\beta}(X^\wedge) \longrightarrow \mathcal{K}^{s-\nu,\beta-\mu}(X^\wedge) \tag{60}$$

parametrized by $t \in \overline{\mathbb{R}}_+$, $\tau \in \mathbb{R}$, satisfying

$$a_{(\nu)}(t,\lambda\tau) = \lambda^\nu \kappa_\lambda a_{(\nu)}(t,\tau)\kappa_\lambda^{-1} \tag{61}$$

for all $\lambda \in \mathbb{R}_+$.

Definition 1.4.15 *Let $\mu,\nu,\beta \in \mathbb{R}$, $\mu-\nu \in \mathbb{N}$, $\mathfrak{g} = (\beta, \beta-\mu, \Theta)$, $\Theta = (-k, 0]$, $k \in \mathbb{N}\setminus\{0\}$. Then*

$$MY^\nu(X^\wedge \times \mathbb{R}_+, \mathfrak{g})$$

is the space of all operators

$$A = A_0 + A_1 + M + G$$

with $M + G \in MY^\nu_{M+G}(X^\wedge \times \mathbb{R}_+, \mathfrak{g})$ (cf. Definition 1.4.12) and $A_i = op_M(a_i)$, $i = 0,1$, with $a_i(t,z)$ being constructed for any $p \in$(49) with the corresponding $f_0(r,t,w,z)$, $f_1(t,z)$ from Theorem 1.4.23 and Theorem 1.4.14 (where here $\alpha = 0$). Finally

$$M\mathcal{Y}^\nu(X^\wedge \times \mathbb{R}_+, \mathfrak{g}; \mathbb{C}^N, \mathbb{C}^M) \tag{62}$$

is the space of all

$$\mathcal{A} = \begin{pmatrix} A & 0 \\ 0 & 0 \end{pmatrix} + \mathcal{G} \tag{63}$$

with $A \in MY^\nu(X^\wedge \times \mathbb{R}_+, \mathfrak{g})$, $\mathcal{G} \in M\mathcal{Y}^\nu_G(X^\wedge \times \mathbb{R}_+, \mathfrak{g}; \mathbb{C}^N, \mathbb{C}^M)$.

Remark 1.4.16 *The restriction of Definition 1.4.15 to (t,z)-dependent $a_i(t,z)$, $m(t,z)$ could be dropped; we might allow (t,t',z)-dependent ones, smooth up to $t,t' = 0$. But it is a property of the calculus that this can be reduced to the (t,z)-dependent case, mod $MY_G^{-\infty}(X^\wedge \times \mathbb{R}_+, \mathfrak{g})$.*

Theorem 1.4.17 *An $\mathcal{A} \in M\mathcal{Y}^\nu(X^\wedge \times \mathbb{R}_+, \mathfrak{g}; \mathbb{C}^N, \mathbb{C}^M)$ induces continuous operators*

$$\mathcal{A} : \mathcal{V}^s_{[comp)}(\mathbb{R}_+, \mathcal{K}^{s,\beta}) \oplus \mathcal{H}^s_{[comp)}(\mathbb{R}_+, \mathbb{C}^N) \longrightarrow \mathcal{V}^{s-\nu}_{[loc)}(\mathbb{R}_+, \mathcal{K}^{s-\nu,\beta-\mu}) \oplus \mathcal{H}^{s-\nu}_{[loc)}(\mathbb{R}_+, \mathbb{C}^M) \tag{64}$$

for all $s \in \mathbb{R}$, and

$$\mathcal{A} : \mathcal{V}^s_{[comp)}(\mathbb{R}_+, \mathcal{K}_P^{s,\beta}) \oplus \mathcal{H}^s_{[comp)}(\mathbb{R}_+, \mathbb{C}^N) \longrightarrow \mathcal{V}^{s-\nu}_{[loc)}(\mathbb{R}_+, \mathcal{K}_Q^{s-\nu,\beta-\mu}) \oplus \mathcal{H}^{s-\nu}_{[loc)}(\mathbb{R}_+, \mathbb{C}^M) \tag{65}$$

or every $P \in As(X, (\beta,\Theta))$ with some $Q \in As(X, (\beta-\mu,\Theta))$ dependent on P and $\mathcal{A}$, not on s. If $\mathcal{A}$ is properly supported with respect to t then we may insert on both sides of (64),(65) [comp) or [loc).

As in the previous section we have two leading symbolic levels. From

$$MY^{\nu}(X^{\wedge} \times \mathbb{R}_{+}, \mathfrak{g}) \subset L^{\nu}_{cl}(X^{\wedge} \times \mathbb{R}_{+})$$

we get the homogeneous principal interior symbolic map

$$\begin{aligned} \sigma^{\nu}_{\psi} : MY^{\nu}(X^{\wedge} \times \mathbb{R}_{+}, \mathfrak{g}) &\longrightarrow r^{-\nu} S^{(\nu)}_{2b}(T^{*}(X^{\wedge} \times \mathbb{R}_{+}) \setminus 0)_{1}, \\ \sigma^{\nu}_{\psi}(A) &:= r^{-\nu} p \end{aligned} \tag{66}$$

with p from Definition 1.4.15. Moreover there is the homogeneous principal Mellin edge symbol of order ν

$$\sigma^{\nu}_{b\wedge} : MY^{\nu}(X^{\wedge} \times \mathbb{R}_{+}, \mathfrak{g}) \longrightarrow \bigcap_{s \in \mathbb{R}} \mathcal{L}^{(\nu)}(T^{*}\overline{\mathbb{R}}_{+} \setminus 0; \mathcal{K}^{s,\beta}, \mathcal{K}^{s-\nu,\beta-\mu}), \tag{67}$$

where

$$\sigma^{\nu}_{b\wedge}(A)(t,\tau) = a_{(\nu)}(t,\tau) + \sigma^{\nu}_{b\wedge}(M)(t,\tau) + \sigma^{\nu}_{b\wedge}(G)(t,\tau), \tag{68}$$

cf. (59), (41), (44). For $\mathcal{A} \in M\mathcal{Y}^{\nu}(X^{\wedge} \times \mathbb{R}_{+}, \mathfrak{g}; \mathbb{C}^{N}, \mathbb{C}^{M})$ we use analogous notations to 1.3.(64),(65).

Theorem 1.4.18 *We have*

$$M\mathcal{Y}^{\nu}(X^{\wedge} \times \mathbb{R}_{+}, \mathfrak{g}; \mathbb{C}^{N}, \mathbb{C}^{M}) \subset \mathcal{Y}^{\nu}(X^{\wedge} \times \mathbb{R}_{+}, \mathfrak{g}; \mathbb{C}^{N}, \mathbb{C}^{M})$$

(cf. Definition 1.3.15) and

$$\sigma^{\nu}_{\wedge}(\mathcal{A})(t,\tau) = \sigma^{\nu}_{b\wedge}(\mathcal{A})(t,\tau_1)|_{\tau_1 = -t\tau} \tag{69}$$

for every $\mathcal{A} \in M\mathcal{Y}^{\nu}(X^{\wedge} \times \mathbb{R}_{+}, \mathfrak{g}; \mathbb{C}^{N}, \mathbb{C}^{M})$ *(cf. the above Proposition 1.4.9).*

In order to introduce (62) globally over $\mathbb{W}$ (cf. 1.2.(25)) we begin with the negligible operators. To this end we shall introduce first the global analogues of the spaces (3),(19),(20). Since we are dealing here with the calculus locally close to a corner, $\mathbb{W}$ is of the form $\mathbb{B} \times \mathbb{R}_{+}$.

The idea will be to glue together the objects on $X^{\wedge} \times \mathbb{R}_{+} \ni (x,r,t)$ close to $r = 0$ and those on $int\mathbb{B} \times \mathbb{R}_{+} \ni (\tilde{x}, t)$ where $int\mathbb{B}$ is regarded as an open subset of some closed compact C^{∞} manifold $\tilde{X}$ ($n+1 = dim\tilde{X}$). We shall fix $\tilde{X}$ once and for all. Then Definition 1.1.1 gives us the spaces $\mathcal{H}^{s,\alpha}(\tilde{X}^{\wedge})$ for every $s, \alpha \in \mathbb{R}$. Let $\tilde{\omega}(r)$ denote a cut-off function supported by $V \cong X \times [0,1)$ (cf. 1.1.(15)). Then $1 - \tilde{\omega}(r)$ vanishes in a collar neighbourhood of $\partial\mathbb{B}$. Set

$$\mathcal{V}^{s}(\mathbb{R}_{+}, \mathcal{H}^{s,\beta}(\mathbb{B})) = [\tilde{\omega}]\mathcal{V}^{s}(\mathbb{R}_{+}, \mathcal{K}^{s,\beta}(X^{\wedge})) + [1 - \tilde{\omega}]\mathcal{H}^{s,(n+1)/2}(\tilde{X}^{\wedge}) \tag{70}$$

Here we have dropped again for abbreviation pull-backs under 1.1.(15). Clearly, $[1-\tilde{\omega}]\mathcal{H}^{s,(n+1)/2}(\tilde{X}^{\wedge})$ is canonically interpreted as a space of distributions over $int\mathbb{B} \times \mathbb{R}_{+}$ with the corresponding Banach space structure. Analogously we set

$$\mathcal{V}^{s}(\mathbb{R}_{+}, \mathcal{H}^{s,\beta}_{P}(\mathbb{B})) = [\tilde{\omega}]\mathcal{V}^{s}(\mathbb{R}_{+}, \mathcal{K}^{s,\beta}_{P}(X^{\wedge})) + [1 - \tilde{\omega}]\mathcal{H}^{s,(n+1)/2}(\tilde{X}^{\wedge}) \tag{71}$$

for any $P \in As(X, (\beta, \Theta))$. In particular we may insert Θ instead of P.

(70) is a Banach space, (71) a Fréchet space, independent of the concrete choice of $\tilde{\omega}$. The operator of multiplication by any $\psi(t) \in C^{\infty}_{0}(\overline{\mathbb{R}}_{+})$ acts continuously on (70),(71). This allows the definition of

$$\mathcal{V}^{s}_{[loc)}(\mathbb{R}_{+}, \mathcal{H}^{s,\beta}(\mathbb{B}))$$

as the subspace of all $u \in \mathcal{D}'(int\mathbb{B} \times \mathbb{R}_{+})$ with $\psi u \in$(70) for every $\psi \in C^{\infty}_{0}(\overline{\mathbb{R}}_{+})$. Moreover

$$\mathcal{V}^{s}_{[comp)}(\mathbb{R}_{+}, \mathcal{H}^{s,\beta}(\mathbb{B}))$$

is the space of all $u \in$(70) with bounded support in t. Analogously we get

$$\mathcal{V}^{s}_{[loc)}(\mathbb{R}_{+}, \mathcal{H}^{s,\beta}_{P}(B)), \mathcal{V}^{s}_{[loc)}(\mathbb{R}_{+}, \mathcal{H}^{s,\beta}_{P}(\mathbb{B})) \tag{72}$$

for any asymptotic type P, all spaces being equipped with the natural locally convex topologies.

Theorem 1.4.19 *For every $s, \beta \in \mathbb{R}$ we have*

$$\mathcal{V}^s(\mathbb{R}_+, \mathcal{H}^{s,\beta}(\mathbb{B})) \subset \mathcal{W}^{s,\beta}_{loc}(\mathbb{W}),$$
$$\mathcal{V}^s(\mathbb{R}_+, \mathcal{H}^{s,\beta}_P(\mathbb{B})) \subset \mathcal{W}^{s,\beta}_{\tilde{P},loc}(\mathbb{W}),$$

(cf. 1.3.(18),(19)) for every $P \in As(X, \mathfrak{a}^\bullet)(\in As(X,\mathfrak{a}))$ with some $\tilde{P} \in As(X,\mathfrak{a}^\bullet)(\in As(X,\mathfrak{a}))$, $\mathfrak{a} = (\beta, \Theta)$.

Definition 1.4.20 *Let $\mu, \beta \in \mathbb{R}$, $\Theta = (\vartheta, 0]$, $-\infty \leq \vartheta < 0$, $\mathfrak{g} = (\beta, \beta - \mu, \Theta)$. Then*

$$M\mathcal{Y}^{-\infty}_G(\mathbb{W}, \mathfrak{g}; \mathbb{C}^N, \mathbb{C}^M) \tag{73}$$

for $\mathbb{W} = \mathbb{B} \times \mathbb{R}_+$ denotes the space of all

$$\mathcal{G} \in \bigcap_{s \in \mathbb{R}} \mathcal{L}(\mathcal{V}^s_{[comp)}(\mathbb{R}_+, \mathcal{H}^{s,\beta}(\mathbb{B})) \oplus \mathcal{H}^s_{[comp)}(\mathbb{R}_+, \mathbb{C}^N), \mathcal{V}^\infty_{[loc)}(\mathbb{R}_+, \mathcal{H}^{\infty,\beta-\mu}(\mathbb{B})) \oplus \mathcal{H}^\infty_{[loc)}(\mathbb{R}_+, \mathbb{C}^M)))$$

which induce continuous operators

$$\begin{aligned} \mathcal{G} : \mathcal{V}^s_{[comp)}(\mathbb{R}_+, \mathcal{H}^{s,\beta}(\mathbb{B})) \oplus \mathcal{H}^s_{[comp)}(\mathbb{R}_+, \mathbb{C}^N) &\longrightarrow \mathcal{V}^\infty_{[loc)}(\mathbb{R}_+, \mathcal{H}^{\infty,\beta-\mu}_P(\mathbb{B})) \oplus \mathcal{H}^\infty_{[loc)}(\mathbb{R}_+, \mathbb{C}^M), \\ \mathcal{G}^* : \mathcal{V}^s_{[comp)}(\mathbb{R}_+, \mathcal{H}^{s,-\beta+\mu}(\mathbb{B})) \oplus \mathcal{H}^s_{[comp)}(\mathbb{R}_+, \mathbb{C}^M) &\longrightarrow \mathcal{V}^\infty_{[loc)}(\mathbb{R}_+, \mathcal{H}^{\infty,-\beta}_Q(\mathbb{B})) \oplus \mathcal{H}^\infty_{[loc)}(\mathbb{R}_+, \mathbb{C}^N) \end{aligned}$$

for all $s \in \mathbb{R}$, for certain $P \in As(X, \mathfrak{a})$ with $\mathfrak{a} = (\beta - \mu, \Theta)$, $Q \in As(X, \mathfrak{b})$ with $\mathfrak{b} \in (-\beta, \Theta)$, dependent on $\mathcal{G}$.

$\mathcal{G}^*$ is the formal adjoint of $\mathcal{G}$ in the sense

$$(\mathcal{G}u, v)_{\mathcal{V}^0(\mathbb{R}_+, \mathcal{H}^0(\mathbb{B})) \oplus \mathcal{H}^0(\mathbb{R}_+, \mathbb{C}^M)} = (u, \mathcal{G}^* v)_{\mathcal{V}^0(\mathbb{R}_+, \mathcal{H}^0(\mathbb{B})) \oplus \mathcal{H}^0(\mathbb{R}_+, \mathbb{C}^M)}$$

for all $u \in C_0^\infty(int\mathbb{B} \times \mathbb{R}_+) \oplus C_0^\infty(\mathbb{R}_+, \mathbb{C}^N)$.

We also write

$$MY^{-\infty}_G(\mathbb{W}, \mathfrak{g}) = l.u.c.M\mathcal{Y}^{-\infty}_G(\mathbb{W}, \mathfrak{g}; \mathbb{C}^N, \mathbb{C}^M).$$

Definition 1.4.21 *$M\mathcal{Y}^\nu(\mathbb{W}, \mathfrak{g}; \mathbb{C}^N, \mathbb{C}^M)$ (with μ, ν, and the weight data $\mathfrak{g}$ being as in Definition 1.4.15 and $\mathbb{W} = \mathbb{B} \times \mathbb{R}_+$) is the space of all*

$$\mathcal{A} = \begin{pmatrix} A & 0 \\ 0 & 0 \end{pmatrix} + \mathcal{G} \tag{74}$$

with $A \in L^\nu_{cl}(int\mathbb{W})$ and $\mathcal{G}$ having kernel in 1.3.(68) where we have to insert $\mathbb{R}_+ = Y$, such that for arbitrary cut-off functions $\omega_1(r), \omega_2(r)$ and $\varphi_1(y), \varphi_2(y) \in C_0^\infty(\overline{\mathbb{R}}_+)$ the operator of the form 1.3.(69) belongs to $M\mathcal{Y}^\nu(X^\wedge \times \mathbb{R}_+, \mathfrak{g}; \mathbb{C}^N, \mathbb{C}^M)$ (cf. Definition 1.4.15) and to $M\mathcal{Y}^{-\infty}_G(\mathbb{W}, \mathfrak{g}; \mathbb{C}^N, \mathbb{C}^M)$ whenever $supp\ \varphi_1 \cap supp\ \varphi_2 = \emptyset$.

For $\Theta = (-\infty, 0]$ we define the corresponding operator spaces as intersections over $k \in \mathbb{N} \setminus \{0\}$ of those for $\Theta_k = (-k, 0]$.

Theorem 1.4.22 *Every $\mathcal{A} \in M\mathcal{Y}^\nu(\mathbb{W}, \mathfrak{g}; \mathbb{C}^N, \mathbb{C}^M)$ induces continuous operators*

$$\mathcal{A} : \mathcal{V}^s_{[comp)}(\mathbb{R}_+, \mathcal{H}^{s,\beta}(\mathbb{B})) \oplus \mathcal{H}^s_{[comp)}(\mathbb{R}_+, \mathbb{C}^N) \longrightarrow \mathcal{V}^{s-\nu}_{[loc)}(\mathbb{R}_+, \mathcal{H}^{s-\nu,\beta-\mu}(\mathbb{B})) \oplus \mathcal{H}^{s-\nu}_{[loc)}(\mathbb{R}_+, \mathbb{C}^M) \tag{75}$$

for all $s \in \mathbb{R}$, further

$$\mathcal{A} : \mathcal{V}^s_{[comp)}(\mathbb{R}_+, \mathcal{H}^{s,\beta}_P(\mathbb{B})) \oplus \mathcal{H}^s_{[comp)}(\mathbb{R}_+, \mathbb{C}^N) \longrightarrow \mathcal{V}^{s-\nu}_{[loc)}(\mathbb{R}_+, \mathcal{H}^{s-\nu,\beta-\mu}_Q(\mathbb{B})) \oplus \mathcal{H}^{s-\nu}_{[loc)}(\mathbb{R}_+, \mathbb{C}^M) \tag{76}$$

for every $P \in As(X, (\beta, \Theta))$ with some $Q \in As(X, (\beta - \mu, \Theta))$ only dependent on $P, \mathcal{A}$, not on s.

The global leading symbolic level also consists of two components, namely

$$\sigma_\psi^\nu : M\mathcal{Y}^\nu(\mathbb{W}, \mathfrak{g}; \mathbb{C}^N, \mathbb{C}^M) \longrightarrow g_0^\nu(r) S_{2b}^{(\nu)}(T^*(int\mathbb{W}) \setminus 0) \tag{77}$$

(cf. 1.1.(28), $g_0^\eta = r^\eta$ close to $\partial\mathbb{B}$, C^∞ and strictly positive elsewhere). (77) acts again only on

$$MY^\nu(\mathbb{W}, \mathfrak{g}) := l.u.c. M\mathcal{Y}^\nu(\mathbb{W}, \mathfrak{g}; \mathbb{C}^N, \mathbb{C}^M). \tag{78}$$

Moreover we have an obvious generalization of (68), the homogeneous principal Mellin edge symbolic map of order ν

$$\sigma_{b\wedge}^\nu : M\mathcal{Y}^\nu(\mathbb{W}, \mathfrak{g}; \mathbb{C}^N, \mathbb{C}^M) \longrightarrow \bigcap_{s\in\mathbb{R}} \mathcal{L}^{(\nu)}(T^*\overline{\mathbb{R}}_+ \setminus 0; \mathcal{K}^{s,\beta} \oplus \mathbb{C}^N, \mathcal{K}^{s-\nu,\beta-\mu} \oplus \mathbb{C}^M) \tag{79}$$

Theorem 1.4.23 *We have*

$$M\mathcal{Y}^\nu(\mathbb{W}, \mathfrak{g}; \mathbb{C}^N, \mathbb{C}^M) \subset \mathcal{Y}^\nu(\mathbb{W}, \mathfrak{g}; \mathbb{C}^N, \mathbb{C}^M)$$

(cf. Definition 1.3.19) and

$$\sigma_\wedge^\nu(\mathcal{A})(t,\tau) = \sigma_{b\wedge}^\nu(\mathcal{A})(t,\tau_1)|_{\tau_1=-t\tau}$$

for every $\mathcal{A} \in M\mathcal{Y}^\nu(\mathbb{W}, \mathfrak{g}; \mathbb{C}^N, \mathbb{C}^M)$.

The material up to now suggests again the standard elements of the calculus. We will content ourselves with the following

Theorem 1.4.24 *Let* $\mathcal{A} \in M\mathcal{Y}^\nu(\mathbb{W}, \mathfrak{g}; \mathbb{C}^N, \mathbb{C}^M)$ *for* $\mathfrak{g} = (\beta, \beta-\mu, \Theta)$, $\mathcal{B} \in M\mathcal{Y}^{\tilde\nu}(\mathbb{W}, \tilde{\mathfrak{g}}; \mathbb{C}^M, \mathbb{C}^L)$ *for* $\tilde{\mathfrak{g}} = (\beta-\mu, \beta-\mu-\tilde\mu, \Theta)$ *with* $\mathcal{A}$ *or* $\mathcal{B}$ *being properly supported with respect to the t-variable. Then* $\mathcal{BA} \in M\mathcal{Y}^{\nu+\tilde\nu}(\mathbb{W}, \mathfrak{g}_0; \mathbb{C}^N, \mathbb{C}^L)$ *for* $\mathfrak{g}_0 = (\beta, \beta-\mu-\tilde\mu, \Theta)$, *and*

$$\begin{aligned} \sigma_\psi^{\tilde\nu+\nu}(\mathcal{BA}) &= \sigma_\psi^{\tilde\nu}(\mathcal{B})\sigma_\psi^\nu(\mathcal{A}), \\ \sigma_{b\wedge}^{\tilde\nu+\nu}(\mathcal{BA}) &= \sigma_{b\wedge}^{\tilde\nu}(\mathcal{B})\sigma_{b\wedge}^\nu(\mathcal{A}). \end{aligned}$$

It remains to look at the ellipticity in the class $M\mathcal{Y}^\mu(\mathbb{W}, \mathfrak{g}; \mathbb{C}^N, \mathbb{C}^M)$, $\mathbb{W} = \mathbb{B} \times \mathbb{R}_+$.

Definition 1.4.25 *An operator* $\mathcal{A} \in M\mathcal{Y}^\mu(\mathbb{W}, \mathfrak{g}; \mathbb{C}^N, \mathbb{C}^M)$ *for* $\mathfrak{g} = (\beta, \beta-\mu, \Theta)$, $\Theta = (-k, 0]$, $k \in \mathbb{N} \setminus \{0\}$, *is called elliptic if*

(i) $\sigma_\psi^\mu(\mathcal{A}) \neq 0$ *on* $T^*(int\mathbb{W}) \setminus 0$ *in the usual sense and if* $p = g_0^\mu(r)\sigma_\psi^\mu(\mathcal{A})$ *in the coordinates* (x, r, t) *close to* $r = 0, t = 0$, *where* $p = p(x, r, t, \xi, r\rho, rt\tau)$, *satisfies* $p(x, r, t, \xi, \rho, \tau) \neq 0$ *up to* $r = 0, t = 0$, *and* $p = p(\tilde x, t, \tilde\xi, t\tau)$ *in the coordinates* $(\tilde x, t) \in int\mathbb{B} \times \mathbb{R}_+$ *close to* $t = 0$ *satisfies* $p(\tilde x, t, \tilde\xi, \tau) \neq 0$ *up to* $t = 0$,

(ii)

$$\sigma_{b\wedge}^\mu(\mathcal{A})(t,\tau) : \begin{array}{c} \mathcal{K}^{s,\beta}(X^\wedge) \\ \oplus \\ \mathbb{C}^N \end{array} \longrightarrow \begin{array}{c} \mathcal{K}^{s-\mu,\beta-\mu}(X^\wedge) \\ \oplus \\ \mathbb{C}^M \end{array} \tag{80}$$

is an isomorphism for all $(t,\tau) \in T^*\overline{\mathbb{R}}_+ \setminus 0$ *for some* $s \in \mathbb{R}$.

If (ii) is satisfied for one $s = s_0$ then automatically for all $s \in \mathbb{R}$.

Definition 1.4.26 *Let* $\mathcal{A} \in M\mathcal{Y}^\mu(\mathbb{W}, \mathfrak{g}; \mathbb{C}^N, \mathbb{C}^M)$, $\mathfrak{g} = (\beta, \beta-\mu, \Theta)$ *and* $\mathcal{P} \in M\mathcal{Y}^{-\mu}(\mathbb{W}, \mathfrak{b}; \mathbb{C}^M, \mathbb{C}^N)$, $\mathfrak{b} = (\beta-\mu, \beta, \Theta)$, $\mathcal{A}$ *or* $\mathcal{B}$ *being properly supported. Then* $\mathcal{P}$ *is called a parametrix of* $\mathcal{A}$ *if*

$$\begin{aligned} \mathcal{PA} - 1 &\in M\mathcal{Y}_G^{-\infty}(\mathbb{W}, \mathfrak{g}_l; \mathbb{C}^N, \mathbb{C}^N), \\ \mathcal{AP} - 1 &\in M\mathcal{Y}_G^{-\infty}(\mathbb{W}, \mathfrak{g}_r; \mathbb{C}^M, \mathbb{C}^M) \end{aligned}$$

with $\mathfrak{g}_l = (\beta, \beta, \Theta)$, $\mathfrak{g}_r = (\beta-\mu, \beta-\mu, \Theta)$, *and* 1 *being the identical operator in the corresponding* $M\mathcal{Y}^0$*-classes.*

Theorem 1.4.27 *Let $\mathcal{A} \in M\mathcal{Y}^{\mu}(\mathbb{W}, \mathfrak{g}; \mathbb{C}^N, \mathbb{C}^M)$ be elliptic, $\mathfrak{g} = (\beta, \beta - \mu, \Theta)$. Then $\mathcal{A}$ has a properly supported parametrix $P \in M\mathcal{Y}^{-\mu}(\mathbb{W}, \mathfrak{b}; \mathbb{C}^M, \mathbb{C}^N)$, $\mathfrak{b} = (\beta - \mu, \beta, \Theta)$. Moreover*

$$\mathcal{A}u = f \in \mathcal{V}^{s}_{[loc)}(\mathbb{R}_+, \mathcal{H}^{s,\beta-\mu}(\mathbb{B})) \oplus \mathcal{H}^{s}_{[loc)}(\mathbb{R}_+, \mathbb{C}^M)$$

for some fixed $s \in \mathbb{R}$, and

$$u \in \mathcal{V}^{-\infty}_{[comp)}(\mathbb{R}_+, \mathcal{H}^{-\infty,\beta}(\mathbb{B})) \oplus \mathcal{H}^{-\infty}_{[comp)}(\mathbb{R}_+, \mathbb{C}^N) \tag{81}$$

imply $u \in \mathcal{V}^{s+\mu}_{[comp)}(\mathbb{R}_+, \mathcal{H}^{s+\mu,\beta}(\mathbb{B})) \oplus \mathcal{H}^{s+\mu}_{[comp)}(\mathbb{R}_+, \mathbb{C}^N)$. Furthermore

$$\mathcal{A}u = f \in \mathcal{V}^{s}_{[loc)}(\mathbb{R}_+, \mathcal{H}^{s,\beta-\mu}_{Q}(\mathbb{B})) \oplus \mathcal{H}^{s}_{[loc)}(\mathbb{R}_+, \mathbb{C}^M) \tag{82}$$

for some $Q \in As(X, (\beta - \mu, \Theta))$ and (81) imply

$$u \in \mathcal{V}^{s+\mu}_{[comp)}(\mathbb{R}_+, \mathcal{H}^{s+\mu,\beta}_{P}(\mathbb{B})) \oplus \mathcal{H}^{s+\mu}_{[comp)}(\mathbb{R}_+, \mathbb{C}^N) \tag{83}$$

for a resulting $P \in As(X, (\beta, \Theta))$, dependent on $\mathcal{A}$ and P.

Remark 1.4.28 *The role of the ellipticity and the elliptic regularity with asymptotics in the set-up of the operator class $M\mathcal{Y}^{\mu}(\mathbb{W}, \mathfrak{g}; \mathbb{C}^N, \mathbb{C}^M)$, $\mathbb{W} = \mathbb{B} \times \mathbb{R}_+$, is to describe the corresponding properties along $K_1 \setminus K_0$ close to K_0 in the final corner ψDO calculus. In this sense (83) gives the edge contribution to the asymptotics close to the corners. The equality was described above by the singular functions (22). In practice we will often have more precise information in the sense of (23).*

2 Mellin operators and corner asymptotics

2.1 Parameter-dependent cone operators

The operators in $MY^{\nu}(\mathbb{W}, \mathfrak{g})$ from 1.4.(78) (where $\mathbb{W} = \mathbb{B} \times \mathbb{R}_+$) can be regarded as ψDO's with operator-valued Mellin symbols, namely

$$A = op_M(a) \ mod \ MY^{-\infty}_{G}(\mathbb{W}, \mathfrak{g})$$

for

$$a(t,z) = \tilde{\omega}(r)\{a_0(t,z) + a_1(t,z) + m(t,z) + g(t,z)\}\tilde{\omega}_0(r) + (1 - \tilde{\omega}(r))f(t,z)(1 - \tilde{\omega}_1(r)), \tag{1}$$

$z \in \Gamma_{1/2}$. Here $a_i(t,z)$ are given by 1.4.(56),(57), $m(t,z)$ by 1.4.(42), $g(t,z) \in R^{\nu}_{G}(\overline{\mathbb{R}}_+ \times \Gamma_{1/2}, \mathfrak{g})$, cf. 1.4.(38), and

$$f(t,z) \in C^{\infty}(\overline{\mathbb{R}}_+, L^{\nu}_{cl}(int\mathbb{B}; \Gamma_{1/2})).$$

The cut-off functions $\tilde{\omega}(r), \tilde{\omega}_i(r), i = 0, 1$, are supported by the fixed collar neighbourhood of $\partial\mathbb{B}$, and we assume

$$\tilde{\omega}\tilde{\omega}_0 = \tilde{\omega}, \quad \tilde{\omega}\tilde{\omega}_1 = \tilde{\omega}_1. \tag{2}$$

An analogous description holds for $M\mathcal{Y}^{\nu}(\mathbb{W}, \mathfrak{g}; \mathbb{C}^N, \mathbb{C}^M)$, cf. Definition 1.4.21. We shall first discuss the left upper corners.

(1) satisfies

$$a(t,z) \in C^{\infty}(\overline{\mathbb{R}}_+ \times \Gamma_{1/2}, C^{\nu}(\mathbb{B}, \mathfrak{g}))$$

with $C^{\nu}(\mathbb{B}, \mathfrak{g})$ as the class of cone operators over $\mathbb{B}$ from Definition 1.2.3 (cf. also Remark 1.2.4). The t independent $a(z)$ will be examples of parameter-dependent cone operator families in the sense of this section. We want to study the parameter-dependent ellipticity for getting order reducing families of operators along $\mathbb{B}$, analogously to 1.1.(6). Furthermore we look at families which are holomorphic in z. The general considerations allow parameters varying in a conical subset of a finite-dimensional vector space. For simplicity here we take a one-dimensional parameter space $\mathbb{R} \ni \tau$. Below we then switch to $\Gamma_\sigma \ni \sigma + i\tau$ for any fixed $\sigma \in \mathbb{R}$.

Definition 2.1.1 *Let $\beta, \delta \in \mathbb{R}$, $\Theta = (\vartheta, 0]$, $-\infty \leq \vartheta < 0$, $\mathfrak{g} = (\beta, \delta, \Theta)$. Then $C_G^{-\infty}(\mathbb{B}, \mathfrak{g}; \mathbb{R})$ is the space of all operator families*

$$g(\tau) \in \bigcap_{s \in \mathbb{R}} C^\infty(\mathbb{R}, \mathcal{L}(\mathcal{H}^{s,\beta}(\mathbb{B}), \mathcal{H}^{\infty,\delta}(\mathbb{B}))$$

such that there are asymptotic types $P \in As(X, (\delta, \Theta))$, $Q \in As(X, (-\beta, \Theta))$ dependent on $\mathfrak{g}$, with

$$g(\tau) \in \bigcap_{s \in \mathbb{R}} \mathcal{S}(\mathbb{R}, \mathcal{L}(\mathcal{H}^{s,\beta}(\mathbb{B}), \mathcal{H}_P^{\infty,\delta}(\mathbb{B})), \tag{3}$$

$$g^*(\tau) \in \bigcap_{s \in \mathbb{R}} \mathcal{S}(\mathbb{R}, \mathcal{L}(\mathcal{H}^{s,-\delta}(\mathbb{B}), \mathcal{H}_Q^{\infty,-\beta}(\mathbb{B})), \tag{4}$$

$g^*(.)$ being the formal adjoint with respect to the fixed $\mathcal{H}^0(\mathbb{B})$ scalar product, taken point-wise.

Remark 2.1.2 *The space*

$$C_G^{-\infty}(\mathbb{B}, \mathfrak{g}; \mathbb{R})_{P,Q}$$

of all $g(\tau) \in C_G^{-\infty}(\mathbb{B}, \mathfrak{g}; \mathbb{R})$ with the asymptotic types P, Q (in the sense of (??),(4)) is Fréchet in a natural way.

Definition 2.1.3 *Let $\beta, \mu, \nu \in \mathbb{R}$, $\mu - \nu \in \mathbb{N}$, $\mathfrak{g} = (\beta, \beta - \mu, \Theta)$ for $\Theta = (\vartheta, 0]$, $-\infty \leq \vartheta < 0$. Then $C_G^\nu(X^\wedge, \mathfrak{g}; \mathbb{R})$ is the space of all*

$$g(\tau) \in \bigcap_{s \in \mathbb{R}} S_{cl}^\nu(\mathbb{R}_\tau; \mathcal{K}^{s,\beta}, \mathcal{K}^{\infty,\beta-\mu})$$

(with $S_{cl}^\nu(\mathbb{R}_\tau; E, \tilde{E})$ being the space of classical symbols $g(\tau)$ on $\mathbb{R}$ with constsant coefficients) such that

$$g(\tau) \in \bigcap_{s \in \mathbb{R}} S_{cl}^\nu(\mathbb{R}_\tau; \mathcal{K}^{s,\beta}, \mathcal{S}_P^{\beta-\mu}), \tag{5}$$

$$g^*(\tau) \in \bigcap_{s \in \mathbb{R}} S_{cl}^\nu(\mathbb{R}_\tau; \mathcal{K}^{s,-\beta+\mu}, \mathcal{S}_Q^{-\beta}) \tag{6}$$

for certain P, Q as in Definition 2.1.1, dependent on $\mathfrak{g}$, and the point-wise formal adjoint g^* with respect to the fixed $\mathcal{K}^0(X^\wedge)$ scalar product (for notations cf. 1.3.(32)).

Furthermore $C_{M+G}^\nu(X^\wedge, \mathfrak{g}; \mathbb{R})$ is the set of all $g(\tau) + m(\tau)$, for $g(\tau) \in C_G^\nu(X^\wedge, \mathfrak{g}; \mathbb{R})$, $m(\tau)$ given by 1.3.(39) with η being replaced by the one-dimensional variable τ and dropped y-dependence.

Note that $C_G^\nu(X^\wedge, \mathfrak{g}; \mathbb{R})$ coincides with the subclass of all y, y'-independent elements of 1.3.(33) for $q = 1$. From 1.4.(41),(44) it follows a homogeneous "Mellin edge symbol" of order ν

$$\sigma_{b\wedge}(m + g)(\tau) \in \bigcap_{s \in \mathbb{R}} \mathcal{L}^{(\nu)}(\mathbb{R}_\tau \setminus \{0\}; \mathcal{K}^{s,\beta}, \mathcal{K}^{s-\nu,\beta-\mu}). \tag{7}$$

Definition 2.1.4 *Let $\beta, \mu, \nu \in \mathbb{R}$, $\mu - \nu \in \mathbb{N}$, $\mathfrak{g} = (\beta, \beta - \mu, \Theta)$ for $\Theta = (-k, 0]$, $k \in \mathbb{N} \setminus \{0\}$. Then $C^\nu(\mathbb{B}, \mathfrak{g}; \mathbb{R})$ is the space of all operator families*

$$a(\tau) = \tilde{\omega}(r)\{a_0(\tau) + a_1(\tau) + m(\tau) + g(\tau)\}\tilde{\omega}_0(r) + (1 - \tilde{\omega}(r))\tilde{f}(\tau)(1 - \tilde{\omega}_1(r)) + g_0(\tau) \tag{8}$$

with cut-off functions $\tilde{\omega}, \tilde{\omega}_i$ satisfying (2), further $m(\tau) + g(\tau) \in C_{M+G}^\nu(X^\wedge, \mathfrak{g}; \mathbb{R})$, $g_0(\tau) \in C_G^{-\infty}(\mathbb{B}, \mathfrak{g};$-$\mathbb{R})$, $\tilde{f}(\tau) \in L_{cl}^\nu(int\mathbb{B}; \mathbb{R}_\tau)$, and

$$\begin{aligned} a_0(\tau) &= \omega(r[\tau])r^{-\nu}op_{M,r}^{\beta-n/2}(f_0)(\tau)\omega_0(r[\tau]), \\ a_1(\tau) &= (1 - \omega(r[\tau]))r^{-\nu}f_1(\tau)(1 - \omega_1(r[\tau])), \end{aligned}$$

with operator families

$$f_0(r,w,\tau) = \underline{h}(r,w,r\tau)$$

for $\underline{h}(r,w,\tau) \in C^\infty(\overline{\mathbb{R}}_+, M_O^\nu(X;\mathbb{R}_\tau))$ *and*

$$f_1(\tau) \in L_{cl}^\nu(X^\wedge;\mathbb{R}_\tau)$$

such that

$$op_{M,r}^{\beta-n/2}(f_0)(\tau) = f_1(\tau) \ mod \ L^{-\infty}(X^\wedge;\mathbb{R}_\tau).$$

Note that we impose here for $a_0(\tau), a_1(\tau)$ a similar structure as for 1.4.(56),(57). Without loss of generality (since negligible remainders only cause a change of $g(\lambda)$) we may assume that

$$f_1(\tau) = \sum_{j=1}^N \varphi_j op_{\psi,(x,r)}(h_j)(\tau)\tilde{\varphi}_j \tag{9}$$

with $\{U_j\}$, $\{\varphi_j\}$, $\{\tilde{\varphi}_j\}$ as in 1.4.(51) and certain

$$h_j(x,r,\xi,\rho,\tau) = \underline{h}_j(x,r,\xi,r\rho,r\tau) \tag{10}$$

with

$$\underline{h}_j(x,r,\xi,\rho,\tau) \in S_{cl}^\nu(U_j \times \overline{\mathbb{R}}_+ \times \mathbb{R}^{n+2}_{\xi,\rho,\tau}).$$

Set

$$C_{M+G}^\nu(\mathbb{B},\mathfrak{g};\mathbb{R}) = \left\{\tilde{\omega}(m(\tau)+g(\tau))\tilde{\omega}_0 + g_0(\tau) : m(\tau)+g(\tau) \in C_{M+G}^\nu(X^\wedge,\mathfrak{g};\mathbb{R}), g_0(\tau) \in C_G^{-\infty}(\mathbb{B},\mathfrak{g};\mathbb{R})\right\} \tag{11}$$

and denote by $C_G^\nu(\mathbb{B},\mathfrak{g};\mathbb{R})$ the subclass where $m(\tau)$ vanishes.

The class $C^\nu(\mathbb{B},\mathfrak{g};\mathbb{R})$ for $\Theta = (-\infty,0]$ is defined as intersection over the corresponding classes containing the finite weight strips $\Theta_k = (-k,0]$, $k \in \mathbb{N}\setminus\{0\}$.

Remark 2.1.5 *$C_G^\nu(\mathbb{B},\mathfrak{g};\mathbb{R})$ is a parameter-dependent version of $C^\nu(\mathbb{B},\mathfrak{g})$ from Definition 1.2.3. The elements $a(\tau) \in C_G^\nu(\mathbb{B},\mathfrak{g};\mathbb{R})$ are a preliminary version of the operator-valued Mellin symbols for corners (cf. the definitions below in Section 2.4.). Note in particular that our special parameter dependence contains an edge degeneracy in τ if τ is regarded as a (Mellin) covariable along the t axis (cf. Section 1.4.).*

The parameter dependent homogeneous principal symbol of order ν of $op_{M,r}^{\beta-n/2}(f_0)(\tau)$ equals

$$p(x,r,\xi,r\rho,r\tau). \tag{12}$$

with $p(x,r,\xi,\rho,\tau)$ being homogeneous of order ν in $(\xi,\rho,\tau)\neq 0$ and C^∞ up to $r=0$. Denote by

$$\tilde{S}^{(\nu)}(T^*\overline{X}^\wedge \times \mathbb{R}_\tau \setminus 0)$$

the space of these p ($\setminus 0$ indicates $(\tilde{\xi},\tau)\neq 0$, $\tilde{\xi}$ being the fibre variable in $T^*(int(\mathbb{B}))$ where

$$p|_{T^*V\times\mathbb{R}_\tau\setminus 0} \in \tilde{S}^{(\nu)}(T^*\overline{X}^\wedge \times \mathbb{R}_\tau \setminus 0)|_{T^*V\times\mathbb{R}_\tau\setminus 0}$$

for the collar neighbourhood V of $\partial\mathbb{B}$ in $\mathbb{B}$, $V \cong X\times[0,1)$.

We have

$$C^\nu(\mathbb{B},\mathfrak{g};\mathbb{R}) \subset L_{cl}^\nu(int\ \mathbb{B};\mathbb{R}).$$

The map σ_ψ^ν that assigns to any $a(\tau) \in L_{cl}^\nu(int\mathbb{B};\mathbb{R})$ the parameter-dependent homogeneous symbol of order ν induces a map

$$\sigma_\psi^\nu : C^\nu(\mathbb{B},\mathfrak{g};\mathbb{R}) \longrightarrow g_0^{-\nu}\tilde{S}^{(\nu)}(T^*\mathbb{B} \times \mathbb{R}_\tau \setminus 0). \tag{13}$$

g_0^η is a strictly positive C^∞ function on $int\mathbb{B}$ which equals r^η close to $\partial\mathbb{B}$ in the coordinates (x,r). If $\tilde{p}(\tilde{x},\tilde{\xi},\tau) \in S^{(\nu)}(T^*(int\mathbb{B}) \times \mathbb{R}_\tau \setminus 0)$ denotes the parameter-dependent homogeneous principal symbol of order ν of $\tilde{f}(\tau)$, then

$$\sigma_\psi^\nu(a) = r^{-\nu}\tilde{\omega}(r)p + (1-\tilde{\omega})\tilde{p},$$

for every $a(\tau) \in C^\nu(\mathbb{B},\mathfrak{g};\mathbb{R})$, where p equals (12).

In order to introduce $\sigma_{b\wedge}^\nu$ on $C^\nu(\mathbb{B},\mathfrak{g};\mathbb{R})$ we set

$$\begin{aligned}(c_\wedge f_0)(r,w,\tau) &= \underline{h}(0,w,r\tau),\\ (c_\wedge f_1)(\tau) &= \sum_{j=1}^N \varphi_j op_{\psi,(x,r)}(c_\wedge h_j)(\tau)\tilde{\varphi}_j\end{aligned}$$

with $(c_\wedge h_j)(x,r,\xi,\rho,\tau) = \underline{h}_j(x,0,\xi,r\rho,r\tau)$ (cf. (9),(10)). Then

$$\begin{aligned}\sigma_{b\wedge}^\nu(a_0)(\tau) &:= \omega(r|\tau|)r^{-\nu}op_M^{\beta-n/2}(c_\wedge f_0)(\tau)\omega_0(r|\tau|),\\ \sigma_{b\wedge}^\nu(a_1)(\tau) &:= (1-\omega(r|\tau|))r^{-\nu}c_\wedge f_1(\tau)(1-\omega_1(r|\tau|))\end{aligned}$$

and for $a(\tau) \in C^\nu(\mathbb{B},\mathfrak{g};\mathbb{R})$, written in the form (1),

$$\sigma_{b\wedge}^\nu(a)(\tau) := \sigma_{b\wedge}^\nu(a_0)(\tau) + \sigma_{b\wedge}^\nu(a_1)(\tau) + \sigma_{b\wedge}^\nu(g+m)(\tau).$$

This gives us the homogeneous Mellin edge symbolic map

$$\sigma_{b\wedge}^\nu : C^\nu(\mathbb{B},\mathfrak{g};\mathbb{R}) \longrightarrow \bigcap_{s\in\mathbb{R}} \mathcal{L}^{(\nu)}(\mathbb{R}_\tau \setminus \{0\};\mathcal{K}^{s,\beta},\mathcal{K}^{s-\nu,\beta-\mu}). \tag{14}$$

Let us now study a number of properties of the operator families of Definition 2.1.4.

Remark 2.1.6 *Without loss of generality we may assume that*

$$\tilde{f}(\tau)|_{intV} = f_1(\tau)|_{int\,V} \ mod\ L^{-\infty}(int\ V;\mathbb{R}_\tau),$$

with V as the above neighbourhood of $\partial\mathbb{B}$. Then the concrete choice of cut-off functions $\tilde{\omega},\tilde{\omega}_i$, ω,ω_i, that are involved in (1) (cf. also 1.4.(56)(57), 1,3,(39)) affects $a(\tau)$ only modulo $C_G^\nu(\mathbb{B},\mathfrak{g};\mathbb{R})$.

Let us set for $\iota \in \mathbb{N}$

$$C^{\nu,-\iota}(\mathbb{B},\mathfrak{g};\mathbb{R}) =$$
$$\left\{a(\tau) \in C^\nu(\mathbb{B},\mathfrak{g};\mathbb{R}) : \tilde{f}(\tau) \in L_{cl}^{\nu-\iota}(int\mathbb{B};\mathbb{R}), \underline{h}(r,w,\tau) \in C^\infty(\overline{\mathbb{R}}_+, M_O^{\nu-\iota}(X_i;\mathbb{R}_\tau)), f_1(\tau) \in L_{cl}^{\nu-\iota}(X^\wedge;\mathbb{R}_\tau)\right\}$$

with the notations of Definition 2.1.4. then

$$C^{\nu,-\iota}(\mathbb{B},\mathfrak{g};\mathbb{R}) \subset L_{cl}^{\nu-\iota}(int\mathbb{B};\mathbb{R})$$

and the parameter-dependent homogeneous principal symbol gives in this case a map

$$\sigma_\psi^{\nu-\iota} : C^{\nu,-\iota}(\mathbb{B},\mathfrak{g};\mathbb{R}) \longrightarrow g_0^{-\nu}\tilde{S}^{(\nu-\iota)}(T^*\mathbb{B} \times \mathbb{R}_\tau \setminus 0). \tag{15}$$

It is clear that

$$C^{\nu,0}(\mathbb{B},\mathfrak{g};\mathbb{R}) = C^\nu(\mathbb{B},\mathfrak{g};\mathbb{R}).$$

Proposition 2.1.7 *(15) is surjective and*

$$C^{\nu,-\iota-1}(\mathbb{B},\mathfrak{g};\mathbb{R}) = ker\sigma_{\psi}^{\nu-\iota}.$$

Proposition 2.1.8 *Let $a_j(\tau) \in C^{\nu,-\iota-j}(\mathbb{B},\mathfrak{g};\mathbb{R})$, $j \in \mathbb{N}$, be an arbitrary sequence. Then there exists an $a(\tau) \in C^{\nu,-\iota}(\mathbb{B},\mathfrak{g};\mathbb{R})$ such that for every $N \in \mathbb{N}$*

$$a(\tau) - \sum_{j=0}^{N} a_j(\tau) \in C^{\nu,-\iota-(N+1)}(\mathbb{B},\mathfrak{g};\mathbb{R}),$$

and $a(\tau)$ is unique mod $C_G^{-\infty}(\mathbb{B},\mathfrak{g};\mathbb{R})$.

For the Mellin ψDO calculus along the corner axis $\mathbb{R}_+ \ni t$ with operator-valued symbols (operating along $\mathbb{B}$) it will be necessary to endow $C^{\nu}(\mathbb{B},\mathfrak{g};\mathbb{R})$ with a natural locally convex topology. Let us begin with $C_G^{\nu}(\mathbb{B},\mathfrak{g};\mathbb{R})$. By definition we have

$$C_G^{\nu}(\mathbb{B},\mathfrak{g};\mathbb{R}) = \tilde{\omega} C_G^{\nu}(X^{\wedge},\mathfrak{g};\mathbb{R})\tilde{\omega}_0 + C_G^{-\infty}(\mathbb{B},\mathfrak{g};\mathbb{R}).$$

Let us denote by

$$C_G^{\nu}(X^{\wedge},\mathfrak{g};\mathbb{R})_{P,Q}$$

the subspace of all $g(\tau) \in C_G^{\nu}(X^{\wedge},\mathfrak{g};\mathbb{R})$ with given fixed asymptotic types P,Q in (5) and (6), respectively. Further denote for abbreviation by $\mathcal{L}_{P,Q}^{(\nu)}$ the subspace of all

$$\sigma(\tau) \in \bigcap_{s\in\mathbb{R}} \mathcal{L}^{(\nu)}(\mathbb{R}_\tau \setminus \{0\}; \mathcal{K}^{s,\beta}, \mathcal{S}_P^{\beta-\mu})$$

for which

$$\sigma^*(\tau) \in \bigcap_{s\in\mathbb{R}} \mathcal{L}^{(\nu)}(\mathbb{R}_\tau \setminus \{0\}; \mathcal{K}^{s,-\beta+\mu}, \mathcal{S}_Q^{-\beta}).$$

The space $\mathcal{L}_{P,Q}^{(\nu)}$ is Fréchet in a canonical way.

$$\sigma_{b\wedge}^{\nu} : C_G^{\nu}(X^{\wedge},\mathfrak{g};\mathbb{R})_{P,Q} \longrightarrow \mathcal{L}_{P,Q}^{(\nu)}$$

is a (surjective) map with

$$ker\sigma_{b\wedge}^{\nu} = C_G^{\nu-1}(X^{\wedge},\mathfrak{g};\mathbb{R})_{P,Q}.$$

It follows for any fixed τ-excision function $\chi(\tau)$

$$g_1(\tau) := g(\tau) - \chi(\tau)\sigma_{b\wedge}^{\nu}(g)(\tau) \in C_G^{\nu-1}(X^{\wedge},\mathfrak{g};\mathbb{R})_{P,Q}$$

for every $g(\tau) \in C_G^{\nu}(X^{\wedge},\mathfrak{g};\mathbb{R})_{P,Q}$. Analogously we can form

$$g_2(\tau) = g_1(\tau) - \chi(\tau)\sigma_{b\wedge}^{\nu-1}(g_1)(\tau) \in C_G^{\nu-2}(X^{\wedge},\mathfrak{g};\mathbb{R})_{P,Q}$$

and so on. This gives us maps

$$C_G^{\nu}(X^{\wedge},\mathfrak{g};\mathbb{R})_{P,Q} \longrightarrow \times_{j=0}^{N} \mathcal{L}_{P,Q}^{(\nu-j)} \tag{16}$$

and

$$C_G^{\nu}(X^{\wedge},\mathfrak{g};\mathbb{R}) \longrightarrow \bigcap_{s\in\mathbb{R}} S^{\nu-(N+1)}(\mathbb{R}_\tau; \mathcal{K}^{s,\beta}, \mathcal{S}_P^{\beta-\mu}), \tag{17}$$

$$C_G^{\nu}(X^{\wedge},\mathfrak{g};\mathbb{R}) \longrightarrow \bigcap_{s\in\mathbb{R}} S^{\nu-(N+1)}(\mathbb{R}_\tau; \mathcal{K}^{s,-\beta+\mu}, \mathcal{S}_Q^{-\beta}), \tag{18}$$

for all $N \in \mathbb{N}$. Here (17) is just given by

$$g(\tau) \longrightarrow g(\tau) - \chi(\tau) \sum_{j=0}^{N} \sigma_{b\wedge}^{\nu-j}(g_j)$$

with $g_0 = g$, and (18) is the adjoint of (17). Now the topology of the projective limit with respect to the mappings (16),(17),(18) for all N turns $C_G^\nu(X^\wedge, \mathfrak{g}; \mathbb{R})_{P,Q}$ to a Fréchet space which is independent of the concrete choice of χ. We obtain also the Fréchet space

$$C_G^\nu(\mathbb{B}, \mathfrak{g}; \mathbb{R})_{P,Q} := [\tilde{\omega}] C_G^\nu(X^\wedge, \mathfrak{g}; \mathbb{R})_{P,Q} + [\tilde{\omega}_0] C_G^{-\infty}(\mathbb{B}, \mathfrak{g}; \mathbb{R})_{P,Q}.$$

(cf. Remark 2.1.2 and the notation 1.1.(1)). It follows

$$C_G^\nu(\mathbb{B}, \mathfrak{g}; \mathbb{R}) = \varinjlim_{P,Q} C_G^\nu(\mathbb{B}, \mathfrak{g}; \mathbb{R})_{P,Q}.$$

Now, we turn to $C_{M+G}^\nu(\mathbb{B}, \mathfrak{g}; \mathbb{R})$. On this space we look at the conormal symbolic maps (only operating on the Mellin terms). First we have the operator-valued homogeneous components

$$\sigma_{b\wedge}^{\nu-\lambda}(m)(\tau) = \omega(r|\tau|) r^{-\nu} \sum_{0 \leq j \leq k-\mu+\nu-1} r^j \sum_{i=0}^{N} op_M^{\beta_{j,i}-n/2}(h_{j,i,j-\lambda}) \tau^{j-\lambda} \tilde{\omega}(r|\tau|)$$

for $\lambda = 0, \ldots, k-\mu+\nu-1$ ($h_{i,j,j-\lambda} = 0$ for $\lambda > j$). Then we can form $\sigma_M^{\nu-j}(\sigma_{b\wedge}^{\nu-\lambda}(m))(w,\tau)$ for $j = 0, \ldots, k-\mu+\nu-1$, namely

$$\sigma_M^{\nu-j}(\sigma_{b\wedge}^{\nu-\lambda}(m))(w,\tau) = \left\{ \sum_{i=0}^{N} h_{j,i,j-\lambda}(w) \right\} \tau^{j-\lambda} \tag{19}$$

which are up to the power of τ in $M_{as}^{-\infty}(X)$. The sequence (19) for $j = 0, \ldots, k-\mu+\nu-1$ determines $\sigma_{b\wedge}^{\nu-\lambda}(m)(\tau)$ up to an element in $\mathcal{L}_{P,Q}^{(\nu-\lambda)}$ for certain asymptotic types P, Q. The sequence of maps (19) for all λ, j is finite and can be interpreted as a sequence

$$\iota_{\lambda,j} : C_{M+G}^\nu(\mathbb{B}, \mathfrak{g}; \mathbb{R}) \longrightarrow M_{as}^{-\infty}(X),$$

$0 \leq \lambda, j \leq k-\mu+\nu-1$. The space $M_{as}^{-\infty}(X)$ can be written as an inductive limit of Fréchet spaces $M_R^{-\infty}(X)$, $R \in \underline{As}(X)$. For every R there are asymptotic types $R_i \in \underline{As}(X)$ with carriers having empty intersections with Γ_{δ_i}, $i = 1,2$, for arbitrary $\delta_1 \neq \delta_2$, such that

$$M_R^{-\infty}(X) = M_{R_1}^{-\infty}(X) + M_{R_2}^{-\infty}(X).$$

Let us fix a choice of δ_1, δ_2 (e.g. $\delta_1 = (n+1)/2 - \beta$, $\delta_2 = (n+1)/2 - \beta + 1/2$) and denote by $C_{M+G}^\nu(\mathbb{B}, \mathfrak{g}; \mathbb{R})_i$ the space of all $m(\tau) + g(\tau) \in C_{M+G}^\nu(\mathbb{B}, \mathfrak{g}; \mathbb{R})$ such that

$$sg(\iota_{\lambda,j}(m(\tau) + g(\tau))) \cap \Gamma_{\delta_i} = \emptyset$$

for all λ, j. Then

$$C_{M+G}^\nu(\mathbb{B}, \mathfrak{g}; \mathbb{R}) = \sum_{i=1,2} C_{M+G}^\nu(\mathbb{B}, \mathfrak{g}; \mathbb{R})_i \tag{20}$$

holds in the sense of vector spaces. It can easily be proved that $C_{M+G}^\nu(\mathbb{B}, \mathfrak{g}; \mathbb{R})_i$ is isomorphic to the cartesian product

$$C_G^\nu(\mathbb{B}, \mathfrak{g}; \mathbb{R}) \times_{j,\lambda} M_{as}^{-\infty}(X)_i \tag{21}$$

where $M_{as}^{-\infty}(X)_i = \{h \in M_{as}^{-\infty}(X) : sg(h) \cap \Gamma_{\delta_i} = \emptyset\}$. But (21) has a canonical locally convex topology (an inductive limit of Fréchet spaces). Then (20) will be endowed with the topology of the (non-direct) sum. In other words $C_{M+G}^\nu(\mathbb{B}, \mathfrak{g}; \mathbb{R})$ is also treated. This was the most subtle part. The full class $C^\nu(\mathbb{B}, \mathfrak{g}; \mathbb{R})$ can be endowed with a topology (an inductive limit of Fréchet spaces) in a natural way, by the cone operator spaces in [S4], Sections 1.3.2. and 1.4.6. For limiting the size of the exposition this simple construction will be postponed to chapter 5 of [S4].

Definition 2.1.9 *Symb* $C^\nu(\mathbb{B},\mathfrak{g};\mathbb{R})$ *denotes the space of all pairs* $(p_\psi,p_\wedge)\in g_0^\nu\tilde{S}^{(\nu)}(T^*\mathbb{B}\times\mathbb{R}_\tau\setminus 0)\times\bigcap_{s\in\mathbb{R}}\mathcal{L}^{(\nu)}(\mathbb{R}_\tau\setminus\{0\};\mathcal{K}^{s,\beta},\mathcal{K}^{s-\nu,\beta-\mu})$ *such that*

$$p_\psi=\sigma_\psi^\nu(a),\quad p_\wedge=\sigma_{b\wedge}^\nu(a) \tag{22}$$

for some $a(\tau)\in C^\nu(\mathbb{B},\mathfrak{g};\mathbb{R})$.

The components of *Symb* $C^\nu(\mathbb{B},\mathfrak{g};\mathbb{R})$ satisfy a natural compatibility condition.

Proposition 2.1.10 *Let* $k-\mu+\nu\geq 1$, *then*

$$(\sigma_\psi^\nu,\sigma_{b\wedge}^\nu):C^\nu(\mathbb{B},\mathfrak{g};\mathbb{R})\longrightarrow Symb\ C^\nu(\mathbb{B},\mathfrak{g};\mathbb{R})$$

is surjective, and

$$ker(\sigma_\psi^\nu,\sigma_{b\wedge}^\nu)=C^{\nu-1}(\mathbb{B},\mathfrak{g};\mathbb{R}).$$

As we have seen in the preceding sections it is necessary also to study matrix-valued families, here denoted by

$$C^\nu(\mathbb{B},\mathfrak{g};\mathbb{C}^N,\mathbb{C}^M;\mathbb{R}) \tag{23}$$

with $C^\nu(\mathbb{B},\mathfrak{g};\mathbb{R})$ as the class of occurring left upper corners. Basically we have only to generalize the Green objects.

Definition 2.1.11 *Let* $\beta,\delta\in\mathbb{R}$, $\Theta=(\vartheta,0]$, $-\infty\leq\vartheta<0$, $\mathfrak{g}=(\beta,\delta,\Theta)$. *Then* $C_G^{-\infty}(\mathbb{B},\mathfrak{g};\mathbb{C}^N,\mathbb{C}^M;\mathbb{R})$ *is the space of all operator families*

$$g(\tau)\in\bigcap_{s\in\mathbb{R}}\mathcal{S}(\mathbb{R},\mathcal{L}(\mathcal{H}^{s,\beta}(\mathbb{B})\oplus\mathbb{C}^N,\mathcal{H}^{\infty,\delta}(\mathbb{B})\oplus\mathbb{C}^M))$$

such that there are asymptotic types $P\in As(X,(\delta,\Theta))$, $Q\in As(X,(-\beta,\Theta))$ *dependent on* g, *with*

$$g(\tau)\ \in\ \bigcap_{s\in\mathbb{R}}\mathcal{S}(\mathbb{R},\mathcal{L}(\mathcal{H}^{s,\beta}(\mathbb{B})\oplus\mathbb{C}^N,\mathcal{H}_P^{\infty,\delta}(\mathbb{B})\oplus\mathbb{C}^M)), \tag{24}$$

$$g^*(\tau)\ \in\ \bigcap_{s\in\mathbb{R}}\mathcal{S}(\mathbb{R},\mathcal{L}(\mathcal{H}^{s,-\delta}(\mathbb{B})\oplus\mathbb{C}^N,\mathcal{H}_Q^{\infty,-\beta}(\mathbb{B})\oplus\mathbb{C}^M)), \tag{25}$$

$g^*(.)$ *being taken here with respect to fixed scalar products in the corresponding spaces* $\mathcal{H}^0(\mathbb{B})\oplus\mathbb{C}^M$ *and* $\mathcal{H}^0(\mathbb{B})\oplus\mathbb{C}^N$, *respectively. Furthermore we set*

$$C_G^\nu(\mathbb{B},\mathfrak{g};\mathbb{C}^N,\mathbb{C}^M;\mathbb{R})\ =$$

$$\left\{\begin{pmatrix}\tilde\omega & 0\\ 0 & 1\end{pmatrix}g(\tau)\begin{pmatrix}\tilde\omega_0 & 0\\ 0 & 1\end{pmatrix}+g_0(\tau):g(\tau)\in\mathcal{C}_G^\nu(\mathbb{B},\mathfrak{g};\mathbb{C}^N,\mathbb{C}^M;\mathbb{R}),g_0(\tau)\in C_G^{-\infty}(\mathbb{B},\mathfrak{g};\mathbb{C}^N,\mathbb{C}^M;\mathbb{R})\right\} \tag{26}$$

with $\tilde\omega,\tilde\omega_0$ *being as in (11) and* $\mathcal{C}_G^\nu(\mathbb{B},\mathfrak{g};\mathbb{C}^N,\mathbb{C}^M;\mathbb{R})$ *defined as the space of all*

$$g(\tau)\in\bigcap_{s\in\mathbb{R}}S_{cl}^\nu(\mathbb{R}_\tau;\mathcal{K}^{s,\beta}\oplus\mathbb{C}^N,\mathcal{S}_P^{\beta-\mu}\oplus\mathbb{C}^M) \tag{27}$$

with

$$g^*(\tau)\in\bigcap_{s\in\mathbb{R}}S_{cl}^\nu(\mathbb{R}_\tau;\mathcal{K}^{s,-\beta+\mu}\oplus\mathbb{C}^M,\mathcal{S}_Q^{-\beta}\oplus\mathbb{C}^N)$$

for certain P,Q *as mentioned, dependent on* g, *and* $g^*(.)$ *being taken with respect to the scalar products of* $\mathcal{K}^0\oplus\mathbb{C}^M$ *and* $\mathcal{K}^0\oplus\mathbb{C}^N$, *respectively.*

The space (23) is defined as the set of all

$$\begin{pmatrix} a(\tau) & 0 \\ 0 & 0 \end{pmatrix} + \tilde{g}(\tau) \tag{28}$$

with $a(\tau) \in C^{\nu}(\mathbb{B}, \mathfrak{g}; \mathbb{R})$, and $\tilde{g}(\tau) \in C^{\nu}_{G}(\mathbb{B}, \mathfrak{g}; \mathbb{C}^N, \mathbb{C}^M; \mathbb{R})$. Moreover we obtain by definition

$$C^{\nu}_{M+G}(\mathbb{B}, \mathfrak{g}; \mathbb{C}^N, \mathbb{C}^M; \mathbb{R})$$

as the set of all (28) with $a(\tau) \in C^{\nu}_{M+G}(\mathbb{B}, \mathfrak{g}; \mathbb{R})$.

The class $C^{\nu}_{G}(\mathbb{B}, \mathfrak{g}; \mathbb{C}^N, \mathbb{C}^M; \mathbb{R})$ for $\Theta = (-\infty, 0]$ is defined as intersection of the classes for $\Theta_k = (-k, 0]$, $k \in \mathbb{N} \setminus \{0\}$.

Remark 2.1.12 *There is an obvious non-parameter dependent analogue*

$$C^{\nu}(\mathbb{B}, \mathfrak{g}; \mathbb{C}^N, \mathbb{C}^M) \tag{29}$$

of (23) which also can be introduced by filling ordinary cone operators $a \in C^{\nu}(\mathbb{B}, \mathfrak{g})$ to matrices

$$\begin{pmatrix} a & c \\ b & d \end{pmatrix} \tag{30}$$

where $b \in \bigcap_{s \in \mathbb{R}} \mathcal{L}(\mathcal{H}^{s,\beta}(\mathbb{B}), \mathbb{C}^N)$ has the property $b^* \in \bigcap_{s \in \mathbb{R}} \mathcal{L}(\mathbb{C}^N, \mathcal{H}^{\infty,-\beta}_{Q}(\mathbb{B}))$ for some $Q \in As(X, (-\beta, \Theta))$, further $c \in \mathcal{L}(\mathbb{C}^M, \mathcal{H}^{\infty,\beta-\mu}_{P}(\mathbb{B}))$ for some $P \in As(X, (\beta - \mu, \Theta))$, $d \in \mathbb{C}^M \otimes \mathbb{C}^N$.

Let

$$C_G(\mathbb{B}, \mathfrak{g}; \mathbb{C}^N, \mathbb{C}^M) \tag{31}$$

denote the set of all (30) with $a \in C_G(\mathbb{B}, \mathfrak{g})$. We set for $a \in$(23)

$$\sigma^{\nu}_{\psi}(a) := \sigma^{\nu}_{\psi}(l.u.c.\ a)$$

(recall that $l.u.c.$ means "left upper corners"). Further if $\tilde{g}(\tau) \in C^{\nu}_{G}(\mathbb{B}, \mathfrak{g}; \mathbb{C}^N, \mathbb{C}^M; \mathbb{R})$ is written as in (26) we have homogeneous principal Mellin edge symbol of order ν by

$$\sigma^{\nu}_{b\wedge}(\tilde{g})(\tau) := \sigma^{\nu}_{b\wedge}(g)(\tau)$$

which is the homogeneous principal part of order ν of (27) namely an operator family in

$$\bigcap_{s \in \mathbb{R}} \mathcal{L}^{(\nu)}(\mathbb{R}_{\tau} \setminus \{0\}; \mathcal{K}^{s,\beta} \oplus \mathbb{C}^N, \mathcal{K}^{s-\nu,\beta-\mu} \oplus \mathbb{C}^M). \tag{32}$$

Here $\mathcal{L}^{(\nu)}(\mathbb{R}_{\tau} \setminus \{0\}; \mathcal{K}^{s,\beta} \oplus \mathbb{C}^N, \mathcal{K}^{s-\nu,\beta-\mu} \oplus \mathbb{C}^M)$ is defined as the set of all

$$b(\tau) \in C^{\infty}(\mathbb{R}_{\tau} \setminus \{0\}, \mathcal{L}(\mathcal{K}^{s,\beta} \oplus \mathbb{C}^N, \mathcal{K}^{s-\nu,\beta-\mu} \oplus \mathbb{C}^M))$$

satisfying

$$b(\lambda\tau) = \lambda^{\nu} \begin{pmatrix} \kappa_{\lambda} & 0 \\ 0 & 1 \end{pmatrix} b(\tau) \begin{pmatrix} \kappa_{\lambda} & 0 \\ 0 & 1 \end{pmatrix}^{-1}$$

for all $\lambda \in \mathbb{R}_+$, $\tau \in \mathbb{R} \setminus \{0\}$.

For (23) we can do the same as for $C^{\nu}(\mathbb{B}, \mathfrak{g}; \mathbb{R})$. The construction of a locally convex topology in (23) is completely analogous as for the left upper corners. The only change concerns the class with subscript G. As a result we obtain

$$C^{\nu}(\mathbb{B}, \mathfrak{g}; \mathbb{C}^N, \mathbb{C}^M; \mathbb{R}) = \bigcup_{\iota \in I} C^{\nu}(\mathbb{B}, \mathfrak{g}; \mathbb{C}^N, \mathbb{C}^M; \mathbb{R})_{\iota} \tag{33}$$

with a natural system of Fréchet subspaces $\mathcal{C}^\nu(\mathbb{B},\mathfrak{g};\mathbb{C}^N,\mathbb{C}^M;\mathbb{R})_\iota$ (described essentially by asymptotic data for every ι from the set I). (33) is endowed with the topology of the inductive limit. For $\Theta = (-\infty,0]$ we pass once again to the corresponding projective limit over the classes for $\Theta_k = (-k,0]$, $k \in \mathbb{N}\setminus\{0\}$.

Moreover we can introduce the set

$$Symb\, \mathcal{C}^\nu(\mathbb{B},\mathfrak{g};\mathbb{C}^N,\mathbb{C}^M;\mathbb{R})$$

of all pairs $(p_\psi, p_\wedge)$ in the Cartesian product of $g_0^{-\nu}\tilde{S}^\nu(T^*\mathbb{B}\times\mathbb{R}_\tau\setminus 0)$ and (32) satisfying (22) for some a in (23). Then

$$(\sigma_\psi^\nu,\sigma_{b\wedge}^\nu): \mathcal{C}^\nu(\mathbb{B},\mathfrak{g};\mathbb{C}^N,\mathbb{C}^M;\mathbb{R}) \longrightarrow Symb\, \mathcal{C}^\nu(\mathbb{B},\mathfrak{g};\mathbb{C}^N,\mathbb{C}^M;\mathbb{R})$$

is again surjective, and

$$ker(\sigma_\psi^\nu,\sigma_{b\wedge}^\nu) = \mathcal{C}^{\nu-1}(\mathbb{B},\mathfrak{g};\mathbb{C}^N,\mathbb{C}^M;\mathbb{R})$$

for $k-\mu+\nu \geq 1$.

Our spaces of operator families have many natural properties with respect to various operations. We will sketch only some of them.

Theorem 2.1.13 *Let $a(\tau) \in \mathcal{C}^0(\mathbb{B},\mathfrak{g};\mathbb{C}^N,\mathbb{C}^M;\mathbb{R})$ for $\mathfrak{g} = (\beta,\beta,\Theta)$. Then*

$$||a(\tau)||_{\mathcal{L}(\mathcal{H}^{0,\beta}(\mathbb{B})\oplus\mathbb{C}^N,\mathcal{H}^{0,\beta}(\mathbb{B})\oplus\mathbb{C}^M)} < c$$

with a constant $c > 0$ independent of $\tau \in \mathbb{R}$.

Clearly there are much more precise norm growth estimates for general $a(\tau) \in \mathcal{C}^\nu(\mathbb{B},\mathfrak{g};\mathbb{C}^N,\mathbb{C}^M;\mathbb{R})$. This will be done in [S8].

Theorem 2.1.14 *$a(\tau) \in \mathcal{C}^\nu(\mathbb{B},\mathfrak{g};\mathbb{C}^N,\mathbb{C}^M;\mathbb{R})$ for $\mathfrak{g} = (\beta,\beta-\mu,\Theta)$ implies $a^*(\tau) \in \mathcal{C}^\nu(\mathbb{B},\mathfrak{g}^*;\mathbb{C}^N,\mathbb{C}^M;\mathbb{R})$ for $g^* = (-\beta+\mu,-\beta,\Theta)$ (where $a^*(\tau)$ is the point-wise formal adjoint wih respect to the scalar products of $\mathcal{H}^0(\mathbb{B})\oplus\mathbb{C}^M$ and $\mathcal{H}^0(\mathbb{B})\oplus\mathbb{C}^N$, respectively) and*

$$\begin{aligned} \sigma_\psi^\nu(a^*) &= \overline{\sigma_\psi^\nu(a)}, && (34)\\ \sigma_{b\wedge}^\nu(a^*) &= \sigma_{b\wedge}^\nu(a)^*, && (35)\end{aligned}$$

where the $*$ on the right of (35) is that for formal adjoints of operator families in the sense of the $\mathcal{L}^{(\nu)}(\ldots)$ classes.

Theorem 2.1.15 *Let $a(\tau) \in \mathcal{C}^\nu(\mathbb{B},\mathfrak{g};\mathbb{C}^N,\mathbb{C}^M;\mathbb{R})$ for $\mathfrak{g} = (\beta,\beta-\mu,\Theta)$, $b(\tau) \in \mathcal{C}^{\tilde\nu}(\mathbb{B},\tilde{\mathfrak{g}};\mathbb{C}^N,\mathbb{C}^M;\mathbb{R})$ for $\tilde{\mathfrak{g}} = (\beta-\mu,\beta-\mu-\tilde\mu,\Theta)$. Then $b(\tau)a(\tau) \in \mathcal{C}^{\nu+\tilde\nu}(\mathbb{B},\mathfrak{g}_0;\mathbb{C}^N,\mathbb{C}^M;\mathbb{R})$ for $\mathfrak{g}_0 = (\beta,\beta-\mu-\tilde\mu,\Theta)$, and*

$$\begin{aligned} \sigma_\psi^{\tilde\nu+\nu}(ba) &= \sigma_\psi^{\tilde\nu}(b)\sigma_\psi^\nu(a),\\ \sigma_{b\wedge}^{\tilde\nu+\nu}(ba) &= \sigma_{b\wedge}^{\tilde\nu}(b)\sigma_{b\wedge}^\nu(a).\end{aligned}$$

Definition 2.1.16 *An operator family $a(\tau) \in \mathcal{C}^\mu(\mathbb{B},\mathfrak{g};\mathbb{C}^N,\mathbb{C}^M;\mathbb{R})$ for $\mathfrak{g} = (\beta,\beta-\mu,\Theta)$, $\Theta = (-k,0]$, $k \in (\mathbb{N}\setminus\{0\})\cup\{\infty\}$ is called elliptic if*

(i) $\sigma_\psi^\mu(a) \neq 0$ on $T^*(int\mathbb{B}\times\mathbb{R}_\tau\setminus 0)$ and if $p = g_0^\mu(r)\sigma_\psi^\mu(a)$ in the coordinates (x,r) close to $r=0$, where $p = p(x,r,\xi,r\rho,r\tau)$, satisfies $p(x,r,\xi,\rho,\tau) \neq 0$ up to $r=0$,

(ii)

$$\sigma^{\mu}_{b\wedge}(a)(\tau): \begin{array}{c}\mathcal{K}^{s,\beta}(X^\wedge)\\ \oplus \\ \mathbb{C}^N\end{array} \longrightarrow \begin{array}{c}\mathcal{K}^{s-\mu,\beta-\mu}(X^\wedge)\\ \oplus \\ \mathbb{C}^M\end{array} \tag{36}$$

is an isomorphism for all $\tau \in \mathbb{R} \setminus \{0\}$.

Similary as above it is clear that when (36) is satisfied for one $s = s_0$ then automatically for all $s \in \mathbb{R}$.

Definition 2.1.16 is basically a sort of parameter-dependent ellipticity of operator families. We hope our shorter notation will not cause confusions.

Definition 2.1.17 *Let* $a(\tau) \in C^{\mu}(\mathbb{B}, \mathfrak{g}; \mathbb{C}^N, \mathbb{C}^M; \mathbb{R})$, $\mathfrak{g} = (\beta, \beta-\mu, \Theta)$, *and* $b(\tau) \in C^{-\mu}(\mathbb{B}, \mathfrak{b}; \mathbb{C}^M, \mathbb{C}^N; \mathbb{R})$ *for* $\mathfrak{b} = (\beta - \mu, \beta, \Theta)$. *Then* $b(\tau)$ *is called a parametrix of* $a(\tau)$ *if*

$$\begin{aligned} b(\tau)a(\tau) - 1 &\in C_G^{-\infty}(\mathbb{B}, \mathfrak{g}_l; \mathbb{C}^N, \mathbb{C}^N; \mathbb{R}), \\ a(\tau)b(\tau) - 1 &\in C_G^{-\infty}(\mathbb{B}, \mathfrak{g}_r; \mathbb{C}^M, \mathbb{C}^M; \mathbb{R}) \end{aligned}$$

with $\mathfrak{g}_l = (\beta, \beta, \Theta)$, $\mathfrak{g}_r = (\beta - \mu, \beta - \mu, \Theta)$, *and 1 being the identical operator family in the corresponding* C^0*-classses.*

Theorem 2.1.18 *Let* $a(\tau) \in C^{\mu}(\mathbb{B}, \mathfrak{g}; \mathbb{C}^N, \mathbb{C}^M; \mathbb{R})$ *be elliptic,* $\mathfrak{g} = (\beta, \beta - \mu, \Theta)$. *Then there exists a constant* $c > 0$ *such that*

$$a(\tau) : \mathcal{H}^{s,\beta}(\mathbb{B}) \oplus \mathbb{C}^N \longrightarrow \mathcal{H}^{s-\mu,\beta-\mu}(\mathbb{B}) \oplus \mathbb{C}^M \tag{37}$$

is an isomorphism for all τ *with* $|\tau| \geq c$, *for all* $s \in \mathbb{R}$.

Remark 2.1.19 *If (37) is an isomorphism for all* $\tau \in \mathbb{R}$ *and fixed* $s = s_0 \in \mathbb{R}$ *then for all* $\tau \in \mathbb{R}, s \in \mathbb{R}$, *and* $a^{-1}(\tau) \in C^{-\mu}(\mathbb{B}, \mathfrak{b}; \mathbb{C}^N, \mathbb{C}^M; \mathbb{R})$.

Theorem 2.1.20 *For every* $\mu, \beta \in \mathbb{R}, \Theta = (-k, 0]$, $k \in \mathbb{N} \setminus \{0\}$, *there exists an elliptic* $a(\tau) \in C^{\mu}(\mathbb{B}, \mathfrak{g}; \mathbb{R})$, $\mathfrak{g} = (\beta, \beta - \mu, \theta)$ *such that*

$$a(\tau) : \mathcal{H}^{s,\beta}(\mathbb{B}) \longrightarrow \mathcal{H}^{s-\mu,\beta-\mu}(\mathbb{B})$$

is bijective for all $s \in \mathbb{R}$ *and all* $\tau \in \mathbb{R}$.

As mentioned τ will be interpreted as $Im\ z$ for $z \in \mathbb{C}$. Let

$$C^{\nu}(\mathbb{B}, \mathfrak{g}; \mathbb{C}^N, \mathbb{C}^M; \Gamma_\sigma) \tag{38}$$

denote the space of all operator families $a(z)$ for $z = \sigma + i\tau$ with fixed $\sigma \in \mathbb{R}$, that belong with respect to τ to $C^{\nu}(\mathbb{B}, \mathfrak{g}; \mathbb{C}^N, \mathbb{C}^M; \mathbb{R}_\tau)$. Analogously we use notations such as $C^{\nu}(\mathbb{B}, \mathfrak{g}; \Gamma_\sigma)$.

Definition 2.1.21 *Let* $a(z) \in \bigcap_{c<\sigma<c'} C^{\nu}(\mathbb{B}, \mathfrak{g}; \mathbb{C}^N, \mathbb{C}^M; \Gamma_\sigma)$ *and*

$$a(z) \in \mathcal{A}(c < Re\ z < c', \mathcal{L}(\mathcal{H}^{s,\beta}(\mathbb{B}) \oplus \mathbb{C}^N, \mathcal{H}^{s-\nu,\beta-\mu}(\mathbb{B}) \oplus \mathbb{C}^M)).$$

We say that $a(z)$ *belongs uniformly to* $C^{\nu}(\mathbb{B}, \mathfrak{g}; \mathbb{C}^N, \mathbb{C}^M; \Gamma_\sigma)$ *with respect to* σ, *for* $c_1 < \sigma < c_1'$, *with certain* $c < c_1, c_1' < c'$, *if* $a(z)|_{\Gamma_\sigma}$ *belongs with respect to* τ *to* $C^{\nu}(\mathbb{B}, \mathfrak{g}; \mathbb{C}^N, \mathbb{C}^M; \Gamma_\sigma)_\iota$ *for all* σ *in the interval with the same* ι *(cf. (33)) and if all semi-norms of* $a(z)|_{\Gamma_\sigma}$ *from* $C^{\nu}(\mathbb{B}, \mathfrak{g}; \mathbb{C}^N, \mathbb{C}^M; \mathbb{R})_\iota$ *are uniformly bounded for* $c_1 < \sigma < c_1'$.

A crucial result of the whole calculus is now the following

Theorem 2.1.22 *For every $a(z) \in C^{\nu}(\mathbb{B}, \mathfrak{g}; \mathbb{C}^N, \mathbb{C}^M; \Gamma_{\sigma_0})$ with fixed $\sigma_0 \in \mathbb{R}$ there exists an*

$$h(z) \in \bigcap_{s \in \mathbb{R}} \mathcal{A}(\mathbb{C}, \mathcal{L}(\mathcal{H}^{s,\beta}(\mathbb{B}) \oplus \mathbb{C}^N, \mathcal{H}^{s-\nu,\beta-\mu}(\mathbb{B}) \oplus \mathbb{C}^M))$$

such that

(i) $h(z)|_{\Gamma_{\sigma_0}} - a(z) \in C_G^{-\infty}(\mathbb{B}, \mathfrak{g}; \mathbb{C}^N, \mathbb{C}^M; \Gamma_{\sigma_{00}})$,

(ii) $h(z)|_{\Gamma_\sigma} \in C^{\nu}(\mathbb{B}, \mathfrak{g}; \mathbb{C}^N, \mathbb{C}^M; \Gamma_\sigma)$

for all $\sigma \in \mathbb{R}$, uniformly in every finite interval $c < \sigma < c'$.

The proof follows by a generalization of kernel cut-off arguments as they have been used systematically in [S1],[S2],[S3],[S4],III. Here τ is treated as $Im\ z$ for the Mellin covariable z with respect to the Mellin transform $M_{t\to z}$.

Remark 2.1.23 *The kernel cut-off technique also allows to choose the order reducing family of Theorem 2.1.20 in the form $a(z)|_{\Gamma_{\sigma_0}}$ for an $a(z) \in \bigcap_{s\in\mathbb{R}} \mathcal{A}(\mathbb{C}, \mathcal{L}(\mathcal{H}^{s,\beta}(\mathbb{B}), \mathcal{H}^{s-\mu,\beta-\mu}(\mathbb{B}))$ with fixed $\sigma_0 \in \mathbb{R}$, and $a(z)|_{\Gamma_\sigma} \in C^{\mu}(\mathbb{B}, \mathfrak{g}; \Gamma_\sigma)$ for all $\sigma \in \mathbb{R}$, uniformly in every finite interval. In addition, for every $c, c' \in \mathbb{R}$ with $c < \sigma_0 < c'$ we can find $a(z)$ in such a way that $a(z)|_{\Gamma_\sigma}$ is order reducing for all $\sigma, c < \sigma < c'$.*

The manipulations for obtaining these results are analogous to those for 3.3.Theorem 1 of [S2].

Remark 2.1.24 *Let $a(z) \in C^{\mu}(\mathbb{B}, \mathfrak{g}; \mathbb{C}^N, \mathbb{C}^M; \Gamma_\sigma)$ be elliptic (where $\tau = Im\ z$) and $h(z)$ be associated with $a(z)$ via Theorem 2.1.22. Then $h(z)|_{\Gamma_\sigma}$ is elliptic for every $\sigma \in \mathbb{R}$. For every $c < c'$ there exists a $d > 0$ such that*

$$h(z) : \mathcal{H}^{s,\beta}(\mathbb{B}) \oplus \mathbb{C}^N \longrightarrow \mathcal{H}^{s-\mu,\beta-\mu}(\mathbb{B}) \oplus \mathbb{C}^M \qquad (39)$$

is an isomorphism for all $z \in \mathbb{C}$ with $c < Re\ z < c'$, $|Im\ z| > d$. Furthermore there exists a sequence $\{p_j\}_{j\in\mathbb{Z}} \subset \mathbb{C}$ with $|Re\ p_j| \to \infty$ as $|j| \to \infty$ such that (39) is an isomorphism for all $z \in \mathbb{C} \setminus \{p_j\}_{j\in\mathbb{Z}}$.

Remark 2.1.25 *Let $h(z)$ be given as in Remark 2.1.24. Then $h^{-1}(z)$, first defined on $\mathbb{C} \setminus \{p_j\}_{j\in\mathbb{Z}}$, extends to a meromorphic operator family in $\mathbb{C}$ with poles at p_j of multiplicities $m_j + 1$, and Laurent expansions*

$$h(z) = \sum_{k=0}^{m_j} \nu_{jk}(z-p_j)^{-(k+1)} + \sum_{k=0}^{\infty} \nu_{jk}^{+}(z-p_j)^k$$

close to p_j, with finite-dimensional $\nu_{jk} \in C_G(\mathbb{B}, \mathfrak{g}; \mathbb{C}^N, \mathbb{C}^M)$ for all $0 \le k \le m_j$, $j \in \mathbb{Z}$ (cf. (31)).

2.2 Corner Sobolev spaces

A further essential ingredient of the calculus are the corner Sobolev spaces. They have been announced already in 1.1.(26). Recall that $\mathbb{K}$ denotes the stretched space belonging to the given "manifold" K with corners $K_0 \subset K$, cf. 1.1.(23). Further we have $\mathbb{W}$, defined by 1.1.(25), which is the stretched space of the manifold $K \setminus K_0$ with edges.

$\mathbb{K}$ is locally close to $p^{-1}(v)$ for any $v \in K_0$ of the form $\mathbb{B} \times [0,1)$ with $\mathbb{B}$ being the (stretched) base of the corner with conical singularities. It will be convenient also to look at the spaces

$$\mathcal{H}^{s,\gamma}(\mathbb{B}^\wedge) \text{ for } s \in \mathbb{R}, \gamma = (\gamma_0, \gamma_1) \in \mathbb{R}^2 \qquad (1)$$

with

$$\mathbb{B}^\wedge = \mathbb{B} \times \mathbb{R}_+, \tag{2}$$

analogously to $\mathcal{H}^{s,\beta}(X^\wedge)$ (cf. Definition 1.1.4). Then the corner Sobolev spaces themselves

$$\mathcal{H}^{s,\gamma}(\mathbb{K}) \text{ for } s \in \mathbb{R}, \gamma = (\gamma_0, \gamma_1) \in \mathbb{R}^2 \tag{3}$$

shall be obtained by glueing together (1) (close to $p^{-1}(v)$) with the corresponding wedge spaces on $\mathbb{W}$, analogously to 1.1.(14).

The various weight shifts that play a role in the following definitions depend on $n = dim X$, X being the base of the cone for the conical singularities of $\mathbb{B}$. These weight conventions will ensure that $\mathcal{H}^{0,(0,0)}(\mathbb{K})$ is the pull back of $L^2(K)$ under 1.1.(23).

For simplicity we shall assume in the following that K_0 consists of a single point v. This can be achieved formally by allowing $\mathbb{B}$ to have several connected components, i.e. $\mathbb{K}$ remains untouched by this assumption.

Let $\gamma = (\gamma_0, \gamma_1) \in \mathbb{R}^2$, $\mu \in \mathbb{R}$. From Theorem 2.1.20 it follows the existence of an operator family

$$b^\mu(z) \in C^\mu(\mathbb{B}, \mathfrak{g}; \Gamma_{(n+2)/2-\gamma_1}) \tag{4}$$

for $\mathfrak{g} = (\gamma_0, \gamma_0 - \mu, \Theta)$, $\Theta = (-k, 0]$ with arbitrary fixed $k \in \mathbb{N} \setminus \{0\}$, such that

$$b^\mu(z) : \mathcal{H}^{s,\gamma_0}(\mathbb{B}) \longrightarrow \mathcal{H}^{s-\mu,\gamma_0-\mu}(\mathbb{B}) \tag{5}$$

is an isomorphism for all $z \in \Gamma_{(n+2)/2-\gamma_1}$.

As usual g_0^β for $\beta \in \mathbb{R}$ will denote a strictly positive function in $C^\infty(int\ \mathbb{B})$ with $g_0^\beta = r^\beta$ in the coordinates $(x, r) \in X \times \mathbb{R}_+$ close to $\partial\mathbb{B} \cong X$. Then

$$||u||_{\mathcal{H}^{s,\beta}(\mathbb{B})} \sim ||g_0^{-\beta} u||_{\mathcal{H}^s(\mathbb{B})}$$

where $\sim$ means equivalence of norms and $\mathcal{H}^s(\mathbb{B}) = \mathcal{H}^{s,0}(\mathbb{B})$. Thus

$$g_0^{s-\beta} b^s(z) : \mathcal{H}^{s,\beta}(\mathbb{B}) \underset{\cong}{\longrightarrow} \mathcal{H}^0(\mathbb{B})$$

for all $s, \beta \in \mathbb{R}$, $z \in \Gamma_{(n+2)2-\gamma_1}$.

Definition 2.2.1 *$\mathcal{H}^{s,\gamma}(\mathbb{B}^\wedge)$ for $s \in \mathbb{R}$, $\gamma = (\gamma_0, \gamma_1) \in \mathbb{R}^2$, is the completion of $C_0^\infty((int\mathbb{B}) \times \mathbb{R}_+)$ with respect to the norm*

$$||u||_{\mathcal{H}^{s,\gamma}(\mathbb{B}^\wedge)} = \left\{ \frac{1}{2\pi i} \int_{\Gamma_{(n+2)/2-\gamma_1}} ||g_0^{s-\gamma_0} b^s(z)(M_{t\to z}u)(z)||^2_{\mathcal{H}^0(\mathbb{B})} dz \right\}^{1/2}. \tag{6}$$

Remark 2.2.2 *If we define (6) with another weight function $\tilde{g}_0^{s-\gamma_0}$ and another order reducing family $\tilde{b}^s(z)$, then we get an equivalent norm. In other words the space $\mathcal{H}^{s,\gamma}(\mathbb{B}^\wedge)$ is independent of the concrete choice of the weight function and the reduction of orders (cf. also Theorem 2.1.13).*

We now regard $\tau = Im\ z \in \mathbb{R}$ as covariable of the one-dimensional Fourier transform

$$(Fv)(\tau) = \int_{\mathbb{R}} e^{-i\tilde{t}\tau} v(\tilde{t}) d\tilde{t}. \tag{7}$$

Let us set $c^s(\tau) = b^s((n+2)/2 - \gamma_1 + i\tau)$ with b^s being given by (4) for $\mu = s$.

Definition 2.2.3 *$\mathcal{Q}^{s,\beta}(\mathbb{B}\times\mathbb{R})$ for $s,\beta\in\mathbb{R}$ is the completion of $C_0^\infty((int\mathbb{B})\times\mathbb{R})$ with respect to the norm*

$$\|v\|_{\mathcal{Q}^{s,\beta}(\mathbb{B}\times\mathbb{R})}=\left\{\int_{\mathbb{R}}\|g_0^{s-\beta}c^s(\tau)(F_{\tilde{t}\to\tau}v)(\tau)\|_{\mathcal{H}^0(\mathbb{B})}d\tau\right\}^{1/2}. \tag{8}$$

Using 1.4.(9) it follows for $\tau=Im\ z$, $z\in\Gamma_{1/2-\alpha}$, $u\in C_0^\infty(\mathbb{R}_+,C_0^\infty(int\mathbb{B}))$,

$$(\Psi M_\alpha u)(\tau)=(F\Phi_\alpha u)(\tau) \tag{9}$$

with the weighted Mellin transform M_α, cf. 1.1.(8).

Thus

$$u\longrightarrow\left\{\int_{\mathbb{R}}\|g_0^{s-\gamma_0}c^s(\tau)(F_{\tilde{t}\to\tau}\Phi_{\tilde{\gamma}_1}u)(\tau)\|^2_{\mathcal{H}^0(\mathbb{B})}d\tau\right\}^{1/2}$$

with $\tilde{\gamma}_1=\gamma_1-(n+1)/2$ is an equivalent norm to (6). Hence $\Phi_{\tilde{\gamma}_1}$ induces an isomorphism

$$\Phi_{\tilde{\gamma}_1}:\mathcal{H}^{s,\gamma}(\mathbb{B}^\wedge)\longrightarrow\mathcal{Q}^{s,\gamma_0}(\mathbb{B}\times\mathbb{R}), \tag{10}$$

$\gamma=(\gamma_0,\gamma_1)$. We obtain

$$\mathcal{H}^{s,\gamma}(\mathbb{B}^\wedge)=\left\{t^{-1/2+\tilde{\gamma}_1}u(\log t):u(\tilde{t})\in\mathcal{Q}^{s,\gamma_0}(\mathbb{B}\times\mathbb{R})\right\}, \tag{11}$$

and consequently

$$\mathcal{H}^{s,\gamma}(\mathbb{B}^\wedge)=t^{\gamma_1}\mathcal{H}^{s,(\gamma_0,0)}(\mathbb{B}^\wedge). \tag{12}$$

In Section 1.4. we have mentioned the spaces $\mathcal{H}^{s,\alpha}(\tilde{X}^\wedge)$ for any fixed closed compact C^∞ manifold $\tilde{X}$ containing $\mathbb{B}$ as a submanifold with boundary, $int\mathbb{B}\subset\tilde{X}$ open, $n+1=dim\tilde{X}$. Let $\tilde{\omega}(r)$ be a cut-off function on $\mathbb{B}$ supported by the fixed collar neighbourhood $V\cong X\times[0,1)$ of $\partial\mathbb{B}\cong X$. As in formula 1.4.(70) the space $[1-\tilde{\omega}]\mathcal{H}^{s,\alpha}(\tilde{X}^\wedge)$ will be canonically identified with a subspace of $\mathcal{D}'(int\ \mathbb{B}^\wedge)$. As usual we have in V the coordinates (x,r). The (stretched) cone $\mathbb{B}^\wedge$ over $\mathbb{B}$ contains the strip $V^\wedge=V\times\mathbb{R}_+$ with the coordinates (x,r,t). In this sense we have $\tilde{\omega}\mathcal{D}'(intV^\wedge)\subset\mathcal{D}'(int\mathbb{B}^\wedge)$.

The following theorem gives a link between the spaces of Definition 2.2.1 and $\mathcal{H}^{s,\alpha}(\tilde{X}^\wedge)$, $\mathcal{V}^{s,\alpha}(\mathbb{R}_+,\mathcal{K}^{s,\beta}(X^\wedge))$, cf. 1.4.(3).

Theorem 2.2.4 *For every $s\in\mathbb{R}$, $\gamma=(\gamma_1,\gamma_2)\in\mathbb{R}^2$, we have*

$$\mathcal{H}^{s,\gamma}(\mathbb{B}^\wedge)=[\tilde{\omega}]\mathcal{V}^{s,\gamma_1-(n+1)/2}(\mathbb{R}_+,\mathcal{K}^{s,\gamma_0}(X^\wedge))+[1-\tilde{\omega}]\mathcal{H}^{s,\gamma_1}(\tilde{X}^\wedge)$$

(cf. 1.1.(1),(2)).

Using 1.4.(13) and the analogous identity for $\mathcal{H}^{s,\alpha}(\tilde{X}^\wedge)$ we get

$$\begin{aligned}\mathcal{H}^{s,\gamma}(\mathbb{B}^\wedge)&=t^{\gamma_1-(n+1)/2}\left\{[\tilde{\omega}]\mathcal{V}^s(\mathbb{R}_+,\mathcal{K}^{s,\gamma_0}(X^\wedge))+[1-\tilde{\omega}]\mathcal{H}^{s,(n+1)/2}(\tilde{X}^\wedge)\right\}\\&=t^{\gamma_1-(n+1)/2}\mathcal{V}^s(\mathbb{R}_+,\mathcal{H}^{s,\gamma_0}(\mathbb{B}))\end{aligned} \tag{13}$$

with the notation 1.4.(70).

Proposition 2.2.5 *For every $s\in\mathbb{R}$, $\gamma=(\gamma_0,\gamma_1)\in\mathbb{R}^2$, we have*

$$\mathcal{H}^{s,\gamma}(\mathbb{B}^\wedge)\subset\mathcal{W}^{s,\gamma_0}_{loc}(\mathbb{V})$$

with the non-compact stretched "wedge" $\mathbb{V}:=int\mathbb{B}\times\mathbb{R}_+$ (cf. the notations from 1.3.(18)).

Definition 2.2.6 *For $s \in \mathbb{R}$, $\gamma = (\gamma_0, \gamma_1) \in \mathbb{R}^2$, we set*

$$\mathcal{H}^{s,\gamma}(\mathbb{K}) = [\omega]\mathcal{H}^{s,\gamma}(\mathbb{B}^\wedge) + [1-\omega]\mathcal{W}^{s,\gamma_0}_{loc}(\mathbb{W}) \tag{14}$$

for any fixed cut-off function $\omega(t)$, (14) being equipped with the topology of the sum.

Clearly we assume $\omega = 0$ for $t > 1-\varepsilon$ with some $0 < \varepsilon < 1$ (remember that a neighbourhood of $p^{-1}(v)$ for $v \in K_0$ is identified with $\mathbb{B} \times [0,1)$).

Remark 2.2.7 *The space (14) is independent of the concrete choice of ω.*

Remark 2.2.8 *Let $\rho \in \mathbb{R}$ and set*

$$g_1^\rho = \omega(t)t^\rho + (1-\omega) \tag{15}$$

for a cut-off function $\omega(t)$, $0 \le \omega \le 1$. Then

$$g_1^\rho \mathcal{H}^{s,\gamma}(\mathbb{K}) = \mathcal{H}^{s,(\gamma_0,\gamma_1+\rho)}(\mathbb{K}), \tag{16}$$

$\gamma = (\gamma_0, \gamma_1) \in \mathbb{R}^2$.

Theorem 2.2.9 *There are canonical continuous embeddings*

$$\mathcal{H}^{s',(\gamma_0',\gamma_1')}(\mathbb{K}) \longrightarrow \mathcal{H}^{s,(\gamma_0,\gamma_1)}(\mathbb{K}) \tag{17}$$

for every $s' \ge s$, $\gamma_i' \ge \gamma_i$, $i = 0,1$. (17) is compact for $s' > s$, $\gamma_i' > \gamma_i$, $i = 0,1$.

Note that the spaces (14) can be endowed with Hilbert norms for which these spaces become Hilbert spaces. We only need someone in

$$\mathcal{H}^0(\mathbb{K}) := \mathcal{H}^{0,(0,0)}(\mathbb{K}). \tag{18}$$

It will be fixed once and for all. A choice follows easily from

$$\mathcal{H}^0(\mathbb{K}) = [\omega]\mathcal{H}^{0,(0,0)}(\mathbb{B}^\wedge) + [1-\omega]\mathcal{W}^{0,0}_{loc}(\mathbb{W})$$

and chosen scalar products in

$$[\omega]\mathcal{H}^{0,(0,0)}(\mathbb{B}^\wedge), [1-\omega]\mathcal{W}^{0,0}_{loc}(\mathbb{W}). \tag{19}$$

In fact the topolgy of a sum $E = E_0 + E_1$ may be represented by $E_0 \oplus E_1/D$ with $D = \{(e,-e) : e \in E_0 \cap E_1)\}$, such that Hilbert structures on E_0, E_1 induce someone on E. Now the scalar product on $\mathcal{H}^{0,(0,0)}(\mathbb{B}^\wedge) := \mathcal{H}^0(\mathbb{B}^\wedge)$ follows by

$$(u,v)_{\mathcal{H}^0(\mathbb{B}^\wedge)} = \frac{1}{2\pi i}\int_{\Gamma_{(n+2)/2-\gamma_1}} (M_{t\to z}u, M_{t\to z}v)_{\mathcal{H}^0(\mathbb{B})}dz.$$

The second space of (19) may be replaced by $[1-\omega]\mathcal{W}^{0,0}_{comp}(\mathbb{W})$ because of the cut-off factor. On $\mathcal{W}^0_{comp}(\mathbb{W}) := \mathcal{W}^{0,0}_{comp}(\mathbb{W})$ we get a scalar product by choosing a Riemannian metric on $K \setminus K_0$ and by defining the adequate scalar products first in local terms and then globally by using a partition of unity and so on. The obvious details will be dropped. Let us only note that 1.3.(7) for $\mathbb{R}_+$ instead of $\mathbb{R}^q$ gives us our scalar product locally close to $r = 0$.

Theorem 2.2.10 *The scalar product $(.,.)_{\mathcal{H}^0(\mathbb{K})}$ extends from $C_0^\infty(int\mathbb{K}) \times C_0^\infty(int\mathbb{K})$ to a non-degenerate sesquilinear pairing*

$$(.,.)_{\mathcal{H}^0(\mathbb{K})} : \mathcal{H}^{s,\gamma}(\mathbb{K}) \times \mathcal{H}^{-s,-\gamma}(\mathbb{K}) \longrightarrow \mathbb{C}, \tag{20}$$

for all $s \in \mathbb{R}$, $\gamma = (\gamma_0, \gamma_1)$, with $-\gamma = (-\gamma_0, -\gamma_1)$ which allows the identification

$$\mathcal{H}^{-s,-\gamma}(\mathbb{K}) = \mathcal{H}^{s,\gamma}(\mathbb{K})'. \tag{21}$$

For any given

$$A \in \bigcap_{s\in\mathbb{R}} \mathcal{L}(\mathcal{H}^{s,\gamma}(\mathbb{K}),\ \mathcal{H}^{s-\nu,\gamma-\delta}(\mathbb{K})), \tag{22}$$

$\delta = (\delta_0, \delta_1) \in \mathbb{R}^2$, $\gamma - \delta = (\gamma_0 - \delta_0, \gamma_1 - \delta_1)$, we obtain by

$$(Au, v)_{\mathcal{H}^0(\mathbb{K})} = (u, A^* v)_{\mathcal{H}^0(\mathbb{K})}, \tag{23}$$

first for $u, v \in C_0^\infty(int\mathbb{K})$, a so-called formal adjoint A^* of A, which induces an

$$A^* \in \bigcap_{s\in\mathbb{R}} \mathcal{L}(\mathcal{H}^{s,-\gamma+\delta}(\mathbb{K}), \mathcal{H}^{s-\nu,-\gamma}(\mathbb{K})). \tag{24}$$

2.3 The corner asymptotics

We turn to the corner asymptotics of distributions belonging to the spaces of Definition 2.2.1, Definition 2.2.6. The asymptotics will consist of two parts, one for $t \to 0$ and another one for $r \to 0$ along the edges emanated from the corners. Let us begin with the corner axis asymptotics. Similary to the cone asymptotics it will be convenient to speak of asymptotic types.

Definition 2.3.1 *Let $\gamma = (\gamma_0, \gamma_1) \in \mathbb{R}^2$, $\mathfrak{a}_i = (\gamma_i, \Theta_i)$ with weight intervals $\Theta_i = (\vartheta_i, 0]$, $-\infty \le \vartheta_i < 0$, $i = 0, 1$. Then $As(\mathbb{B}, \mathfrak{a})$ for $\mathfrak{a} := (\mathfrak{a}_0, \mathfrak{a}_1^\bullet)$ and $\vartheta_1 = -\infty$ is the set of all sequences*

$$P = \{(p_j, m_j, L_j)\}_{j\in\mathbb{N}} \tag{1}$$

whereas for $\vartheta_1 > -\infty$ it is the set of all

$$P = \{(p_j, m_j, L_j)\}_{0\le j\le N}, \tag{2}$$

with $N = N(P) < \infty$, where $p_j \in \mathbb{C}$, $m_j \in \mathbb{N}$,

$$(n+2)/2 - \gamma_1 + \vartheta_1 < Re\ p_j < (n+2)/2 - \gamma_1 \tag{3}$$

for all j, with $Re\ p_j \to -\infty$ as $j \to \infty$ when $\vartheta_1 = -\infty$, and L_j being finite-dimensional subspaces of $\mathcal{H}_D^{\infty,\gamma_0}(\mathbb{B})$ for some $D = D(P) \in As(X, \mathfrak{a}_0)$, independent of j (cf. 1.2.(8),1.2.(14)).

Set

$$\pi_{\mathbb{C}}P = \{p_j\}_{j\in\mathbb{N}} \text{ for } \vartheta_1 = -\infty, \tag{4}$$

$$\pi_{\mathbb{C}}P = \{p_j\}_{0\le j\le N(P)} \text{ for } \vartheta_1 > -\infty. \tag{5}$$

Further the correspondence $P \to D(P)$ will be denoted by π_X, i.e. we have a map

$$\pi_X : As(\mathbb{B}, \mathfrak{a}) \longrightarrow As(X, \mathfrak{a}_0). \tag{6}$$

Recall that in 1.2. we have distinguished between the discrete and the continuous cone asymptotics, the first ones being indicated by upper dots at the weight data. Here for the corner the notation $\mathfrak{a} = (\mathfrak{a}_0, \mathfrak{a}_1^\bullet)$ means that the asymptotics with respect to the corner axis variable t are discrete, those for the cone axis variable r continuous. Also the other combinations make sense, for instance $As(\mathbb{B}, \mathfrak{a})$ for $\mathfrak{a} = (\mathfrak{a}_0^\bullet, \mathfrak{a}_1^\bullet)$ would mean that the above $D(P) = \pi_X P$ belongs to $As(X, \mathfrak{a}_0^\bullet)$.

Clearly the reader who wants first to study discrete asymptotics may look at the latter case, anyway. It is contained in Definition 2.3.1. By definition $D \in As(X, \mathfrak{a}_0^\bullet)$ (e.g. for $\vartheta_0 = -\infty$) consists of a sequence

$$D = \{(q_i, n_i, N_i)\}_{i\in\mathbb{N}} \tag{7}$$

with $q_i \in \mathbb{C}$, $Re\, q_i < (n+1)/2 - \gamma_0$, $Re\, q_i \to -\infty$ as $i \to \infty$, $n_i \in \mathbb{N}$, and finite-dimensional subspaces $N_i \subset C^\infty(X)$ (cf. 1.2.(1),(2)).

According to 1.2.(3) the space $\mathcal{H}_D^{s,\gamma_0}(\mathbb{B})$ with D being of the form (7) consists of all $u \in \mathcal{H}^{s,\gamma_0}(\mathbb{B})$ such that for every real β there exists an $M = M(\beta)$ with

$$\tilde{\omega}(r)\left\{u - \sum_{i=0}^{M}\sum_{k=0}^{n_i} \nu_{ik} r^{-q_i} \log^k r\right\} \in \mathcal{H}^{s,\beta}(\mathbb{B})$$

for certain $\nu_{ik} \in N_i$, $0 \le k \le n_i$, and any cut-off function $\tilde{\omega}(r)$.

Let $\Theta_1 = (\vartheta_1, 0]$, $\vartheta_1 > -\infty$ be the weight interval for the t asymptotics ("on the left of the weight line $\Gamma_{(n+2)/2-\gamma_1}$") and set

$$\mathcal{H}_{\Theta_1}^{s,\gamma}(\mathbb{B}^\wedge) = \varprojlim_{\epsilon>0} \left\{[\omega]\mathcal{H}^{s,(\gamma_0,\gamma_1-\vartheta_1-\epsilon)}(\mathbb{B}^\wedge) + [1-\omega]\mathcal{H}^{s,\gamma}(\mathbb{B}^\wedge)\right\} \tag{8}$$

for any cut-off function $\omega(t)$.

Definition 2.3.2 *Let $s \in \mathbb{R}$, $\gamma = (\gamma_0, \gamma_1) \in \mathbb{R}^2$, and $P \in As(\mathbb{B}, \mathfrak{a})$ as in Definition 2.3.1. Then, $\mathcal{H}_P^{s,\gamma}(\mathbb{B}^\wedge)$ for $\vartheta_1 > -\infty$ denotes the subspace of all $u \in \mathcal{H}^{s,\gamma}(\mathbb{B}^\wedge)$ with*

$$\omega\left\{u - \sum_{j=0}^{N}\sum_{k=0}^{m_j} \zeta_{jk} t^{-p_j} \log^k t\right\} \in \mathcal{H}_{\Theta_1}^{s,\gamma}(\mathbb{B}^\wedge) \tag{9}$$

for certain $\zeta_{jk} \in L_j$, $0 \le k \le m_j$, $0 \le j \le N = N(P)$, with any cut-off function $\omega(t)$. For $\vartheta_1 = -\infty$ the space $\mathcal{H}_P^{s,\gamma}(\mathbb{B}^\wedge)$ consists of all $u \in \mathcal{H}^{s,\gamma}(\mathbb{B}^\wedge)$ with

$$\omega\left\{u - \sum_{j=0}^{N}\sum_{k=0}^{m_j} \zeta_{jk} t^{-p_j} \log^k t\right\} \in t^\beta \mathcal{H}^{s,\gamma}(\mathbb{B}^\wedge) \tag{10}$$

where $\zeta_{jk} \in L_j$, $0 \le k \le m_j$, for all j and all every real β with some $N = N(\beta)$.

If Θ_1 is finite, $P \in As(\mathbb{B}, \mathfrak{a})$, we set

$$\mathcal{E}_P(\mathbb{B}^\wedge) = \left\{\sum_{j=0}^{N}\sum_{k=0}^{m_j} \zeta_{jk} t^{-p_j} \log^k t\, \omega(t) : \zeta_{jk} \in L_j, 0 \le k \le m_j\right\}$$

Here $N = N(P)$ and the cut-off function ω is fixed. Then

$$\mathcal{E}_P(\mathbb{B}^\wedge) \subset \mathcal{H}_P^{\infty,\gamma}(\mathbb{B}^\wedge) \tag{11}$$

is a finite-dimensional subspace, and we have

$$\mathcal{H}_P^{s,\gamma}(\mathbb{B}^\wedge) = \mathcal{H}_{\theta_1}^{s,\gamma}(\mathbb{B}^\wedge) + \mathcal{E}_P(\mathbb{B}^\wedge) \tag{12}$$

where the sum (12) is direct. This gives a natural Banach space structure in (12).

If $\vartheta_1 = -\infty$ we choose a sequence $\{\vartheta_{1,(m)}\}_{m\in\mathbb{N}}$ with $\vartheta_{1,(m+1)} < \vartheta_{1,(m)} < 0$ for all m and $\vartheta_{1,(m)} \to -\infty$ as $m \to \infty$, If $\mathfrak{a}_{1,(m)} = (\gamma_1, \Theta_{1,(m)})$ with $\Theta_{1,(m)} = (\vartheta_{1,(m)}, 0]$, then

$$\begin{aligned} P_{(m)} &= \left\{(p, m, L) \in P : Re\, p > (n+2)/2 - \gamma_1 + \vartheta_{1,(m)}\right\} \\ &\in As(\mathbb{B}, \mathfrak{a}_{(m)}) \text{ for } \mathfrak{a}_{(m)} = (\mathfrak{a}_0, \mathfrak{a}_{1,(m)}). \end{aligned} \tag{13}$$

This yields a sequence of Banach spaces $\mathcal{H}^{s,\gamma}_{P_{(m)}}(\mathbb{B}^\wedge)$, $m \in \mathbb{N}$, with continuous embeddings

$$\mathcal{H}^{s,\gamma}_{P_{(m+1)}}(\mathbb{B}^\wedge) \subset \mathcal{H}^{s,\gamma}_{P_{(m)}}(\mathbb{B}^\wedge)$$

for all m. We endow

$$\mathcal{H}^{s,\gamma}_{P}(\mathbb{B}^\wedge) = \bigcap_{m \in \mathbb{N}} \mathcal{H}^{s,\gamma}_{P_{(m)}}(\mathbb{B}^\wedge)$$

with the Fréchet topology of the projective limit.

The Mellin calculus in Sobolev spaces requires a description of the asymptotics under the Mellin transform. Let $f(z)$ be a $\mathcal{H}^{s,\beta}(\mathbb{B})$-valued function on $\Gamma_{(n+2)/2-\gamma_1}$. Set for $\gamma = (\gamma_0, \gamma_1)$

$$||f||_{s,\gamma} = \left\{ \frac{1}{2\pi i} \int_{\Gamma_{(n+2)/2-\gamma_1}} ||g_0^{s-\gamma_0} b^s(z)(M_{t\to z}u)(z)||^2_{\mathcal{H}^0(\mathbb{B})} dz \right\}^{1/2}$$

with the operator family $b^s(z)$ from 2.2.(4). By shifting $b^s(z)$ to any other weight line $\Gamma_{(n+2)/2-\sigma}$ we get in the same way the expressions $||f||_{s,(\gamma_0,\sigma)}$.

Set

$$S^{\gamma_1}_{\Theta_1} = \{z \in \mathbb{C} : (n+2)/2 - \gamma_1 + \vartheta_1 < Re\, z \leq (n+2)/2 - \gamma_1\},$$

$\Theta_1 = (\vartheta_1, 0]$.

Definition 2.3.3 *Let $s \in \mathbb{R}$, $\gamma = (\gamma_0, \gamma_1) \in \mathbb{R}^2$, and $P \in As(\mathbb{B}, \mathfrak{a})$ for $\mathfrak{a} = (\mathfrak{a}_0, \mathfrak{a}_1^\bullet)$, $\mathfrak{a}_i = (\gamma_i, \gamma_i - \mu, \Theta_i)$, P being written as (1). Then $\mathcal{A}^{s,\gamma}_P(\mathbb{B})$ denotes the space of all $f(z) \in \mathcal{A}(int S^{\gamma_1}_{\Theta_1} \setminus \pi_{\mathbb{C}} P, \mathcal{H}^{s,\gamma_0}(\mathbb{B}))$ which are meromorphic in $int S^{\gamma_1}_{\Theta_1}$ with poles at all p_j of multiplicities $m_j + 1$ and Laurent expansions*

$$f(z) = \sum_{k=0}^{m_j} \zeta_{jk}(z - p_j)^{-(k+1)} + \sum_{k=0}^{\infty} \zeta^+_{jk}(z - p_j)^k$$

close to p_j, with $\zeta_{jk} \in L_j$, $0 \leq k \leq m_j$, for all j, and further

$$\sup_\sigma ||\chi(z) f(z)||_{s,(\gamma_0,\sigma)} < \infty \tag{14}$$

where for $\vartheta_1 > -\infty$ the supremum (14) is taken over $\gamma_1 < \sigma < \gamma_1 - \vartheta_1 - \varepsilon$ for every $\varepsilon > 0$ with $-\vartheta_1 > \varepsilon$, and for $\vartheta_1 = -\infty$ over $\gamma_1 < \sigma < \gamma_1 + \delta$ for arbitrary $\delta > 0$. Here χ is an arbitrary $\pi_{\mathbb{C}} P$-excision function.

Note that the coefficients ζ_{jk} are uniquely determined by f. Thus there is a system of linear operators

$$\mathcal{A}^{s,\gamma}_P(\mathbb{B}) \longrightarrow L_j, \; f \longrightarrow \zeta_{jk}, \tag{15}$$

$0 \leq k \leq m_j$ for all j. The space $\mathcal{A}^{s,\gamma}_P(\mathbb{B})$ is Fréchet in the semi-norms system induced from $\mathcal{A}(int S^{\gamma_1}_{\Theta_1} \setminus \pi_{\mathbb{C}} P, \mathcal{H}^{s,\gamma_0}(\mathbb{B}))$ together with those from (15) and (14) for all ε (δ).

For empty P we get $\mathcal{A}^{s,\gamma}_{\Theta_1}(\mathbb{B}) \subset \mathcal{A}(int S^{\gamma_1}_{\Theta_1}, \mathcal{H}^{s,\gamma_0}(\mathbb{B}))$. Moreover, we have

$$\mathcal{A}^{\infty,\gamma}_P(\mathbb{B}) = \bigcap_{s \in \mathbb{R}} \mathcal{A}^{s,\gamma}_P(\mathbb{B}).$$

Then

$$\mathcal{A}^{s,\gamma}_P(\mathbb{B}) = \mathcal{A}^{s,\gamma}_{\Theta_1}(\mathbb{B}) + \mathcal{A}^{\infty,\gamma}_P(\mathbb{B})$$

holds as a sum of Fréchet spaces (cf. 1.1.(2)).

Theorem 2.3.4 *The weighted Mellin transform*

$$M_{\gamma_1-(n+1)/2} : C_0^\infty(int\mathbb{B}\times\mathbb{R}_+) \longrightarrow \mathcal{S}(\Gamma_{(n+2)/2-\gamma_1}, C_0^\infty(int\mathbb{B}))$$

extends from $[\omega]C_0^\infty(int\mathbb{B}\times\mathbb{R}_+)$ *to a continuous operator*

$$M_{\gamma_1-(n+1)/2} : [\omega]\mathcal{H}_P^{s,\gamma}(\mathbb{B}^\wedge) \longrightarrow \mathcal{A}_P^{s,\gamma}(\mathbb{B}) \tag{16}$$

for every cut-off function $\omega(t)$. *An element* $u \in \mathcal{H}^{s,\gamma}(\mathbb{B}^\wedge)$ *belongs to* $\mathcal{H}_P^{s,\gamma}(\mathbb{B}^\wedge)$ *iff* $M_{\gamma_1-(n+1)/2}(\omega u) \in \mathcal{A}_P^{s,\gamma}(\mathbb{B})$ *for any* ω. *The Fréchet topology of* $\mathcal{H}_P^{s,\gamma}(\mathbb{B}^\wedge)$ *can equivalently be defined by the projective limit with respect to (16) and* $\mathcal{H}_P^{s,\gamma}(\mathbb{B}^\wedge) \subset \mathcal{H}^{s,\gamma}(\mathbb{B}^\wedge)$.

Next we consider the corner axis asymptotics (9) or (10) together with the edge asymptotics for $r \to 0$. Analogously to 1.4.(70),(71) we use the spaces

$$\begin{aligned} \mathcal{V}^{s,\alpha}(\mathbb{R}_+, \mathcal{H}^{s,\beta}(\mathbb{B})) &= t^\alpha \mathcal{V}^s(\mathbb{R}_+, \mathcal{H}^{s,\beta}(\mathbb{B})), \\ \mathcal{V}^{s,\alpha}(\mathbb{R}_+, \mathcal{H}_D^{s,\beta}(\mathbb{B})) &= t^\alpha \mathcal{V}^s(\mathbb{R}_+, \mathcal{H}_D^{s,\beta}(\mathbb{B})) \end{aligned}$$

for arbitrary $\alpha, \beta \in \mathbb{R}$, $D \in As(X, (\beta, \Theta_0))$. Similary to (8) we set

$$\begin{aligned} &\mathcal{V}_{\Theta_1}^{s,\gamma_1-(n+1)/2}(\mathbb{R}_+, \mathcal{H}^{s,\gamma_0}(\mathbb{B})) \\ = {} &\varprojlim_{\epsilon>0} \left\{[\omega]\mathcal{V}^{s,\gamma_1-(n+1)/2-\vartheta_1-\epsilon}(\mathbb{R}_+, \mathcal{H}^{s,\gamma_0}(\mathbb{B})) + [1-\omega]\mathcal{V}^{s,\gamma_1-(n+1)/2}(\mathbb{R}_+, \mathcal{H}^{s,\gamma_0}(\mathbb{B}))\right\}. \end{aligned} \tag{17}$$

In the same way we get

$$\mathcal{V}_{\Theta_1}^{s,\gamma_1-(n+1)/2}(\mathbb{R}_+, \mathcal{H}_D^{s,\gamma_0}(\mathbb{B})) \tag{18}$$

by replacing $\mathcal{H}^{s,\gamma_0}(\mathbb{B}))$ in (17) by $\mathcal{H}_D^{s,\gamma_0}(\mathbb{B})$.

Remark 2.3.5 *Let* $P \in As(\mathbb{B}, \mathfrak{a})$ *for* $\mathfrak{a} = (\mathfrak{a}, \mathfrak{a}_1^\bullet)$, $\mathfrak{a}_i = (\gamma_i, \Theta_i)$, *and* $D = \pi_X P \in As(X, \mathfrak{a}_0)$, *cf. (6). Then*

$$\mathcal{E}_P(\mathbb{B}^\wedge) \subset \mathcal{V}^{\infty,\gamma_1-(n+1)/2}(\mathbb{R}_+, \mathcal{H}_D^{\infty,\gamma_0}(\mathbb{B})). \tag{19}$$

Definition 2.3.6 *Let* $P = (P_0, P_1) \in As(X, \mathfrak{a}_0) \times As(\mathbb{B}, \mathfrak{a})$ *for* $\mathfrak{a} = (\mathfrak{a}_0, \mathfrak{a}_1^\bullet)$, $\mathfrak{a}_i = (\gamma_i, \Theta_i)$, $i = 0, 1$, $\gamma = (\gamma_0, \gamma_1)$, P_1 *being written as in Definition 2.3.1,* $s \in \mathbb{R}$. *Then* $\mathcal{H}_P^{s,\gamma}(\mathbb{B}^\wedge)$ *denotes the subspace of all* $u \in \mathcal{H}^{s,\gamma}(\mathbb{B}^\wedge)$ *such that*

(i) $u \in \mathcal{H}_{P_1}^{s,\gamma}(\mathbb{B}^\wedge)$ *in the sense of Definition 2.3.2,*

(ii)$_1$ in case $\vartheta_1 > -\infty$ *we have*

$$u - \omega(t)\sum_{j=0}^{N}\sum_{k=0}^{m_j} \zeta_{jk} t^{-p_j} \log^k t \in \mathcal{V}_{\Theta_1}^{s,\gamma_1-(n+1)/2}(\mathbb{R}_+, \mathcal{H}_{P_0}^{s,\gamma_0}(\mathbb{B}))$$

with $N = N(P)$,

(ii)$_\infty$ in case $\vartheta_1 = -\infty$ *there is for every* $\alpha \in \mathbb{R}$ *an* $N \in \mathbb{N}$ *with*

$$u - \omega(t)\sum_{j=0}^{N}\sum_{k=0}^{m_j} \zeta_{jk} t^{-p_j} \log^k t \in \mathcal{V}^{s,\alpha}(\mathbb{R}_+, \mathcal{H}_{P_0}^{s,\gamma_0}(\mathbb{B}))$$

where (both for (ii)$_1$ and (ii)$_\infty$) $\zeta_{jk} \in L_j$, $0 \le k \le m_j$, *for all* j, *and an arbitrary cut-off function* $\omega(t)$.

For $\vartheta_1 > -\infty$ we can write

$$\mathcal{H}_P^{s,\gamma}(\mathbb{B}^\wedge) = \mathcal{V}_{\Theta_1}^{s,\gamma_1-(n+1)/2}(\mathbb{R}_+, \mathcal{H}_{P_0}^{s,\gamma_0}(\mathbb{B})) + \mathcal{E}_{P_1}(\mathbb{B}^\wedge) \tag{20}$$

which is a direct sum, $P = (P_0, P_1)$. This turns $\mathcal{H}_P^{s,\gamma}(\mathbb{B}^\wedge)$ to a Fréchet space, since both summands are Fréchet.

For $\vartheta_1 = -\infty$ we form $P_{(m)} = (P_0, P_{1,(m)})$ where $P_{1,(m)}$ is defined in the same way as (13) above, now for P_1. Then we have continuous embeddings

$$\mathcal{H}_{P_{(m+1)}}^{s,\gamma}(\mathbb{B}^\wedge) \subset \mathcal{H}_{P_{(m)}}^{s,\gamma}(\mathbb{B}^\wedge)$$

for all m, and we endow

$$\mathcal{H}_P^{s,\gamma}(\mathbb{B}^\wedge) = \bigcap_{m\in\mathbb{N}} \mathcal{H}_{P_{(m)}}^{s,\gamma}(\mathbb{B}^\wedge)$$

with the Fréchet topology of the projective limit. Now we define the subspaces of the global spaces 2.2.(14) over $\mathbb{K}$ with asymptotics.

Definition 2.3.7 *Let $\mathbb{K}$ be the stretched manifold belonging to the given space K with corners. Then, for $P = (P_0, P_1) \in As(X, \mathfrak{a}) \times As(\mathbb{B}, \mathfrak{a})$, $\mathfrak{a} = (\mathfrak{a}_0, \mathfrak{a}_1^\bullet)$, with P_0 satisfying the shadow condition (cf. 1.3.(20)) we set*

$$\mathcal{H}_P^{s,\gamma}(\mathbb{K}) = [\omega]\mathcal{H}_P^{s,\gamma}(\mathbb{B}^\wedge) + [1-\omega]\mathcal{W}_{P_0,loc}^{s,\gamma_0}(\mathbb{W}), \tag{21}$$

for $s \in \mathbb{R}$, $\gamma = (\gamma_0, \gamma_1) \in \mathbb{R}^2$, for any fixed cut-off function $\omega(t)$ (cf. 1.3.(19)). The space (21) will be endowed with the topology of the sum.

Remark 2.3.8 *(21) is independent of the concrete choice of ω, cf. Theorem 1.4.7. If we fix ω and form*

$$[\omega]\mathcal{H}_P^{s,\gamma}(\mathbb{B}^\wedge) + [1-\omega]\mathcal{W}_{P_2,loc}^{s,\gamma_0}(\mathbb{W}) \tag{22}$$

with arbitrary $P = (P_0, P_1) \in As(X, \mathfrak{a}_0) \times As(\mathbb{B}, \mathfrak{a})$, $P_2 \in As(X, \mathfrak{a}_0)$ then there is a resulting asymptotic type Q_0 such that (22) equals $\mathcal{H}_Q^{s,\gamma}(\mathbb{K})$ with $Q = (Q_0, P_1)$.

2.4 Corner Mellin symbols

The next step for establishing the class of ψDO's on manifolds with corners (analogously to Definition 1.2.3 for conical singularities is to define the spaces of operator-valued Mellin symbols, here operating globally along the base $\mathbb{B}$ of the corner.

Definition 2.4.1 *Let $\mu, \nu, \beta \in \mathbb{R}$, $\mu - \nu \in \mathbb{N}$, $\mathfrak{g} = (\beta, \beta - \mu, \Theta)$ for $\Theta = (-k, 0]$, $k \in (\mathbb{N}\setminus\{0\})\cup\{\infty\}$. Then $\mathcal{M}_O^\nu(\mathbb{B}, \mathfrak{g}; \mathbb{C}^N, \mathbb{C}^M)$ is the space of all*

$$h(z) \in \bigcap_{s\in\mathbb{R}} \mathcal{A}(\mathbb{C}, \mathcal{L}(\mathcal{H}^{s,\beta}(\mathbb{B}) \oplus \mathbb{C}^N, \mathcal{H}^{s-\nu,\beta-\mu}(\mathbb{B}) \oplus \mathbb{C}^M))$$

with

$$h(z)|_{\Gamma_\sigma} \in C^\nu(\mathbb{B}, \mathfrak{g}; \mathbb{C}^N, \mathbb{C}^M; \Gamma_\sigma)$$

for all $\sigma \in \mathbb{R}$, uniformly in every finite interval $c < \sigma < c'$ (cf. Definition 2.1.21). Furthermore we set

$$M_O^\nu(\mathbb{B}, \mathfrak{g}) = l.u.c.\ \mathcal{M}_O^\nu(\mathbb{B}, \mathfrak{g}; \mathbb{C}^N, \mathbb{C}^M) \tag{1}$$

(i.e. the class of left upper corners).

Observe the relation of Definition 2.4.1 to Theorem 2.1.22.

The corner calculus will employ not only holomorphic but meromorphic $C^\nu(\mathbb{B},\mathfrak{g};\mathbb{C}^N,\mathbb{C}^M)$-valued Mellin symbols (cf. Remark 2.1.12). The poles and Laurent expansions will be described in terms of asymptotic types.

Definition 2.4.2 $\underline{As}(\mathbb{B},\mathfrak{g};\mathbb{C}^N,\mathbb{C}^M)^\bullet$ *for* $\mathfrak{g}=(\beta,\beta-\mu,\Theta)$ *denotes the set of all sequences*

$$P=\{(p_j,m_j,N_j)\}_{j\in\mathbb{Z}} \tag{2}$$

with $p_j\in\mathbb{C}$, $|Re\ p_j|\to\infty$ *as* $|j|\to\infty$, $m_j\in\mathbb{N}$, *and finite-dimensional subspaces* N_j *of finite-dimensional operators* $C_G(\mathbb{B},\mathfrak{g};\mathbb{C}^N,\mathbb{C}^M)$ *(cf. 2.1.(31)), where the asymptotic types of the operators in* N_j *are independent of* j. *For* $N=M=0$ *we set by definition*

$$\underline{As}(\mathbb{B},\mathfrak{g})^\bullet. \tag{3}$$

In this case $N_j\subset C_G(\mathbb{B},\mathfrak{g})$. *We set*

$$\pi_{\mathbb{C}}P=\{p_j\}_{j\in\mathbb{N}}.$$

A $\pi_{\mathbb{C}}P$*-excision function is a* $\chi(z)\in C^\infty(\mathbb{C})$, $0\le\chi\le1$, *with* $\chi(z)=1$ *for* $dist(z,\pi_{\mathbb{C}}P)>\varepsilon$ *for some* $\varepsilon>0$, $\chi(z)=0$ *in an open neighbourhood of* $\pi_{\mathbb{C}}P$.

Definition 2.4.3 $\mathcal{M}_P^{-\infty}(\mathbb{B},\mathfrak{g};\mathbb{C}^N,\mathbb{C}^M)$ *for* $P\in\underline{As}(\mathbb{B},\mathfrak{g};\mathbb{C}^N,\mathbb{C}^M)^\bullet$, $\mathfrak{g}=(\beta,\beta-\mu,\Theta)$, *is the space of all*

$$g(z)\in\bigcap_{s\in\mathbb{R}}\mathcal{A}(\mathbb{C}\setminus\pi_{\mathbb{C}}P,\mathcal{L}(\mathcal{H}^{s,\beta}(\mathbb{B})\oplus\mathbb{C}^N,\mathcal{H}^{\infty,\beta-\mu}(\mathbb{B})\oplus\mathbb{C}^M))$$

which are meromorphic with poles at all p_j *of multiplicities* m_j+1 *and Laurent expansions*

$$g(z)=\sum_{k=0}^{m_j}\gamma_{jk}(z-p_j)^{-(k+1)}+\sum_{k=0}^{\infty}\gamma_{jk}^{+}(z-p_j)^k$$

close to p_j, *with* $\gamma_{jk}\in N_j$, $0\le k\le m_j$, *for all* $j\in\mathbb{Z}$, *and further*

$$(\chi g)(\sigma+i\tau)\in C_G^{-\infty}(\mathbb{B},\mathfrak{g};\mathbb{C}^N,\mathbb{C}^M;\Gamma_\sigma)$$

uniformly in σ *in finite intervals, for every* $\pi_{\mathbb{C}}P$*-excision function* χ.

Set

$$M_P^{-\infty}(\mathbb{B},\mathfrak{g})=l.u.c.\ \mathcal{M}_Q^{-\infty}(\mathbb{B},\mathfrak{g};\mathbb{C}^N,\mathbb{C}^M) \tag{4}$$

for $P\in\underline{As}(\mathbb{B},\mathfrak{g})^\bullet$ being the left upper corner of Q.

Remark 2.4.4 *Both* $\mathcal{M}_O^\nu(\mathbb{B},\mathfrak{g};\mathbb{C}^N,\mathbb{C}^M)$ *and* $\mathcal{M}_P^{-\infty}(\mathbb{B},\mathfrak{g};\mathbb{C}^N,\mathbb{C}^M)$ *are inductive limits of Fréchet spaces in a canonical way.*

We can write

$$\mathcal{M}_P^\nu(\mathbb{B},\mathfrak{g};\mathbb{C}^N,\mathbb{C}^M)=\mathcal{M}_O^\nu(\mathbb{B},\mathfrak{g};\mathbb{C}^N,\mathbb{C}^M)+\mathcal{M}_P^{-\infty}(\mathbb{B},\mathfrak{g};\mathbb{C}^N,\mathbb{C}^M) \tag{5}$$

in the topology of the (non-direct) sum, and

$$\mathcal{M}_P^\nu(\mathbb{B},\mathfrak{g})=\mathcal{M}_O^\nu(\mathbb{B},\mathfrak{g})+\mathcal{M}_P^{-\infty}(\mathbb{B},\mathfrak{g}) \tag{6}$$

for P in $\underline{As}(\mathbb{B},\mathfrak{g};\mathbb{C}^N,\mathbb{C}^M)^\bullet$ and $\underline{As}(\mathbb{B},\mathfrak{g})^\bullet$, respectively.

Set

$$sg(a) = \pi_{\mathbb{C}} P \tag{7}$$

if $a(z)$ belongs to (5),(6).

Define

$$\mathcal{M}^{\nu}_{as}(\mathbb{B}, \mathfrak{g}; \mathbb{C}^N, \mathbb{C}^M)^{\bullet} = \lim_{\overrightarrow{P}} \mathcal{M}^{\nu}_{P}(\mathbb{B}, \mathfrak{g}; \mathbb{C}^N, \mathbb{C}^M), \tag{8}$$

$$M^{\nu}_{as}(\mathbb{B}, \mathfrak{g})^{\bullet} = \lim_{\overrightarrow{P}} M^{\nu}_{P}(\mathbb{B}, \mathfrak{g}). \tag{9}$$

Every

$$a(z) \in \mathcal{M}^{\nu}_{P}(\mathbb{B}, \mathfrak{g}; \mathbb{C}^N, \mathbb{C}^M)$$

gives rise to a Mellin ψDO

$$op^{\alpha}_{M}(a)u(t) = (t^{\alpha} op_M (T^{-\alpha}a) t^{-\alpha} u)(t),$$

$\alpha \in \mathbb{R}$, for $sg(a) \cap \Gamma_{1/2-\alpha} = \emptyset$; $(T^{-\alpha}a)(z) = a(z-\alpha)$. $u(t)$ is regarded as vector-valued function of $t \in \mathbb{R}_+$, and we assume first $u(t) \in C^{\infty}_0(\mathbb{R}_+, C^{\infty}_0(int\mathbb{B}) \oplus \mathbb{C}^N)$. Then

$$op^{\alpha}_{M}(a) : C^{\infty}_0(\mathbb{R}_+, C^{\infty}_0(int\mathbb{B}) \oplus \mathbb{C}^N) \longrightarrow C^{\infty}(\mathbb{R}_+, C^{\infty}(int\mathbb{B}) \oplus \mathbb{C}^M) \tag{10}$$

is continuous.

Theorem 2.4.5 *Let $a(z) \in \mathcal{M}^{\nu}_{as}(\mathbb{B}, \mathfrak{g}_0; \mathbb{C}^N, \mathbb{C}^M)^{\bullet}$, $\mathfrak{g}_0 = (\gamma_0, \gamma_0 - \mu, \Theta_0)$, $sg(a) \cap \Gamma_{(n+2)/2-\gamma_1} = \emptyset$, $\gamma = (\gamma_0, \gamma_1)$. Then $\omega_1 op^{\tilde{\gamma}_1}_{M}(a)\omega_2$ for $\tilde{\gamma}_1 = \gamma_1 - (n+1)/2$ (with arbitrary cut-off functions $\omega_i(t), i = 1,2$) extends to a continuous operator*

$$\omega_1 op^{\tilde{\gamma}_1}_{M}(a)\omega_2 : \mathcal{H}^{s,\gamma}(\mathbb{B}^{\wedge}) \oplus \mathcal{H}^{s,\tilde{\gamma}_1}(\mathbb{R}_+, \mathbb{C}^N) \longrightarrow \mathcal{H}^{s-\nu,(\gamma_0-\mu,\gamma_1)}(\mathbb{B}^{\wedge}) \oplus \mathcal{H}^{s-\nu,\tilde{\gamma}_1}(\mathbb{R}_+, \mathbb{C}^M) \tag{11}$$

for all $s \in \mathbb{R}$. Moreover for every $P = (P_0, P_1) \in As(X, \mathfrak{a}_0) \times As(\mathbb{B}, \mathfrak{a})$ (cf. Definition 2.3.6), $\tilde{P} \in As(\tilde{\mathfrak{a}}^{\bullet})$ for $\tilde{\mathfrak{a}} = (\tilde{\gamma}_1, \Theta_1)$ there exist asymptotic types $Q = (Q_0, Q_1) \in As(X, \mathfrak{b}_0) \times As(\mathbb{B}, \mathfrak{b})$ for $\mathfrak{b} = (\mathfrak{b}_0, \mathfrak{a}^{\bullet})$, $\mathfrak{b}_0 = (\gamma_0 - \mu, \Theta_0)$, and $\tilde{Q} \in As(\mathfrak{a}^{\bullet})$ such that (11) restricts to continous operators

$$\omega_1 op^{\tilde{\gamma}_1}_{M}(a)\omega_2 : \mathcal{H}^{s,\gamma}_{P}(\mathbb{B}^{\wedge}) \oplus \mathcal{H}^{s,\tilde{\gamma}_1}_{\tilde{P}}(\mathbb{R}_+, \mathbb{C}^N) \longrightarrow \mathcal{H}^{s-\nu,(\gamma_0-\mu,\gamma_1)}_{Q}(\mathbb{B}^{\wedge}) \oplus \mathcal{H}^{s-\nu,\tilde{\gamma}_1}_{\tilde{Q}}(\mathbb{R}_+, \mathbb{C}^M) \tag{12}$$

for all $s \in \mathbb{R}$.

Remark 2.4.6 *The operators*

$$\omega_1 op^{\tilde{\gamma}_1}_{M}(a)\omega_2 \tag{13}$$

for $\tilde{\gamma}_1 = \gamma_1 - (n+1)/2 = 0$, $a \in \mathcal{M}^{\nu}_{as}(\mathbb{B}, \mathfrak{g}; \mathbb{C}^N, \mathbb{C}^M)^{\bullet}$, belong to $M\mathcal{Y}^{\nu}(\mathbb{W}, \mathfrak{g}_0; \mathbb{C}^N, \mathbb{C}^M)$ (cf. Definition 1.4.21) and Theorem 2.4.5 is in this case a more precise version of Theorem 1.4.22 (close to $t = 0$).

3 Pseudo-differential operators on manifolds with corners

3.1 The corner operator algebra

The operators of the corner algebra on $\mathbb{K}$ (cf. the notations around 1.1.(23)) are expected to be matrix-valued with the "proper ψDO's" in the left upper corners and the additional trace and potential conditions along the edges. They can be introduced at once by concise notations. But we prefer here to look first at the left upper corners separately. The general class will then follow in Section 3.2.

As above we assume for notational convenience that K_0 consists of a single corner v; the base $\mathbb{B}$ is allowed to have several connected components.

Let us start with the Green operators.

Definition 3.1.1 *Let* $\mu \in \mathbb{R}$, $\gamma = (\gamma_0, \gamma_1) \in \mathbb{R}^2$, $\mathfrak{g}_i = (\gamma_i, \gamma_i - \mu, \Theta_i)$, $\Theta_i = (\vartheta_i, 0]$, $i = 0, 1$, $\mathfrak{g} = (\mathfrak{g}_0, \mathfrak{g}_1^\bullet)$. *Then* $C_G(\mathbb{K}, \mathfrak{g})$ *is the space of all operators*

$$G \in \bigcap_{s \in \mathbb{R}} \mathcal{L}(\mathcal{H}^{s,\gamma}(\mathbb{K}), \mathcal{H}^{\infty,\gamma-\mu}(\mathbb{K})),$$

$\gamma - \mu := (\gamma_0 - \mu, \gamma_1 - \mu)$, *which induce continuous operators*

$$G : \mathcal{H}^{s,\gamma}(\mathbb{K}) \longrightarrow \mathcal{H}_P^{\infty,\gamma-\mu}(\mathbb{K}),$$
$$G^* : \mathcal{H}^{s,-\gamma+\mu}(\mathbb{K}) \longrightarrow \mathcal{H}_Q^{\infty,-\gamma}(\mathbb{K})$$

for all $s \in \mathbb{R}$, *with asymptotic types* $P = (P_0, P_1) \in As(X, \mathfrak{a}_0) \times As(\mathbb{B}, \mathfrak{a})$, $\mathfrak{a} = (\mathfrak{a}_0, \mathfrak{a}_1^\bullet)$, $\mathfrak{a}_i = (\gamma_i - \mu, \Theta_i)$, $Q = (Q_0, Q_1) \in As(X, \mathfrak{b}_0) \times As(\mathbb{B}, \mathfrak{b})$, $\mathfrak{b} = (\mathfrak{b}_0, \mathfrak{b}_1^\bullet)$, $\mathfrak{b}_i = (-\gamma_i, \Theta_i)$, $i = 0, 1$, *dependent on* G *(not on* s*). Here* G^* *is the formal adjoint of* G *with respect to the scalar product of* $\mathcal{H}^0(\mathbb{K})$ *(cf. 2.2.(23)).*

Here and in the sequel we shall automatically assume that the first components of asymptotic types of distributions on *int*$\mathbb{K}$ satisfy $R(P_0) = P_0$, $R(Q_0) = Q_0$ (cf. Definition 2.3.7).

Definition 3.1.2 *Let* $\mu, \nu \in \mathbb{R}$, $\mu - \nu \in \mathbb{N}$, $\mathfrak{g} = (\mathfrak{g}_0, \mathfrak{g}_1^\bullet)$ *for* $\mathfrak{g}_i = (\gamma_i, \gamma_i - \mu, \Theta_i)$, $\Theta_i = (-k_i, 0]$, $k_i \in \mathbb{N} \setminus \{0\}$, $\gamma_i \in \mathbb{R}$, $i = 0, 1$. *Then* $C^\nu(\mathbb{K}, \mathfrak{g})$ *is the space of all operators*

$$A = A_0 + A_1 + M + G \tag{1}$$

with

$$A_0 = \omega t^{-\nu} op_M^{\tilde{\gamma}_1}(h) \omega_0, \; h(t, z) \in C^\infty(\overline{\mathbb{R}}_+, M_O^\nu(\mathbb{B}, \mathfrak{g}_0)) \tag{2}$$

(cf. 2.4.(1)) for $\tilde{\gamma}_1 = \gamma_1 - (n+1)/2$, $n + 1 = dim\mathbb{B}$,

$$A_1 = (1 - \omega) P (1 - \omega_1), P \in Y^\nu(\mathbb{W}, \mathfrak{g}_0) \tag{3}$$

(cf. 1.3.(70) with $\mathbb{W}$ *being given by 1.1.(25)),*

$$M = \omega_2 \left\{ t^{-\nu} \sum_{0 \le j \le k_1 - \mu + \nu - 1} t^j op_M^{\tilde{\delta}_j}(h_j) \right\} \omega_3 \tag{4}$$

where $h_j \in M_{\mathfrak{a}_s}^{-\infty}(\mathbb{B}, \mathfrak{g}_0)^\bullet$, *(cf. 2.4.(9)),* $\tilde{\delta}_j = \delta_j - (n+1)/2$,

$$\gamma_1 - (\mu - \nu) - j \le \delta_j \le \gamma_1 \tag{5}$$

$sg(h_j) \cap \Gamma_{(n+2)/2 - \delta_j} = \emptyset$ *for all* j, *and finally* $G \in C_G(\mathbb{K}, \mathfrak{g})$. *Here* $\omega(t), \omega_i(t)$ *are arbitrary cut-off functions supported by* $[0, 1)$.

Note that 3.1.2 is analogous to Definition 1.2.3.

We have $M = 0$ for $k_1 - \mu + \nu < 1$.

Recall from Section 1.1. that there is a neighbourhood V of $v \in K_0$ with $p^{-1}(V) \cong \mathbb{B} \times [0, 1)$. In other words there is a neighbourhood $\mathbb{V}$ of $p^{-1}(v)$ in $\mathbb{K}$ and a fixed diffeomorphism

$$\varphi : \mathbb{V} \setminus p^{-1}(v) \longrightarrow \mathbb{B} \times (0, 1). \tag{6}$$

Similary to 1.2.(33) it is allowed to assume that

$$\varphi_*(P|_{\mathbb{V} \setminus p^{-1}(v)}) \sim t^{-\nu} op_M^{\tilde{\gamma}_1}(h)|_{\mathbb{B} \times (0,1)} \tag{7}$$

where $\sim$ means equality $mod Y_G^{-\infty}(\mathbb{B}\times(0,1),\mathfrak{g}_0)$ (cf. 1.3.(67). If we suppose (7) and assume

$$\omega\omega_0=\omega\ ,\ \omega\omega_1=\omega_1 \tag{8}$$

then any other choice of ω,ω_1,ω_2 will modify (1) only by some element in $C_G(\mathbb{K},\mathfrak{g})$.

Also the concrete choice of ω_2,ω_3 and of the weights δ_j (under the mentioned conditions) only affects $A\ mod\ C_G(\mathbb{K},\mathfrak{g})$. The latter property follows by the same technique as for [S4],I,1.1.4. Theorem 20 (cf. also [S1]).

In (1) we have dropped for abbreviation the pull-backs of Mellin operators under (6) to $\mathbb{K}$ itself. Also below this will often be omitted if identifications of objects with their pull-backs are obvious.

Theorem 3.1.3 *An operator $A\in C^\nu(\mathbb{K},\mathfrak{g})$ (with $\mathfrak{g}$ being as in Definition 3.1.2) induces continuous operators*

$$A:\mathcal{H}^{s,\gamma}(\mathbb{K})\longrightarrow\mathcal{H}^{s-\nu,\gamma-\mu}(\mathbb{K}) \tag{9}$$

for all $s\in\mathbb{R}$. Here $\gamma=(\gamma_0,\gamma_1)$, $\gamma-\mu=(\gamma_0-\mu,\gamma_1-\mu)$. Moreover for every $P=(P_0,P_1)\in As(X,\mathfrak{a}_0)\times As(\mathbb{B},\mathfrak{a})$, $\mathfrak{a}=(\mathfrak{a}_0,\mathfrak{a}_1^\bullet)$, $\mathfrak{a}_i=(\gamma_i,\Theta_i)$ (cf. Definition 2.3.7) there exists a $Q=(Q_0,Q_1)\in As(X,\mathfrak{b}_0)\times As(\mathbb{B},\mathfrak{b})$, $\mathfrak{b}=(\mathfrak{b}_0,\mathfrak{b}_1^\bullet)$, $\mathfrak{b}_i=(\gamma_i-\mu,\Theta_i)$, such that (9) induces continous operators

$$A:\mathcal{H}_P^{s,\gamma}(\mathbb{K})\longrightarrow\mathcal{H}_Q^{s-\mu,\gamma-\mu}(\mathbb{K})$$

for all $s\in\mathbb{R}$.

Remark 3.1.4 *The inclusions*

$$C^\nu(\mathbb{K},\mathfrak{g})\subset L^\nu_{cl}(\Omega) \tag{10}$$

with $\Omega=K\setminus K_1$ (cf. the notations in the beginning),

$$C^\nu(\mathbb{K},\mathfrak{g})\subset Y^\nu(\mathbb{W},\mathfrak{g}_0)$$

(cf. 1.3.(70)),

$$t^{-\widetilde{\gamma}_1+\nu}\varphi_*\left\{C^\nu(\mathbb{K},\mathfrak{g})|_{\mathbb{V}\setminus p^{-1}(v)}\right\}t^{\widetilde{\gamma}_1}\subset MY^\nu(\mathbb{B}\times(0,1),\mathfrak{g}_0)$$

(cf. 1.4.(78)). Here $\widetilde{\gamma}_1=\gamma_1-(n+1)/2$, give rise to the symbolic levels over $C^\nu(\mathbb{W},\mathfrak{g})$ from the corresponding larger classes, namely

$$C^\nu(\mathbb{K},\mathfrak{g})\ni A\longrightarrow\sigma_\psi^\nu(A)\in S^{(\nu)}(T^*\Omega\setminus 0), \tag{11}$$

$$C^\nu(\mathbb{K},\mathfrak{g})\ni A\longrightarrow\sigma_\psi^\nu(A)\in\bigcap_{s\in\mathbb{R}}\mathcal{L}^{(\nu)}(T^*\Omega_1\setminus 0;\mathcal{K}^{s,\gamma_0}(X^\wedge),\mathcal{K}^{s-\nu,\gamma_0-\mu}(X^\wedge)) \tag{12}$$

$\Omega_1:=K_1\setminus\{v\}$ (cf. 1.3.(71))

$$C^\nu(\mathbb{K},\mathfrak{g})\ni A\longrightarrow\sigma_{b\wedge}^\nu(\widetilde{A})\in\bigcap_{s\in\mathbb{R}}\mathcal{L}^{(\nu)}(T^*([0,1))\setminus 0;\mathcal{K}^{s,\gamma_0}(X^\wedge),\mathcal{K}^{s-\nu,\gamma_0-\mu}(X^\wedge)) \tag{13}$$

$\widetilde{A}=t^\nu\varphi_*(A|_{\mathbb{V}\setminus p^{-1}(v)})$ (cf. 1.4.(79)).

Remark 3.1.5 *(12),(13) satisfy the compatibility condition*

$$\sigma_\wedge^\nu(A)(t,\tau)=t^{-\nu}\sigma_{b\wedge}^\nu(\widetilde{A})(t,\tau_1)|_{\tau_1=-t\tau}. \tag{14}$$

(cf. 1.1.49)).

The features of (11) (the homogeneous principal symbols of order ν) over the various parts of $\mathbb{K}$ (namely close to $p^{-1}(v)$ or close to $p^{-1}(K_0 \setminus \{v\})$) are those from the interior symbols of the classes MY^ν and Y^ν, respectively.

Denote by

$$S_{2b}^{(\nu)}(T^*\Omega \setminus 0) \tag{15}$$

(cf. 1.1.(24)) the subspace of all $p \in S^{(\nu)}(T^*\Omega \setminus 0)$ which are (i) close to the corner near the edges (i.e. for small t and r) in the coordinates (x, r, t) of the form

$$p_0(x, r, t, \xi, r\rho, rt\tau)$$

with $p_0(x, r, t, \xi, \rho, \tau)$ being C^∞ in r, t up to $r = 0, t = 0$, (ii) close to the corner far from the edges (i.e. for small t) in the coordinates $(\tilde{x}, t)$ of the form

$$\tilde{p}\,(\tilde{x}, t, \tilde{\xi}, t\tau)$$

with $\tilde{p}\,(\tilde{x}, t, \tilde{\xi}, t\tau)$ being C^∞ in t up to $t = 0$, (iii) close to the edges far from the corner (i.e. for small r in the coordinates (x, r, t) of the form

$$p_1(x, r, t, \xi, \rho, r\tau)$$

with $p_1(x, r, t, \xi, \rho, \tau)$ being C^∞ in r up to $r = 0$.

Let g_2^η for $\eta \in \mathbb{R}$ denote a strictly positive function in $C^\infty(\Omega)$ (a weight function) with the properties 1.1.(32),(33),(34). It is clear that then

$$\sigma_\psi^\nu(A) \in g_2^{-\nu} S_{2b}^{(\nu)}(T^*\Omega \setminus 0) \tag{16}$$

for every $A \in C^\nu(\mathbb{K}, \mathfrak{g})$.

Remark 3.1.6 *It may happen that the true interior order of an $A \in C^\nu(\mathbb{K}, \mathfrak{g})$ as well as the edge order is less than ν, i.e. $\nu - \iota$ for some $\iota \in \mathbb{N}$. Set*

$$C^{\nu}(\mathbb{K}, \mathfrak{g}) = C^\nu(\mathbb{K}, \mathfrak{g}) \cap Y^{\nu-\iota}(\mathbb{W}, \mathfrak{g}_0) \tag{17}$$

(cf. (10)). For every $A \in C^{\nu,\iota}(\mathbb{K}, \mathfrak{g})$ we then have a homogeneous principal interior symbol

$$\sigma_\psi^{\nu-\iota}(A) \in g_2^{-\nu} S_{2b}^{(\nu-\iota)}(T^*\Omega \setminus 0) \tag{18}$$

and a homogeneous principal edge symbol

$$\sigma_\wedge^{\nu-\iota}(A) \in \bigcap_{s \in \mathbb{R}} \mathcal{L}^{(\nu)}(T^*\Omega_1 \setminus 0; \mathcal{K}^{s,\gamma_0}(X^\wedge), \mathcal{K}^{s-\nu-\iota,\gamma_0-\mu}(X^\wedge)).$$

Definition 3.1.7 *(16) is called the homogeneous principal interior symbol of order ν, (12) the homogeneous principal edge symbol of order ν and (13) the homogeneous principal Mellin edge symbol of order ν of A. For $k_1 - \mu + \nu \geq 1$ we also define*

$$\sigma_M^\nu(A)(z) := h(0, z) + h_0(z) \in M_{as}^\nu(\mathbb{B}, \mathfrak{g}_0)^\bullet \tag{19}$$

called the principal conormal symbol of A of (conormal) order ν ($h(t, z)$ and $h_0(z)$ were given in (2) and (4), respectively).

(19) will be regarded as family of operators

$$\sigma_M^\nu(A)(z) : \mathcal{H}^{s,\gamma_0}(\mathbb{B}) \longrightarrow \mathcal{H}^{s-\nu,\gamma_0-\mu}(\mathbb{B}) \tag{20}$$

for any $s \in \mathbb{R}$, parametrized by $z \in \Gamma_{(n+2)/2-\gamma_1}$.

Remark 3.1.8 *The operator classes of Definition 3.1.2 have natural properties with respect to formal adjoints and compositions. We shall postpone this to the following section where they will be formulated for the matrices with $A \in C^{\nu}(\mathbb{K}, \mathfrak{g})$ in the left upper corners.*

Let us now look at some examples. The typical differential operators A from the Introduction and Section 1.1. belong to the class $C^{\mu}(\mathbb{K}, \mathfrak{g})$ for $\mathfrak{g} = (\gamma_i, \gamma_i - \mu, (-\infty, 0])$ with arbitrary $\gamma_0, \gamma_1 \in \mathbb{R}$. Remember that A was written close to $t = 0$ in the form 1.1.(36).

Set

$$h(t,z) = \tilde{\omega}(r) r^{-\mu} \sum_{j+k\leq\mu} a_{jk}(r,t)(-r\frac{\partial}{\partial r})^j(-rz)^k + (1 - \tilde{\omega}(r)) \sum_{k=0}^{\mu} a_k(t) z^k \tag{21}$$

with the coefficients a_{jk}, a_k of 1.1.(37),(38). Let us check that

$$h(t,z) \in C^{\infty}(\overline{\mathbb{R}}_+, M_O^{\mu}(\mathbb{B}, \mathfrak{g}_0)) \tag{22}$$

for $\mathfrak{g}_0 = (\gamma_0, \gamma_0 - \mu, (-\infty, 0])$ with arbitrary $\gamma_0 \in \mathbb{R}$. The operator A is then of the form (2) (with $\mu = \nu$) close to $t = 0$. If we set

$$\begin{aligned} f_1(t,z) &= \tilde{\omega}_0(r) \sum_{j+\mu\leq k} a_{jk}(r,t)(-r\frac{\partial}{\partial r})^j(rz)^k, \\ f_0(r,t,w,z) &= \tilde{\omega}_0(r) \sum_{j+\mu\leq k} a_{jk}(r,t) w^j (rz)^k \end{aligned}$$

for a cut-off function $\tilde{\omega}_0(r)$ (say with $\tilde{\omega}(r)\tilde{\omega}_0(r) = \tilde{\omega}(r)$) then we can form

$$\begin{aligned} a_0(t,z) &= \omega(r[\tau]) r^{-\mu} op_{M,r}^{\gamma_0 - n/2}(f_0)(t,z)\omega_0(r[\tau]), \\ a_1(t,z) &= (1 - \omega(r[\tau])) r^{-\mu} f_1(t,z)(1 - \omega_1(r[\tau])) \end{aligned}$$

with cut-off functions $\omega, \omega_0, \omega_1$, $\omega\omega_0 = \omega$, $\omega\omega_1 = \omega_1$ (c.f. 1.4.(56),(57)). It follows

$$\tilde{\omega}(r) r^{-\mu} \sum_{j+k\leq\mu} a_{jk}(r,t)(-r\frac{\partial}{\partial r})^j(-rz)^k = \tilde{\omega}(r)\{a_0(t,z) + a_1(t,z)\}.$$

Thus

$$\begin{aligned} h(t,z) &= \tilde{\omega}(r)\{a_0(t,z) + a_1(t,z)\} + (1 - \tilde{\omega}(r)) f(t,z) \\ &= \tilde{\omega}(r)\{a_0(t,z) + a_1(t,z)\} + (1 - \tilde{\omega}(r)) f(t,z)(1 - \tilde{\omega}_1(r)) \end{aligned}$$

with

$$f(t,z) = \sum_{k=0}^{\mu} a_k(t) z^k$$

and arbitrary cut-off functions $\tilde{\omega}_0(r), \tilde{\omega}_1(r)$ satisfying $\tilde{\omega}\tilde{\omega}_0 = \tilde{\omega}$, $\tilde{\omega}\tilde{\omega}_1 = \tilde{\omega}_1$. We get $h(t,z)$ in the form 2.1.(1) which was the model for Definition 2.1.4. In other words it follows (22), indeed. It can easily be verified that A belongs to $Y^{\mu}(\mathbb{W}, \mathfrak{g}_0)$ over $\mathbb{W}$ (cf. Definition 1.3.19, 1.3.(70)). This can be red off immediately from 1.1.(41). In 1.1.(44) we have already expressed the leading (operator-valued) edge symbol $\sigma_{\wedge}^{\mu}(A)(t,\tau)$. It shows a typical relation between the various symbolic levels. Since $\sigma_{\wedge}^{\mu}(A)(t,\tau)$ is a $C^{\mu}(X^{\wedge}, \mathfrak{g}_0)$-valued symbol (cf. 1.3.(34)), we can look at the leading conormal symbol of $\sigma_{\wedge}^{\mu}(A)(t,\tau)$ with respect to the r-variable (cf. 1.1.(19), Remark 1.2.5)). It equals in our case

$$\sigma_{M,r}^{\mu} \sigma_{\wedge}^{\mu}(A)(t,w) = \sum_{j=0}^{\mu} c_{j0}(0,t) w^j \tag{23}$$

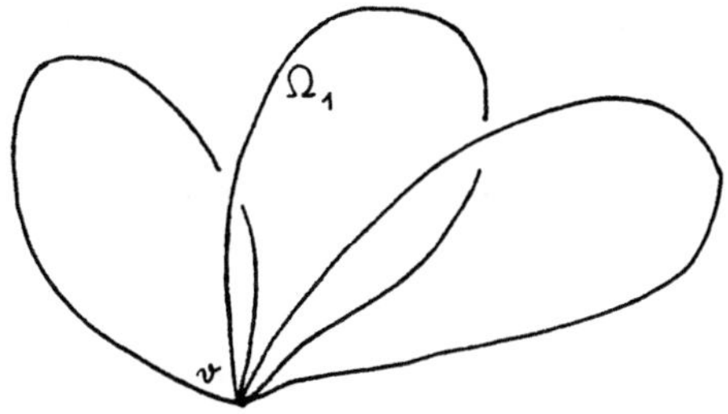

Figure 3:

(cf. 1.1.(44)). We see in particular that it is independent of τ. On the other hand we have for the leading conormal symbol with respect to t

$$\sigma^{\mu}_{M,t}(A)(z) = h(0,z)$$

with the above (21), which is a $C^{\mu}(\mathbb{B},\mathfrak{g}_0)$-valued Mellin symbol. As such it also has a leading conormal symbol with respect to the r-variable close to $\partial\mathbb{B}$, namely

$$\sigma^{\mu}_{M,r}\sigma^{\mu}_{M,t}(A)(w) = \sum_{j=0}^{\mu} a_{jo}(0,0)w^j. \tag{24}$$

Using 1.1.(43) it follows

$$t^{\mu}\sigma^{\mu}_{M,r}\sigma^{\mu}_{\wedge}(A)(t,w)|_{t=0} = \sigma^{\mu}_{M,r}\sigma^{\mu}_{M,t}(A)(w). \tag{25}$$

This is, of course, a crucial relation for understanding the interaction between corner and edge contribution to the asymptotics of solutions. The use of the variable t is allowed, since (25) concerns a neighbourhood of $t=0$. Otherwise we could also take the function g_1^{μ} (cf. 2.2.(15)).

An analogous relation holds in general, i.e. we have the

Proposition 3.1.9 *Let* $A \in C^{\nu}(\mathbb{K},\mathfrak{g})$, $\mathfrak{g} = (\mathfrak{g}_0, \mathfrak{g}_1^{\bullet})$, $\mathfrak{g}_i = (\gamma_i, \gamma_i - \mu, \Theta_i)$, $k_i = -\mu + \nu \geq 1$, $i = 0,1$. *Then*

$$t^{\nu}\sigma^{\nu}_{M,r}\sigma^{\nu}_{\wedge}(A)(t,w)|_{t=0} = \sigma^{\nu}_{M,r}\sigma^{\nu}_{M,t}(A)(w). \tag{26}$$

3.2 The corner algebra with trace and potential conditions

As announced we now complete the operators of Definition 3.1.2 to matrices, where two of the entries have the meaning of trace and potential operators, respectively, with respect to the edges, and the right lower corners are operators analogous to those from Definition 1.2.3, acting along the "branched space" formed by the edges. These edges

$$\Omega_1 = K_1 \setminus K_0 \tag{1}$$

are diffeomorphic to a system of open intervals, cf. Figure 3, i.e. to a disjoint union of components $\Omega_{1,(m)} \cong (0,1)\,, m = 1,\ldots,l$. Sometimes we will need the closure

$$\overline{\Omega}_1 = \bigcup_{m=1}^{l} \overline{\Omega}_{1,(m)} \tag{2}$$

where the right hand side of (2) means the disjoint union, such that $\overline{\Omega}_1$ does not necessarily coincide with K_1. For every $s, \alpha \in \mathbb{R}$ we have the space

$$\begin{aligned} \mathcal{H}^{s,\alpha}\left(\Omega_{1,(m)}, \mathbb{C}^{N_m}\right) &= \mathcal{H}^{s,\alpha}\left(\Omega_{1,(m)}\right) \otimes \mathbb{C}^{N_m}, \\ \mathcal{H}^{s,\alpha}\left(\Omega_{1,(m)}\right) &\subset H^s_{loc}\left(\Omega_{1,(m)}\right), \\ \mathcal{H}^{s,\alpha}\left(\Omega_{1,(m)}\right) &\cong \mathcal{H}^{s,\alpha}\left((0,1)\right), \end{aligned}$$

where

$$\begin{aligned} \mathcal{H}^{s,\alpha}\left((0,1)\right)|_{(0,1-\varepsilon)} &= \mathcal{H}^{s,\alpha}\left(\mathbb{R}_+\right)|_{(0,1-\varepsilon)}, \\ \mathcal{H}^{s,\alpha}\left((0,1)\right)|_{(\varepsilon,1)} &= \mathcal{H}^{s,\alpha}\left((-\infty,1)\right)|_{(\varepsilon,1)}, \end{aligned}$$

$0 < \varepsilon < 1$, with $\mathcal{H}^{s,\alpha}(\mathbb{R}_+)$ being the completion of $C_0^\infty(\mathbb{R}_+)$ with respect to the norm

$$||u||_{\mathcal{H}^{s,\alpha}(\mathbb{R}_+)} = \left\{ \frac{1}{2\pi i} \int_{\Gamma_{1/2-\alpha}} \left(1+|z|^2\right)^\alpha |M_\alpha u(z)|^2 dz \right\}^{1/2},$$

and $u(t) \in \mathcal{H}^{s,\alpha}((-\infty,1)) \leftrightarrow u(\tilde{t}) \in \mathcal{H}^{s,\alpha}(\mathbb{R}_+)$ for $\tilde{t} = 1-t$.

Let us set

$$\mathcal{H}^{s,\alpha}(\Omega_1, J) = \bigoplus_{m=1}^{l} \mathcal{H}^{s,\alpha}\left(\Omega_{1,(m)}, \mathbb{C}^{N_m}\right) \tag{3}$$

and analogously

$$\mathcal{H}^{s,\alpha}(\Omega_1, L) = \bigoplus_{m=1}^{l} \mathcal{H}^{s,\alpha}\left(\Omega_{1,(m)}, \mathbb{C}^{M_m}\right) \tag{4}$$

for another choice of dimensions M_m. The discrete asymptotic types of distribution in $\mathcal{H}^{s,\alpha}(\Omega_1, J)$ close to v will be a straigthtforward generalization from $\mathcal{H}^{s,\alpha}\left(\mathbb{R}_+, \mathbb{C}^N\right)$. The latter ones form the set $As(\mathfrak{a}^\bullet)$, $\mathfrak{a} = (\alpha, \Theta_i)$, consisting of all sequences

$$P = \{(p_j, m_j)\} \subset \mathbb{C} \times \mathbb{N}$$

which are infinite for infinite $\Theta_1 = (-\infty, 0]$ (with $Re\, p_j \to -\infty$ as $j \to \infty$) and finite for $\Theta_1 = (\vartheta_1, 0]$ finite, with the usual position of p_j in the complex Mellin plane, namely $Re\, p_j < 1/2 - \alpha$ for all j. The asymptotics of $u(t) \in \mathcal{H}^{s,\alpha}_P\left(\mathbb{R}_+, \mathbb{C}^N\right)$ are of the form

$$u(t) \sim \sum_j \sum_{m=0}^{m_j} \zeta_{jk} t^{-p_j} \log^k t \text{ as } t \to 0.$$

Now the asymptotic types for $u \in \mathcal{H}^{s,\alpha}(\Omega_1, J)$ will obviously consist of sequences $P \in \times_{j=1}^{N(J)} As(\mathfrak{a}^\bullet)$, $N(J) = \sum_{m=1}^{l} 2N_m$, with the analogous meaning of data. In other words we get the space $\mathcal{H}^{s,\alpha}_P(\Omega_1, J)$. We could, of course, distinguish between the weight data on different branches and allow them to be different. But we prefer the simpler form.

The matrix-valued analogue of Definition 3.1.1 is as follows.

Definition 3.2.1 *Let $\mu \in \mathbb{R}$, $\gamma = (\gamma_0, \gamma_1) \in \mathbb{R}^2$, $\mathfrak{g}_i = (\gamma_i, \gamma_i - \mu, \Theta_i)$, $\Theta_i = (\vartheta_i, 0]$, $i = 0, 1$, $\mathfrak{g} = (\mathfrak{g}_0, \mathfrak{g}_1^\bullet)$, and $J, L \in Vect(\Omega_1)$. Then $C_G(\mathbb{K}, \mathfrak{g}; J, L)$ denotes the space of all*

$$\mathcal{G} \in \bigcap_{s \in \mathbb{R}} \mathcal{L}\left(\mathcal{H}^{s,\gamma}(\mathbb{K}) \oplus \mathcal{H}^{s,\tilde{\gamma}_1}(\Omega_1, J), \mathcal{H}^{\infty,\gamma-\mu}(\mathbb{K}) \oplus \mathcal{H}^{\infty,\tilde{\gamma}_1-\mu}(\Omega_1, L)\right),$$

$\gamma-\mu=(\gamma_0-\mu,\gamma_1-\mu)$, $\tilde{\gamma}_1=\gamma_1-(n+1)/2$, $n+1=dim\mathbb{B}$, *which induce continous operators*

$$\mathcal{G}:\mathcal{H}^{s,\gamma}(\mathbb{K})\oplus\mathcal{H}^{s,\tilde{\gamma}_1}(\Omega_1,J)\longrightarrow\mathcal{H}_P^{\infty,\gamma-\mu}(\mathbb{K})\oplus\mathcal{H}_{\tilde{P}}^{\infty,\tilde{\gamma}_1-\mu}(\Omega_1,L),\tag{5}$$

$$\mathcal{G}^*:\mathcal{H}^{s,-\gamma+\mu}(\mathbb{K})\oplus\mathcal{H}^{s,-\tilde{\gamma}_1+\mu}(\Omega_1,L)\longrightarrow\mathcal{H}_Q^{\infty,-\gamma}(\mathbb{K})\oplus\mathcal{H}_{\tilde{Q}}^{\infty,-\tilde{\gamma}_1}(\Omega_1,J)\tag{6}$$

for all $s\in\mathbb{R}$, *with asymptotic types* $P=(P_0,P_1)$, $Q=(Q_0,Q_1)$ *as in Definition 3.1.1 and* $\tilde{P}\in\times_{j=1}^{N(L)}As(\tilde{\mathfrak{a}}^\bullet)$, $\tilde{\mathfrak{a}}=(\tilde{\gamma}_1,\Theta_1)$, $\tilde{Q}\in\times_{j=1}^{N(J)}As(\tilde{\mathfrak{b}}^\bullet)$, $\tilde{\mathfrak{b}}=(-\tilde{\gamma}_1,\Theta_1)$, *dependent on* $\mathcal{G}$ *(not on* s*). Here* $\mathcal{G}^*$ *is the formal adjoint of* $\mathcal{G}$ *in the sense*

$$(\mathcal{G}u,v)_{\mathcal{H}^0(\mathbb{K})\oplus\mathcal{H}^0(\Omega_1,L)}=(u,\mathcal{G}^*v)_{\mathcal{H}^0(\mathbb{K})\oplus\mathcal{H}^0(\Omega_1,J)}$$

for all $u\in C_0^\infty(\Omega)\oplus C_0^\infty(\Omega_1,J)$, $v\in C_0^\infty(\Omega)\oplus C_0^\infty(\Omega_1,L)$ *(*$\Omega=K\setminus K_1$ *being regarded as open dense subset of* $\mathbb{K}$*), and the scalar products of* $\mathcal{H}^0(\Omega_1,J)$, $\mathcal{H}^0(\Omega_1,L)$ *refer to a fixed Riemannian metric on* Ω_1 *(* C^∞ *up to the end points of the intervals* $\Omega_{1,(m)}$, $m=1,\ldots,l$*)* .

Definition 3.2.2 *Let* $\mu,\nu\in\mathbb{R}$, $\mu-\nu\in\mathbb{N}$, $\mathfrak{g}=(\mathfrak{g}_0,\mathfrak{g}_1^\bullet)$ *for* $\mathfrak{g}_i=(\gamma_i,\gamma_i-\mu,\Theta_i)$, $\Theta_i=(-k_i,0]$, $k_0\in(\mathbb{N}\setminus\{0\})\cup\{\infty\}$, $k_1\in\mathbb{N}\setminus\{0\}$, $\gamma_i\in\mathbb{R}$, $i=0,1$. *Then* $C^\nu(\mathbb{K},\mathfrak{g};J,L)$ *denotes the space of all operators*

$$\mathcal{A}=\mathcal{A}_0+\mathcal{A}_1+\mathcal{M}+\mathcal{G}\tag{7}$$

with

$$\mathcal{A}_0=\omega t^{-\nu}op_M^{\tilde{\gamma}_1}(h)\,\omega_0,\ h(t,z)\in C^\infty\left(\overline{\mathbb{R}}_+,\mathcal{M}_O^\nu\left(\mathbb{B},\mathfrak{g}_0;\mathbb{C}^{N(J)},\mathbb{C}^{N(L)}\right)\right)\tag{8}$$

(cf. Definition 2.4.1) for $\tilde{\gamma}_1=\gamma_1-(n+1)/2$, $n+1=dim\mathbb{B}$,

$$\mathcal{A}_1=(1-\omega)\,\mathcal{P}\,(1-\omega_1)\,,\mathcal{P}\in\mathcal{Y}^\nu(\mathbb{W},\mathfrak{g}_0;J,L)\tag{9}$$

(cf. Remark 1.3.23 with $\mathbb{W}$ *being given by 1.1.(25))* ,

$$\mathcal{M}=\omega_2\left\{t^{-\nu}\sum_{0\le j\le k_1-\mu+\nu-1}t^j op_M^{\tilde{\delta}_j}(h_j)\right\}\omega_3\tag{10}$$

where $h_j(z)\in\mathcal{M}_{\mathfrak{a}_\bullet}^{-\infty}\left(\mathbb{B},\mathfrak{g}_0;\mathbb{C}^{N(J)},\mathbb{C}^{N(L)}\right)^\bullet$ *(cf. 2.4.(9))* $\tilde{\delta}_j=\delta_j-(n+1)/2$,

$$\gamma_1-(\mu-\nu)-j\le\delta_j\le\gamma_1\tag{11}$$

$sg(h_j)\cap\Gamma_{(n+1)/2-\delta_j}=\emptyset$ *for all* j, *and finally* $\mathcal{G}\in C_G(\mathbb{K},\mathfrak{g};J,L)$. *Here* $\omega(t),\omega_i(t)$ *are arbitrary cut-off functions supported by* $[0,1)$.

For $k_1=\infty$ we define $C^\nu(\mathbb{K},\mathfrak{g};J,L)$ as the intersection of the corresponding classes for $\Theta_{1,(m)}=(-m,0]$ over $m\in\mathbb{N}\setminus\{0\}$.

By definition it follows

$$C^\nu(\mathbb{K},\mathfrak{g})=l.u.c.\ C^\nu(\mathbb{K},\mathfrak{g};J,L)\,,\tag{12}$$

(cf. Definition 3.1.2) . We define

$$C^{\nu,\iota}(\mathbb{K},\mathfrak{g};J,L)=C^\nu(\mathbb{K},\mathfrak{g};J,L)\cap\mathcal{Y}^{\nu-\iota}(\mathbb{K},\mathfrak{g}_0;J,L)\,.\tag{13}$$

for $\iota\in\mathbb{N}$. Then $\mathcal{A}\in$(13) implies $l.u.c.\ \mathcal{A}\in C^{\nu,\iota}(\mathbb{K},\mathfrak{g})$ (cf. 3.1.(17)) .

Let us now look at the leading symbolic levels. The homogeneous principal interior symbol of order ν of $\mathcal{A}\in C^\nu(\mathbb{K},\mathfrak{g};J,L)$ is defined by the left upper corner

$$\sigma_\psi^\nu(\mathcal{A}):=\sigma_\psi^\nu(l.u.c.\ \mathcal{A})\,;\tag{14}$$

thus

$$\sigma_{\psi}^{\nu}(\mathcal{A}) \in g_2^{-\nu} S_{2b}^{(\nu)}(T^*\Omega \setminus 0). \tag{15}$$

The homogeneous principal edge symbol of order ν of $\mathcal{A}$ follows from

$$\mathcal{C}^{\nu}(\mathbb{K}, \mathfrak{g}; J, L) \subset \mathcal{Y}^{\nu}(\mathbb{W}, \mathfrak{g}_0; J, L) \tag{16}$$

(cf. Remark 1.3.23) . Analogously to Remark 3.1.4 we have

$$t^{-\tilde{\gamma}_1+\nu}\varphi_* \mathcal{C}^{\nu}(\mathbb{K}, \mathfrak{g}; J, L)\,|_{\mathbb{V}\setminus p^{-1}(v)} t^{\tilde{\gamma}_1} \subset M\mathcal{Y}^{\nu}(\mathbb{B} \times (0,1), \mathfrak{g}_0; J, L) \tag{17}$$

(cf. Definition 1.4.21) , $\tilde{\gamma}_1 = \gamma_1 - (n+1)/2$.

Denote by g_1^{η} for $\eta \in \mathbb{R}$ a strictly positive C^{∞} function on Ω_1 which is on the component $\Omega_{1(m)} \cong (0,1)$

$$g_1^{\eta} = \begin{cases} t^{\eta} \text{ close to } t = 0, \\ (1-t)^{\eta} \text{ close to } t = 1, \end{cases} \tag{18}$$

$m = 1, \ldots, l$. (16) and 1.3.(71) induce the homogeneous edge symbolic map

$$\sigma_{\wedge}^{\nu} : \mathcal{C}^{\nu}(\mathbb{K}, \mathfrak{g}; J, L) \longrightarrow g_1^{-\nu} \bigcap_{s \in \mathbb{R}} \mathcal{L}^{(\nu)}\left(T^*\overline{\Omega}_1 \setminus 0; \mathcal{K}^{s,\gamma_0} \oplus J_t, \mathcal{K}^{s-\nu,\gamma_0-\mu} \oplus L_t\right), \tag{19}$$

(cf. (2)) where $J_t\,(L_t)$ indicate the "fibres" of $J\,(L)$ over the point $t \in \Omega_1$. Note that

$$J_t \cong \mathbb{C}^{N_m}, L_t \cong \mathbb{C}^{M_m} \text{ for } t \in \Omega_{1,(m)}. \tag{20}$$

Analogously (17) and 1.4.(79) lead to

$$\sigma_{b\wedge}^{\nu}\varphi^* : \mathcal{C}^{\nu}(\mathbb{K}, \mathfrak{g}; J, L)\,|_{\mathbb{V}\setminus p^{-1}(v)} \longrightarrow t^{-\nu} \bigcap_{s \in \mathbb{R}} \mathcal{L}^{(\nu)}\left(T^*[0,1) \setminus 0; \mathcal{K}^{s,\gamma_0} \oplus J_t, \mathcal{K}^{s-\nu,\gamma_1-\mu} \oplus L_t\right). \tag{21}$$

Then, for $\tilde{\mathcal{A}} = t^{\nu}\varphi_*\left(\mathcal{A}|_{\mathbb{V}\setminus p^{-1}(v)}\right)$, it follows

$$\sigma_{\wedge}^{\nu}(\mathcal{A})(t,\tau) = t^{-\nu}\sigma_{b\wedge}^{\nu}\left(\tilde{\mathcal{A}}\right)(t,\tau_1)\,|_{\tau_1 = -t\tau}. \tag{22}$$

Moreover we get analogously to 3.1.(19) for $k_1 - \mu + \nu \geq 1$

$$\sigma_M^{\nu}(\mathcal{A})(z) := h(0,z) + h_0(z) \in \mathcal{M}_{a_s}^{\nu}\left(\mathbb{B}, \mathfrak{g}_0; \mathbb{C}^{N(J)}, \mathbb{C}^{N(L)}\right)^{\bullet}, \tag{23}$$

called the principal conormal symbol of $\mathcal{A}$ of (conormal) order ν. Here $h(t,z)$ and $h_0(z)$ are the Mellin symbols from Definition 3.2.2. (23) is regarded as family of operators

$$\sigma_M^{\nu}(\mathcal{A})(z) : \mathcal{H}^{s,\gamma_0}(\mathbb{B}) \oplus \mathbb{C}^{N(J)} \longrightarrow \mathcal{H}^{s-\nu,\gamma_0-\mu}(\mathbb{B}) \oplus \mathbb{C}^{N(L)} \tag{24}$$

for any $s \in \mathbb{R}$, parametrized by $z \in \Gamma_{(n+2)/2-\gamma_1}$.

For dealing with the asymptotic information under actions of operators $\mathcal{A} \in \mathcal{C}^{\nu}(\mathbb{K}, \mathfrak{g}; J, L)$ it is also interesting to have the conormal symbols of $\mathcal{A}$ of lower orders, namely

$$\sigma_M^{\nu-j}(\mathcal{A})(z) = \frac{1}{j!}\frac{\partial^j h}{\partial t^j}(0,z) + h_j(z) \tag{25}$$

for $j = 0, \ldots, k_1 - \mu + \nu - 1$ $(\Theta_1 = (-k_1, 0])$. They disappear for $j > k_1 - \mu + \nu - 1$.

Definition 3.2.3 *Let $k_1 - \mu + \nu \geq 1$. Denote by*

$$Symb\, C^{\nu}(\mathbb{K}, \mathfrak{g}; J, L) \tag{26}$$

the space of all

$$\begin{aligned} (p_{\psi}, p_{\wedge}, p_M) &\in g_2^{-\nu} S_{2b}^{(\nu)}(T^*\Omega \setminus 0) \\ &\times\ g_1^{-\nu} \bigcap_{s \in \mathbb{R}} \mathcal{L}^{(\nu)}(T^*\overline{\Omega}_1 \setminus 0; \mathcal{K}^{s,\gamma_0} \oplus J_t, \mathcal{K}^{s-\nu,\gamma_0-\mu} \oplus L_t) \\ &\times\ \mathcal{M}_{\mathfrak{a}_s}^{\nu}(\mathbb{B}, \mathfrak{g}_0; \mathbb{C}^{N(J)}, \mathbb{C}^{N(L)})^{\bullet} \end{aligned} \tag{27}$$

such that for some $\mathcal{A} \in C^{\nu}(\mathbb{K}, \mathfrak{g}; J, L)$

$$p_{\psi} = \sigma_{\psi}^{\nu}(\mathcal{A}), p_{\wedge} = \sigma_{\wedge}^{\nu}(\mathcal{A}), p_M = \sigma_M^{\nu}(\mathcal{A}).$$

By definition we get a symbolic map

$$(\sigma_{\psi}^{\nu}, \sigma_{\wedge}^{\nu}, \sigma_M^{\nu}) : C^{\nu}(\mathbb{K}, \mathfrak{g}; J, L) \longrightarrow Symb\, C^{\nu}(\mathbb{K}, \mathfrak{g}; J, L). \tag{28}$$

Proposition 3.2.4 *Let $k_i - \mu + \nu \geq 1$, $i = 0, 1$. Then*

$$ker\,(\sigma_{\psi}^{\nu}, \sigma_{\wedge}^{\nu}, \sigma_M^{\nu}) = C^{\nu-1}(\mathbb{K}, \mathfrak{g}; J, L).$$

Theorem 3.2.5 *An $\mathcal{A} \in C^{\nu}(\mathbb{K}, \mathfrak{g}; J, L)$ induces continuous operators*

$$\mathcal{A} : \mathcal{H}^{s,\gamma}(\mathbb{K}) \oplus \mathcal{H}^{s,\tilde{\gamma}_1}(\Omega_1, J) \longrightarrow \mathcal{H}^{s-\nu,\gamma-\mu}(\mathbb{K}) \oplus \mathcal{H}^{s-\nu,\tilde{\gamma}_1-\mu}(\Omega_1, L) \tag{29}$$

for every $s \in \mathbb{R}$. Here $\gamma = (\gamma_0, \gamma_1)$ are the weights, involved in $\mathfrak{g}$, and $\gamma - \mu = (\gamma_0 - \mu, \gamma_1 - \mu)$, $\tilde{\gamma}_1 = \gamma_1 - (n+1)/2$. Moreover for every $P = (P_0, P_1) \in As(X, \mathfrak{a}_0) \times As(\mathbb{B}, \mathfrak{a})$, $\mathfrak{a} = (\mathfrak{a}_0, \mathfrak{a}_1^{\bullet})$, $\mathfrak{a}_i = (\gamma_i, \Theta_i)$, $\tilde{P} \in \times_{j=1}^{IN} As(\tilde{\mathfrak{a}}^{\bullet})$, $\tilde{\mathfrak{a}} = (\tilde{\gamma}_1, \Theta_1)$ there exists asymptotic types $Q = (Q_0, Q_1) \in As(X, \mathfrak{b}_0) \times As(\mathbb{B}, \mathfrak{b})$, $\mathfrak{b} = (\mathfrak{b}_0, \mathfrak{b}_1^{\bullet})$, $\mathfrak{b}_i = (\gamma_i - \mu, \Theta_i)$, $\tilde{Q} \in \times_{j=1}^{IM} As(\tilde{\mathfrak{b}}^{\bullet})$, $\tilde{\mathfrak{b}} = (\tilde{\gamma}_1 - \mu, \Theta_1)$, such that (29) restricts to a continuous operator

$$\mathcal{A} : \mathcal{H}_P^{s,\gamma}(\mathbb{K}) \oplus \mathcal{H}_{\tilde{P}}^{s,\tilde{\gamma}_1}(\Omega_1, J) \longrightarrow \mathcal{H}_Q^{s-\nu,\gamma-\mu}(\mathbb{K}) \oplus \mathcal{H}_{\tilde{Q}}^{s-\nu,\tilde{\gamma}_1-\mu}(\Omega_1, L) \tag{30}$$

for all $s \in \mathbb{R}$.

Theorem 3.2.6 *Let $\mathcal{A} \in C^{\nu-1}(\mathbb{K}, \mathfrak{g}; J, L)$ and $k_i - \mu + \nu \geq 1$, $i = 0, 1$. Then (29) is a compact operator.*

Theorem 3.2.7 *Let $\mathcal{A}_j \in C^{\nu,j}(\mathbb{K}, \mathfrak{g}; J, L)$, $j \in \mathbb{N}$, be an arbitrary sequence. Then there exists an $\mathcal{A} \in C^{\nu}(\mathbb{K}, \mathfrak{g}; J, L)$ with*

$$\mathcal{A} - \sum_{j=0}^{N} \mathcal{A}_j \in C^{\nu,N+1}(\mathbb{K}, \mathfrak{g}; J, L)$$

for every $N \in \mathbb{N}$, and any other $\tilde{\mathcal{A}} \in C^{\nu}(\mathbb{K}, \mathfrak{g}; J, L)$ with this property satisfies

$$\mathcal{A} - \tilde{\mathcal{A}} \in C_G(\mathbb{K}, \mathfrak{g}; J, L).$$

If an operator $\mathcal{A}$ is continuous in the sense of (29) for all $s \in \mathbb{R}$ then we can form the formal adjoint

$$\mathcal{A} : \mathcal{H}^{s,-\gamma+\mu}(\mathbb{K}) \oplus \mathcal{H}^{s,-\gamma_1-(n+1)/2+\mu}(\Omega_1, L) \longrightarrow \mathcal{H}^{s-\nu,-\gamma}(\mathbb{K}) \oplus \mathcal{H}^{s,-\gamma_1-(n+1)/2}(\Omega_1, J) \tag{31}$$

in the sense

$$(\mathcal{A}u,v)_{\mathcal{H}^0(\mathbb{K})\oplus\mathcal{H}^{0,-(n+1)/2}(\Omega_1,L)} = (u,\mathcal{A}^*v)_{\mathcal{H}^0(\mathbb{K})\oplus\mathcal{H}^{0,-(n+1)/2}(\Omega_1,J)} \tag{32}$$

first for all $u \in C_0^\infty(\Omega) \oplus C_0^\infty(\Omega_1, J)$, $v \in C_0^\infty(\Omega) \oplus C_0^\infty(\Omega_1, L)$, which extends then to continuous operators (31) for all $s \in \mathbb{R}$.

The weights on the second component of the scalar products in (32) are motivated by the following observation, say on $\mathbb{R}_+$. We have

$$\mathcal{H}^{s,\alpha}(\mathbb{R}_+) = t^\alpha \mathcal{H}^s(\mathbb{R}_+)$$

with $\mathcal{H}^s(\mathbb{R}_+) := \mathcal{H}^{s,0}(\mathbb{R}_+)$, and $\mathcal{H}^0(\mathbb{R}_+) = L^2(\mathbb{R}_+)$. The $L^2(\mathbb{R}_+)$-scalar product

$$\int_0^\infty u\overline{v}dt = (u,v)_0$$

extends from $C_0^\infty(\mathbb{R}_+) \times C_0^\infty(\mathbb{R}_+)$ to a non-degenerate sesqui-linear pairing

$$\mathcal{H}^{s,\rho}(\mathbb{R}_+) \times \mathcal{H}^{-s,-\rho}(\mathbb{R}_+) \longrightarrow \mathbb{C} \quad , s,\rho \in \mathbb{R}.$$

Thus the formal adjoint A_0^* of an $A \in \bigcap_{s\in\mathbb{R}} \mathcal{L}\left(\mathcal{H}^{s,\lambda}(\mathbb{R}_+), \mathcal{H}^{s-\nu,\lambda-\mu}(\mathbb{R}_+)\right)$ with respect to $(.,.)_0$ belongs to $\bigcap_{s\in\mathbb{R}} \mathcal{L}\left(\mathcal{H}^{s,-\lambda+\mu}(\mathbb{R}_+), \mathcal{H}^{s,-\lambda}(\mathbb{R}_+)\right)$. Let us calculate A_α^*, the formal adjoint of a with respect to $\mathcal{H}^{0,\alpha}(\mathbb{R}_+) = t^\alpha L^2(\mathbb{R}_+)$ which has the scalar product

$$(u,v)_\alpha = \int t^{-2\alpha} u\overline{v}dt.$$

It follows

$$\int t^{-2\alpha} Au\overline{v}dt = \int t^{-2\alpha} t^{2\alpha} u\overline{A_0^* t^{-2\alpha} v}dt.$$

Thus

$$A_\alpha^* = t^{2\alpha} A_0^* t^{-2\alpha}.$$

We obtain

$$A_\alpha^* \in \bigcap_{s\in\mathbb{R}} \mathcal{L}\left(\mathcal{H}^{s,-\lambda+\mu+2\alpha}(\mathbb{R}_+), \mathcal{H}^{s-\nu,-\lambda+2\alpha}(\mathbb{R}_+)\right).$$

Since in the present case of (31) we have in the first components to interchange the weights with negative sign, we need in the second component again $-\gamma_1 - (n+1)/2 = (-\gamma_1)^\sim$ instead of $-\gamma_1$. This follows just by taking $\alpha = -(n+1)/2$.

Theorem 3.2.8 *$\mathcal{A} \in \mathcal{C}^\nu(\mathbb{K}, \mathfrak{g}; L, J)$ for $\mathfrak{g} = (\mathfrak{g}_0, \mathfrak{g}_1^*)$, $\mathfrak{g}_i = (\gamma_i, \gamma_i - \mu, \Theta_i)$, $i = 0,1$, implies $\mathcal{A}^* \in \mathcal{C}^\nu(\mathbb{K}, \mathfrak{g}^*; J, L)$ for $\mathfrak{g}^* = (\mathfrak{g}_0^*, \mathfrak{g}_1^{**})$, $\mathfrak{g}_i^* = (-\gamma_i + \mu, -\gamma_i, \Theta_i)$, $i = 0,1$, and*

$$\sigma_\psi^\nu(\mathcal{A}^*) = \overline{\sigma_\psi^\nu(\mathcal{A})}, \tag{33}$$
$$\sigma_\wedge^\nu(\mathcal{A}^*) = \sigma_\wedge^\nu(\mathcal{A})^*, \tag{34}$$

where the $*$ on the right side of (33) refers to the formal adjoint of operator families, taken point-wise, as they occur as edge symbols, further for $k_1 - \mu + \nu \geq 1$

$$\sigma_M^\nu(\mathcal{A}^*) = T^\nu \sigma_M^\nu(\mathcal{A})^* \tag{35}$$

where the $*$ on the right of (35) is to be interpreted as

$$h^*(z) = h^{(*)}(n+2-\overline{z}),$$

with $(*)$ being the point-wise formal adjoint of the values of h.

Theorem 3.2.9 *Let $\mathcal{A} \in C^\nu\left(\mathbb{K}, \mathfrak{g}; J, \tilde{L}\right)$ for $\mathfrak{g} = (\mathfrak{g}_0, \mathfrak{g}_1^\bullet)$, $\mathfrak{g}_i = (\gamma_i, \gamma_i - \mu, \Theta_i)$, $\mathcal{B} \in C^{\tilde{\nu}}\left(\mathbb{K}, \tilde{\mathfrak{g}}; \tilde{L}, M\right)$ for $\tilde{\mathfrak{g}} = (\tilde{\mathfrak{g}}_0, \tilde{\mathfrak{g}}_1^\bullet)$, $\tilde{\mathfrak{g}}_i = (\gamma_i - \mu, \gamma_i - \mu - \tilde{\mu}, \Theta_i)$, $i = 0, 1$. Then $\mathcal{BA} \in C^{\nu+\tilde{\nu}}(\mathbb{K}, \mathfrak{c}; J, M)$ for $\mathfrak{c} = (\mathfrak{c}_0, \mathfrak{c}_1^\bullet)$, $\mathfrak{c}_i = (\gamma_i, \gamma_i - \mu - \tilde{\mu}, \Theta_i)$, $i = 0, 1$, and*

$$\begin{aligned}\sigma_\psi^{\nu+\tilde{\nu}}(\mathcal{BA}) &= \sigma_\psi^{\tilde{\nu}}(\mathcal{B})\,\sigma_\psi^{\nu}(\mathcal{A}),\\ \sigma_\wedge^{\nu+\tilde{\nu}}(\mathcal{BA}) &= \sigma_\wedge^{\tilde{\nu}}(\mathcal{B})\,\sigma_\wedge^{\nu}(\mathcal{A}),\end{aligned}$$

and further for $k_1 - \mu + \nu \geq 1$, $k_1 - \tilde{\mu} + \tilde{\nu} \geq 1$

$$\sigma_M^{\nu+\tilde{\nu}}(\mathcal{BA}) = \left(T^\nu \sigma_M^{\tilde{\nu}}(\mathcal{B})\right)\sigma_M^{\nu}(\mathcal{A}).$$

Remark 3.2.10 *The conormal symbols of lower orders of $\mathcal{BA}$ follow by the Mellin translation product*

$$\sigma_M^{\nu+\tilde{\nu}-d}(\mathcal{BA})(z) = \sum_{p+q=d}\left\{T^{\nu-q}\sigma_M^{\tilde{\nu}-p}(\mathcal{B})(z)\right\}\sigma_M^{\nu-q}(\mathcal{A})(z) \tag{36}$$

(cf. (25)) for $d = 0, \ldots, k_1 - \mu + \nu - 1$.

Remark 3.2.11 *Let g_1^ρ be the weight function along Ω_1 defined above (cf. (18)), regarded now also as a function on $\mathbb{K}$ (cf. 2.2.(15)). Then, for every $\rho \in \mathbb{R}$,*

$$g_1^\rho C^\nu(\mathbb{K}, \mathfrak{g}; J, L)\, g_1^{-\rho} = C^\nu(\mathbb{K}, \mathfrak{g}_{(\rho)}; J, L)$$

for $\mathfrak{g}_{(\rho)} = \left(\mathfrak{g}_0, \mathfrak{g}_{1,(\rho)}^\bullet\right)$ with $\mathfrak{g}_{1,(\rho)} = (\gamma_1 + \rho, \gamma_1 + \rho - \mu, \Theta_1)$, and

$$\begin{aligned}\sigma_\psi^\nu\left(g_1^\rho \mathcal{A} g_1^{-\rho}\right) &= \sigma_\psi^\nu(\mathcal{A}),\\ \sigma_\wedge^\nu\left(g_1^\rho \mathcal{A} g_1^{-\rho}\right) &= \sigma_\wedge^\nu(\mathcal{A}),\\ \sigma_M^{\nu-j}\left(g_1^\rho \mathcal{A} g_1^{-\rho}\right) &= T^\rho \sigma_M^{\nu-j}(\mathcal{A}),\quad j = 0, \ldots, k_1 - \mu + \nu - 1.\end{aligned}$$

3.3 Ellipticity, parametrices within the class, elliptic regularity with asymptotics

The ellipticity of the operators of Definition 3.2.2 will refer to $\nu = \mu$.

Definition 3.3.1 *Let $\mu \in \mathbb{R}$, $(\gamma_0, \gamma_1) \in \mathbb{R}^2$, $\mathfrak{g} = (\mathfrak{g}_0, \mathfrak{g}_1^\bullet)$ for $\mathfrak{g}_i = (\gamma_i, \gamma_i - \mu, \Theta_i)$, $i = 0, 1$. An operator $\mathcal{A} \in C^\mu(\mathbb{K}, \mathfrak{g}; J, L)$ is called elliptic if*

(i)

$$\sigma_\psi^\mu(\mathcal{A}) \neq 0 \text{ on } T^*\Omega \setminus 0$$

and if $p = g_2^\mu \sigma_\psi^\mu(\mathcal{A})$ in the coordinates (x, r, t) close to $r = 0, t = 0$, written as $p = p(x, r, t, \xi, r\rho, rt\tau)$, satisfies $p(x, r, t, \xi, \rho, \tau) \neq 0$ up to $r = 0, t = 0$, in the coordinates $(\tilde{x}, t)$ close to $t = 0$, p being written as $p(\tilde{x}, t, \tilde{\xi}, t\tau)$, satisfies $p(\tilde{x}, t, \tilde{\xi}, \tau) \neq 0$ up to $t = 0$, further in the coordinates (x, r, t) close to $r = 0$, p being written as $p(x, r, t, \xi, \rho, r\tau)$, satisfies $p(x, r, t, \xi, \rho, \tau) \neq 0$ up to $r = 0$ (cf. the definition of 3.1.(15))

(ii)

$$\sigma_\wedge^\mu(\mathcal{A})(t, \tau) : \begin{array}{c}\mathcal{K}^{s,\gamma_0}(X^\wedge)\\ \oplus\\ J_t\end{array} \longrightarrow \begin{array}{c}\mathcal{K}^{s-\mu,\gamma_0-\mu}(X^\wedge)\\ \oplus\\ L_t\end{array} \tag{1}$$

is an isomorphism for all $(t, \tau) \in T^*\Omega_1 \setminus 0$ and some $s \in \mathbb{R}$ (cf. 3.2.(20)), and $g_1^\mu \sigma_\wedge^\mu(\mathcal{A})(t, \tau)$ remains an isomorphism up to $t = 0$,

(iii)

$$\sigma_M^{\mu}(\mathcal{A})(z): \begin{array}{c}\mathcal{H}^{s,\gamma_0}(\mathbb{B})\\ \oplus \\ \mathbb{C}^{N(J)}\end{array} \longrightarrow \begin{array}{c}\mathcal{H}^{s-\mu,\gamma_0-\mu}(\mathbb{B})\\ \oplus \\ \mathbb{C}^{N(L)}\end{array} \tag{2}$$

is an isomorphism for all $z \in \Gamma_{(n+2)/2-\gamma_1}$ *and some* $s \in \mathbb{R}$ *(cf. 3.2.(23)).*

Remark 3.3.2 *If the condition (1) is satisfied for an* $s = s_0$ *then for all* $s \in \mathbb{R}$. *The same is true of (2).*

Definition 3.3.3 *Let* $\mathcal{A} \in \mathcal{C}^{\nu}(\mathbb{K}, \mathfrak{g}; J, L)$ *for* $\mathfrak{g} = (\mathfrak{g}_0, \mathfrak{g}_1^{\bullet})$, $\mathfrak{g}_i = (\gamma_i, \gamma_i - \mu, \Theta_i)$, $i = 0, 1$. *An operator* $\mathcal{P} \in \mathcal{C}^{-\mu}(\mathbb{K}, \mathfrak{b}; L, J)$ *with* $\mathfrak{b} = (\mathfrak{b}_0, \mathfrak{b}_1^{\bullet})$, $\mathfrak{b}_i = (\gamma_i - \mu, \gamma_i, \Theta_i)$, $i = 0, 1$, *is called a parametrix of* $\mathcal{A}$ *if*

$$\mathcal{P}\mathcal{A} - 1 \in \mathcal{C}_G(\mathbb{K}, \mathfrak{g}_l; J, J), \tag{3}$$

$$\mathcal{A}\mathcal{P} - 1 \in \mathcal{C}_G(\mathbb{K}, \mathfrak{g}_r; L, L) \tag{4}$$

with $\mathfrak{g}_l = (\mathfrak{g}_{l,0}, \mathfrak{g}_{l,1}^{\bullet})$, $\mathfrak{g}_{l,i} = (\gamma_i, \gamma_i, \Theta_i)$, $\mathfrak{g}_r = (\mathfrak{g}_{r,0}, \mathfrak{g}_{r,1}^{\bullet})$, $\mathfrak{g}_{r,i} = (\gamma_i - \mu, \gamma_i - \mu, \Theta_i)$, $i = 0, 1$, *and* 1 *being the identical operator in the corresponding* $\mathcal{C}^0$*-classes.*

Theorem 3.3.4 *Let* $\mathcal{A} \in \mathcal{C}^{\nu}(\mathbb{K}, \mathfrak{g}; J, L)$, $\mu \in \mathbb{R}$, $\mathfrak{g} = (\mathfrak{g}_0, \mathfrak{g}_1^{\bullet})$, $\mathfrak{g}_i = (\gamma_i, \gamma_i - \mu, \Theta_i)$, $i = 0, 1$, $\gamma = (\gamma_0, \gamma_1) \in \mathbb{R}^2$. *Then the following conditions are equivalent*

(i)

$$\mathcal{A}: \mathcal{H}^{s,\gamma}(\mathbb{K}) \oplus \mathcal{H}^{s,\widetilde{\gamma}_1}(\Omega_1, J) \longrightarrow \mathcal{H}^{s-\mu,\gamma-\mu}(\mathbb{K}) \oplus \mathcal{H}^{s-\mu,\widetilde{\gamma}_1-\mu}(\Omega_1, L) \tag{5}$$

is a Fredholm operator for some $s \in \mathbb{R}$, *where* $\gamma - \mu = (\gamma_0 - \mu, \gamma_1 - \mu)$, $\widetilde{\gamma}_1 = \gamma_1 - (n+1)/2$,

(ii) $\mathcal{A}$ *is elliptic.*

If $\mathcal{A}$ *is elliptic there exists a parametrix* $\mathcal{P} \in \mathcal{C}^{-\mu}(\mathbb{K}, \mathfrak{b}; J, L)$ *of* $\mathcal{A}$, *with* $\mathfrak{b}$ *from Definition 3.3.3. Moreover*

$$\mathcal{A}u = f \in \mathcal{H}_Q^{s,\gamma-\mu}(\mathbb{K}) \oplus \mathcal{H}_{\widetilde{Q}}^{s,\widetilde{\gamma}_1-\mu}(\Omega_1, L) \tag{6}$$

for some $s \in \mathbb{R}$ *and certain* $Q = (Q_0, Q_1) \in As(X, \mathfrak{c}_0) \times As(\mathbb{B}, \mathfrak{c})$ *for* $\mathfrak{c} = (\mathfrak{c}_0, \mathfrak{c}_1^{\bullet})$, $\mathfrak{c}_i = (\gamma_i - \mu, \Theta_i)$, *and* $\widetilde{Q} \in \times_{j=1}^{lM} As(\widetilde{\mathfrak{c}}^{\bullet})$, $\widetilde{\mathfrak{c}} = (\widetilde{\gamma}_1 - \mu, \Theta_1)$, *and*

$$u \in \mathcal{H}^{-\infty,\gamma}(\mathbb{K}) \oplus \mathcal{H}^{-\infty,\widetilde{\gamma}_1}(\Omega_1, J) \tag{7}$$

imply

$$u \in \mathcal{H}_P^{s+\mu,\gamma}(\mathbb{K}) \oplus \mathcal{H}_{\widetilde{P}}^{s+\mu,\widetilde{\gamma}_1}(\Omega_1, J) \tag{8}$$

for resulting asymptotic types $P = (P_0, P_1) \in As(X, \mathfrak{a}_0) \times As(\mathbb{B}, \mathfrak{a})$ *for* $\mathfrak{a} = (\mathfrak{a}_0, \mathfrak{a}_1^{\bullet})$, $\mathfrak{a}_i = (\gamma_i, \Theta_i)$, *and* $\widetilde{P} \in \times_{j=1}^{lN} As(\widetilde{\mathfrak{a}})$, $\widetilde{\mathfrak{a}} = (\widetilde{\gamma}_1, \Theta_1)$, *dependent on* $\mathcal{A}$ *and* $Q, \widetilde{Q}$, *not on* s. *Finally*

$$\mathcal{A}u = f \in \mathcal{H}^{s,\gamma-\mu}(\mathbb{K}) \oplus \mathcal{H}^{s,\widetilde{\gamma}_1-\mu}(\Omega_1, L) \tag{9}$$

and (7) imply

$$u \in \mathcal{H}^{s+\mu,\gamma}(\mathbb{K}) \oplus \mathcal{H}^{s+\mu,\widetilde{\gamma}_1}(\Omega_1, J). \tag{10}$$

Remark 3.3.5 *Let $\mathcal{A} \in C^\mu(\mathbb{K}, \mathfrak{g}; J, L)$ be elliptic. Then (5) is a Fredholm operator for all $s \in \mathbb{R}$ and ind $\mathcal{A}$ (the index of $\mathcal{A}$) is independent of s. The kernel ker$\mathcal{A}$ is a finite-dimensional subspace of $\mathcal{H}_P^{\infty,\gamma}(\mathbb{K}) \oplus \mathcal{H}_{\tilde{P}}^{\infty,\tilde{\gamma}_1}(\Omega_1, J)$ for certain asymptotic types $P, \tilde{P}$. If for $\mathcal{A}_0, \mathcal{A}_1 \in C^\mu(\mathbb{K}, \mathfrak{g}; J, L)$*

$$\sigma_\psi^\mu(\mathcal{A}_0) = \sigma_\psi^\mu(\mathcal{A}_1), \sigma_\wedge^\mu(\mathcal{A}_0) = \sigma_\wedge^\mu(\mathcal{A}_1), \sigma_M^\mu(\mathcal{A}_0) = \sigma_M^\mu(\mathcal{A}_1),$$

then ind $\mathcal{A}_0 =$ ind $\mathcal{A}_1$.

Theorem 3.3.6 *Let $\mathcal{A} \in C^\mu(\mathbb{K}, \mathfrak{g}; J, L)$ and assume that (5) is an isomorphism for an $s \in \mathbb{R}$. Then the inverse $\mathcal{A}^{-1}$ (taken in the sense of operators between the corresponding Sobolev spaces) belongs to $C^{-\mu}(\mathbb{K}, \mathfrak{b}; L, J)$, with $\mathfrak{b}$ being as in Definition 3.3.3.*

Theorem 3.3.4 is the main result of our corner operator theory. Therefore we want to sketch the main ideas of the proof. First observe that (ii)$\Rightarrow$(i) follows from the existence of a parametrix $\mathcal{P}$ in the asserted operator class, since the left hand sides of (3),(4) are compact operators in $\mathcal{H}^{s,\gamma}(\mathbb{K}) \oplus \mathcal{H}^{s,\tilde{\gamma}_1}(\Omega_1, J)$ and $\mathcal{H}^{s-\mu,\gamma-\mu}(\mathbb{K}) \oplus \mathcal{H}^{s-\mu,\tilde{\gamma}_1-\mu}(\Omega_1, L)$, respectively (cf. Theorem 3.2.6). In other words we have to construct $\mathcal{P}$. This has to be done, of course, by inverting the symbols. It employs the following

Proposition 3.3.7 *Let $\mathcal{A} \in C^\mu(\mathbb{K}, \mathfrak{g}; J, L)$ be elliptic. Then, there exists a $\mathcal{P}_0 \in C^{-\mu}(\mathbb{K}, \mathfrak{b}; L, J)$ with*

$$\sigma_\psi^{-\mu}(\mathcal{P}_0) = \sigma_\psi^\mu(\mathcal{A}^{-1}), \sigma_\wedge^{-\mu}(\mathcal{P}_0) = \sigma_\wedge^\mu(\mathcal{A}^{-1}), \sigma_M^{-\mu}(\mathcal{P}_0)^{-1} = T^{-\mu}\sigma_M^\mu(\mathcal{A}^{-1}).$$

The construction of $\mathcal{P}_0$ is rather complex, because of the various symbolic levels. But we can proceed step by step. First we construct an elliptic $\mathcal{P}_\psi \in C^{-\mu}(\mathbb{K}, \mathfrak{b}; L, J)$ with $\sigma_\psi^{-\mu}(\mathcal{P}_\psi) = \sigma_\psi^\mu(\mathcal{P}^{-1})$. Then

$$\mathcal{A}_0 := \mathcal{P}_\psi \mathcal{A} \in C^0(\mathbb{K}, \mathfrak{g}_l; J, J)$$

is also elliptic and $\sigma_\psi^0(\mathcal{A}_0) = 1$. Next we construct an elliptic $\mathcal{P}_\wedge \in C^0(\mathbb{K}, \mathfrak{g}_l; J, J)$ with $\sigma_\psi^0(\mathcal{P}_\wedge) = 1$, $\sigma_\wedge^0(\mathcal{P}_\wedge) = \sigma_\wedge^0(\mathcal{A}_0^{-1})$. Then

$$\mathcal{A}_1 := \mathcal{P}_\wedge \mathcal{P}_\psi \mathcal{A} \in C^0(\mathbb{K}, \mathfrak{g}_l; J, J)$$

is elliptic and $\sigma_\psi^0(\mathcal{A}_1) = 1$, $\sigma_\wedge^0(\mathcal{A}_1)$ is a family of identical operators. It remains to construct an elliptic $\mathcal{P}_M \in C^0(\mathbb{K}, \mathfrak{g}_l; J, J)$ with $\sigma_M^0(\mathcal{P}_M) = \sigma_M^0(\mathcal{A}_1)^{-1}$. Then we may set $\mathcal{P}_0 = \mathcal{P}_M \mathcal{P}_\wedge \mathcal{P}_\psi$. The details of the construction need the technicalities of the Mellin ψDO calculus, where the scheme is to large extent analogous to the usual cone and edge theories (cf. [S1],[S4]).

From $\mathcal{P}_0$ we now pass to $\mathcal{P}$ itself by a formal Neumann series argument (cf. ,e.g. , analogously [S1], Section 1.2.2.Theorem 12). Here we also use Theorem 3.2.7.

Now, the elliptic regularity with asymptotics expressed by (6), (7)$\Rightarrow$(8), is a consequence of the existence of $\mathcal{P}$ which is needed here as a left parametrix. In fact

$$\mathcal{P}\mathcal{A} = 1 + \mathcal{R},\ \mathcal{R} \in C_G^0(\mathbb{K}, \mathfrak{g}_l; J, J)$$

imply for (7)

$$\mathcal{A}\mathcal{P}u = \mathcal{P}f = (1 + \mathcal{R})u.$$

Using Theorem 3.2.5 we get the asserted quality of $\mathcal{P}f$, whereas the regularity of $\mathcal{R}u$ is just expressed by Definition 3.2.1.

Finally the necessity of the ellipticity for the Fredholm property of (5) follows from a combination of corresponding conclusions from the standard ψDO calculus, the wedge and the cone calculus. This means that we conclude first the bijectivity of $\sigma_\psi^\mu(\mathcal{A})$, since *l.u.c.*$\mathcal{A}$ as operator in $L_{cl}^{-\mu}(\Omega)$, then we conclude the bijectivity of $\sigma_\wedge^\mu(\mathcal{A})$, since $\mathcal{P}$ corresponds to a parametrix of $\mathcal{A}$ in the sense of $\mathcal{Y}^{-\mu}(\mathbb{W}, \mathfrak{b}_0; L, J)$, and then we find the bijectivity of $\sigma_M^\mu(\mathcal{A})$, by the same methods as for the usual cone (cf. [S1],2.2.1 Theorem 14).

The proof of 3.3.6 employs first that $\mathcal{A}$ is necessarily elliptic. The mentioned steps for constructing $\mathcal{P}_0$ in terms of $\mathcal{P}_\psi, \mathcal{P}_\wedge, \mathcal{P}_M$ may the carried out in such a way that $\mathcal{P}_\psi, \mathcal{P}_\wedge, \mathcal{P}_M$ are invertible once $\mathcal{A}$ is invertible (this uses in particular Remark 3.3.5). Thus $\mathcal{P}_0$ is also bijective in the Sobolev space sense as well as

$$\mathcal{P}_0\mathcal{A} = 1 + \mathcal{R}_0, \quad \mathcal{R}_0 \in C_G^0(\mathbb{K}, \mathfrak{g}_l; J, J).$$

Then an immediate consequence of Definition 3.2.1 is that

$$(1 + \mathcal{R}_0)^{-1} = 1 + \mathcal{R}_1$$

for another $\mathcal{R}_1 \in C_G^0(\mathbb{K}, \mathfrak{g}_l; J, J)$. Thus

$$(1 + \mathcal{R}_1)\mathcal{P}_0\mathcal{A} = 1$$

and hence $(1 + \mathcal{R}_1)\mathcal{P}_0 = \mathcal{A}^{-1} \in C^{-\mu}(\mathbb{K}, \mathfrak{b}; L, J)$ (cf. Theorem 3.2.9).

References

[A1] M.S. Agranovič, M.I. Višik
Elliptic problems with parameter and parabolic problems of general type. Uspechi Mat. Nauk 19, 3 (1964) 53 - 161

[A1] M.F. Atiyah, I.M. Singer
The index of elliptic operators, I. Ann. of Math. 87 (1968) 484 - 530

[B1] L. Boutet de Monvel
Boundary problems for pseudo-differential operators. Acta Math. 126 (1971) 11 - 51

[C1] H.O. Cordes
A global parametrix for pseudo-differential operators over $\mathbb{R}^n$ with applications. SFB 72 Reprints, Bonn 1976

[D1] M.Dauge
Elliptic boundary value problems on corner domains. Springer Lecture Notes in Math. 1341 1988

[G1] B. Gramsch
Inversion von Fredholmfunktionen bei stetiger und holomorpher Abhängigkeit von Parametern. Math. Ann. 214 (1975) 95 - 147

[H1] T. Hirschmann
Functional analysis in cone and edge Sobolev spaces. Ann. Glob. Anal. and Geometry 8, 2 (1990) 167 - 192

[H2] L. Hörmander
The analysis of linear partial differential operators. Springer, New York, Vols. 1 and 2, 1983, Vols. 3 and 4, 1985

[K1] V.A. Kondrat'ev
Boundary value problems for elliptic equations in domains with conical points. Trudy Mosk. Mat. Obšč. 16 (1967) 209 - 292

[M1] V.G. Maz'ja, B.A. Plamenevskij
Elliptic boundary problems on manifolds with singularities. Problemy mat. analiza 6, published by the Leningrad University 1977, pp. 85 - 142

[M2] R. Melrose, G. Mendoza
Elliptic operators of totally characteristic type. Reprint MSRI 047 - 83 Berkeley 1983

[M3] R. Melrose
Pseudo-differential operators on manifolds with corners. Manuscript MIT, Boston 1987

[P1] B.A. Plamenevskij
Algebras of pseudo-differential operators. Moscow, Nauka 1986

[R1] S. Rempel, B.-W. Schulze
Complete Mellin and Green symbolic calculus in spaces with conormal asymptotics. Ann. Glob. Analysis and Geometry 4,2 (1986) 137 - 224

[R2] J. Roßmann
Gewichtete Sobolev-Slobodezkij Räume und Anwendungen auf elliptische Randwertaufgaben in Gebieten mit Kanten. Doktor Diss. Rostock 1988

[R3] S. Rempel, B.-W. Schulze
Asymptotics for elliptic mixed boundary problems (Pseudo-differential and Mellin operators in spaces with conormal singularity). Math. Research 50, Akademie-Verlag Berlin 1989

[S1] E. Schrohe
Spaces of weighted symbols and weighted Sobolev spaces on manifolds. In "Pseudo-Differential Operators", Springer Lecture Notes in Math. 1256, Berlin, Heidelberg, 1987, 360 - 377

[S2] B.-W. Schulze
Pseudo-differential operators on manifolds with edges. Symposium "Partial Differential Equations" Holzhau 1988, Teubner-Texte Math. 112, 259 - 288, Leipzig 1989

[S3] B.-W. Schulze
Corner Mellin operators and reductions of orders with parameters. Ann. sc. Norm. Sup. Pisa 16,1 (1989) 1 - 81

[S4] B.-W. Schulze
Mellin representations of pseudo-differential operators on manifolds with corners. Preprint P-Math. 28/89, Karl-Weierstrass-Institute for Math., Berlin; Ann. Glob. Anal. Geom. 8, 3 (1990) 261-287

[S5] B.-W. Schulze
Pseudo-differential operators and asymptotics on manifolds with corners. Reports R-Math Karl-Weierstrass-Institute for Math., I : 07/89, II : 02/90, III : 04/90 IV : 06/90, VI : 08/90 , VII : 01/91, VIII : 02/91 ,IX : 03/91 Berlin (further chapters in preparation)

[S6] B.-W. Schulze
Pseudo-differential operators on manifolds with singularities. North Holland 1991

[S7] B.-W. Schulze
Mellin operators with corner asymptotics (in preparation)

[S8] B.-W. Schulze, G. Wildenhain
Methoden der Potentialtheorie für elliptische Differentialgleichungen beliebiger Ordnung. Akademie-Verlag, Berlin 1977; Birkhäuser-Verlag Basel, Stuttgart 1977

A Lefschetz Fixed Point Formula in Reduced L^2-Cohomologies on Manifolds with Cylinders

M.A. Shubin
Moscow, USSR

1 Introduction.

In this paper we shall introduce the Lefschetz number in reduced L^2-cohomologies for proper maps of a manifold with cylinders (or cylindrical ends), which are linear with respect to the axis coordinate near infinity. We prove the homotopy invariance of this Lefschetz number when the topology is taken in the class of the maps of the same sort. Then we calculate the Lefschetz number for the case when the bases of all the cylinders are spheres.

In an earlier paper [S-S] which was a predecessor of this work the same results were obtained in case when the compact manifold with boundary to which the cylinders are attached is mapped into itself. This condition considerably simplifies the situation and the answer was given in [S-S] in terms of the map of this compact part. Here the answer involves the asymptotic of the map near infinity.

Another Lefschetz type theorem in reduced L^2-cohomologies was proved by A. Efremov ([E]), but there von Neumann traces and dimensions were used which is not necessary here because the reduced L^2-cohomologies are finite dimensional in our siuation as was noticed in [A-P-S].

2 Preliminaries and formulation of the main results.

We shall remind some results from[S-S] in slightly different notations.

Let M be a noncompact Riemannian manifold which has the form $Y \times [0,\infty)$ with a compact Y near infinity, $dimM = N$. More exactly we suppose that there exists a compact part $X_0 \subset M$ such that X_0 is diffeomorphic to a manifold X with the boundary Y, which is a $(n-1)$-dimensional compact closed manifold, and $M = X_0 \cup (Y \times [0,\infty))$, ∂X_0 being identified with $Y \times \{0\}$. We always suppose that the Riemannian metric on $Y \times [0,\infty)$ is a product metric of the form $dy^2 \oplus dt^2$ where dy^2 is a fixed Riemannian metric on on Y and t is the standard coordinate in $[0,\infty)$. Of course X_0 is not unique (e.g. it can be replaced by $X_{t_0} = X_0 \cup (Y \times [0,t_0]))$ but we fix such a compact part X_0 for the sake of simplicity of notations. In this situation we shall say thet M is a manifold with cylinders (or with cylindrical ends).

Let $\Lambda^k(M)$ be the space of all smooth exterior k-forms on M, $L^2\Lambda^k(M)$ be the Hilbert space of all square integrable k-forms on M, $C_0^\infty\Lambda^k(M)$ be the space of all forms from $\Lambda^k(M)$ having a compact support. Let $d : \Lambda^k(M) \to \Lambda^{k+1}(M)$ denote the usual De Rham differential, $\delta = d^*$ be the formally adjoint operator with respect to the scalar product in $L^2\Lambda^k M$. We shall also consider the maximal extension $d_{\max}$ of d to $L^2\Lambda^k(M)$ which is a closed unbounded operator from $L^2\Lambda^k(M)$ to $L^2\Lambda^{k+1}(M)$, its null-space will be denoted by $Ker_k d_{\max}$. It is a closed subspace in $L^2\Lambda^k M$. The reduced L^2-cohomologies of M are defined as

$$L^2H^k(M) = Ker_k d_{\max}/\overline{dC_0^\infty\Lambda^{k-1}(M)} \tag{1}$$

where the closure is taken in $L^2\Lambda^k(M)$. Let us defin also the space of all harmonic L^2 forms

$$\mathcal{H}^k(M) = \{\omega : \omega \in \Lambda^k(M) \cap L^2\Lambda^k(M), d\omega = \delta\omega = 0\}.$$

Then there is an orthogonal Kodaira decomposition [K]

$$L^2\Lambda^k(M) = \overline{dC_0^\infty\Lambda^{k-1}(M)} \oplus \mathcal{H}^k(M) \oplus \overline{\delta C_0^\infty\Lambda^{k+1}(M)} \tag{2}$$

and in this decomposition

$$Ker_k d_{\max} = \overline{dC_0^\infty\Lambda^{k-1}(M)} \oplus \mathcal{H}^k(M) \tag{3}$$

(see e.g. [C]). It follows that the natural inclusion $\mathcal{H}^k(M) \subset L^2\Lambda^k(M)$ induces an isomorphism

$$J_k : \mathcal{H}^k(M) \xrightarrow{\sim} L^2H^k(M).$$

It was proved in [A-P-S] that $dim\mathcal{H}^k(M) < \infty$ for all $k = 0, 1, \ldots, n$. Moreover, the components of the forms $\omega \in \mathcal{H}^k(M)$ and their derivatives in natural product coordinates decay exponentially as $t \to \infty$ i.e. there exists $\varepsilon > 0$ such that

$$|\partial^\alpha \omega(y,t)| = 0(\exp(-\varepsilon t)),\ t \to +\infty,$$

the derivative ∂^α being taken in product coordinates for every component of the form ω and absolute value may be understood e.g. as a usual norm of a vector.

The space $\mathcal{H}^k(M)$ can be also described as the space of all forms $\omega \in L^2\Lambda^k(M)$ such that $\Delta\omega = 0$ where $\Delta = d\delta + \delta d$ is the Laplace-Beltrami operator on k-forms.

We shall consider proper C^∞-maps $f : M \to M$ which are linear with respect to the axis variable t near infinity i.e. such that in product coordinates

$$f(y,t) = (g(y), a(y)t + b(y)), y \in Y, t \geq t_0, \tag{4}$$

if $t_0 > 0$ is sufficiently large. We would like to define a natural action of f on $L^2H^k(M)$. It can not be done in a straightforward manner because the map $f^* : C_0^\infty H^k(M) \to C_0^\infty\Lambda^k(M)$ can be extended to a bounded linear operator $f^* : L^2\Lambda^k(M) \to L^2\Lambda^k(M)$ if and only if f is a diffeomorphism. To avoid this difficulty we shall use $\mathcal{H}^k(M)$ instead of $L^2H^k(M)$ with the image included actually in

$$Ker\{d : \Lambda^k(M) \longrightarrow \Lambda^{k+1}(M)\} \bigcap L^2\Lambda^k(M) = Ker_k d_{\max} \bigcap \Lambda^k(M).$$

Now let P_k denote the orthogonal projection on $L^2\Lambda^k(M)$ with the image $\mathcal{H}^k(M)$ in the Kodaria decomposition 1.(2). Then we can define a map

$$P_k f^* : \mathcal{H}^k(M) \longrightarrow \mathcal{H}^k(M)$$

which is a linear map of finite-dimensional spaces. Using the isomorphism J_k we can define a linear map $f_k^* : L^2H^k(M) \to L^2H^k(M)$ via the following commutative diagram

$$\begin{array}{ccc} \mathcal{H}^k(M) & \xrightarrow{P_k f^*} & \mathcal{H}^k(M) \\ J_k \downarrow & & J_k \downarrow \\ L^2H^k(M) & \xrightarrow{f_k^*} & L^2H^k(M) \end{array}$$

Proposition 1 *The map f_k^* depends only on the uniform conformal class of the Riemannian metric i.e. if g, g' are two Riemannian metrics on M such that there exists $C > 0$ such that $C^{-1}g \leq g' \leq Cg$ then the corresponding maps f_k^* coincide for all $k = 0, 1, \ldots, n$,*

Corollary 2 *The trace* $trP_kf^* = trf_k^*$ *does not depend on the choice of a Riemannian metric in the same uniform conformal class.*

The proof of Proposition 1 will be given in Section 2. Now we can define the Lefschetz number

$$L(f) = \sum_{0\le k\le n} (-1)^k trf_k^* = \sum_{0\le k\le n} (-1)^k trP_kf^*,$$

which also depends only on the uniform conformal class of the Riemannian metrics.

An important property of the maps f_k^* is their homotopy invariance with respect to appropriate homotopies. Namely we shall consider homotopies which are proper C^∞-maps

$$F : M \times [0,1] \longrightarrow M$$

such that for every $\tau \in [0,1]$ the map $F : (\cdot,\tau)M \to M$ has the form 1.(4) near infinity. In this case we shall say that F is an admissible homotopy and the maps $F(\cdot,0)$ and $F(\cdot,1)$ are admissibly homotopic.

Proposition 3 *Let the maps* $f, f' : M \to M$ *be admissibly homotopic. Then they induce the same maps in the reduced* L^2*-cohomologies i.e.* $f_k^* = f_k'^*$, $k = 0,1,\ldots,n$.

The proof will be also given in Section 2.

Corollary 4 *The traces* $trP_kf^* = trf_k^*$ *do not change if the map* f *is replaced by an admissibly homotopic map. The same is true for the Llefschetz number* $L(f)$.

Now we turn to the connection between the Lefschetz number and the fixed points.

Let us consider a fixed point x of a C^∞-map $f : M \to M$ i.e. $x \in M$ and $f(x) = x$. The derivative map at x is then a linear map of the tangent space TM into itself

$$df_x : T_xM \to T_xM.$$

We shall say that a fixed point x is simple or non-degenerate if $det(1 - df_x) \neq 0$ or, equivalently, if 1 is not an eigenvalue of df_x (another equivalent formulation is that the graph $\Gamma_f = \{(x, f(x))|x \in M\} \subset M \times M$ is transversal to the diagonal in $M \times M$). All simple fixed points of f are isolated so the set of all simple fixed points is locally finite. We shall always suppose that f satisfies the following condition

(C1) f is a proper C^∞-map which has the form 1.(4) near infinity, all the fixed points of f and g are simple and f has no fixed points near infinity i.e. outside some compact part $X_{t_0} \subset M$.

It follows that the sets of all fixed points of f in M and of g in Y are finite. We shall denote these sets F^0 and F_b respectively. It follows also that for every fixed point $y \in F_b$ one and only one of the following two possibilities holds:

(i) $a(y)t + b(y) < t$ if $t > t_0$ (in this case we shall write $y \in F_+$);

(ii)$a(y)t + b(y) > t$ if $t > t_0$ (in this case we shall write $y \in F_-$).

In this way the set F_b is represented as a disjoint union $F_b = F_+ \cup F_-$, $F_+ \cap F_- = \emptyset$, $F_\pm$ being finite sets in Y.

The map $g : Y \to Y$ is thus defined by the asymptotics of f near infinity. This asymptotics also defines the classification of the fixed points of g as the points from F_+ or F_-.Roughly speaking F_+

is the set of asymptotic fixed points which are attracted by X_0 and repulsed by infinity and F_- is vice versa attracted by infinity and repulsed by X_0

The same situation appeared in [S-S] that f defines a map

$$f : (X_0, \partial X_0) \longrightarrow (X_0, \partial X_0)$$

i.e. $f(X_0) \subset X_0$ and $f(\partial X_0) \subset \partial X_0$. Then it was proved that all the traces $tr(P_k f^*)$ depend only on the restriction $f|X_0$. after a homotopy it was then possible to achieve a situation when there are no fixed points in $M \setminus X_0$ and f has the form 1.(4) outside X_0 with the additional condition $b \equiv 0$. In this case g can be identified with the restriction $f|\partial X_0$ and F_+ (resp. F_-) correspond to attractive (resp. repulsive) boundary fixed points of $f|X_0$ due to the classification given in [B-S].

Now we shall introduce the contributions of the fixed points

$$\begin{aligned} \nu(x) &= \text{sgn det } (1 - df_x), \ x \in F^0; \\ \nu_b(x) &= \text{sgn det } (1 - dg_y), \ x \in F_b. \end{aligned}$$

We shall use the notation $degf$ for the degree $degf$ of $f : M \to M$ in case when M is orientable and connected. It can be understood e.g. as the sum of ± 1 over the points $x \in f^{-1}(x_0)$ for a generic point $x_0 \in M$ (such that $df_x : T_x M \to T_{x_0} M$ is an isomorphism for every $x \in f^{-1}(x_0)$; the choice of the sign + or − depends on wether df_x conserves a chosen orientation of M or not).

Now let $Y_1, \ldots, Y_l$ be all connected components of Y and $Y_1, \ldots, Y_r$ be the fixed ones (i.e. those wich are mapped by g into themselves). Then the restriction of g to Y_j is a map $g_j : Y_j \to Y_j$, $j = 1, \ldots, r$. The degree $degg_j$ is the obviously defined.

The following theorem gives the Lefschetz type theorem in a special case of the described situation.

Theorem 5 *Let M be an orientable connected manifold with cylinders based on Y which is a non-empty disjoint union of spheres $Y_j \cong S^{n-1}$, $j = 1, \ldots, l$. (Here S^{n-1} is the standard unit sphere in $\mathbb{R}^n$ and $Y_j \cong S^{n-1}$ means that Y_j is diffeomorphic to S^{n-1}). Let $f : M \to M$ satisfy (C1). Then*

$$\begin{aligned} L(f) &= \sum_{x \in F^0} \nu(x) - \sum_{y \in F_+} \nu_b(y) + (-1)^{n+1} degf + r - 1 \\ &= \sum_{x \in F^0} \nu(x) + \sum_{y \in F_-} \nu_b(y) + (-1)^{n+1} degf + (-1)^n \sum_{j=1}^{r} degg_j - 1. \end{aligned} \qquad (5)$$

This theorem generalizes the result of [S-S] to the case when f has no invariant compact part like X_0. Of course it is very easy to generalize Theorem 5 to the case when M is not necessarily connected by summing over all fixed connected components.

The proof of Theorem 5 will be given in section 3.

3 Basic properties of the action on L^2-cohomologies.

Proof of Proposition 1 Let g, g' be Riemannian metrics on M which are in the same uniform conformal class. Let us denote by $\mathcal{H}^k(M)$, $\mathcal{H}'^k(M)$ the corresponding spaces of L^2-harmonic forms; J_k, J'_k the corresponding isomorphisms of $\mathcal{H}^k(M)$, $\mathcal{H}'^k(M)$ and $L^2 H^k(M)$ (see Section 1); P_k, P'_k the orthogonal projections of $L^2\Lambda^k(M)$ to $\mathcal{H}^k(M)$, $\mathcal{H}'^k(M)$ respectively where the scalar product is defined by the corresponding metric. We have to prove that the following diagram

$$\begin{array}{ccccc} & \mathcal{H}^k(M) & \overset{J_k}{\underset{\sim}{\longrightarrow}} & L^2H^k(M) & \overset{J_k'}{\underset{\sim}{\longleftarrow}} & \mathcal{H}'^k(M) & \\ P_kf^* & \downarrow & & & & \downarrow & P_k'f^* \\ & \mathcal{H}^k(M) & \overset{J_k}{\underset{\sim}{\longrightarrow}} & L^2H^k(M) & \overset{J_k'}{\underset{\sim}{\longleftarrow}} & \mathcal{H}'^k(M) & \end{array}$$

is commutative i.e. that the diagram

$$\begin{array}{ccccc} & \mathcal{H}^k(M) & \overset{(J_k')^{-1}J_k}{\longrightarrow} & \mathcal{H}'^k(M) & \\ P_kf^* & \downarrow & & \downarrow & P_k'f^* \\ & \mathcal{H}^k(M) & \overset{(J_k')^{-1}J_k}{\longrightarrow} & \mathcal{H}'^k(M) & \end{array}$$

is commutative. To do this let us remark first that the arguments which were given in [S-S] and used Dodziuk Kodaira decomposition in Sobolev spaces [Dod] show that forms $\omega_1, \omega_2 \in \mathcal{H}^k(M)$ coincide if and only if all their periods (i.e. integrals over finite simplificial cycles) coincide. The same arguments show that if $\omega \in \mathcal{H}^k(M)$, $\omega' \in \mathcal{H}'^k(M)$ then the periods of ω and ω' coincide if and only if ω, ω' define the same class in $L^2H^k(M)$ i.e. $J_k\omega = J_k'\omega'$.

Now let us choose $\omega \in \mathcal{H}^k(M)$. We have proved that the form $\omega' = (J_k')^{-1}J_k\omega$ has the same periods ass ω. Therefore the same is true for the forms $f^*\omega$, $f^*\omega'$. Again using the Dodziuk decomposition we obtain that $P_kf^*\omega$ and $f^*\omega$ and $f^*\omega$ have the same periods (and similarly for $P_k'f^*\omega'$ and $f^*\omega'$). Hence $P_kf^*\omega$ and $P_k'f^*\omega'$ have the same periods, therefore $J_kP_kf^*\omega = J_k'P_k'f^*\omega'$ which proves the required commutativity. □

Proof of Proposition 3 Let us consider an admissible homotopy $F : M \times [0,1] \to M$ of two maps $f_j = F(\cdot, j) : M \to M$, $j = 0, 1$. Let us construct in the usual way a cochain homotopy $T : \Lambda^k(M) \to \Lambda^{k-1}(M)$ given by

$$T\omega = \int_0^1 (\partial/\partial\tau \rfloor F^*\omega)d\tau$$

so that

$$F_1^*\omega - f_0^* = (Td + dT)\omega, \ \omega \in \Lambda^k(M).$$

Now let $\omega \in \mathcal{H}^k(M)$. Then $d\omega = 0$, hence $f_1^*\omega - f_0^*\omega = d(T\omega)$. The form $T\omega \in \Lambda^{k-1}(M)$ decreases exponentially as $t \to +\infty$ with all the derivates. Using some appropriate cut-off functions we easily obtain that $d(T\omega) \in \overline{dC_0^\infty\Lambda^{k-1}(M)}$ where the bar means closure in $L^2\Lambda^k(M)$. Hence 1.(3) shows that $P_kf_1^*\omega = P_kf_0^*\omega$. □

Remark. It is easily seen from the proofs that Propositions 1 and 3 (and hence Corollaries 2 and 4) are true in a much more general situation e.g. on coverings of compact manifolds for maps which behave reasonably at infinity.

4 The case of spherically based cylinders.

The proof of Theorem 5 will be based on the following idea. Let $\overline{M}$ be the compactification of M obtained by adding a point z_j at infinity of each of the cylinders $Y_j \times [0, \infty)$, $j = 1, \dots, l$. Then $\overline{M}$ has a natural structure of a C^∞-manifold such that there is a natural inclusion $M \subset \overline{M}$ with the image $\overline{M} \setminus \{z_1, \dots, z_l\}$.

Our idea is to make a homotopy of f to a map which acts simpler near infinity and then extend the obtained map to a map of $\overline{M}$ in order to be able to apply the classical Lefschetz - Hopf theorem on $\overline{M}$.

Lemma 6 *There exists a C^∞-map $\overline{f}: M \to M$ such that the following conditions are satisfied:*

(i) $\overline{f}$ is admissibly homotopic to f;

(ii) $\overline{f}$ coincides with f on on X_{t_0} where t_0 is so chosen that f has the form 1.(4) on $M \setminus X_{t_0}$ and f has no fixed points outside X_{t_0};

(iii) $\overline{f}$ has the form $\overline{f}(y,t) = (g(y), \alpha(y,t))$ outside X_{t_0} and $\alpha(y,t) = 2t$ outside X_{t_1} for some $t_1 > t_0$ with the same map $g: Y \to Y$ as in 1.(4) for f;

(iv) $\overline{f}$ has only non-degenerate fixed points;

(v) the set F'_+ of all fixed points of $\overline{f}$ in $M \setminus X_{t_0}$ is in a one-one correspondence with F_+; more precisely, $F'_+ = \{(y, t(y)) | y \in F_+\}$ and $\partial\alpha(y,t)/\partial t > 1$ if $y \in F_+$ and $t = t(y)$.

Proof. Let us begin by constucting the function α from (iii). It should be chosen so that the following requirements are satisfied:

(a) $\alpha \in C^\infty(Y \times [t_0, \infty))$;

(b) $\alpha(y,t) = a(y)t + b(y)$ if $t \in [t_0, t_0 + \varepsilon]$ where $\varepsilon > 0$ and a, b are the same as in 1.(4);

(c) $\alpha(y,t) = 2t$ if $t \geq t_1$ for some $t_1 > t_0$;

(d) if $y \in F_-$ then $\alpha(y,t) > t$ for all $t \geq t_0$;

(e) if $y \in F_+$ then a point $t = t(y) \in (t_0, t_1)$ such that $\alpha(y,t) = t$, is unique (note that it necessarily exists due to the choice of t_0 and t_1 : see (i) in Section 1 and (c) here); moreover

$$\partial\alpha(y,t)/\partial t > 1 \text{ if } t = t(y),\ y \in F_+.$$

It follows that $(y, t(y))$ will be a non-degenerate fixed point of the map $(y,t) \to (g(y), \alpha(y,t))$ and its contribution equals $-\nu_b(y)$.

Let us choose a small neighbourhood U of F_+ such that $U \cap F_- = \emptyset$. Then it is easy to construct a function $\alpha_+ \in C^\infty(U \times [t_0, +\infty))$ such that α_+ satisfies the conditions (b)-(e) above. Now let V be an open set in Y such that $V \cap F_+ = \emptyset$ and $U \cup V = Y$. Let us set $\alpha_-(y,t) = \alpha(y)trb(y)$ in $V \times [0, +\infty)$¿ Let us choose $h \in C^\infty(Y)$ such that *supp* $h \subset U$, $0 \leq h \leq 1$ and $h = 1$ in a neighbourhood of F_+. Then we can take

$$\alpha(y,t) = h(y)\alpha_+(y,t) + (1 - h(y))\alpha_-(y,t)$$

and all the conditions (a)-(e) will be satisfied.

Now we can define $\overline{f}$ by (ii) and (iii) in the statement of Lemma 6 with α constructed above. Then (iv), (v) are satisfied due to (c)-(e). It remains to prove (i). But the desired homotopy

$$F : [0,1] \times M \longrightarrow M$$

may be chosen constant on X_{t_0} with respect to the homotopy parameter $\tau \in [0,1]$ i.e. $F(\tau, x) = f(x)$ for all $\tau \in [0,1]$, and linear with respect to τ outside X_{t_0} i.e.

$$\begin{aligned} F(\tau, y, t) &= (g(y), [a(y)t + b(y)](1 - \tau) + \tau\alpha(y,t)), \\ & \quad t \geq t_0,\ \tau \in [0,1],\ y \in Y. \end{aligned}$$

Clearly it is an admissible homotopy which proves the Lemma. □

Proof of Theorem 5. There is a unique extension of $\overline{f}$ to a continuous map of $\overline{M}$ into itself which we shall still denote by $\overline{f}$. Clearly $\overline{f}$ performs a permutation of the points $z_1, \ldots z_l$; among them $z_1, \ldots, z_r$ remain fixed.

We shall use natural isomorphisms ([R])

$$\mathcal{H}^k(M) = \begin{cases} H^k(\overline{M}) & \text{if } \ 0 < k < n, \\ 0 & \text{if } \quad k = 0, n. \end{cases}$$

It follows thet if $L(\overline{f})$ is the usual Lefschetz number of $\overline{f}$ on $\overline{M}$ then

$$L(\overline{f}) = L(f) + 1 + (-1)^n tr \overline{f}^* | H^n(\overline{M}) = L(f) + 1 + (-1)^n deg f.$$

Now due to the general Lefschetz-Hopf theorem (e.g. for CW-complexes- see [Dol], Sect. 7.6) $l(\overline{f})$ may be calculated as the sum of contributions of all fixed points. The sums of contributions of all fixed points in X_{t_0} and in $M \setminus X_{t_0}$ equal $\sum_{x \in F^0} \nu(x)$ and $- \sum_{y \in F_+} \nu_b(y)$, respectively, due to Lemma 6.

It remains to calculate the contributions of the fixed points z_j, $j = 1, \ldots, r$. Using the homotopy invariance we can locally reduce $\overline{f}$ to the constant map, hence the contributions of z_j equals 1. Therefore

$$L(\overline{f}) = \sum_{x \in F^0} \nu(x) - \sum_{y \in F_+} \nu_b(y) + r,$$

which implies the first equality in 1.(5). To prove the second one we can act as in the proof of Lemma 6 to construct a map $\overline{f}$ with repulsive fixed points at infinity (e.g. with $\alpha(y,t) = t/2$ near infinity). Again using local homotopy we can reduce $Id - \overline{f}$ near z_j to the map $-\overline{f}$. Hence the contribution of z_j will be equal to the trace of $(-\overline{f})^*$ on $H^n(\overline{M}, \overline{M} \setminus \{z_j\}) \cong H^n(B_j \setminus \{z_j\})$ where B_j is a small ball centered at z_j with respect to a Riemannian metric on M. Passing to the exact sequence of the pair $(B_j, B_j \setminus \{z_j\})$ we easily obtain that the contribution of z_j is equal the trace of $(-\overline{f})^*$ on $H^{n-1}(B_j \setminus \{z_j\}) \cong H^{n-1}(\partial B_j)$. This trace is equal to the trace induced by the composition of g_j and the antipodal map of $Y_j = B_j$ in $H^{n-1}(\partial B_j)$. Thus the contribution of z_j will be equal to $(-1)^n deg g_j$. Summing up we obtain the second equality in 2.(5). On the other hand it can be also deduced from the first one and the classical Lefschetz formulas on the spheres Y_j. □

References

[A-P-S] M.F. Atiyah, V.K. Patodi, I.M. Singer. Spectral asymmetry and Riemannian geometry I. Proc. Cambridge Phil. Soc., 77,43-69 (1975)

[B-S] A.V. Brenner, M.A. Shubin. The Atiyah - Bott - Lefschetz fixed point theorem for manifolds with boundary. Funct. Analysis and its Applications, 15, no.4, 67-68 (1981)

[C] J. Cheeger. On the Hodge theory of Riemannian pseudomanifolds. Proc. Symp. Pure Math. 36, 91-146 (1980)

[Dod] J. Dodziuk. De Rham - Hodge theory for L^2-cohomology of infinite coverings. Topology, 16, 157-165 (1977)

[Dol] A. Dold. Lectures on algebraic topology. Springer - Verlag, Berlin e.a. 1972.

[E] A.V. Efremov. TheAtiyah - Bott - Lefschetz theorem for Hilbert bundles. Vestnik Moscovskogo Univ., Ser. Mat. - Mech., No.4 (1988), 90-93

[K] K. Koidara. Harmonic fields in Riemannian manifolds Ann. of Math., 50, 587-665 (1949)

[R] S. Rosenberg. Harmonic forms and L^2-cohomology on manifolds with cylinders. Indiana Univ. Math. Journal, 34, 335-373 (1985)

[S-S] S. Seifarth, M.A. Shubin. A Lefschetz fixed point formula for manifolds with cylindrical ends. Annals of Global Analysis and Geometry, 1990

M.A. Shubin
Moscow State University
Department of Mathematics
Moscow V-234, USSR

A Mixed Problem for Evolution Equations

L. R. Volevich
Keldysh Institute of Applied Mathematics, Ac. Sc. USSR

In this lecture we present some general remarks about classes of differential operators with variable coefficients stable under perturbations of lower (in some sense) terms for which Cauchy and the mixed problem are correctly posed. The results presented here are obtained in collaboration with S.G. Gindikin and are published in [1]-[4].

1

We introduce coordinates (x,t), $x = (x_1, \dots, x_n) \in \mathbb{R}^n$, $t \in \mathbb{R}$ is regarded as time, Ω is a domain in $\mathbb{R}^n$. We introduce partial differential operators $A, B_1, \dots, B_\mu$,

$$A(x,t,\mathcal{D}_x,\mathcal{D}_t) = \mathcal{D}_t^k + \sum_{j\geq 1} A_j(x,t,\mathcal{D}_x)\mathcal{D}_t^{k-j}$$

and consider the mixed problem

$$A(x,t,\mathcal{D}_x,\mathcal{D}_t)u(x,t) = f(x,t), \ x \in \Omega, \ t \geq 0; \tag{1}$$

$$B_j(x,t,\mathcal{D}_x,\mathcal{D}_t)u(x,t)|_{\partial\Omega} = g_j(x',t), \ x' \in \partial\Omega, \ t \geq 0, \ j = 1,\dots,\mu \tag{2}$$

$$\mathcal{D}_t^j u(x,t)|_{t=0} = \varphi_j(x), \ j = 0,\dots,k-1. \tag{3}$$

There are two remarkable classes of operators for which the problem (1), (2), (3) was treated in great generality. These classes are :

(I) q-parabolic operators, $q = 2b$ (or $2b$-parabolic in sense of Petrovsky), the simplest example of these operators is heat operator

$$\frac{\partial}{\partial t} - \sum \frac{\partial^2}{\partial x_j^2}.$$

(II) Strictly hyperbolic operators (or hyperbolic in sense of Petrovsky), the simplest example of these operators is the wave operator

$$\frac{\partial^2}{\partial t^2} - \sum \frac{\partial^2}{\partial x_j^2}.$$

(III) q-hyperbolic operators, the simplest example of these operators is the operator

$$\frac{\partial}{\partial t} \pm \sum \frac{\partial^3}{\partial x_j^3}.$$

Our aim is to describe the main properties of classes (I), (II), (III) permitting to treat the problem (1), (2), (3) in the general situation and to explain that in some sense other general classes do not exist.

2

If we want to treat a mixed problem in the domain Ω of general nature we must expect that the class of operators under investigation is invariant under diffeomorphisms of $\mathbb{R}^n$. If we want to treat operators with variable coefficients we have to provide that our classes are stable under variations of lower terms. To define the lower terms as in the case of parabolic equations we attach the "weight q" to the time variable and the weight 1 to the space variable and set

$$deg_q \mathcal{D}^{\alpha_1}_{x_1} \dots \mathcal{D}^{\alpha_n}_{x_n} = \alpha_2 + \dots + \alpha_n + qj.$$

By linearity we define $deg_q C$ for arbitrary differential operators C. In the same way we define $deg_q C(\xi, \tau)$ for polynomials of variables dual to x and t.

Note that the problem (1), (2), (3) cover the Cauchy problem as a special case corresponding to $\Omega = \mathbb{R}^n$. It is reasonable to investigate the case of unique solvability of the Cauchy problem. In the case of operators $A(D_x, D_t)$ with constant coefficients there exists a necessary and sufficient condition for the Cauchy problem to be correctly posed. It is the

Petrovsky condition :

$$\exists \gamma_0 \in \mathbb{R} \text{ with } A(\xi,\tau) \neq 0, \ \forall \xi \in \mathbb{R}^n, \ Im\, \tau \leq \gamma_0.$$

If we want the Cauchy problem to be correctly posed under arbitrary variation of lower terms of the operator A we must impose the

Stable Petrovsky condition :

$$\forall Q(\xi,\tau), \ deg_q Q < deg_q A, \ A(\xi,\tau) + Q(\xi,\tau) \neq 0, \ \forall \xi \in \mathbb{R}^n, \ Im\, \tau \leq \gamma_0.$$

The following proposition gives us an algebraic description of the class of polynomials satisfying the stable Petrovsky condition. Without loss of generality we can confine to the case of q-homogeneous polynomials:

$$A(\lambda\xi, \lambda^q \tau) = \lambda^m A(\xi,\tau), \ \lambda > 0 \ (m = deg_q A).$$

Proposition 1 *Let $Q \geq 1$ and let $A(\xi,\tau)$ be a q-homogeneous polynomial. The following conditions are equivalent:*

(i) stable Petrovsky condition ;

(ii) existence of the estimate

$$c|Im\, \tau|^{1/q}(|\tau|^{1/q} + |\xi|)^{m-1} \leq |A(\xi,\tau)|, \ \xi \in \mathbb{R}^n, \ Im\, \tau \leq 0;$$

(iii) q equals 1 or is an even number and

- *(a) in the case $q = 1$ the polynomial $A(\xi,\tau)$ is strictly hyperbolic, i.e. the roots $\tau_j(\xi)$, $j = 1,\dots,k$ are real and different for $\xi \neq 0$,*
- *(b) in the case $q = 2b$ the polynomial $A(\xi,\tau)$ is $2b$-parabolic i.e. $\exists \kappa > 0$ so that $Im\, \tau_j(\xi) > \kappa|\xi|^{2b}$, $j = 1,\dots k$.*

Pareseval's identity permits us to rewrite Proposition 1 as someone about differential operators. Denote by $H_\gamma^{(s/q)}$ the space of functions (distributions) with finite norm

$$||u||_{(s/q),\gamma} = \left(\int\int_{Im\ \tau=\gamma} (|\tau|^{2/q} + |\xi|^2)^s |\hat{u}(\xi,\tau)|^2 d\xi d\tau, \right.$$

where $\hat{u}(\xi,\tau)$ is Fourier-Laplace transform of $u(x,t)$. For $s = 0$ the norm (4) is the usual L^2-norm with weight $\exp(\gamma t)$. Denote by $H_\gamma^{0(s/q)}$ the subspace of functions equal 0 when $t < 0$.

Proposition 2 *Let $q \geq 1$. For the differential operator A with q-homogeneous symbol $A(\xi, \tau)$ the following conditions are equivalent.*

(i') $\forall Q(\mathcal{D}_x, \mathcal{D}_t)$ $deg_q Q < deg_q A = m$ $\exists \gamma_0 = \gamma_0(Q)$ so that for $\forall s \in \mathbb{R}$ and $\forall \gamma \leq \gamma_0$ the equation

$$(A + Q)(\mathcal{D}_x, \mathcal{D}_t)u = f$$

$\forall f \in H_\gamma^{0(s/q)}$ has unique solution $u \in H_\gamma^{0(s+M-1/q)}$;

(ii') existence of the estimate

$$C|\gamma|^{1/q}||u||_{(m-1+s/q),\gamma} \leq ||Au||_{(s/q),\gamma}\ \forall \gamma \leq 0;$$

(iii) the same condition as in Proposition 1.

It is easy to reformulate (not to prove!) the proposition for the case of variable coefficients and even for the case of the mixed problem. These results will be presented below in a more general framework.

3

Propositions 1, 2 show that if we restrict ourselves to traditional (for problems of evolution) $H^{(s)}$-spaces with weight $\exp(\gamma t)$ then only hyperbolic and parabolic operators satisfy the conditions of unique solvability of the Cauchy or mixed problems under arbitrary lower terms. To extend the class of operators under investigation we have to treat our problems in a wider class of functional spaces. We consider $H^{(s)}$-spaces with weight

$$\exp(\gamma t + \nu < N, x >),\ \ \nu \geq 0, \tag{4}$$

where $N \in \mathbb{R}^n, < , >$- scalar product in $\mathbb{R}^n$. Because of the equality

$$P(\mathcal{D}_x, \mathcal{D}_t)[v \exp(\gamma t + \nu < N, x >)] = \exp(\gamma t + \nu < N, x >)P(\mathcal{D}_x + i\nu N, \mathcal{D}_t + i\gamma)v$$

Petrovsky's condition in the spaces with weight (4) has the form

$$A(\zeta, \tau) \neq 0,\ \zeta = \xi + i\nu N,\ Im\ \tau \leq \gamma_0(\nu) \tag{5}$$

and the following theorem is an exact generalization of Proposition 1.

Theorem 1 *Let $q \geq 1$, $N \in \mathbb{R}^n$ and $A(\zeta, \tau)$ be a q- homogeneous polynomial. The following conditions are equivalent.*

(i) stable Petrovsky condition with weight (4), i.e. $\forall Q(\zeta, \tau)$, $deg_q Q < deg_q A = m$ $\exists \nu_0(Q)$ so that

$$A(\zeta, \tau) + Q(\zeta, \tau) \neq 0,\ \zeta = \xi + i\nu N,\ \nu \geq \nu_0(Q),\ Im\ \tau \leq \gamma_0(\nu, Q);$$

(ii) existence of the estimate

$$C\nu(|\tau|^{1/q} + |\xi|)^{m-1} \leq |A(\xi + i\nu N, \tau)|,\ \nu > 0,\ Im\ \tau < -R\nu^q;$$

(iii) q is an integer and

(a) if $q = 1$ then A is strictly hyperbolic and N is an arbitrary element of $\mathbb{R}^n$;

(b) if q is even then A is q-parabolic and N is arbitrary;

(b) if q is odd then A is q-hyperbolic (in the sense of definition below) and N belongs to the open cone of q-hyperbolicity of A

Definition. *Let q be a positive odd integer and let*

$$A(\xi,\tau) = \tau^k + \sum_{j>0} A_j(\xi)\tau^{k-j}$$

is a q-homogeneous polynomial. A is called q-hyperbolic if the following conditions are satisfied:

(i) the roots $\tau_j(\xi), j = 1,\ldots,k$ are real and different for $\xi \neq 0$ (and consequently smooth for $\xi \neq 0$),

(ii) there exists an $N \in \mathbb{R}^n$ such that

$$< \nabla\tau_j(\xi), N > > \ 0, \ j = 1,\ldots,k, \xi \neq 0.$$

It can be shown that the directions $N \in \mathbb{R}^n$ in (ii) form an open convex cone in $\mathbb{R}^n$ and this cone does not contain straight lines. That cone is called the cone of q-hyperbolicity of A.

If A is a polynomial of general form, then q-hyperbolicity of A means q-hyperbolicity of the principal q-homogeneous part of A.

Example. $A(\xi,\tau) = \tau - \sum a_j\xi_j^3$ *is 3- hyperbolic if and only if $a_j, j = 1,\ldots,n$ are real and $a_j \neq a_k$ for $j \neq k$.*

4

To formulate the extension of Proposition 2 we introduce the space $H^{(s/q)}_{\gamma,\nu}$ with the norm

$$||u||_{(s/q),\gamma,\nu} = (\int_{Im\ \tau=\gamma}\int_{Im\ \zeta=\nu N} (|\tau|^{2/q} + |\zeta|^2)^s|\hat{u}(\zeta,\tau)|^2 d\zeta d\tau)^{1/2},$$

here $\hat{u}(\zeta,\tau)$ is the Fourier-Laplace transform. Denote by $H^{0(s/q)}_{\gamma,\nu}$ the subspace of elements equal zero for $t < 0$.

Theorem 2 *Let $q \geq 1$, $N \in \mathbb{R}^n$ and A be a q-homogeneous polynomial (symbol). The following conditions are equivalent.*

(i) $\forall Q$, $deg_q Q < deg_q A = m$, $\forall s \in \mathbb{R}\ \exists \nu_0, R_1, R_2$ such that for $\nu \geq \nu_0$, $\gamma < -R_1\nu^q - R_2$ the equation

$$(A+Q)(\mathcal{D}_x,\mathcal{D}_t)u = f$$

for $\forall f \in H^{0(s/q)}_{\gamma,\nu}$ has a unique solution $u \in H^{(s+m-1/q)}_{\gamma,\nu}$;

(ii) $\forall s \in \mathbb{R}\ \exists \nu_0, R_1, R_2$ such that the following estimate holds

$$C\nu||u||_{(s+m-1/q),\gamma,\nu} \leq ||Au||_{(s/q),\gamma,\nu}\ \forall u \in H^{0(\infty)}_{\gamma,\nu}\ \nu \geq \nu_0,\ \gamma < -R_1\nu^q - R_2;$$

(iii) the condition (iii) of Theorem 1 is satisfied.

Remark. *The theorem above can be reformulated for the case of partial differential operators with smooth coefficients. In this case in (i) we must replace $A(D_x,D_t)$, $Q(D_x,D_t)$ by $A(t,x,D_x,D_t)$ and $Q(x,t,D_x,D_t)$ and condition (iii) must be reformulated in such form that the polynomials $A_{(x,t)}(\xi,\tau) = A(x,t,\xi,\tau)$ satisfy condition (ii) of Theorem 1 uniformly in the parameters $(x,t) \in \mathbb{R}^{n+1}$. In particular in the q-hyperbolic case there exist $\kappa, \kappa_1 > 0$ such that*

$$|\tau_j(x,t,\xi) - \tau_k(x,t,\xi)| \ \geq \ \kappa|\xi|^q, \ j \neq k,$$
$$< \nabla_\xi\tau_j(x,t,\xi), N > \ \geq \ \kappa_1|\xi|^{q-1}, \ j = 1,\ldots,k$$

and N belongs to the intersection of cones of q-hyperbolicity of the polynomials $A_{(x,t)}$.

5

We shall consider the mixed problem (1), (2), (3) in the half-space $< x, N > > 0$. In other words

$$A(\mathcal{D}_x, \mathcal{D}_t)u(x,t) = f(x,t), \quad < x, N > > 0, \tag{1'}$$
$$B_j(\mathcal{D}_x, \mathcal{D}_t)u(x,t)|_{<x,N>=0} = g_j(x',t), \quad j = 1, \ldots, \mu, \tag{2'}$$

where x' are the points of the plane $< x, N >= 0$. In the case of zero Cauchy data we can extend functions u, f, g_i in (1'), (2') as zeros for $t < 0$ and treat (1'), (2') as equations for all $t \in \mathbb{R}$. In this case (3) is replaced by

$$u(x,t) = f(x,t) = 0, \quad g_j(x',t) = 0, \quad t < 0. \tag{3'}$$

We denote by $\xi' \in \mathbb{R}^{n-1}$ the variables dual to x'. Then for every $\xi \in \mathbb{R}^n$ there exists such λ that $\xi = \lambda N + \xi'$ and we pose $\xi = (\lambda, \xi')$.

For every positive integer l and $s \in \mathbb{R}$ we define the space $H^{(l,s/q)}_{\gamma,\nu}$ with the norm

$$||u||_{(l,s/q),\gamma,\nu} = (\sum_{j=0}^{l} \int_{Im\tau=\gamma} \int_{Im\zeta=\nu N} |\lambda + i\nu|^{2j} \times (|\tau|^{2/q} + |\xi'|^2)^{s+l-j} |\hat{u}(\zeta,\tau)|^2 d\zeta d\tau)^{1/2}.$$

If $\mathbb{R}^{n+1}_N$ is the product of half-space $< x, N) > 0$ and straight line $\mathbb{R}$, then by $H^{(l,s/q)}_{\gamma,\nu}(\mathbb{R}^{n+1}_N)$ we denote the space of restrictions of functions from $H^{(l,s/q)}_{\gamma,\nu}$ to $\mathbb{R}^{n+1}_N$. Denote by $H^{(s/q)}_{\gamma}(\mathbb{R}^n)$ the space introduced above in the case of functions on $\mathbb{R} \times \{x, < x, N >= 0\}$ and let $\{\}_{(s/q),\gamma}$ be the norm in this space.

We suppose that $A, B_1, \ldots, B_\mu$ are q-homogeneous polynomials of degrees $m, \beta_1, \ldots, \beta_\mu$, respectively, and $m > \beta_1, \ldots, \beta_\mu$.

Theorem 3 *Let $q > 1$ and $A, B_1, \ldots, B_\mu$ be as above. The following conditions are equivalent.*

(i) For arbitrary operators $Q, R_1, \ldots, R_\mu$, $deg_q Q < m$, $deg_q R_j < \beta_j$, $j = 1, \ldots, \mu$, $\forall s \in \mathbb{R}\ \exists \gamma_0, R$ so that for $\nu \geq \nu$ and $\gamma < -R\nu^q$ and $\forall f \in H^{0(0,s+1/q)}_{\gamma,\nu}$ and $\forall g_j \in H^{0(s+m-\beta_j/q)}_j(\mathbb{R}^n)$, $j = 1, \ldots, \mu$ the problem (1'), (2'), (3') where A and B_j are replaced by $A + Q$, $B_j + R_j$ has a unique solution $u \in H^{0(m,s/q)}_{\gamma,\nu}(\mathbb{R}^{n+1}_N)$ and

$$< N, D >^j u|_{<x,N>=0} \in H^{(m+s-j/q)}(\mathbb{R}^n), \ j = 0, \ldots m-1$$

(ii) $\forall s \in \mathbb{R}\ \exists \nu_0, R$ such that the estimate holds

$$C\left[\nu||u||^2_{(m,s/q),\gamma,\nu} + \sum_{j=0}^{m-1} \{< N, D >^j u\}_{(m+s-j/q),\gamma}\right]$$
$$\leq \frac{1}{\nu}||Au||^2_{(0,s+1/q),\ \gamma,\nu} + \sum_{j=1}^{\mu} \{B_j u\}_{(m+s-\beta_j),\gamma}.$$

(iii) Following conditions are satisfied:

(a) condition (iii) of Theorem 2,

(b) the direction of the normal to the plane $< x, N >= 0$ is non-characteristic for A. In other words

$$A(\lambda) := A(\xi' + \lambda N, \tau) = a_0\lambda^m + 0((\lambda^{m-1}),$$

(c) μ in condition (2') equals to the number of roots of $A(\lambda) = 0$ in half-plane $Im\ \lambda > 0$

(d) uniform Lopatinsky condition is satisfied (see [2], [4]).

Remark. *As in the case of Theorem 2 we can reformulate the theorem above for the mixed problem with variable coefficients. This formulation and detailed proofs can be found in [4].*

References

[1] S. Gindikin, L. Volevič. Sur le problème mixte puor des opérateurs différentielles $2(b+1)$-hyperboliques. - C.R. Acad. Sci. Paris 1979, v.289, p.A437-A439.

[2] S. Gindikin, L. Volevič. Sur le problème mixte pour des opérateurs différentielles a partie principale quasi homogène. - C.R. Acad. Sci. Paris, 1980, v.291, pp.A107-A110.

[3] L.R. Volevich, S.G. Gindikin. A mixed problem for $(2b+1)$-hyperbolic equations, Trans. Moscow Mat. Soc.,1983 ,v.43, pp.215-283.

[4] L.R. Volevich, S.G. Gindikin. Method of energy estimates in the mixed problem, Uspechi Mat.Nauk, 1980, v.35,N 5, pp.53-120.

L.R. Volevich
Moscow State University
Department of Mathematics
Moscow V-234, USSR

LIST OF LECTURES

ABD EL-RADY
Uniqueness theorem for Neumann problems of elliptic equations in an unbounded domain.

ALI MEHMETI
Interaction problems and propagation of singularities.

AZZAM
On asymptotic expansions of solutions of the first boundary value problem for elliptic equations near corners.

BAGIROV
On solutions of degenerate elliptic boundary value problems.

BRENNER
Fixed-point formulas on strictly pseudo-convex manifolds.

COSTABEL
A collocation method for a screen problem in $\mathbb{R}^3$.

DAUGE
Edge asymptotics on a skew cylinder, from a work with M. Costabel.

EGOROV, KONDRAT'EV
On the negative spectrum of elliptic operators.

ELSCHNER
The double layer potential operator over polyhedral domains; solvability in weighted Sobolev spaces and spline approximation.

GOLDSTEIN
Three-dimensional elasticity crack problems with small parameters or variations of the crack shape.

GRAMSCH
On closed geodesics in the manifold of idempotent elements in some classes of pseudo-differential operators.

IL'IN
On a bisingular problem for a nonlinear equation.

JENTSCH
On a bimetal problem for heat conductions.

JUHL
Dynamical Lefschetz formulas and the singularities of the Ruelle zeta-functions.

KONDRAT'EV
On the boundedness of the Schrödinger operator in weighted spaces.

KONDRAT'EV, OLEĬNIK
On Korn's inequalities for domains depending on parameters.

KOZLOV
Spectral properties of operator pencils associated with boundary value problems in a cone and in an angle.

LEOPOLD
On function spaces of variable and generalized smoothness.

MAZZEO
Edge operators, scalar curvature and index theory.

MEISTER
On a class of singular integral and integro-differential equations on the half-line.

MENDOZA
On solvability for systems of vector fields.

NAZAROV
Asymptotics of solutions at an edge. Radiation conditions in an edge. Scattering operator on an edge.

NICAISE
Exact controllability of a pluridimensional coupled problem.

OLEĬNIK
On smoothness of solutions of some problems of mathematical physics in nonsmooth domains.

PANEYAKH
On degenerate elliptic boundary value problems.

PLAMENEVSKIĬ
On the index formula for pseudo-differential operators with discontinuities in symbolds.

RABINOVICH
Spectral and scattering theory for acoustic operators in perturbed fluids.

ROĬTBERG
The Cauchy problem, boundary value and mixed problems for hyperbolic systems in the complete scale of Sobolev type spaces.

ROĬTBERG, SHEFTEL'
On the density of the solutions of elliptic boundary value problems in the sense of Petrovsky systems in functional spaces on manifolds.

ROSSMANN
On the Agmon-Miranda maximum principle for strongly elliptic differential operators in non-smooth domains.

SANCHEZ-PALENCIA
Crack phenomena in heterogeneous media.

SÄNDIG
Calculation of 3D-elasticity-singularities in a cone.

SCHMITZ
On Dirichlet-Signorini problems.

SCHMUTZLER
About the structure of branching asymptotics for elliptic boundary value problems in domains with edges.

SCHOMBURG
Singularity functions of solution of the two dimensional floating body problem.

SCHROHE
Boutet de Monvel's algebra for boundary value problems on non-compact manifolds.

SCHULZE
Mellin operators and elliptic regularity with asymptotics for second order corners.

SHUBIN
On Neumann spectra and topology of non-simply connected manifolds.

VOLEVICH
Mixed problems for evolution equations.

INTERNATIONAL WORKSHOP
ANALYSIS IN DOMAINS AND ON MANIFOLDS WITH SINGULARITIES

April 30 - May 5, 1990 in Breitenbrunn, Saxony, Germany

List of participants

Abd El-Rady, Qena
Ali Mehmeti, Mainz
Apel, Chemnitz
Azzam, Kuwait
Bagirov, Moscow
Baranowski, Berlin
Berger, Leipzig
Brenner, Karaganda
Bunke, Greifswald
Costabel, Darmstadt
Dauge, Nantes
Doppel, Berlin
Egorov, Moscow
Elschner, Berlin
Föllinger, Berlin
Goldstein, Moscow
Gramsch, Mainz
Gürlebeck, Chemnitz
Hanke, Chemnitz
Hochmuth, Berlin
Hirschmann, Berlin
Il'in, Sverdlovsk
Jentsch, Chemnitz
Juhl, Berlin
Kondrat'ev, Moscow
Kozlov, Leningrad
Langer, Chemnitz
Leopold, Jena
Lorenz, Chemnitz
Mazzeo, Standford
Meister, Darmstadt
Mendoza, Caracas
Mirschinka, Chemnitz
Nazarov, Leningrad
Nicaise, Lille
Oleĭnik, Moscow
Paneyakh, Moscow
Plamenevskiĭ, Leningrad
Rabinovich, Rostov
Rempel, Berlin
Roĭtberg, Chernigov
Roßmann, Rostock
Sanchez-Palencia, Paris
Seifarth, Berlin
Schmidt, Berlin
Schmitz, Stuttgart
Schmutzler, Chemnitz
Schomburg, Berlin
Schrohe, Mainz
Schulze, Berlin
Sheftel', Chernigov
Shubin, Moscow
Smaiel, Homs
Sprößig, Freiberg
Umnuß, Chemnitz
Volevich, Moscow

GPSR Compliance

The European Union's (EU) General Product Safety Regulation (GPSR) is a set of rules that requires consumer products to be safe and our obligations to ensure this.

If you have any concerns about our products, you can contact us on ProductSafety@springernature.com

In case Publisher is established outside the EU, the EU authorized representative is:

Springer Nature Customer Service Center GmbH
Europaplatz 3
69115 Heidelberg, Germany

Batch number: 10577878

Printed by Printforce, the Netherlands